Daniel K Harder
Berkeley - March 1989

Daniel K Harder
Berkeley - March 1989

Handbook of LEGUMES of World Economic Importance

Handbook of LEGUMES of World Economic Importance

James A. Duke
United States Department of Agriculture
Beltsville, Maryland

Plenum Press • New York and London

Library of Congress Cataloging in Publication Data

Main entry under title:

Handbook of legumes of world economic importance.

Includes index.
1. Legumes. I. Duke, James A., 1929-
SB317.L43H36   633.3   80-16421
ISBN 0-306-40406-0

First Printing — January 1981
Second Printing — October 1983

© 1981 Plenum Press, New York
A Division of Plenum Publishing Corporation
233 Spring Street, New York, N.Y. 10013

All rights reserved

No part of this book may be reproduced, stored in a retrieval system, or transmitted, in any form or by any means, electronic, mechanical, photocopying, microfilming, recording, or otherwise, without written permission from the Publisher

Printed in the United States of America

Contents

Credits vii
Contributors ix
Introduction 1
    Format 1
    Text Abbreviations 3
    Collaborators 4

Legume Species 5
Appendix 311
    TABLE 1. Legume "Toxins": Their Toxicity and Generic
        Distribution 311
    TABLE 2. Legume Genera and Their Toxins 314
    TABLE 3. Ecosystematic Attributes of Legumes 317
    TABLE 4. Economic Legumes: Their Tolerances, Yields, Centers of
        Diversity, and Ecocenters 321
    TABLE 5. Recommended Inoculants for Various Legumes 327
    TABLE 6. Zero-Moisture Nutritional Analysis of Legumes 328
    TABLE 7. Amino Acid Composition of Various Legumes
        (g/ 16 g N) 334

General References 341

Credits

Acad. Repubc. Pop. Rom. **5,** Bucharest, Rumania, 1957: Fig. 115.

Ali, S. I., *Flora of West Pakistan,* Third Annual Report, Karachi, Pakistan: Fig. 26.

Ali, S. I., *Flora of West Pakistan* **100,** 1977: Figs. 51, 134, 136, 137.

Ann. Jard. Bot. Buitenzorg **44,** 1935: Fig. 29.

Aublet, J. B. C. F., *Histoire des Plantes de la Guiane Française,* Vol. 4, Didot, London/Paris, 1775: Figs. 61, 108.

Basu, B. D., *Indian Medicinal Plants,* Allahabad, India, 1918 (reprinted 1975): Figs. 24, 103, 142.

Bentham, G., *Handbook of the British Flora,* Vol. 1, Reeve, London, 1865: Fig. 125.

Berg. Char. Pflanzen Gattungen, Gaertner, Berlin, 1861: Fig. 41.

Chuang, C.-C., and Huang, C., *The Leguminosae of Taiwan for Pasture and Soil Improvement,* Taipei, Taiwan, 1965: Figs. 70, 83, 139, 141.

Constantin, P., *Marvelles de la Nature. Les Plantes,* Baillera, Paris, 1894–1896: Fig. 3.

Corner, E. J. H., and Watambe, K., *Illustrated Guide to Tropical Plants,* Hirohawa, Tokyo, 1969: Fig. 19.

Coste, H. J., *Flora France,* Vol. 1, 1901: Fig. 79.

Curtis. Bot. Mag. **3,** 1789: Fig. 105.

Curtis. Bot. Mag. **139,** 1913: Fig. 38.

Dodson, C. H., and Gentry, A. H., *Flora of the Rio Palenque Science Center, Los Rios, Ecuador,* The Marie Selby Botanical Gardens, Sarasota, Florida, 1978: Figs. 40, 90.

Duke, P. (tracings of xerocopies): Figs. 8, 13, 20, 34, 44, 54, 69, 88, 107, 111, 120.

Engler, H. G. A., *Pflanzenfam:* Figs. 11, 15, 27, 52, 82.

Engler, H.G.A., and Drude, C. G. O., *Veget. der Erde,* 1911: Fig. 17.

Farmers Bull. **693,** 1915: Fig. 73.

Faul, K. (drawing): Fig. 85.

Flora of Taiwan, Vol. 3, Epoch, Taiwan, 1977: Figs. 1, 9, 14, 16, 18, 25, 28, 30, 33, 35, 36, 43, 49, 50, 57, 58, 59, 60, 63, 67, 68, 75, 81, 89, 97, 99, 101, 102, 110, 122, 123, 145.

Freeman, G. F., *Bot. Gaz.* **56,** 1913: Fig. 92.

Garcke, C. A. F., *Illustrated Flora Deutschland,* 17th ed., 1895: Fig. 10.

Grieve, M., and Leyel, C. F., *A Modern Herbal,* Vol. 1, 1937: Fig. 86.

Gunn, C. R., Classification of *Medicago sativa* L. using legume characters and flower colors, *U.S. Dep. Agric., Tech. Bull.* **1574,** 1978: Fig. 74.

Hegi, G., *Illus. Flora Mittel. Europ.,* Munich, 1923–1924: Fig. 129.

Hogg, P., and Johnson, C. W., *Wild Flowers of Great Britain,* Vol. 8, London, 1874: Fig. 80.

Humboldt, A., Bonpland, A., and Kunth, C. S., *Nova Genera et Species Plantarum,* 1823: Fig. 109.

Iowa Geol. Surv. Bull. **4,** Weed flora of Iowa, Des Moines, 1913: Fig. 46.

Irvine, F. R., *Woody Plants of Ghana,* Oxford University Press, London, 1961: Figs. 5, 12, 48, 91.

J. Linn. Soc. London, Bot. **13,** 1871–1873: Fig. 72.

Karsten, H., *Dtsch. Flora. Pharm. Med. Botanik,* Spaeth, Berlin, 1880–1883: Fig. 6.

Kew Bull., 1912: Fig. 71.

Killip, E. E., and Smith, A. C., The identity of South American fish poisons, *J. Wash. Acad. Sci.* **20**(5), 1930: Fig. 62.

Kotschy, T., *De plantis nilotico—Acthiopicis Knoblecheriana,* 1864–1865: Fig. 32.

Lindley, J., *Med. Econom. Bot.,* 1849: Fig. 45.

Maiden, J. H., *The Forest Flora of New South Wales,* Vol. 3, Government Printers, Sydney, Australia, 1907: Fig. 2.

Maiden, J. H., *Wattles and Wattles Barks,* 2nd ed., Government Printers, Sydney, Australia, 1891: Fig. 4.

Moris, G. G., *Flora Sardoa Iconographic,* 1837: Fig. 76.

Notizblatt Bot. Gart. Mus. Berlin. Dahlem., 1909: Figs. 98, 100.

Notizblatt Bot. Gart. Mus. Berlin. Dahlem., 1911: Fig. 106.

Reed, C. F., *Selected Weeds,* 1970: Fig. 135.

Reichenbach, H. G., and Beck, G., *Icon. Fl. Germ. Helv.* **22,** 1903: Figs. 42, 47, 65, 87, 113, 116, 117, 127.

Schlechtendahl, D. F. L. von, *et al., Flora von Deutschland,* 5th ed., Bd. 23, 1880: Fig. 121.

Schlechtendahl, D. F. L. von, *et al., Flora von Deutschland,* Bd. 24, 1880: Figs. 130, 132, 133.

Smartt, J., *Tropical Pulses,* Longman Group Ltd., London, 1976: Figs. 143, 146.

Syme, J. T., *et al., English Botany,* Vol. 3, 3rd ed., Hardwicke, London, 1864: Fig. 64.

Taylor, T. M. C., The pea family of British Columbia, *B.C. Prov. Mus. Handb.* **32,** 1974: Fig. 126.

Credits

Trans. Linn. Soc. N.Y. **29**, 1873–1875: Fig. 112.
U.S. Dep. Agric., Bull. **119**, 1914: Fig. 138.
Westphal, E., *Pulses in Ethiopia. Their Taxonomy and Agricultural Significance,* Centre for Agricultural Publishing and Documentation, Wageningen, 1974: Figs. 53, 55, 56, 66, 84, 93, 94, 95, 96, 128, 131, 140, 144.
Wight, R., *Icon. Pl. Ind. Orientalis* **1**, 1840: Figs. 23, 37.

Wight, R., *Icon. Pl. Ind. Orientalis* **3**, 1843–1845: Fig. 21.
Wilbur, R. L., *The Leguminous Plants of North Carolina,* The North Carolina Agricultural Experiment Station, 1963: Figs. 78, 104, 118.
Zohary, M., PL-480 Project A-10-CR-11, 1967: Figs. 114, 119, 124.

Contributors

Dr. J. C. Baudet
Faculte des Sciences
Agronomique de L'Etat
5800 Gembloux, Belgium

Dr. T. E. Boswell
Texas A. & M. University
Plant Disease Research Station
Yoakum, Texas 77995

Dr. J. L. Brewbaker
University of Hawaii
Honolulu, Hawaii 96822

Dr. A. H. Bunting
Plant Science Laboratory, University of
 Reading
White Knights
Reading, England
RG62AF

Dr. W. V. Campbell
Associate Professor of Insect Pathology
The School of Agriculture and Life
 Sciences and Forest Resources
North Carolina State University
Raleigh, North Carolina 27607

Dr. S. Chandra
Senior Plant Breeder
Division of Genetics and Plant
 Physiology
Central Soil Salinity Research Institute
Karnal, 132001
India

Dr. C. S. Cooper
Plant Science Research Division
United States Department of Agriculture
Bozeman, Montana 59715

Dr. W. A. Cope
Research Leader
USDA–SEA–FR–SR
North Carolina State University
P.O. Box 5155
Raleigh, North Carolina 27607

Dr. S. Dana
Bidhan Chandra Krishi
Vistva Vidyalaya
Faculty of Agriculture
Department of Genetics and Plant
 Breeding
Kalyani 741235
West Bengal, India

Dr. T. E. Devine
Room 218, Building 001, BARC-West
Department of Agriculture
Beltsville, Maryland 20705

Dr. Tricia Dodd
Plant Science Laboratory, University of
 Reading
White Knights
Reading, England
RG62AF

Dr. J. A. Duke
Chief
Economic Botany Laboratory
Building 265, BARC/East
United States Department of Agriculture
Beltsville, Maryland 20705

Dr. N. R. Farnsworth
College of Pharmacy
University of Illinois
833 S. Wood Street
Chicago, Illinois 60612

Dr. M. B. Forde
Plant Introduction Office
Grasslands Division
DSIR, Private Bag
Palmerson North, New Zealand

Dr. W. Foulds
Science Department
Claremont Teachers College
West Australia

Mr. A. Fyson
Department of Biological Sciences
The University
Dundee, United Kingdom

Dr. P. B. Gibson
Agronomy Department
Clemson University
Clemson, South Carolina 29631

Dr. M. A. Golden
Room 159, Building 011-A
United States Department of Agriculture
Beltsville, Maryland 20705

Dr. H. J. Gorz
University of Nebraska
Lincoln, Nebraska 68503

Dr. Peter H. Graham
CIAT, Apartado Aereo 6713
Cali, Colombia

Dr. W. C. Gregory
Professor of Genetics
The School of Agriculture
North Carolina State University
Raleigh, North Carolina 27607

Dr. Ray O. Hammons
Research Leader
USDA–SEA–FR–SR
Georgia Coastal Plains Experiment
 Station
Tifton, Georgia 31794

Contributors

Dr. R. Hegnauer
Laboratorium and Voor
Experimenele Plantensystematiele
Schelpenkade 14a
2131-2T Leiden
Netherlands

Dr. J. M. Hopkinson
Agricultural Branch
Department of Primary Industries
Brisbane Old
Australia 4000

Dr. C. S. Hoveland
Auburn University
Auburn, Alabama 36830

Dr. E. M. Hutton
Chief of Tropical Crops and Pastures
Commonwealth Scientific and Industrial
  Research Organization
St. Lucia, Queensland, Australia

Dr. T. Hymowitz
University of Illinois at
  Urbana–Champaign
Urbana, Illinois 60801

Dr. R. F. Keeler
Utah State University
UMC 48, Agricultural Science Building
Logan, Utah 84322

Dr. T. N. Khan
Department of Agriculture
Jarrah Road
South Perth, 6151
Western Australia

Dr. A. E. Kretschmer, Jr.
University of Florida
Institute of Food, Agricultural Sciences
Agricultural Research Center
Fort Pierce, Florida 33450

Dr. J. Langenheim
University of California
Botany Department
Santa Cruz, California 95053

Dr. R. M. Lantican
University of the Philippines at
  Los Banos
College of Agriculture
College, Laguna, 3720
Philippines

Dr. J. M. Lenne
University of Florida
Institute of Food, Agricultural Sciences
Agricultural Research Center
Fort Pierce, Florida 33450

Dr. R. L. Lynch

Dr. L. 't Mannetje
CSIRO
Mill Road
St. Lucia
QLD. Australia 4067

Dr. R. Marechal
Faculte Des Sciences
Agronomiques De L'Etat
5800 Gembloux
Belgium

Dr. J. D. Miller
Agronomy Department
Georgia Coastal Plains Experiment
  Station
Tifton, Georgia 31794

Dr. A. J. Norden
Institute of Food and Agriculture
  Sciences
College of Agriculture
University of Florida
304 Newell Hall
Gainesville, Florida 32611

Dr. B. N. Okigbo
Director
International Institute for Tropical
  Agriculture
Ibadan, Nigeria

Dr. B. P. Pandya
Agricultural Station
Pantnager, India

Ms. Hazel Pollard
DSAD, National Agricultural Library
Beltsville, Maryland 20705

Dr. K. O. Rachie
CIAT, Apartado Aereo 6713
Cali, Valle, Colombia

Dr. C. F. Reed
Reed Herbarium
10105 Harford Road
Baltimore, Maryland 21234

Dr. Peter P. Rotar
Department of Agronomy and Soil
  Science, CTA
University of Hawaii at Manoa
3190 Maile Way
Honolulu, Hawaii 96822

Dr. D. G. Roux
Department of Chemistry
University of The Orange Free State
P.O. Box 339, Bloemfontein 9300
Republic of South Africa

Dr. N. H. Shaw
CSIRO
Division of Tropical Crops and Pastures
St. Lucia, Queensland
Australia

Dr. C. E. Simpson
Texas A. & M. University
Agricultural Research and Extension
  Center
Stephenville, Texas 76401

Dr. R. P. Sinha

Dr. A. E. Slinkard
Senior Research Scientist
Crop Science Department
University of Saskatchewan
Saskaton, Saskatchewan
Canada S7N 0W0

Dr. J. Smartt
Department of Biology
University of Southampton
509 5NH
United Kingdom

Dr. Donald H. Smith
Associate Professor
Texas A. & M. University
Plant Disease Research Station
Yoakum, Texas 77995

Dr. Olin D. Smith
Associate Professor of Peanut Breeding
Soil and Crop Science
Texas A. & M. University
College Station, Texas 77843

Dr. R. R. Smith
USDA, ARS, North Central Region
Department of Agronomy
University of Wisconsin
Madison, Wisconsin 53706

Dr. J. I. Sprent
Department of Biological Sciences
The University
Dundee, United Kingdom

Dr. R. J. Summerfield
Plant Environmental Laboratory
Shinfield Grange
Cutbush Lane, Shinfield
Reading, Berks
United Kingdom

Dr. N. L. Taylor
University of Kentucky
Agronomy Department
College of Agriculture
University of Kentucky
Lexington, Kentucky 40506

Contributors

Dr. C. E. Townsend
Crops Research Laboratory
Colorado State University
Bay Road south of West Prospect
Fort Collins, Colorado 80523

Dr. L. J. G. van der Maesen
Pulse Germplasm Botanist
1-11-256 Begumpet
Hyderabad, 500016 A. P.
India

Dr. Jurgen K. P. Weder
Institut FUR Lebensmittelchemie
Technische Universitat Munchen
8 Munchen 2
Lothstrasse 17
Nese Ruf-NR. 2105-4246
West Germany

Dr. E. B. Whitty
Associate Professor of Field Crop
  Management
Agronomy Department, 304 Newell Hall
University of Florida
Gainesville, Florida 32611

Dr. E. C. Williams

Dr. M. C. Williams
University of Agriculture and Applied
  Sciences
Logan, Utah 84322

Dr. J. C. Wynne
Crop Science Department
North Carolina State University
Raleigh, North Carolina 27607

Dr. Clyde T. Young
Food Science Department
Georgia Agricultural Experiment Station
Experiment, Georgia 30212

Dr. D. M. Zohary
Department of Genetics
Hebrew University
Jerusalem, Israel

Introduction

In 1971, Dr. Quentin Jones, now of the National Program Staff, SEA, USDA, suggested that the Plant Taxonomy Laboratory devise a format for concise write-ups on 1,000 economic plants (Duke and Terrell, 1974; Duke et al., 1975). Dr. C. F. Reed was contracted to search the literature on these economic plants, which included 146 species of legumes. From 1971 through 1974, Dr. Reed prepared rough drafts of write-ups on the 1,000 species. It was my responsibility to establish the format and monitor the write-ups, to ensure that they would answer many questions on legumes directed to the USDA by our taxpaying public. Since then, a computerized system alerts me to new publications on legumes. I have ordered for our files copies of the more promising documents. With the evolution of the manual, many later references supplanted earlier ones. Some repetitive information was dropped by condensing the write-ups on secondary species within the genera.

Informally in 1975 and formally in 1976, Dr. Roger Polhill, Curator of Legumes, Royal Botanical Gardens, Kew, joined by Dr. A. S. Bunting, Professor, The University of Reading, and Dr. J. P. M. Brenan, Director, Royal Botanical Gardens, Kew, conceived the splendid idea of an International Legume Conference at Kew. We agreed that this would be a good forum for presentation of the *Handbook of Legumes*. Beginning in 1976, I circulated second drafts of the individual legume write-ups to a mailing list of contributors. Responses ranged from obloquy to praise; many contributors went over the drafts carefully.

I also benefited from attending a National Academy of Science Legume Conference in Maui, Hawaii, where an international panel convened to discuss and assemble information on underexploited tropical legumes. Conversations at that meeting and subsequent correspondence with the participants also yielded new information on some of the tropical legumes.

Finally in 1978, 100 copies of the writeups were delivered to the International Legume Conference at Kew, July 24th–August 4, and all were given to potential cooperators before my lecture on the manual (July 31st). New information presented in lectures at that conference and personal communications behind the scenes have also been used to update and embellish the write-ups so that they are more than a bibliographic echo.

On May 7, 1978, I was transferred from USDA's Plant Taxonomy Laboratory to the Medicinal Plant Resources Laboratory (now the Economic Botany Laboratory). Since then, I have included the Folk Medicine paragraph in the final format. My computerized ecosystematic program (Duke, 1978b) includes data not only on legumes, but on economic plants in general, many of which are intercropped with legumes to good advantage.

FORMAT

Except for beginning each write-up under the heading of scientific name, there was little unanimity about what should be included or excluded, and in what order the included paragraphs should be arranged. I have, without sacrificing clarity, adopted a terse style, often leaving out the verbs in descriptive sentences. Furthermore, I have gener-

Introduction

ously sprinkled the text with abbreviations that are identified in the following section titled Text Abbreviations. It should be noted that these abbreviations may differ from those specifically defined in tables found in the Appendix. I finally elected to follow the format below, at least for the first species in each genus.

NOMENCLATURE: All agreed that the scientific name, authority, and recent significant synonyms be included. Because Anglicans consistently use the word "lucerne" for the plant Americans call "alfalfa," and because this manual was intended for international usage, common names in a few major languages are included. However, the large number of common names dictated that most of them be omitted.

USES: Following the Nomenclature Section, important general uses are reported. Some of the uses that may seem trivial in developed countries are important in developing countries, and vice versa.

FOLK MEDICINE: Folk medicine is separated from general uses, even though many collaborators believe that the information is parascientific at best. Frequently, however, folk medical usage is related to chemical characteristics. Certainly the USDA does not recommend folk medicine, and discourages self-medication. Still, I quote Dr. R. J. McCracken from Perdue and Hartwell (1975). "We have heard much lately about the use of plants and the products derived from them, in meeting world food problems. But of equal concern and importance is human health and in this respect plants have much to contribute that is of therapeutic value." In the cancer-screening program of the National Institutes of Health, several compounds found in legumes have proved active against experimental cancers. *Cassia obtusa* has been reported to give the active colubrinol; *Cercidium microphyllum* has given an active; *Coronilla varia* gave hyrcanoside; *Crotalaria spectabilis,* monocrotaline; *Derris trifoliata,* rotenone; *Entada phaseoloides,* an active; *Lonchocarpus urucu,* rotenone; and *Piscidia erythrina,* rotenone (Perdue and Hartwell, 1975). Folklore on medicinal plants is a good guide in our quest for new medicinal chemicals.

CHEMISTRY: I did not include detailed analytical data on the chemistry of legumes but included nutritional data, where available. Often approximate nutritional composition varied widely among sources. In the individual chapters I have reported the nutritional data from the various sources listed in the references. Many of these data were recently published (Duke 1977a). To facilitate comparison, I tabulated nutrients (zero-moisture basis) in Table 6 and amino acid compositions in Table 7 of the Appendix. I also list chemurgic chemicals and/or those cited as "toxic" (NIOSH, 1975). There has been discussion in the United States urging that plants containing toxins be so labeled. Consequently they were included in Appendix Tables 1 and 2, derived from my Phytotoxin Tables (Duke, 1977b). Many correspondents questioned the merit of including mere generic lists. I have initiated a computer program to accommodate the quantitative distribution of chemicals within the species and the location in the plant, rather than in the genera. That should prove even more useful than the Phytotoxin Tables.

DESCRIPTION: The manual was designed more specifically for agronomists than for taxonomists, although I hope both groups will find it useful. Simple terms were sought for the descriptions. Line drawings have been borrowed from the open literature where possible.

GERMPLASM: Space limited the discussion of germplasm to the more important cultivars, comments on the centers of origin, breeding mechanisms, tolerances, and chromosome counts.

DISTRIBUTION: Modern distribution of many economically important legumes bears little relation to their historic distribution. In this section I tried to summarize facts about current and historic distribution.

ECOLOGY: For more than 5 years, I have solicited local floral checklists associated with ecological data. The resultant data complement other published data. Ecological data from my computerized ecosystematic data base (Duke, 1978a,b,c) on legumes are tabulated in the Appendix (Table 3). If Table 3 represented the entire world rather than a small biased sample, the climate and pH of a remote area might be estimated by a study of its native legumes. The concluding sentence of the Ecology Section with data on life zones, annual precipitation, annual temperature, and pH is derived from my ecosystematic program; the data were confirmed by my computer program. Some data may differ from those published but each datum was based on a case history that is retrievable by computer.

CULTIVATION: Here I included information for developed "mechanized" societies, for developing "manual" societies, or for both, on land preparation, planting, fertilization, intercropping, cultivation and maintenance of the legume until harvest. Some data on field spacing and seed rates were often included, but these vary so much from system to system, that I reported only a few.

HARVESTING: Methods for harvesting were reported, often with notes on processing the harvested plant. Data on optimum stage of growth and conditions for harvest also were included, where available.

YIELDS AND ECONOMICS: Conventional data for yields were reported. Data for biomass, nitrogen fixation, and any other available nonconventional estimates of yield also were included. Dollar figures, where possible, are tied to dates so that this manual would not immediately lose its comparative value.

BIOTIC FACTORS: Pests, at least some of the more important ones, were listed. I consulted the historic file maintained in the U.S. National Fungus Collection at Beltsville, and was assisted editorially by many of my colleagues in other disciplines at Beltsville. However they can be held responsible only for the correct names. I found no one with time enough to check out all the names recorded in the National Fungus Collection, hence many synonyms have crept in. I would rather have a species listed under an outdated name than not listed at all. For conservation of space, the authors of the species names of the various pests have been deleted. Occasionally, I have listed pesticides in this section. This section is strictly bibliographic, and the lists do not constitute a recommendation or endorsement of these pesticides by the USDA. As a matter of fact, some of them are prohibited in the US and many are dangerous, unless used strictly according to the instructions.

SPECIFIC REFERENCES: Many correspondents have pointed out several recent references, too many for inclusion. General familial or generic references consulted in early drafts or in final revisions have been placed in the reference citations of the section titled General References. Even though these works were consulted in preparing individual write-ups, they are listed only in the final general collection of references. For inclusion with each write-up, preference has been given to specific references or those dated 1975 or later, especially review papers. General references were listed only at the end of the manual; a few specific references were listed at the end of most chapters.

TEXT ABBREVIATIONS

ann	annum, annual
Apr.	April
Aug.	August
avg.	average
AVRDC	The Asian Vegetable Research and Development Center
B	boron
B.C.	before Christ
BP	before present
Br	bromine
bu	bushel
BYMV	bean yellow mosaic virus
°C	degrees centigrade (Celsius)
Ca	calcium
ca.	about, nearly
cf.	compare with
CIAT	Centro Internacional de Agricultura Tropical
Cl	chlorine
cm	centimeter
Cu	copper
cu	cubic
cv	cultivar
cvgr	cultigroups
cwt	hundredweight
Dec.	December
DM	dry matter
dm	decimeters
EC	electrical conductivity
e.g.	for example
et al.	and others
etc.	and so forth
FAO	Food and Agricultural Organization of the United Nations
FDA	Federal Drug Administration
Fe	iron
Feb.	February
Fig.	figure

Introduction

Fl.	flower
Fr.	fruit
g	gram
ha	hectare
HCN	hydrogen cyanide
hr	hour
I	iodine
i.e.	that is
IITA	International Institute of Tropical Agriculture
IU	International Unit
Jan.	January
K	potassium
kg	kilogram
km	kilometer
lb	pounds
LAI	Leaf Area Index
m	meter
MD	Maryland
Mar.	March
max.	maximum
Mg	magnesium
mg	milligram
ml	milliliter
mm	millimeter
min	minute
min.	minimum
Mn	manganese
Mo	molybdenum
mos.	months
MT	metric ton
N	nitrogen
n as in $2n$	number (specifically chromosome number when given $2n =$)
Na	sodium
NC	cultigroups
Ni	nickel
NIOSH	National Institute of Occupational Safety and Health
N.H.	Northern Hemisphere
No.	number
Nov.	November
NPK	nitrogen: phosphorus: potassium
Oct.	October
p.	page
pp.	pages
P	phosphorus
PA	Pennsylvania
PI	plant introduction
ppm	parts per million
prec	precipitation
R	reproductive
S	sulfur
SA	South America
Sep.	September
sp.	species (singular)
spp.	species (plural)
sq	square
ssp.	subspecies (singular)
sspp.	subspecies (plural)
temp.	temperature
µg	microgram
U.K.	United Kingdom
USDA	United States Department of Agriculture
USSR	Russia
V	vegetative
var	variety
WARF	Wisconsin Alumni Research Foundation
WI	West Indies
WM	wet matter (as opposed to DM)
WOI	Wealth of India
wt.	weight
yr	year

Collaborators: Perhaps the most important item in each chapter is the list of contributors. Many specialists have contributed and some lists may be incomplete. I wish, however, to thank R. F. Barnes, L. R. Batra, T. E. Boswell, J. L. Brewbaker, K. R. Bromfield, A. H. Bunting, J. C. Burton, W. V. Campbell, S. Chandra, W. A. Cope, S. Dana, T. C. Davidson, T. E. Devine, D. Ellington, N. R. Farnsworth, M. B. Forde, A. Fyson, M. Golden, H. J. Gotz, P. H. Graham, W. C. Gregory, C. R. Gunn, R. O. Hammons, R. Hegnauer, C. C. Heyn, E. M. Hoover, C. S. Hoveland, E. M. Hutton, T. Hymowitz, T. J. Khan, R. F. Keeler, L. Knutson, A. E. Kretschmer, Jr., J. Langenheim, R. M. Lantican, R. L. Lynch, R. Marechal, J. D. Miller, A. J. Norden, B. N. Okigbo, B. P. Pandya, R. M. Polhill, H. Pollard, K. O. Rachie, C. F. Reed, P. P. Rotar, D. G. Roux, C. E. Simpson, A. E. Slinkard, J. Smartt, D. H. Smith, O. D. Smith, J. I. Sprent, R. J. Summerfield, R. A. Taber, N. L. Taylor, E. E. Terrell, C. E. Townsend, L. J. G. van der Maesen, J. K. P. Weder, E. B. Whitty, E. C. Williams, M. C. Williams, J. C. Wynne, C. T. Young, and D. Zohary.

Legume Species

Acacia farnesiana (L.) Willd.

FAMILY: Mimosaceae
COMMON NAMES: Cassie, Huisache
SYNONYM: *Mimosa farnesiana* L.

USES: Cassie perfume distilled from the flowers. Cassie absolute used in preparation of violet bouquets, extensively used in European perfumery. Cassie pomades manufactured in Uttar Pradesh and the Punjab. Pods contain 23% tannin, a glucoside of ellagic acid, and are used for tanning leather. Bark also used for tanning and dyeing leather in combination with iron ores and salts. In Bengal and West Indies pods used for a black leather dye. Gummy substance from pods used in Java as cement for broken crockery. Gum exuding from trunk considered superior to gum arabic in arts. In Ivory Coast trees used as ingredient in arrow poison; elsewhere used as fences and to check erosion. Wood is hard and durable underground, used for wooden plows and for pegs. Often planted as an ornamental.

FOLK MEDICINE: Bark is astringent and demulcent, and with leaves and roots is used for medicinal purposes. Woody branches used in India as toothbrushes; gummy roots chewed for sore throat. Said to be used as alterative, antispasmodic, aphrodisiac, astringent, demulcent, antidiarrhetic, febrifuge, antirheumatic, and stimulant.

CHEMISTRY: Dried seeds of one *Acacia* sp. reported to contain per 100 g: 377 calories, 7.0% moisture, 12.6 g protein, 4.6 g fat, 72.4 g carbohydrate, 9.5 g fiber, and 3.4 g ash. Raw leaves of *Acacia* contain per 100 g: 57 calories, 81.4% moisture, 8.0 g protein, 0.6 g fat, 9.0 g carbohydrate, 5.7 g fiber, 1.0 g ash, 93 mg Ca, 84 mg P, 3.7 mg Fe, 12,255 µg β-carotene equivalent, 0.20 mg thiamine, 0.17 mg riboflavin, 8.5 mg niacin, and 49 mg ascorbic acid. Reporting 55% protein on a dryweight basis, Van Etten *et al.* (1963) break down the amino acids as follows: lysine, 4.7 (g/16 g N); methionine, 0.9;

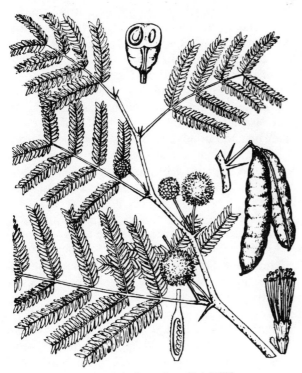

Figure 1. *Acacia farnesiana* (L.) Willd.

Acacia farnesiana (L.) Willd.

arginine, 9.2; glycine, 3.4; histidine, 2.3; isoleucine, 3.5; leucine, 7.5; phenylalanine, 3.5; tyrosine, 2.8; threonine, 2.5; valine, 3.9; alanine, 4.3; aspartic acid, 8.8; glutamic acid, 12.6; hydroxyproline, 0.0; proline, 5.1; serine, 4.1; with 76% of the total nitrogen as amino acids. Cassie has been reported to contain anisaldehyde, benzoic acid, benzyl alcohol, butyric acid, coumarin, cresol, cuminaldehyde, decyl aldehyde, eicosane, eugenol, farnesol, geraniol, hydroxyacetophenone, methyleugenol, methyl salicylate, nerolidol, palmitic acid, salicylic acid, and terpineol. The genus *Acacia* is reported to contain: anhaline, anisaldehyde, anthraquinone, benzaldehyde, benzyl alcohol, butyraldehyde, cresol, cuminic aldehyde, decanal, dimethyl tryptamine, eugenol methyl ether, gallic acid, heptanoic acid, hydrocyanic acid, indole, isobutyraldehyde, linalool, methyl salicylate, nicotine, palmitic acid, phenethylamine, quercitin, rutin, saponin, tannic acid, terpineol, trigonelline, tyramine (for relative toxicities, see Appendix Table 2). Seeds of the genus *Acacia* reported to contain trypsin inhibitors and chymotrypsin inhibitors.

DESCRIPTION: Thorny bush or small tree, 1.5 m tall; bark, light brown, rough; branches glabrous or nearly, purplish to gray, with very small glands; stipules spinescent, usually short, up to 1.8 cm long, rarely longer, never inflated; leaves twicepinnate, with a small gland on petiole and sometimes one on the rachis near top of pinnae; pinnae 2–8 pairs, leaflets 10–12 pairs, minute, 2–7 mm long, 0.75–1.75 mm wide, glabrous, leathery; flowers in axillary pedunculate heads, calyx and corolla glabrous, scented; pod indehiscent, straight or curved, 4–7.5 cm long, about 1.5 cm wide, subterete and turgid, dark brown to blackish, glabrous, finely longitudinally striate, pointed at both ends; seeds chestnut brown, in 2 rows, embedded in a dry spongy tissue, 7–8 mm long, about 5.5 mm broad, smooth, elliptic, thick, only slightly compressed; areole 6.5–7 mm long, 4 mm wide.

GERMPLASM: Both *A. farnesiana* and its var *cavenia* are extensively cultivated in and around Cannes, southern France, the center for production of the perfume; *cavenia* seems more resistant to drought and frost. Assigned to the South American Center of Diversity, cassie or cvs thereof is reported to exhibit tolerance to drought, high pH, heat, low pH, salt, sand, slope, and savanna. ($2n = 52, 104$).

DISTRIBUTION: Probably native to tropical America, but widely naturalized and cultivated—Africa (Rhodesia, Mozambique) and Australia. Planted in coastal areas of Ghana and elsewhere in tropical Africa. Grown throughout India, often in gardens.

ECOLOGY: Thrives in dry localities and on loamy or sandy soils; may serve as a sand binder. Grows on loose, sandy soil of river beds, on pure sand in plains of Punjab. Requires a dry tropical climate. Ranging from Warm Temperate Dry through Tropical Desert to Moist Forest Life Zones, cassie is reported to tolerate annual precipitation of 6.4–40.3 dm (mean of 20 cases = 14.0 dm), annual mean temperature of 14.7–27.8°C (mean of 20 cases = 24.1°C), and pH of 5.0–8.0 (mean of 15 cases = 6.8).

CULTIVATION: Propagated mainly from seed and cuttings. Seeds germinate readily and plants grow rapidly. Plants do not require much cultivation, watering, or care.

HARVESTING: Trees begin to flower from the third year, mainly November–March. Perfume is extracted from the flowers in form of concrete or pomade. Macerated flowers held several hours in melted purified natural fat, then repeatedly replaced by fresh flowers until the fat is saturated with perfume. Fat melted, strained, and cooled to yield pomade. Odor is that of violets but more intense. For absolute, pomade is mixed with alcohol (2–3 kg to ca. 4 liters) and held for 3–4 weeks at ca. −5°C. The alcohol is separated by distillation. The extract is an olive-green liquid with strong odor of cassie flowers.

YIELDS AND ECONOMICS: Mature trees yield up to 1 kg of flowers per season. Southern France (Cannes and Grasse) is main production center for cassie flower perfume. India and other Eastern countries produce much for local use.

BIOTIC FACTORS: Fungi reported on this plant include: *Camptomeris albizziae, Clitocybe tabescens, Hypocrea borneensis, Lenzites palisoti, L. repanda, Phyllachora acaciae, Phymatotrichum omnivorum, Polystictus flavus, Ravenelia austris, R. hieronymi, R. siliquae, R. spegazziniana, Schizophyllum commune, Systingophora hieronymi, Tryblidiella rufula,* and *Uromycladium notabile.* It may also be parasitized by the flowering plants *Dendrophthoe falcata* and *Santalum album.*

CONTRIBUTORS: J. A. Duke, C. F. Reed, J. K. P. Weder.

Acacia mearnsii de Wild.

FAMILY: Mimosaceae

COMMON NAMES: Black wattle, Acacia negra, Acacia noir, Schwarze akazie, Gomboom

SYNONYMS: *Acacia mollissima* auct., not Willd. *Acacia decurrens* var *mollis* Lindl.

USES: Tree of economic importance in South and East Africa, Rhodesia, India, and Rio Grande do Sul area of South America and environs for fuel and for tanning of soft leather. Dried bark contains 30–54% tannin. Trees furnish fuel and lumber in some areas and add nitrogen and organic material to the soil. Bark is used for wood adhesives and flotation agents.

FOLK MEDICINE: Products are often used in folk medicine as styptics or astringents.

CHEMISTRY: Black wattle bark contains (−)-robinetinidol and (+)-catechin; the biflavonoids (−)-fisetinidol-(+)-catechin (2 diastereoisomers), (−)-robinetinidol-(+)-catechin and (−)-robinetinidol-(+)-gallocatechin; triflavonoids and condensed tannins. The heartwood is rich in (+)-leucofisetinidin (mollisacacidin) together with (−)-fisetinidol, (+)-fustin, butin, fisetin, butein, and biflavonoid condensates (tannins).

DESCRIPTION: Tree 6–20 m tall, 10–60 cm in diameter; crown conical or rounded; all parts except flowers usually pubescent or puberulous; stems without spines or prickles; leaves bipinnate, on petioles 1.5–2.5 cm long, with a gland above; rachis 4–12 cm long with numerous raised glands all along its upper side; pinnae in 8–30 pairs, pinnules in 16–70 pairs, linear–oblong, 1.5–4 mm long,

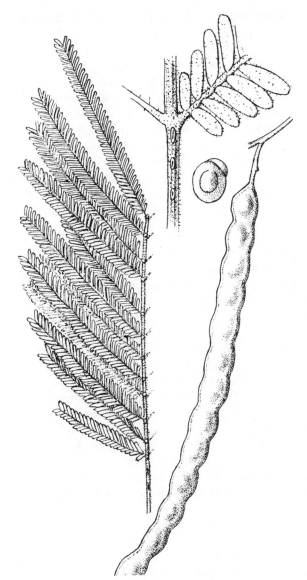

Figure 2. *Acacia mearnsii* de Wild.

0.5–0.75 mm wide; flowers in globose heads 5–8 mm in diameter, borne in panicles or racemes, on peduncles 2–6 mm long; pale yellow and fragrant; pods gray-puberulous, or sometimes glabrous, almost moniliform, dehiscing, usually 3–10 cm long, 0.5–0.8 cm wide, with 3–14 joints; seeds black, smooth, elliptic or compressed ovoid, 3–5 mm long, 2–3.5 mm wide; caruncle conspicuous; areole 3.5 mm long, 2 mm wide. Seeds 66,000–110,000/kg.

GERMPLASM: Can be crossed with *Acacia decurrens;* hybrids show more sterility than parents.

Acacia mearnsii **de Wild.**

Meiosis is regular, with no gross cytological abnormalities, and sterility may be due to gene differentiation between species. There is little geographic overlap in the native Australian ranges of the species, and there are differences in phenology (flowering; seedset). Most of the characters that vary among species are quantitative. The development of black wattle strains or of hybrids with enhanced vigor, better quality bark, outstanding stem form, or resistance to insect pests and disease would benefit the wattle industry. Assigned to the Australian Center of Diversity, black wattle or cvs thereof is reported to exhibit tolerance to drought, laterite, and poor soil. ($2n = 26$.)

DISTRIBUTION: Native to Southeast Australia (Victoria to New South Wales and southern Queensland) and Tasmania. Introduced and cultivated widely for afforestations. See Sherry (1971) for details.

ECOLOGY: In Kenya grows on or near Equator at altitudes of 2,000–2,800 m; is well adjusted to the climate of East Africa. Grows well at 30°S Lat. in South America on rolling terrain at altitudes of 50–70 m. Thrives on poor, dry soils but favors deep, moist, fertile soils. In Australia, black wattle may occur on soils derived from shales, mudstones, sandstones, conglomerates, and alluvial deposits; in Kenya on podsols, krasnozems, sandy hills, lava flows or on mixtures of lava and contemporaneous volcanic tuffs and breccias. In South America, grows on red clay or sandy soils that have suffered from severe erosion and soil depletion (ferruginous clay loams with little or no free silica). In East Africa grows where annual rainfall is 1,041–1,321 mm, (about 75% between April and September). On the equator where black wattle is grown in South America, the rain pattern is nearly opposite, mean annual temperature range is 17–23°C; there is little seasonal variation, but considerable diurnal variation. At higher altitudes in South America, frost is a risk and heavy snows may break tree limbs. Tannin content varies inversely with precipitation. Ranging from Warm Temperate Dry through Tropical Thorn to Tropical Moist Forest Life Zones, black wattle is reported to tolerate annual precipitation of 6.6–22.8 dm (mean of 6 cases = 12.6), annual mean temperature of 14.7–27.8°C (mean of 6 cases = 22.6°C), and pH of 5.0–7.2 (mean of 5 cases = 6.5).

CULTIVATION: Propagation by seed is easy. Seeds retain their viability for several years. For germination seeds are covered with boiling water and allowed to stand until cool. This cracks the hard outer coat and facilitates germination. Seeds may be broadcast or sown in rows on any barren site. Usually they are sown about 5 cm apart in seedbeds, and are transplanted after 3–6 months. In South America, fields are usually plowed and harrowed in April or May. Seedlings are set out May–November, but usually in winter, June–August, after a rain. Plants are spaced 2 m each way, at rate of 2,500/ha. Propagation by cuttings is almost impossible without mist. Air layering is more promising. Two types of farmers grow acacia: the tanner or business person plants 200 ha or so entirely to black wattle, usually one section at a time so that seeds can be planted and harvested within the same year and continue year after year; the farmer plants half or less of the land to black wattle and the rest to crops such as corn, beans, manioc, sugarcane, other vegetables, or pasture. The farmer plants 2–6 ha of acacia each year and thus evenly distributes work and production. Oxen may be useful for plowing, but most work is manual, using only plows and hoes. Intercrops may be grown the first year during which trees grow about 4–5 m in height, and about 2.5 cm in diameter.

HARVESTING: Trees provide bark 5–10 years after seeding (ave. 7). Bark is stripped from lower part of tree, then tree is felled, the remaining bark removed, and tree and bark are cut into 1-m lengths. Thoroughly dried bark is arranged in bales of 75–80 kg when ready for transportation. Tanning power improves by 10–15% in bark carefully stored for a season. Percent tannin does not differ between barks harvested in dry and wet seasons. However, the amount of bark on trees may be less on poor than on rich soils. Tannin runs about 25–35% of dried bark, on either poor or rich soil. Acacia bark may be sold as baled bark or bark powder. Dried bark may go first to commercial bark processors where it is ground or shredded in a hammermill, then sold in 40-kg sacks. Bark powder is sold in 60-kg sacks. Liquid extract is sold in 300-kg wooden barrels. In Rio Grande do Sul an estimated 5,000 MT of liquid extract is produced annually.

YIELDS AND ECONOMICS: Like some mangrove species, black wattle in pure stand produces more tannin per hectare than most tanniniferous plants. In South Africa well-managed stands have produced the equivalent of 3 MT/ha tannin, about twice the average, when grown in rotations in excess of 12 years. One 7-yr-old tree produces 3–5 kg dried bark; 12 trees produce 1 cu m of firewood. The wood of debarked trees is dried and used for mine timbers, pulpwood, and fuel. Moisture loss is rapid in first 4 weeks after felling, then much slower. Wood weighs 708.7 kg/cu m. One tree can produce up to 10 cwt of bark or about 5 cwt stripped. One ton of black wattle bark is sufficient to tan 2,530 hides, best adapted for sole leather and other heavy goods; the leather is fully as durable as that tanned with oak bark. One ton of bark yields 4 cwt of extract tar. Destructive distillation of the wood yields 33.2% charcoal, 9.5% lime acetate, and 0.81 methyl alcohol. As a source of vegetable tannin, black wattle shares with quebracho and chestnut a large portion of the world market. Plantation-grown wattle in South Africa, Rhodesia, Tanzania, Kenya, and Brazil supplied much of world demand for tannin (Sherry, 1971). In South Africa, the largest producer, annual output was 72,000 MT of ca. 120,000 MT on the world market. *Eucalyptus grandis* produces more wood than wattle, but it is inferior for fuel and charcoal. At one time in South Africa, 56% of the proceeds from wattle was from bark, the balance from timber.

BIOTIC FACTORS: The most serious disease is dieback, caused by *Phoma herbarum*. Other fungi attacking black wattle include *Chaetonium cochliodes, Daldinia* sp., and *Trichodesma viride*. In Rio Grande do Sul, disease and insects cause about 20% loss of trees. Principal insects attacking Brazilian wattle are *Molippa sabina, Achryson surinamum, Placosternus crinicornis, Eburodacrys dubitata, Neoclytus pusillus, Oncideres impluviata, Oncideres saga,* and *Trachyderes thoracica*. Ants, termites, and borers are the most damaging. The sauva ant, which attacks the leaves, is fought constantly with arsenicals and carbon disulfide.

REFERENCES:
Dougherty, J. L., 1946, Acacia negra industry in Rio Grande do Sul, *Agric. Am.* **6**(9):139–141, 147, illus.
Heiberg-Iurgensen, K., 1967, A contribution to the economics of the growing of black wattle *Acacia mearnsii* de Wild. in Natal, Ph.D. thesis, University of Natal, Pietermaritzburg.
Olembo, T. W., 1972, *Phoma herbarum* Westend.: A pathogen of *Acacia mearnsii* de Wild. in Kenya, *East Afr. Agric. For. J.* Oct. 1972:201–206.
Saayman, H. M., and Roux, D. G., 1975, The origins of tannins and flavonoids in black-wattle barks and heartwoods, and their associated "non-tannin" components, *Biochem. J.* **97**:794.
Schonau, A. P. G., 1969, A site evaluation study in black wattle (*Acacia mearnsii* de Wild), *Ann. Univ. Stellenbosch Ser. A* **44**(No. 2).
Schonau, A. P. G., 1973, Height growth and site index curves for *Acacia mearnsii* on the Uasin Gishu Plateau of Kenya, *Commonw. For. Rev.* **52**(3):245–253.
Sherry, S. P., 1971, *The Black Wattle (Acacia mearnsii* de Wild.), University of Natal Press, Pietermaritzburg, 402 pp.
Stubbings, J. A., and Schonau, A. P. G., 1972, Density and air-drying rate of the timber of black wattle *(Acacia mearnsii)*, *Univ. Natal Wattle Res. Inst., Rep.* **25**:36–38.

CONTRIBUTORS: J. A. Duke, C. F. Reed, D. G. Roux.

Acacia nilotica (L.) Del.

FAMILY: Mimosaceae
COMMON NAMES: Babul, Egyptian mimosa, or Thorn
SYNONYM: *Mimosa nilotica* L.

USES: Inner bark contains 18–23% tannin, used for tanning and dyeing leather black. Young pods produce a very pale tint in leather, notably goat hides (Kano leather). Pods were used by the ancient Egyptians. Young bark used as fiber, esteemed for toothbrushes (chewsticks). Trees tapped for gum

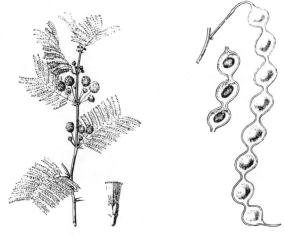

Figure 3. *Acacia nilotica* (L.) Del.

Acacia nilotica (L.) Del.

arabic. Tender pods and shoots used as vegetable and as forage for camels, sheep, and goats, especially in Sudan, where said to improve milk from these animals. Seeds are a valuable cattle food. Roated seed kernels, sometimes used for flavoring and when crushed provide dye for black strings worn by Nankani women. Trees useful in Sudan for afforestation of inundated areas. Sapwood is yellowish-white, heartwood reddish-brown, hard, heavy, durable, difficult to work, but takes a high polish. Because of its resins, resists insects and water; timber used for boat-making, posts, buildings, water pipes, well-planking, plows, cabinetwork, wheels, mallets, and other implements. Wood yields excellent firewood and charcoal.

FOLK MEDICINE: Bark, gum, leaves, and pods used medicinally in West Africa. Sap or bark, leaves, and young pods are strongly astringent due to tannin, and are chewed in Senegal as antiscorbutic; also used for dysentery. Bark decoction drunk for intestinal pains and diarrhea. Other preparations used for coughs, gargle, toothache, ophthalmia, and syphilitic ulcers. Also said to be used for antitussive, colds, demulcent, smallpox, stimulant, and tuberculosis.

CHEMISTRY: Babul reported to contain l-arabinose, catechol, galactan, galactoaraban, galactose, N-acetyldjenkolic acid, N-acetyldjenkolic acid, sulfoxide, pentosan, saponin, tannin.

DESCRIPTION: Small tree, 2.5–14 m tall, quite variable in many aspects; bark of twigs not flaking off, gray to brown; branches spreading, with flat or rounded crown; bark thin, rough, fissured, deep red-brown; branchlets purple-brown, shortly or densely gray-pubescent, with lenticels; spines gray-pubescent, slightly recurved, up to 3 cm long; leaves often with 1–2 petiolar glands and other glands between all or only the uppermost pinnae; pinnae 2–11 (–17) pairs; leaflets 7–25 (–30) pairs, 1.5–7 mm long, 0.5–1.5 mm wide, glabrous or pubescent, apex obtuse; peduncles clustered at nodes of leafy and leafless branchlets; flowers bright yellow, in axillary heads 6–15-mm diam.; involucel from near the base to about halfway up the peduncle, rarely somewhat higher; calyx 1–2 mm long, subglabrous to pubescent; corolla 2.5–3.5 mm long, glabrous or pubescent outside; pods especially variable, linear, indehiscent, 8–17 (–24) cm long, 1.3–2.2 cm broad, straight or curved, glabrous or gray-velvety, turgid, blackish, about 12-seeded; seeds deep blackish-brown, smooth, subcircular, compressed, areole 6–7 mm long, 4.5–5 mm wide. Fl. Oct.–Dec.; Fr. Mar.–June.

GERMPLASM: *Acacia nilotica* ssp. *kraussiana* (Benth.) Brenan is the most common form in east tropical Africa. Young branches more or less densely pubescent; pods not necklacelike, 1–1.8-cm wide, oblong, more or less pubescent all over at first with raised parts over seeds becoming glabrescent, shining and black when dry, margins shallowly crenate. Exhibits wide range of altitudinal and habitat requirements. Found in Botswana, Zambia, Rhodesia, Malawi, Tanzania, Angola, Mozambique, Transvaal, and Natal. *A. nilotica* var *tomentosa* A. F. Hill (*A. arabica* var *tomentosa* Benth.) has pods straight, constricted between seeds and densely tomentose; found in Senegal and northern Nigeria, to Sudan, Arabia, and India. *A. nilotica* var *adansonii* (Guill. et Perr.) Kuntze is a tree up to 17 m with dark reddish-brown bark, deeply fissured, tomentose, reddish-brown twigs and gray fruits; commonest variety in West Africa, from Senegal to Nigeria and widespread in northern parts of Tropical Africa. Assigned to the African Center of Diversity, babul or cvs thereof is reported to exhibit tolerance to high pH, salt, savanna, and waterlogging. ($2n = 52$.)

DISTRIBUTION: Native from Egypt south to Mozambique and Natal; apparently introduced to Zanzibar, Pemba, India, Arabia.

ECOLOGY: Various woodlands, wooded grasslands, scrub and thickets, Thrives in dry areas, but endures floods. Grows 10–1,340 m altitude, in a wide range of conditions. Ranging from Subtropical Desert to Subtropical Dry through Tropical Desert to Tropical Dry Forest Life Zones, babul is reported to tolerate annual precipitation of 3.8–22.8 dm (mean of 12 cases = 12.0 dm), annual mean temperature of 18.7–27.8°C (mean of 12 cases = 24.1°C), and pH of 5.0–8.0 (mean of 10 cases = 6.9).

CULTIVATION: Trees propagated in forest by seeds. Not formally cultivated.

HARVESTING: There are other sources of gum arabic, but trees are still tapped for the gum by removing a bit of bark 5–7.5 cm wide and bruising the surrounding bark with mallet or hammer. The resulting reddish gum, almost completely soluble and tasteless, is formed into balls, used in commerce to some extent but is inferior to other forms of gum arabic, with which it is sometimes mixed.

YIELDS AND ECONOMICS: Various products of the tree are used locally in tropical Africa, but none enter international markets.

BIOTIC FACTORS: Following fungi have been reported on this plant: *Cytospora acaciae, Diatryphe acaciae, Diplodia acaciae, Fomes badius, F. endotheius, F. fastuosus, F. rimosus, Fusicoccum indicum, Phyllactinia acaciae, Ravenelia acaciae-arabicae, Septogloeum acaciae, Septoria mortolensis, Sphaerostilbe acaciae.* Trees are also parasitized by *Dendrophthoe falcata* and *Loranthus globiferus* var *verrucosus.*

REFERENCE:
Theresa, Y. M., and Nayudamma, Y., 1972, A comparative study of Babul pod tannins of Indian *(Acacia arabica)* and Sudanese *(Acacia nilotica)* origin, *Leather Sci. (Madras)* **19**:341–348.

CONTRIBUTORS: J. A. Duke, C. F. Reed.

Acacia pycnantha Benth.

FAMILY: Mimosaceae
COMMON NAMES: Golden wattle, Broadleaved wattle
SYNONYM: *Acacia petiolaris* Lehm.

USES: Trees exude a gum, containing mainly arabogalactin, called wattle gum or Australian gum. Red color is due to tannins. Extract of bark is said to equal the best Indian catechu. Bark contains 33–36% tannins. Flowers used for perfume, plants as sod-binder.

FOLK MEDICINE: No data available.

CHEMISTRY: Golden wattle reported to contain β-acetylpropionic acid, β-diaminopropionic acid, albizziine, phenethylamine, S-carboxyethylcysteine, S-carboxyethylcysteine sulfoxide, and tannin.

DESCRIPTION: Shrub or small tree; branchlets more or less pendulous; phyllaries pinniveined, oblong-

Figure 4. *Acacia pycnantha* Benth.

lanceolate to falcate-lanceolate or broadly obovate, 6.5–15 cm long, 2–3-cm wide, the single vein more or less excentric; gland 1.3–2 cm from base; racemes simple or compound, with 50–80 large flowers in head (weight of bloom often bends tree) with peduncles 0.3 cm long; sepals 5, ciliate, almost as long as petals; petals showy, pods varying 5–12 cm long, 5–7 mm wide. contracted and slightly constricted between seeds; funicle (aril) whitish, club-shaped, not folded, half as long as seed or occasionally folded and transverse the seed. Fl. Feb.–Mar.; Fr. Aug.–Oct.

GERMPLASM: Assigned to the Australian Center of Diversity, golden wattle or cvs thereof is reported to exhibit tolerance to drought.

DISTRIBUTION: Native to South Australia (southern districts to Flinders Range, Yorke Peninsula, Kangaroo Islands, Murraylands); experimental plantations grown in the Nilgiris, India.

ECOLOGY: Golden wattle is less hardy in South Africa than other species, and does not tolerate poorly drained locations. Requires a temperate climate, similar to that in southern Australia. Ranging

Acacia pycnantha Benth.

from Warm Temperate Dry to Warm Temperate Moist through Tropical Thorn to Tropical Dry Forest Life Zones, golden wattle is reported to tolerate annual precipitation of 6.6–22.8 dm (mean of 6 cases = 13.0 dm), annual mean temperature of 12.8–27.8°C (mean of 6 cases = 19.9°C), and pH of 5.0–7.2 (mean of 4 cases = 6.4).

CULTIVATION: Propagated mainly by seed. Plants grow rapidly. If borers are removed as soon as noticed, trees live for many years with little attention. Established plants may be pruned by cutting back the 1-year-old wood immediately after flowering; should be watered after pruning. Seedling leaves sometimes 12.5 cm long and 10 cm wide, hence the name "broadleaved wattle."

HARVESTING AND ECONOMICS: Valuable locally in South Australia for its tanning bark and gum. No data on yields or economic value.

BIOTIC FACTORS: Following fungi have been reported on this plant: *Coniothyrium acacide, C. pycnanthae, Coryneum acaciae, Pleospora herbarum, Pseudocommis vitis, Uromyces fusisporus, Uromycladium simplex,* and *U. tepperianum.*

CONTRIBUTORS: J. A. Duke, C. F. Reed.

Figure 5. *Acacia senegal* (L.) Willd.

Acacia senegal (L.) Willd.

FAMILY: Mimosaceae

COMMON NAMES: Gum arabic, Senegal gum, Sudan gum arabic, Kher, Kumta

SYNONYM: *Acacia verek* Guill. et Perr.

USES: Tree yields commercial gum arabic, used extensively in pharmaceutical preparations, inks, pottery pigments, water colors, wax polishes, and liquid gum; for dressing fabrics, giving luster to silk and crepe; for thickening colors and mordants in calico-printing; in confections and sweetmeats. Strong rope made from bark fibers. White wood used for tool handles, black heartwood for weaver's shuttles. The long flexible strands of surface roots provide one of the strongest of local fibers, used for cordage, well-ropes, fishing nets, horsegirdles, foot-ropes, and so forth. Young foliage makes good forage. Plants useful for afforestation of arid tracts and soil reclamation.

FOLK MEDICINE: The demulcent, emollient gum is used internally in inflammation of intestinal mucosa, and externally to cover inflamed surfaces, as burns, sore nipples, and nodular leprosy. Also said to be used for antitussive, astringent, catarrh, colds, coughs, diarrhea, dysentery, expectorant, gonorrhea, hemorrhage, sore throat, typhoid, urinary tract.

CHEMISTRY: Main component of gum arabic is arabic acid, a polysaccharide composed of L-arabinose, L-rhamnose, D-galactose, and D-glucuronic acid in different molar ratios (depending on the species of source). Gum arabic is reported to contain trypsin inhibitors.

DESCRIPTION: Savanna shrub or tree, up to 20 m tall, over 1.3 m in girth, spiny; bark gray to brown or blackish, scaly, rough; young branchlets densely to sparsely pubescent, soon glabrescent, crown dense; stipules not spinescent; prickles just below the nodes, either in threes up to 7-mm long, with the middle one hooked downward and the lateral ones curved upward, or solitary with the laterals absent; leaves bipinnate, up to 2.5-cm long; leafaxis finely downy with 2 glands; pinnae 6–20 pairs; leaflets small, 7–25 pairs, rigid, leathery, glabrous, linear to elliptic-oblong, ciliate on margins, pale glaucous-green, apex obtuse to subacute; flowers in spikes 5–10 cm long, not very dense, on peduncles 0.7–2 cm long, normally produced with the leaves; calyx bell-shaped, glabrous, deeply toothed; corolla white to yellowish, fragrant, sessile; pod straight or slightly curved, strap-shaped, 7.5–18 cm long, 2.5 cm wide, thin, light brown or gray, papery or woody, firm, indehiscent, glabrous, 5–6 (–15) seeded; seeds greenish-brown. Fl. Jan.–Mar.; Fr. Jan.–Apr., July, Aug., or Oct.

GERMPLASM: Typical *A. senegal* is a tree with a single central stem and a dense flat-topped crown, bark without any papery peel, rough, gray or brown, with pubescent, rarely glabrous inflorescence, and pods variable in size, rounded to somewhat pointed but not rostrate or acuminate at apex. Variety *rostrata* Brenan is a shrub, branching at or close to base, or a small tree, with a single stem, 1–6 m tall, with dense flattened crown, bark normally with a flaking papery peel, creamy-yellow to yellow-green or gray-brown, inflorescence axis always pubescent and pods 2–3.5 times as long as wide, rostrate or acuminate at apex. Variety *leiorhachis* Brenan, is always a tree with central stem, and rounded or irregular with straggling branches; bark with conspicuous yellow papery peel, and inflorescence axis always glabrous. Variety *pseudoglaucophylla* occurs on fixed sand dunes in Africa. Assigned to the African Center of Diversity, gum arabic is reported to exhibit tolerance to alkali, drought, fire, high pH, poor soil, sand, and slope. ($2n = 26$.)

DISTRIBUTION: Widespread in tropical Africa from Mozambique, Zambia to Somalia, Sudan, Ethiopia, Kenya, and Tanzania.

ECOLOGY: Thrives on dry rocky hills, in low-lying dry savannas, and areas where annual rainfall is 25–36 cm. This hardy species survives many adverse conditions, and seems to be favored by low rainfall and absence of frost. Ranging from Warm Temperate Thorn through Tropical Thorn to Tropical Dry Forest Life Zones, gum arabic is reported to tolerate annual precipitation of 3.8–22.8 dm (mean of 9 cases = 12.4 dm), annual mean temperature of 16.2–27.8°C (mean of 9 cases = 23.8°C), and pH of 5.0–7.7 (mean of 7 cases = 6.4).

CULTIVATION: In Sudan, trees are cultivated over a very large area. Best propagated from seeds, which are produced once every few years, grown in Sudan, in special "gum gardens." Elsewhere, collected from wild trees.

HARVESTING: Gum exudes from cracks in bark of wild trees, mostly in the dry season, with little or none in the rainy season during flowering. In some areas, a long strip of bark is torn off and the gum allowed to exude. In Africa, it is regularly tapped from 6-year-old trees by making narrow transverse incisions in bark in February and March. In about a month, tears of gum form on surface and are gathered. Trees begin to bear between 4–18 years of age and are said to yield only when they are in unhealthy state due to poor soil, lack of moisture, or damaged. Attempts to improve conditions tend to reduce yield. Gum from wild trees is variable and somewhat darker colored than that from cultivated plants. Collected gum is carefully freed of extraneous matter, sorted and sometimes ripened in sun before export. Gum arabic is odorless with a bland taste, yellowish, and some tears are vermiform in shape. Ripened or bleached gum occurs in rounded or ovoid tears over 2.5 cm in diameter, and in broken fragments. Tears are nearly white or pale yellow and break readily with a glassy fracture. Gum is almost completely soluble in an equal volume of water and gives a translucent, viscous, slightly acid solution, but is insoluble in 90% alcohol. Kordofan (Sudan) Gum is yellow or pinkish, has fewer cracks and is more transparent.

YIELDS AND ECONOMICS: Annual yields from young trees may range from 188 to 2856 g (avg 900 g), from older trees, 379 to 6754 g (avg 2,000 g). Gum arabic is important export from some areas in tropical Africa and Mauritania. From Africa some genuine gum is shipped to India then to Europe and America. Between 1940 and 1950, United States imports were 3,179–8,989 MT.

Acacia senegal (L.) Willd.

BIOTIC FACTORS: Fungi reported on this crop are *Cladosporium herbarum*, *Fusarium* sp., *Ravenelia acaciae-senegalae*, and *R. acaciocola*. Many insect visitors mimic the plant, the buffalo treehopper, *Stictocephala bubalus*, being a good example. Spiders (*Cyclops* sp.) may completely cover the young growing apex.

REFERENCE:

Cheema, M. S. Z. A., and Qadir, S. A., Autecology of *Acacia senegal* (L.) Willd., *Vegetatio* 27(1–3):131–162.

CONTRIBUTORS: J. A. Duke, C. F. Reed, J. K. P. Weder.

Acacia seyal Del.

FAMILY: Mimosaceae

COMMON NAMES: Shittim wood, White whistling wood

USES: Tree yields a good quality gum, inferior to that of *A. senegal*. Systematic tapping has improved color and flavor of product. Bark contains tannin, yields a red liquid extract. Wood is white to yellow-brown, finely striated with dark lines, coarse-grained, soft, easy to work, polishes well, but discolors easily with mold and is susceptible to insect attack. The gum is said to be edible. Leaves are important for forage and the wood for fuel where trees are abundant. The pods are sold, especially for fattening sheep.

FOLK MEDICINE: The slightly acid gum is believed to be somewhat aphrodisiac. The bark decoction is used for dysentery and leprosy. The gum is used as emollient and astringent for colds, diarrhea, hemorrhage, and ophthalmia.

CHEMISTRY: This species has been reported to contain 18–20% tannin.

DESCRIPTION: Tree 3–12 m tall, crown flat-topped; bark powdery, white to greenish-yellow or orange-red; sparsely branched, the branches horizontal or ascending; young branchlets with sparse hairs or almost glabrous, with numerous reddish sessile glands; epidermis of twigs becoming reddish and shed annually; leaves often with a large gland on petiole and between the top 1–2 pairs of pinnae; stipules spinescent, up to 8 cm long, ant-galls present or absent; pinnae usually 3–7 pairs, the leaflets in 11–20 pairs, 3–8 cm long, 0.75–1 mm wide, sparingly ciliolate or glabrous; lateral veins invisible beneath; flowers bright yellow, in axillary, pedunculate heads 10–13 mm across, borne on terminal or short lateral shoots of current season; involucel in lower half of peduncle 2–4 mm long; apex of bracteoles rounded to elliptic, sometimes pointed; calyx 2–2.5 mm long, puberulous in upper part; corolla 3.5–4 mm long, glabrous outside; pods 7–20 cm long, 0.5–0.9 cm in diameter, dehiscent, falcate, constricted between seeds, glabrous except for sessile glands, 6–9-seeded; seeds elliptic, 7–9 mm long, 4.5–5 mm wide, compressed, minutely wrinkled, olive-brown to olive; areole 5–6 mm long, 2.5–3.5 mm wide. Fl. July–Oct.

GERMPLASM: Species has several botanical varieties. The two main ones are: *A. seyal* var *fistula* (Schweinf.) Oliv. (*A. fistula* Schweinf.), is white-barked, with some pairs of spines fused at base into 'ant-galls,' 0.8–3 cm in diameter, grayish or whitish, often marked with sienna-red and with longitudinal furrows down center, more or less 2-lobed. Found

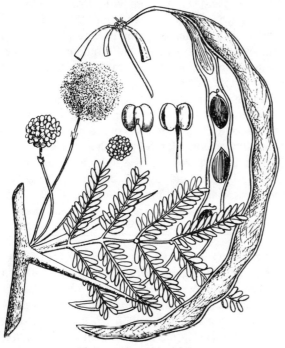

Figure 6. *Acacia seyal* Del.

in Zambia, Malawi, and Mozambique. *A. seyal* var *multijuga* Schweinf. ex Baker f. (*A. stenocarpa* Oliv., pro partem), a shrub or tree, usually less than 5 m tall, sometimes up to 13 m, flattened crown; bark on main stem greenish-brown, peeling in papery rolls; bark on branchlets red-brown, thorns straight, weak, usually less than 2.5 cm long, sometimes absent; pinnae 4–12 pairs, leaflets 10–20 pairs; flowers golden-yellow; pod narrow-linear, strongly curved, up to 10 cm long, 0.6 cm wide, dehiscing on tree. Common in overgrazed pastures and widely distributed in East Africa. Hybrids, *A. seyal* var *fistula* × *A. xanthophloea* Benth., are known from woodlands on black clay loams on lake flood plains in Malawi. Pods are conspicuously irregular, 4–11 cm long, 6–10 mm wide, ill-formed, and curved. Assigned to the Africa Center of Diversity, shittim wood or cultivars thereof is reported to exhibit tolerance to high pH, heavy soil, mycobacteria, salt, poor soil, slope, savanna, and waterlogging. ($2n = 26$.)

DISTRIBUTION: Widespread in tropical Africa, northward to Sudan and Somalia.

ECOLOGY: Trees thrive in *Sclerocarya caffra* woodlands, wooded grasslands and especially on seasonally flooded black-cotton soils along watercourses. Requires a heavy clay-alluvium, but will grow on stony ground at base of hills. Grows 20–1,220 m altitude. A gregarious savanna tree, ranging from Subtropical Desert to Dry through Tropical Desert to Very Dry Forest Life Zones, shittim wood is reported to tolerate annual precipitation of 8.7–22.8 dm (mean of 7 cases = 15.0 dm), annual mean temperature of 18.7–27.8°C (mean of 7 cases = 24.0°C), and pH of 5.0–8.0 (mean of 5 cases = 6.9).

CULTIVATION: Propagated from seed. No formal cultivation of plants noted.

HARVESTING: Pods, bark, or wood are harvested in season from trees or shrubs in native habitats. Gum also obtained from native plantings, in manner similar to that for other gum arabic plants.

YIELDS AND ECONOMICS: Gum and other products of some local importance in East Africa, but do not enter international trade.

BIOTIC FACTORS: Following fungi reported on this plant: *Fomes rimosus, Ganoderma lucidum, Leveillula taurica, Ravenelia volkensii, Trametes meyenii,* and *Uromyces schweinfurthii.*

CONTRIBUTORS: J. A. Duke, C. F. Reed.

Alysicarpus vaginalis (L.) DC.

FAMILY: Fabaceae
COMMON NAMES: Alyceclover, Buffalo clover
SYNONYM: *Hedysarum vaginale* L.

USES: Mainly grown as a forage crop, for hay and soil improvement, also makes good pasturage. In Malaya, grown as cover crop in rubber plantations. Hay appears to be equal to other legume hays in feeding value. Used both as dry and green forage in India, for grazing in Florida, and described as the best forage legume in Philippine lowlands.

Figure 7. *Alysicarpus vaginalis* (L.) DC.

Alysicarpus vaginalis (L.) DC.

FOLK MEDICINE: Decoction of roots said to be used in Java for coughs.

CHEMISTRY: Reported to have nutritive value equivalent to alfalfa (Bailey and Bailey, 1976). Reported to have significantly higher leaf nitrogen and phosphorus content than Townsville Stylo. Green foliage contains ca. 84% water and 40% protein.

DESCRIPTION: Erect or spreading procumbent summer perennial herb (self-regenerating summer annual in Florida), woody at base; stems 10–60 cm tall, densely to sparsely pubescent, becoming glabrous, often rooting at nodes; leaves unifoliate, the leaflet ovate, oblong to lanceolate, 0.5–6.5 cm long, 0.3–2.6 cm broad, acute to emarginate and mucronulate at apex, subcordate at base, finely puberulous and with longer hairs, or almost glabrous above, ciliate; reticulation prominent on both surfaces; petioles 0.4–1.5 cm long, the petiolule 1–2 mm long; stipules lanceolate, 0.7–1.7 cm long; inflorescence terminal and opposite a leaf, 2–13 cm long, mostly dense, the internodes mostly shorter than the flowers, glabrous or usually densely pubescent; rachis 2–13 cm long; peduncle 0.5 cm long; pedicles 0.5–2 mm long; primary bracts ovate-lanceolate, to lanceolate, 4–8 mm long, 1.5–2.5 mm broad, acuminate, more or less glabrous, deciduous; secondary bracts similar, 2.5–3.5 mm long; calyx pubescent, teeth narrowly triangular, 3–4 mm long, 0.5–1 mm broad, acuminate, not imbricate; standard orange, pinkish-buff or purple, often paler than other petals, 4–6 mm long, 3–4 mm broad; wings mauve, keel magenta, mauve or green with purple tip; pods 1.2–2.5 cm long, not constricted between articles; articles 4–7, subcylindrical, 2.5–3 mm long, 1.5–3 mm broad, with raised reticulate ridges, puberulous; seeds yellowish, speckled brown or entirely yellow-brown, ellipsoidal, slightly compressed, longest dimension 1.7 mm, shorter 1.5 mm, about 1-mm thick. Fl. Aug.–Feb. (Taiwan). Seeds 661,500/kg; wt. kg/hl = 77.

GERMPLASM: Alyceclover has been variously assigned to the African, Hindustani, and Indochina–Indonesia centers of diversity. Many cvs (e.g., 'Clarence,' 'Cooper,' 'Malawi,' 'Tinaroo') and strains have been developed, one having been developed in California and introduced to Taiwan as forage crop. In var *villosa* Verdc., whole plant is covered with long pale brownish hairs over 1-mm long. Alyceclover or cvs thereof is reported to exhibit tolerance to heavy soil, low pH, poor soil, and weeds. ($2n = 16, 20$.)

DISTRIBUTION: Widespread throughout the Old World Tropics, India, Philippine Islands, Malaya, Java, Ceylon, East Africa (Uganda, Kenya, Tanzania, Zanzibar). Occurs in Tanzania and Kenya, 200–420-m altitude. Introduced to West Indies, Florida, Guiana, and tropical America. Uncommon in West Africa.

ECOLOGY: Adapted to areas near Gulf of Mexico; requires a tropical to subtropical climate, and will grow from sea level to 1,380-m altitude in many tropical areas. Grows in most cultivated soils, but favors clay loams and sandy soils. Found growing in dry grassy or rocky places, on old cultivations, roadsides, bushland, and open woodlands. Ranging from Subtropical Moist through Tropical Dry to Wet Forest Life Zone, this species has been reported to tolerate annual precipitation of 9.1–42.9 dm (mean of 11 cases = 15.7 dm), annual mean temperature 18.3–27.4°C (mean of 11 cases = 24.6), and pH of 5.5–8.7 (mean of 6 cases = 7.0), with 9–12 consecutive frostfree months, each with at least 60 mm rainfall. Tolerates lower soil pH than most leguminous forage plants. Tolerates high rainfall, but minimum rainfall requirement is as low as 3 dm over 4 mos.

CULTIVATION: Easily propagated from scarified seed. Seeded in late spring for seed to mature in same season. Seeding rate 11–13 kg/ha broadcast, 2kg/ha for pasture. Inoculation, unnecessary. Stems branch in thin stands but plants grow 1-m tall with little branching in thick stands. In Taiwan, seeded in grass pastures in April–May. If fruit matures and shatters, crop is reseeded in succeeding years. Requires little cultivation or care after seeding. Like Townsville Stylo, this reponds well to phosphorus.

HARVESTING: Crop sets seed abundantly in late fall, and seed must be harvested before the fruits shatter. Plants used for green forage through summer and cut for dry forage in fall.

YIELDS AND ECONOMICS: In the 1950s, seed from about 2,000 ha was harvested—sufficient to plant ca 30,000 ha. Yield, about 45 T green fodder/ha with 16.21% dry matter and 3.88% crude protein on fresh basis. A very important tropical legume used for forage in many lands, as Taiwan, Philippines, Southeast Asia, India, Malaya. Some planted in California and other subtropical areas.

BIOTIC FACTORS: Following fungi have been reported on this plant: *Acanthostigma heterochaete, Erysiphe polygoni, Ophiogene philippinensis, Parodiella perisporioides, Pellicularia filamentosa, Synchytrium alysicarpi*. It is also attacked by the bacterium, *Xanthomonas alysicarpi* var *vaginalidis*. It is sensitive to nematode attacks, especially on heavy, poorly drained soils.

REFERENCE:
Martin, T. J., and Torssell, W. R., 1974, Buffalo clover [*Alysicarpus vaginalis* (L.) DC]: A pasture legume in Northern Australia, *J. Aust. Inst. Agric. Sci.* **40**(3):232–234.

CONTRIBUTORS: J. A. Duke, A. E. Kretschmer, Jr., C. F. Reed.

Anthyllis vulneraria L.

FAMILY: Fabaceae
COMMON NAME: Kidneyvetch

USES: Cultivated as a forage plant in the temperate regions of Europe and North Africa. Especially good for goats and sheep.

FOLK MEDICINE: Considered astringent, depurative, diuretic, and vulnerary; used for dressing wounds.

CHEMISTRY: Russian analyses indicate: 86% dry matter (= ? 14% water) 8.6% protein, 2.7–3.2% fat, 26.7–32.9% crude cellulose, 37.2—40.4% N-free extract, 2.1–2.3% nitrogen, 6.0–7.8% ash. The ash breaks down as 4.4–9.2% phosphoric acid, 12.3–25.4% K, 25–34% Ca, and 2.2–3.0% Mg. This species has been reported to contain malonic acid, raffinose, saponins, and tannins. Seeds are reported to contain trypsin inhibitors.

DESCRIPTION: Annual, biennial, or perennial; stems up to 90-cm tall, erect, ascending or decumbent, much or little branched, plant herbaceous; leaves

Figure 8. *Anthyllis vulneraria* L.

compound with 1–9 leaflets, the lowermost leaves reduced to a terminal leaflet; upper leaves imparipinnate; stipules small, caducous; flowers in dense, rounded, usually paired heads, subtended and half-encircled by 2 dissected leafy bracts borne close beneath the flowers; flowers 12–15 mm broad; calyx inflated at anthesis, constricted at the apex, with 5 equal teeth, the mouth oblique, with dense white

Anthyllis vulneraria L.

woolly hairs, distinctly constricted at the throat; corolla yellow, red, purple, orange, white, or multicolored; legume 1–2-seeded; seeds 396,900/kg; wt. kg/hl = 77. Fl. summer, June–Aug.; Fr. fall. In warmer climates Fl. April–Aug.

GERMPLASM: A polymorphic species with about 30 infraspecific taxa, about 24 of them in Europe. Taxa fall into two rather distinct groups, but intermediates occur and make identification difficult. The main differences are width of calyx, color of calyx, height of stem, habit of plant, and amount of branching. In Europe, *A. v.* ssp. *carpatica* (Pant.) Nyman is cultivated for forage and is naturalized; *A. v.* ssp. *polyphylla* (DC.) Blocki, a subglabrous variant, is cultivated as forage plant. Assigned to the Eurosiberian and Mediterranean Centers of Diversity, kidneyvetch or cvs thereof is reported to exhibit tolerance to high pH, poor soil, sand, and slope. ($2n = 12$.)

DISTRIBUTION: Native throughout Europe from the Atlantic Coast to the Baltic region, Turkey, south into North Africa (Algeria, Morocco, Ethiopia). Also in Caucasus and Iran; Iceland.

ECOLOGY: A warm-season crop adapted to many types of temperate climate, often naturalized and found in dry fields, meadows, alpine pastures, on hillsides, and screes. Not particular as to soil type or structure. Thrives equally well from sealevel to 2,230-m elevations in the high mountainous regions of Pyrenees and Alps; a deep-rooted herb, of some value on poor light sandy soils of upper latitudes. Ranging from Boreal Wet through Warm Temperate Thorn Forest Life Zone, kidneyvetch is reported to tolerate annual precipitation of 4.4–13.6 dm (mean of 18 cases = 7.0 dm), annual mean temperature of 6.6–18.6°C (mean of 18 cases = 9.1°C), and pH of 4.8–8.0 (mean of 17 cases = 6.8).

CULTIVATION: Propagated by seeds broadcast at rate of 17–22 kg/ha. Cultivated much like clover and requiring about the same fertilizer and care.

HARVESTING: Established crop cut for forage nearly anytime from July until frost; usually mowed like alfalfa or clover. Commonly used for sheep and goats.

YIELDS AND ECONOMICS: Russia reports yields of 4–20 MT/ha, all in one cut. An important forage plant in temperate Europe, the Mediterranean region and North Africa, especially used as food for sheep and goats.

BIOTIC FACTORS: The following fungi have been reported on this plant: *Cercospora radiata, Erysiphe martii, Microdiplodia microsporella, Mycosphaerella vulnerariae, Phoma anthyllidis, Pleospora anthyllidis,* and *Uromyces anthyllidis.* Also the nematode *Ditylenchus dipsaci* has been isolated from the plant. Entirely cross-pollinated, mostly by bees.

REFERENCE:
Sterk, A. A., 1975, Demographic studies of *Anthyllis vulneraria* L. in the Netherlands, *Acta Bot. Neerl.* **24**(3/4):315–337.

CONTRIBUTORS: J. A. Duke, C. F. Reed.

Arachis hypogaea L.

FAMILY NAME: Fabaceae
COMMON NAMES: Groundnut, Peanut

USES: Seeds yield a nondrying, edible oil, used in cooking, margarines, salads, canning, for deep-frying, for shortening in pastry and bread, in pharmaceutical industry and for soaps, cold creams, pomades and lubricants, emulsions for insect control, vegetable ghee, and fuel for diesel engines. The cake is a high-protein livestock feed and may be used as flour, for humans. Other products include peanut milk, ice cream, paints, dyes, and massage oil. Seeds are used whole-roasted and salted, or chopped in confectioneries, or ground into peanut butter. Young pods may be consumed as a vegetable. Peanut hulls are used for furfural, fuel, filler for fertilizers, livestock feed, or sweeping compounds. Foliage provides silage and forage. Hogs may glean the fields following the harvester. Most peanuts produced in the USA are marketed as peanut butter (50%), salted peanuts (21%) and confectionery (16.5%). Elsewhere peanuts are processed mainly for oil.

FOLK MEDICINE: Peanuts play a small role in various folk pharmacopoeias.

CHEMISTRY: Shelled, uncooked seeds are reported to contain approximately per 100 g: over 500 calo-

Arachis hypogaea L.

Figure 9. *Arachis hypogaea* L.

ries, 4–13 g moisture, 21.0–36.4 g protein, 35.8–54.2 g fat, 6.0–24.9 g total carbohydrate, 1.2–4.3 g fiber, 1.8–3.1 g ash, 49 mg Ca, 409 mg P, 3.8 mg Fe, 15 µg β-carotene equivalent, 0.79 mg thiamine, 0.14 mg riboflavin, 15.5 mg niacin, and 1 mg ascorbic acid. Roasted seeds contain approximately per 100 g: 595 calories, 1.8 g moisture, 23.2 g protein, 50.9 g fat, 21.7 g total carbohydrate, 3.2 g fiber, 2.4 g ash, 42 mg Ca, 354 mg P, 0.45 mg thiamine, 0.11 mg riboflavin, and 15.3 mg niacin. Boiled seeds contain per 100 g: 235 calories, 44.6 g moisture, 16.8 g protein, 8.3 g fat, 26.3 g total carbohydrate, 6.1 g fiber, 4.0 g ash, 45 mg Ca, 260 mg P, 5.1 mg Fe, 0.44 mg thiamine, 0.16 mg riboflavin, and 1.4 mg niacin. Raw leaves contain per 100 g: 69 calories, 78.5 g moisture, 4.4 g protein, 0.6 g fat, 14.9 g total carbohydrate, 4.6 g fiber, 1.6 g ash, 262 mg Ca, 82 mg P, 4.2 mg Fe, 7,735 µ β-carotene equivalent, 0.23 mg thiamine, 0.58 mg riboflavin, 1.6 mg niacin, and 98 mg ascorbic acid.

DESCRIPTION: Annual ascending (Guaranian and sequential Peruvian) to somewhat longer-lived ascending, decumbent, or prostrate (Bolivian and Amazonan), geocarpic, glabrate to hirsute herbs, with upright main or *n*-axes. The two principal, *n* + 1 order, vegetative axes arise in the axils of the alternately arranged cotyledons (cataphylls). The first two nodes (or sometimes three) of every vegetative branch are also subtended by cataphylls. The longer-lived Bolivian and Amazonan subspecies *A. hypogaea hypogaea** produces a pair (or three) of *n* + 2 order vegetative (*V*) axes, from the cataphyllar nodes of the *n* + 1 axes, followed by a pair (or three) of *n* + 2 order reproductive (*R*) axes, to establish along the length of the *n* + 1 laterals an alternating pattern of paired *V* with paired *Rn* + 2 axes. With rare exception, all *n* + 1 order axes are *V* and no flowering axes occur along the main stem. In axes of higher order (*n* + 4, *n* + 5), *R/V* tends to rise. The Guaranian and sequential Peruvian subspecies *A. hypogaea fastigiata* produce *n* + 2*R* axes from their cataphyllar nodes, and produce *n* + 1 order *R* axes along the *n* axis above the fourth node. The *n* + 2*R* axes succeed each other along the *n* + 1*V* axes uninterrupted by an *n* + 2*V* axis (sequential branching pattern) to terminate all further branching in *A. hypogaea fastigiata fastigiata,* or occur in large sequential runs, interrupted by shorter runs of *n* + 2*V* axes in *A. hypogaea fastigiata vulgaris*. The *R* axes in both subspecies are small replicas of the *V* axes with scalelike leaves, in the axils of which, either the very short peduncle bearing a single pedicel of a flower is borne (simple inflorescence) or a peduncle bearing secondary peduncles (compound inflorescence) is produced. In *A. hypogaea hypogaea* the *R* axes are simple and expand only moderately during maturation. In *A. hypogaea fastigiata vulgaris* they may be compound but expand only moderately during maturation. In Peruvian *A. hypogaea fastigiata fastigiata,* the internodes of the simple *R* axis may elongate during maturation to form a conspicuous branch, which in late season may become vegetative at its tip, bearing secondary inflorescences in the axils of its new green leaves. These *R* axes can be confused with *V* axes, but may be distinguished by their numerous scale-leaf basal nodes and their much smaller diameters. Tap root with four series of spirally arranged lateral roots with abundant branching and usually heavily supplied with nodules. Root tips without epidermis and without root hairs. Leaves stipulate, pinnate with two opposite

---

* Although taxonomists usually avoid the use of trinomials and quadrinomials, they are here used to conform with the usages now prevalent among peanut specialists.

Arachis hypogaea L.

pairs of leaflets, alternately arranged in a 2/5 phyllotaxy on the *n* axis; distichous on *n* + 1 and higher-order branches. Flowers pealike, enclosed between two bracts, one simple, subtending a very short peduncle, the other bifid, subtending the pedicel; sessile, but appear to be stalked after growth of a tubular hypanthium just before anthesis. The ovary is surrounded by the base of the hypanthium (perigynous), on the distal end of which are inserted two calyx lobes, one awllike opposite the keel and the other broad and four-notched opposite the back of the standard. Petals orange, yellow, cream or rarely white; inserted between the calyx lobes and the fused bases of the anther filaments (staminal column). The standard is orange with red veins marking the more yellow central face or brick-red by extension of the red-veined area. Wings yellow, or yellow at base and orange apically, to brick-red; keel colorless to faintly yellow, clasping the staminal column and bending at right angles with it about halfway along its length. Stamens 10, sterile filaments usually 2, anthers 8 (sometimes 9, rarely 10), 4 globose, uniloculate, alternating with 4 oblong, 3 of which are biloculate and 1, opposite the standard, uniloculate. The tip of the ovary, bearing from 1–5 ovules, grows out from between the floral bracts, bearing with it the dried petals, calyx lobes, and hypanthium; creating a unique floral structure—the peg. The peg quickly turns down toward the soil and thrusts its tip with its ovules several centimeters into the soil, where the tip turns horizontally and develops into the pod. Fruit an indehiscent legume up to 10 cm long; seeds 1–5, from less than 1 cm long × 0.5 cm thick to 3.5 cm × 1.5 cm weighing from less than 0.2 g to over 2.0 g. Testa thin, colors pink, red, purple, tan, brown, yellow, white or red and white, pink and white, brown and white, purple and white, or marked with small purple dashes or splashes on a base color. ($2n = 4\times = 40$.)

GERMPLASM: More than 4,000 entries in the germplasm bank in the United States have arisen from the Guaranian (Spanish, Valencia, Natal Common, Barberton, Manyema, Tatu, Pollachi and numerous other locally named cvs), the Bolivian and Amazonan (Virginia Bunch, Virginia and Georgia Runners, Matevere, Overo, Mani Pintado, etc.) and Peruvian (Tinga Maria, Chinese) gene centers and their extensions into North America, Africa, Europe, and Asia. Their classification:

	A. hypogaea L. (1753)	
	ssp. *hypogaea* Krap. & Rig. (1960)	
Alternate: Bolivian and Amazonan	var. *hypogaea* Krap. (1968)	
Coastal Peruvian	var. *hirsuta* Kohler (1898)	
	ssp. *fastigiata* Waldron (1919)	
Sequential: Peruvian Selva and Guaranian	var. *fastigiata* Krap. (1968)	
Guaranian	var. *vulgaris* Harz (1885)	

Cultivar distinction within botanical cvs is based on pod and seed. In addition to the cultivated peanut, wild *Arachis* species known to be cross-compatible with cultivated peanuts and have resistance to pests and diseases, including early and late leafspot and spidermites. Cultivars resistant to diseases: 'Schwarz 21,' resistant to slime disease, *Pseudomonas solanacearum;* 'Tarapota' and PI's 314817 and 315608, resistant to rust, *Puccinia arachidis;* 'NC 3033,' resistant to black rot, *Cylindrocladium crotalariae;* IRHO Nos. 56–369 and 'H32,' resistant to rosette virus; Valencia PI's 337394F and 337409, resistant to *Aspergillus flavus;* 'Tarapoto' and PI 109839, resistant to early leafspot, *Cercospora arachidicola;* NC 2, resistant to stem rot, *Sclerotium rolfsii;* PI's 295233 and 290606 resistant to lesion nematode, *Pratylenchus brachyurus;* and 'Natal Common' and 'Kumawu Erect,' resistant to root knot nematode, *Meloidogyne arenaria.* Cultivars resistant to insects: 'Southeastern Runner 56-15,' resistant to fall armyworm, *Spodoptora frugiperda;* 'NC 6,' resistant to the southern corn rootworm, *Diabrotica undecimpunctata howardi;* 'Spancross' to leaf feeding; 'NC 10247,' 'NC 10272,' 'NC 15729' and 'NC 15745,' resistant to the potato leafhopper, *Empoasca fabae.* Assigned to the South American and African Centers of Diversity, peanut or cvs thereof is reported to exhibit tolerance to aluminum, disease, drought, frost, fungus, high pH, heat, insects, laterite, limestone, low pH, sand, smog, savanna, ultraviolet, and virus.

DISTRIBUTION: Native to South America; now widely cultivated in warm countries throughout the world. Introduced in pre-Columbian times to West Indies and Mexico, in early post-Columbian times to Africa and eastern Asia and during the colonial period to Atlantic North America.

ECOLOGY: Suitable for tropics, subtropics, and warm temperate regions, grown from 40°S to 40°N latitude. Growing period 3½–5 months ('Chico' matures in 80 days in South Texas). Frost sensitive. Thrives with 5 dm water in the growing season with most in mid-one-third of season. Grows on light, friable, well-drained sandy loams, but will grow in heavier soils. Ranging from Cool Temperate Moist through Tropical Thorn to Wet Forest Life Zones, peanut is reported to tolerate annual precipitation of 3.1–41.0 dm (mean of 162 cases = 13.8 dm), annual mean temperature of 10.5°C–28.5°C (mean of 161 cases = 23.5°C), and pH of 4.3–8.7 (mean of 90 cases = 6.5).

CULTIVATION: All commercial peanuts are propagated from seed. Virginia-type (alternately branched) peanuts have a dormancy period; Spanish-Valencia types (sequentially branched) have little or no dormancy. Seedbed should be prepared, either on the flat, or widely ridged. Seed often treated with antifungal dressing before planting. In countries of advanced agriculture peanuts are often grown in monoculture and by mechanized means. In many countries they are cultivated by hand and sometimes in mixed culture. The spacing and seed rate vary with growth habit and production methods. Stands of 250,000 plants/ha are sought in machine-drilled planting. For hand planting rates may be much lower. Weeds are controlled by cultivation and by pre- and postplanting applications of selective herbicides. Short season cvs in semiarid regions of West Africa respond to early application of N. Phosphorous (P) is added on tropical red earths. Less P is added on temperate sandy soils where other crops in the rotation receive P. Roots and fruits absorb nutrients. Calcium (Ca) supply in the pegging zone is essential for high yield of good quality peanuts in large-podded, alternate types. Seeds produced on Ca-deficient soil often have poor germination and poor seedling growth. In tropical red soils of Africa, addition of S may be beneficial.

HARVESTING: Although flowering may commence in 30 days, 80–150 days or more are required for fruit maturation. In hand-harvest plants are pulled and turned over on the ground or stacked or placed on racks to cure. Pods are picked and allowed to finish drying in depths of 5 cm or less on trays, or spread in the sun in the dry season tropics. In fully mechanized harvesting a single operation pulls, inverts and windrows the plants where they remain a few days for preliminary drying. The pods are removed by combines and elevated into baskets attached to the combine or blown directly into trailing "drying wagons" that, when full, are towed to a drying station where warm or ambient air is forced through the load of peanuts. In Argentina the combines pick and shell the pods in one operation; the crop is marketed as dried seeds instead of dried pods.

YIELDS AND ECONOMICS: Yields have increased remarkably in the USA and other countries since 1951 and now range from 2,000 to 6,000 kg/ha, depending on cv, agronomic practice and weather. With poorer conditions and cvs, yields range from 400 to 1,500 kg/ha. Shelling percentage: 75–80% (sequential types) and 60–80% (alternate types). World production with shell in 1975 from 19,384,000 ha was 19,117,000 MT (avg. 986 kg/ha). Asia produced 11,128,000 MT (avg. 866 kg/ha). Africa produced 5,116,000 MT (avg. 743 kg/ha). North America produced 1,936,000 MT (avg. 2559 kg/ha); South America, 879,000 MT (avg. 1128 kg/ha); Oceania, 35,000 MT (avg. 1228 kg/ha), and Europe, 23,000 Mt (avg. 2202 kg/ha). Production was highest in India, 6,600,000 MT; second in China, 2,791,000 MT; third in United States, 1,750,000 MT; fourth in Senegal, 1,130,000 MT; and fifth in South Africa, 1,100,000 MT.

BIOTIC FACTORS: Self-pollinating, occasionally outcrossed by bees. Fungal diseases: *Ascochoyta arachidis* (leafspot), *Aspergillus flavus* (yellow mold), *A. niger* (crown rot), *A. pulverulentus* (crown rot), *Botrytis cinerea* (blight), *Cercospora arachidicola* (early leafspot), *Cercospora canescens*, *Cercosporidium personatum* (late leafspot), *Colletotrichum arachidis* (anthracnose), *C. dematium* (anthracnose), *C. mangenoti* (anthracnose), *Diplodia arachidis* (collar rot), *D. gossypina* (collar rot), *Dothiorella arachidis* (stem disease), *Fusarium moniliforme*, *F. oxysporum*, *F. roseum*, *F. solani* var *martii*, *Leptosphaerulina crassiasca* (pepper spot and leaf scorch), *Macrophomina phaseoli* (wilt, root rot, and stem rot), *Oidium arachidis* (powdery mildew), *Pestalotiopsis arachidis* (leafspot), *Phyllosticta arachidis* and *Ph. hypogaeae* (leafspot), *Puccinia arachidis* (rust), *Pythium debaryanum* (pod rot), *P. myriotylum* (pod rot), *P. ultimum*, *Rhizoctonia solani* (root rot), *Rhizopus arrhizus*, *R. oryzae*, *R. stolonifer*, *Rhizoctonia solani* (all cause seed and preemergence seedling rot), *Sclerotinia arachidis*,

Arachis hypogaea L.

S. minor (root and pod rot), *S. sclerotiorum* (root and pod rot), *Sclerotium rolfsii* (stem rot), *Verticillium dahliae* and *V. albo-atrum* (wilt and pod rot), *Sphaceloma arachidis* (scab), *Cylindrocladium crotalariae* (black rot of roots, pegs, and pods), *Phomopsis sojae* (leaf and stem diseases), *Diaporthe sojae*, *Phomopsiodes arachidis* (stem diseases), *Chalara elegans* (black hull), *Phoma arachidicola* (web blotch), *Cristulariella pyramidalis* (zonate leafspot). Some strains of *Aspergillus flavus* and *A. parasiticus*, soilborne pathogens, may enter pods and kernels and produce toxic and carcinogenic aflatoxins. Bacterial diseases: *Bacterium solanacearum*, *Phytomonas solanacearum*, *Xanthomonas solanacearum*, *Pseudomonas solanacearum*, and brown bacterial leafspot. Viruses: abutilon mosaic, alfalfa mosaic, bean chlorotic ringspot, bean mosaic, bean necrosis, bean yellow mosaic, Brazilian tobacco streak, bunchy plant, chlorotic rosette, Euphorbia mosaic, Kromnek disease, leaf curl, marginal chlorosis, mosaic rosette, ringspot and mottle, *Arachis* virus I, rugose leafcurl, tobacco mosaic, southern sunnhemp mosaic, tomato spotted wilt, turnip mosaic, white clover mosaic and witches' broom, peanut stunt. Bud necrosis (TSWV) is a serious disease in India, and rosette can be devastating in Africa. Nematodes: *Belonolaimus longicaudatus*, *Meloidogyne arenaria*, *M. hapla*, *Pratylenchus brachyurus*. Of lesser importance are: *Aphasmatylenchus straturratus*, *Aphelenchoides arachidis*, *Criconemoides* spp., *Helicotylenchus* spp., *Hemicycliohora* spp., *Hoplolaimus* spp., *Longidorus* spp., *Meloidogyne javanica*, *Radopholus similis*, *Scutellonema* spp., *Telotylenchus* spp., *Trichodorus* spp., *Tylenchorhynchus* spp., and *Ziphinema* spp. Insects: (1) Soil insects: lesser cornstalk borer, *Elasmopalpus lignosellus* Zeller; southern corn rootworm, *Diabrotica undecimpunctata howardi* Barker and also *D. balteata;* whitefringed beetles, *Graphognathus* spp.; burrowing bug, *Pangeaus bilineatus* Say and *P. congruus;* white grub, *Strigoderma arboricola* Fabricius; bahiagrass borer, *Derobrachus brevicollis* Audinet-Serville; and wireworms *Conoderus*, *Melanotus*, *Heterodires* and *Cebria*. (2) Foliage insects: corn earworm, *Heliothis zea* Boddie; tobacco budworm, *H. virescens* Fabricius; fall armyworm, *Spodoptera frugiperda* J. E. Smith; beet armyworm, *S. exigua* Hubner; granulate cutworm, *Feltia subterranea* Fabricius; velvetbean caterpillar, *Anticarsia gemmatalis* Hubner; rednecked peanutworm, *Stegasta bosqueella* Chambers; the salt marsh caterpillar, *Estigmene acrea;* green cloverworm, *Platypena scabra* Fabricius; cabbage looper, *Trichoplusia ni* Hubner; tobacco thrips, *Frankliniella fusca* Hinds; potato leafhopper, *Empoasca fabae* Harris; three-cornered alfalfa hopper, *Spissistilus festinus* Say; and the arachnid spidermites, *Tetranychus urticae*, *T. cinnabarinus,* and *T. desertorum* Koch. (3) Storage insects: Indian meal moth, *Plodia interpunctella* Hubner; Mediterranean flour moth, *Anagasta kuehniella* Zeller; almond moth, *Cadra cautella* Walker (Ephestis); sawtoothed grain beetle, *Oryzaephilus surinamensis* L.; red flour beetle, *Tribolium castaneum* Herbst; and the confused flour beetle, *T. confusum* (duVal). Other insects: *Aphis craccivora* Koch vector of rosette and other viruses (worldwide), *Holotrichia* sp., white grubs (India), *Amsacta* sp. (India), *Peridontopyge, Entermes, Anoplocnemis,* and *Halticus* (Senegal). Dicotyledonous parasites: *Alectra abyssinica*, *A. senegalensis* var *arachidis*, *A. vogelii*, *Striga asiatica*, *S. gesneriodies*, *S. hermonthica*, *S. lutea* and *S. senegalensis*. Weeds: *Ageratum conyzoides*, *Cenchrus echinatus*, *Cynodon dactylon*, *Cyperus rotundus*, *Digitaria longiflora*, *Digitaria sanguinalis*, *Echinochloa colonum*, *Eleusine indica*, *Portulaca oleracea*, *Rottboellia exaltata*, *Setaria pallidefusca*, *Sorghum halepense*, *Tribulus terrestris*, *Tridax procumbens*.

REFERENCES:

Wilson, C. T. (ed.), 1973, *Peanuts—Culture and Uses*, American Peanut Research and Education Association. Stillwater, Oklahoma, 684 pp.

Woodroof, J. G. (ed.), 1973, *Peanuts: Production, Processing, Products*. Avi Publishing, Westport, Conn., 330 pp.

CONTRIBUTORS: T. E. Boswell, A. H. Bunting, W. V. Campbell, Tricia Dodd, J. A. Duke, W. C. Gregory, R. O. Hammons, R. L. Lynch, A. J. Norden, H. Pollard, C. F. Reed, C. E. Simpson, J. Smartt, Donald H. Smith, Olin D. Smith, E. B. Whitty, J. C. Wynne, Clyde T. Young.

Astragalus cicer L.

FAMILY: Fabaceae
COMMON NAME: Cicer milkvetch

USES: Cicer milkvetch is a valuable forage legume for range and pasture plantings and for erosion control.

Astragalus cicer L.

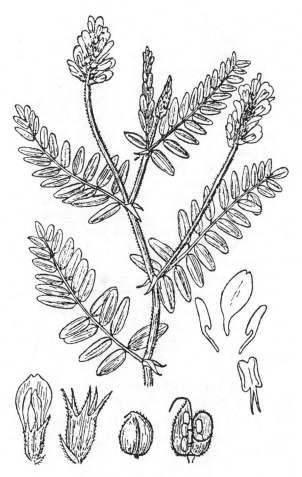

Figure 10. *Astragalus cicer* L.

CHEMISTRY: This species has not been reported to cause bloat in the grazing animal. It does not contain toxic levels of nitro compounds, tannins, oxalates, alkaloids, or selenium as do certain other *Astragalus* species. Quality of forage as measured by crude protein, *in vitro* dry matter digestibility, cell wall constituents, lignin, hemicellulose, cellulose, and silica is equivalent to that of the more commonly grown forage legumes.

DESCRIPTION: A long-lived, rhizomatous perennial with a branched taproot, up to 1.3 m tall; stems ascending or prostrate, sparsely vested with appressed short hairs; stipules 8–10 mm long, connate at base, oblong to triangular-oblong with short white hairs; leaves compound, 7–21 cm long; leaflets 8–17 pairs, lanceolate to lance-oblong, round-tipped obtuse or acute, mucronulate, 15–30 mm long, short appressed hairs on both sides; racemes compact, many-flowered, 4–6 cm long; bracts linear, about 5–7 mm long, covered with short black and white hairs; calyx 7–9 mm long; corolla ochroleucous; standard 14–16 mm long; pods sessile, ovoid-spherical or spherical, inflated, 10–14 mm long, membranous, densely villous with short black and predominately long white appressed hairs.

GERMPLASM: Two released cvs, 'Lutana' from the United States and 'Oxley' from Canada. Germplasm is available from breeding programs in the United States and Canada and from native stands in Europe and Asia. Assigned to the Eurosiberian Center of Diversity, cicer milkvetch is reported to tolerate drought, frost, and limestone. ($2n = 64$.)

DISTRIBUTION: A European–Asiatic species, distributed from the Caucasus Mountains through southern Europe to Spain. Variability within the species apparently maximum in the Mediterranean and sub-Mediterranean zones of Europe.

ECOLOGY: In native habitat, common in both damp and dry meadows, orchards, and woodlands. In the United States and Canada, is well-adapted to a wide range of soil types, especially those derived from limestone. Rhizome development is best in moderately coarse to coarse-textured soils containing adequate moisture. It is very winter-hardy and moderately drought resistant and apparently adapted to regions receiving more than 40 cm of annual precipitation. Also adapted to subirrigated sites.

CULTIVATION: Propagated from seeds but can be propagated from rhizomes. Generally grown in association with cool season forage grasses and used as pasture, but can be used for hay. Under favorable conditions, one of the most persistent forage legumes.

YIELDS AND ECONOMICS: When harvested for hay under irrigation cicer milkvetch yields nearly as much (75–80%) as alfalfa (*Medicago sativa* L.). When clipped frequently, however, it yields more and persists better than alfalfa. Is less persistent than alfalfa in areas receiving less than 4 dm annual precipitation. Relatively poor seedling vigor and stand establishment have limited acceptance but have been improved. Livestock producers appreciate the nonbloating character.

BIOTIC FACTORS: Cross-pollinated and predominately self-incompatible.

Astragalus cicer L.

REFERENCES:

Johnston, A., Smoliak, S., Hanna, M. R., and Hironaka, R., 1975, *Cicer Milkvetch for Western Canada,* Agriculture Canada Publication 1536, 16 pp.

Smoliak, S., Johnston, A., and Hanna, M. R., 1972, Germination and seedling growth of alfalfa, sainfoin and cicer milkvetch, *Can. J. Plant Sci.* **52:**757–762.

Stroh, J. R., Carleton, A. E., and Seamands, W. J., 1972, Management of Lutana cicer milkvetch for hay, pasture, seed, and conservation uses, *Mont. Agric. Exp. Stn., Bull.* **666:**16 pp.

Townsend, C. E., 1974, Selection for seedling vigor in *Astragalus cicer* L., *Agron. J.* **66:**241–245.

Townsend, C. E., 1977, Recurrent selection for high seed weight in cicer milkvetch, *Crop Sci.* **17:**473–476.

Townsend, C. E., Christensen, D. K., and Dotzenko, A. D., 1978, Yield, quality, and persistence of cicer milkvetch as influenced by cutting frequency, *Agron. J.* **70:**109–113.

CONTRIBUTOR: C. E. Townsend.

Astragalus gummifer Labill.

FAMILY: Fabaceae

COMMON NAME: Tragacanth

USES: Perhaps man first used tragacanth as a survival food. Ants, goats, and sheep appear to relish the sweeter gums. Tragacanth gum is one of the oldest natural emulsifiers known to man. The gum consists of two major parts: (1) tragacanthin, which is water soluble, and (2) bassorin, a complex carbohydrate that swells but is insoluble in water. Used to give body to liqueurs. Best grades used in pharmacy as a demulcent and emulsifier, and as an adhesive agent and excipient for pills and tablets and for suspensions of insoluble powders. Other grades are used in calico printing. Films made from aqueous solutions of gum tragacanth are unaffected by organic solvents (e.g., ethanol, carbon tetrachloride, acetone, xylene, or toluene). Painted with solutions of basic or neutral lead acetate, these films become insoluble in water. Gum is also used in salad dressings, sauces, ice creams, confections, syrups, milk powder stabilizers, citrus oil emulsions, cheeses. Medicinally, it is used in cod-liver oil, mineral oil, or paraffin oil emulsions, toothpastes, diabetic syrups, lubricating jellies, and as an emulsifier for poorly soluble substances, such as steroids, cardiac glycosides, barbiturates, and fat-soluble vitamins. Importance of the wood as fuel might jeopardize many natural stands in Iran.

FOLK MEDICINE: Mucilaginous, demulcent, emollient, and laxative, tragacanth is occasionally used as a remedy for cough or diarrhea, where demulcents are indicated. Aphrodisiac qualities are ascribed to the gum (Gentry, 1957).

CHEMISTRY: This species has been reported to contain barium, canavanine, glycyrrhizin (2), malonic acid, saponin, *S*-methylcysteine, and trigonelline. Gum tragacanth is reported to contain tragacanthin, bassorin (60–70%) and small amounts of cellulose and starch. Tragacanthin is a soluble polysaccharide composed of D-galacturonic acid, D-galactose, L-fucose, D-xylose, and L-arabinose. Bassorin is an insoluble methylated acidic polysaccharide, probably methylated tragacanthin. Seed of some *Astragalus* species are reported to contain trypsin inhibitors.

DESCRIPTION: Low umbraciform shrub, up to 1 m tall, thorny and branching; leaf-rachis spiny, 2–5 cm long, straight, glabrous or sparsely pubescent; leaves compound, the leaflets 5–8 (–10) mm long, elliptic, usually mucronate, glabrescent, 4–7-paired; stipules 7–9-mm long, triangular-ovate, glabrous or white-hairy; flowers 2–3 per leaf-axil, in lax, ovoid to oblong, 8–20-flowered inflorescences;

Figure 11. *Astragalus gummifer* Labill.

Astragalus gummifer Labill.

bracts 4–6 (–8) mm long, ovate-orbicular, navicular, bilobed, glabrous or white-hairy; bracteoles absent; calyx 5–7 mm long, white-pilose to the base; calyx-lobes divided almost to base; standard 10–12 mm long.

GERMPLASM: *Astragalus gummifer* is not the only gum-producing species. Gentry (1957) includes: *Astragalus adscendens* Boiss. & Haussk., *A. brachycentrus* Fisch., *A. cerasocrenus* Bunge, *A. echidnaeformis* Sirjaev, *A. elymaiticus* Boiss. & Haussk., *A. geminanus* Boiss. & Haussk., *A. globiflorus* Boiss., *A. gossypinus* Fisch., *A. microcephalus* Willd., *A. myriacanthus* Boiss., *A. senganensis* Bunge, *A. brachycalyx* Fisch., *A. creticus* Lam., *A. cylleneus* Boiss. & Heldr., *A. eriostylus* Boiss. & Haussk., *A. heratensis* Bunge, *A. leiocladus* Boiss., *A. pycnocladus* Boiss., *A. strobiliferus* Royle, *A. stromatodes* Bunge and *A. verus* Oliv. Astragalus is a difficult genus with many useful and some poisonous species. Some of the American species are listed as endangered. This may pose a quandary. The endangered species act may be interpreted as dictating that government agencies are required to nurture endangered species or species that might be confused with endangered species. Many poisonous range species may be confused with some endangered species. Assigned to the Middle Eastern Mediterranean and Near Eastern Centers of Diversity, gum tragacanth or cvs thereof is reported to exhibit tolerance to drought, limestone, poor soil, and slope.

DISTRIBUTION: Native to the Eastern Mediterranean region (Greece, Turkey) and eastward (Kurdistan, Iraq, NW India, Syria, and Lebanon), Russia. Common in sections of Asia Minor, and in semidesert and mountainous regions in Iraq, Syria, and Turkey.

ECOLOGY: In Iran, tragacanths occupy arid slopes between 1,200 and 3,050 m, being partial to lime, well-drained slopes, and foothills. Frost is common throughout the area and temperatures may fall to −17°C. It grows in a Mediterranean climate, or with winter–spring precipitation, rainless summers. Plants require water during the growing season, and a relatively dry climate during collection time. *Astragalus gummifer* ranges from Cool Temperate Moist through Warm Temperate Thorn Forest Life Zone, with pH of 7.2–7.8, annual precipitation of 1.5–6 dm and annual mean temperature of 10–19°C.

CULTIVATION: Propagated from seeds.

HARVESTING: The gum exudes spontaneously from shrubs and hardens into ribbons, which are long, flat, flexible, curled, almost opaque and 5–10 cm long, or into flakes which are oval, thick, brittle, ca. 1.5 cm in diameter. To increase the supply of gum, natives make systematic incisions in shrubs during early spring. Ribbon tragacanth is collected between April and July; flakes, from late September to November. Excessive rains and wind storms during collection season cause discoloration. After collection, gum is put in sheds and visually sorted into 5 grades. The better grades are the lighter colored of the longer ribbons or larger flakes. After sorting, ribbons are packed in cloth-lined boxes and flakes in double burlap bags.

YIELDS AND ECONOMICS: Yield was estimated as 15 g gum per plant, or ca. 375 kg/ha with about 25,000 plants per hectare by Gentry, an average yield one-fifth as large by Schery (1972). When plants are injured, the cell walls of the pith and then of the medullary rays are gradually transformed into gum. The gum absorbs water and produces pressure inside the stem that forces the gum to the surface of the incision or injury. As water evaporates, the gum hardens and then is collected. The amount of gum so exuded varies among plants. Most commercial tragacanth gum is exported from Kurdistan, Iraq, and Persia. Best grades are preferred in pharmacy trade and come from Smyrna and other ports along the Persian Gulf (known as Persian or Syrian Tragacanth). In the 1950s annual Iranian export of gum tragacanth averaged about 4,000 MT of flakes and 400 MT of ribbons. United States and Great Britain normally import the most (about 50%) followed by West Germany, France, Italy, Russia, and Japan. Around 1958 prices per kg were as follows: No. 1, $7.70 to $10.35; No. 2, $7.05 to $8.40; No. 3, $5.85 to $7.50; USP, $2.30 to $3.05. By 1977, better tragacanth commanded $32 to $33 per kg. Between 1945 and 1954, the United States imported about 1 million kg/yr with an annual value of about $1,800,000.

Astragalus gummifer Labill.

BIOTIC FACTORS: I have no reports of any serious diseases or insects that affect tragacanth.

REFERENCES:
Ferri, C. M., 1959, in: *Industrial Gums* (R. L. Whittier and J. N. BeMiller, eds.), pp. 511–515, Academic, New York.
Gentry, H. S., 1957, Gum tragacanth in Iran, *Econ. Bot.* 11(1): 40–63.

CONTRIBUTORS: J. A. Duke, C. F. Reed, J. K. P. Weder.

Baphia nitida Lodd.

FAMILY: Fabaceae

COMMON NAME: Camwood

SYNONYMS: *Baphia haematoxylon* Hook. f.
Carpolobia versicolor G. Don
Podalyria haemtatoxylon Schum.

USES: Formerly planted and cultivated as a dyewood, now largely replaced by American Logwood. Camwood was at one time much exported, chiefly from Sierra Leone. Wood is yellowish-white when cut, only slightly darker when dry. The small dark red heart is the source of the dye. Dye, insoluble in water, readily soluble in alkali, used for wool and for red bandanas. Natives use for a red cosmetic. Durable wood used for houseposts and rafters, pestles for rice-mortars, axe handles, farm implements, walking sticks. Wood is close-grained, of fine texture, planes smoothly and was formerly exported to Europe for turnery. Twigs sometimes used for chewsticks. Plants have potential for stockfeed; produce good supplies of high-protein leaves and grow under diverse conditions. In villages often planted for hedge or living fence.

FOLK MEDICINE: Camwood paste made up with shea butter is applied to sprains or swollen joints. Leaves or sap from heated leaves applied for parasitic skin diseases. Leaves along with other medicines applied to sores. Roots pounded and macerated give a red extract that is mixed with palm oil to treat foot diseases. Bark and leaves prepared as an enema for constipation.

CHEMISTRY: This species has been reported to contain homopterocarpin, pterocarpin, and santal.

Figure 12. *Baphia nitida* Lodd.

DESCRIPTION: Large erect shrub, 2.6–3.3 m tall; branches slender, terete, glabrous; petioles 1.3–2.5 cm long on flowering branches; leaves simple, ovate or oblong, 10–12.5 cm long, about half as broad, acute, rounded at base, subcoriaceous, glabrous; flowers 1–4, on the main branches, on slender glabrous pedicels 0.6 cm long; bracteoles small, round, spreading, united at base; calyx membranous, glabrous, 0.6 cm deep; corolla white, 1.3 cm deep, the standard round and about 1.3 cm broad; pod linear, straight, 10 cm long, 1.3–1.5 cm broad, valves turgid, coriaceous, glabrous; seeds 2–4 per pod; stamens free.

GERMPLASM: Assigned to the African Center of Diversity, camwood or cvs thereof is reported to exhibit tolerance to slope, savanna, and virus. ($2n = 44$.)

DISTRIBUTION: Native to West Tropical Africa (Upper Guinea, Sierra Leone, Ghana, Fernando Po, Guinea).

ECOLOGY: Common in coastal regions, often at edge of villages and on old farms. Distributed over a wide range of conditions from favorable sites in dry open tree savanna to moist closed secondary forest, with slightly acid soil. Ranging from Subtropical Moist through Tropical Dry to Tropical Moist Forest Life Zones, camwood is reported to tolerate annual precipitation of 13.6–40.3 dm (mean of 4 cases = 23.8 dm), annual mean temperature of 23.5–26.6°C (mean of 4 cases = 25.4°C), pH of 5.0–5.3 (mean of 3 cases = 5.1).

CULTIVATION: Trees grow easily but slowly from seeds or cutting. Often grown as a fence or hedge.

HARVESTING: Very little camwood is now harvested. Main harvest is of the lush, palatable evergreen leaves used for stockfeed in areas of adaptation in West Tropical Africa.

YIELDS AND ECONOMICS: Potentially valuable stockfeed in West Tropical Africa. In 1946 Kamdye was exported chiefly to the United States and was priced in London at £5–8/ton.

BIOTIC FACTORS: Fungi reported on camwood: *Asterina baphiae, Chaconia baphiae, Hypoxylon annulatum, Meliola baphiae-nitidae, Micropeltis ugandae, Periconia byssoides, Tryblidiella rufula, Uraecium africanum, Uredo baphiae.*

REFERENCE:
Harms, H., 1911, *Baphia* spp. in Cameroons, *Notizbl. App.* **21**(2):66.

CONTRIBUTORS: J. A. Duke, C. F. Reed.

Bauhinia esculenta Burchell

FAMILY: Caesalpiniaceae
COMMON NAME: Camel's foot
SYNONYM: *Bauhinia burkeana* Benth. ex Harv.

USES: Pods are an important food for natives of South Africa. Seeds sometimes used like peas or garbanzos.

FOLK MEDICINE: Said to be astringent.

CHEMISTRY: Raw leaves are reported to contain per 100 g, 82 calories, 73.3% moisture, 4.3 g protein, 0.1 g fat, 19.9 g carbohydrate, 6.8 g fiber, 2.4 g ash, 436 mg Ca, 86 mg P, and 68 mg ascorbic acid. Baked or dried seeds contain 554 calories, 1.1% moisture, 29.2 g protein, 42.4 g fat, 24.1 g carbohydrate, 20.3 g fiber, 3.2 g ash, 194 mg Ca, 474 mg P, 6.5 mg Fe, 0.08 mg thiamine, 0.97 mg riboflavin, and 1.9 mg niacin. The genus *Bauhinia* has been reported to contain the following: hydrocyanic acid, quercitin, and rutin (for relative toxicity, see Appendix). Seeds of *Bauhinia* are reported to contain trypsin inhibitors and chymotrypsin inhibitors.

DESCRIPTION: Woody, slender vine or climber, several meters long; young parts thinly villous-pubescent; stems angular; leaves of 2 partially connate leaflets, glabrous, deeply divided, 7.5–10 cm broad, each leaflet 3.5–5 cm long, 5–6.5 cm broad, veins branching; peduncles tomentose, many-flowered, some abortive and tendril-bearing; calyx-lobes 1.3 cm long, spreading, separate, lanceolate; stamens 2, fertile, exserted; petals not much longer than the calyx, yellow, striate; vexillum with a very prominent channeled callus; staminodes 5, spatulate, 3 broader than the others; ovary glabrous, stipitate; legume, 1-seeded, dark brown, about 3 cm long and broad, nearly round;

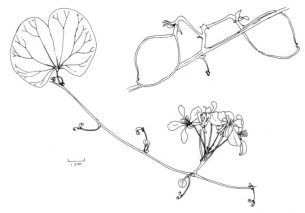

Figure 13. *Bauhinia esculenta* Burchell

seed dark brown, about 1-cm diameter. Fl. and Fr. Dec.–Feb.

GERMPLASM: Assigned to the African Center of Diversity, camel's foot or cvs thereof is reported to exhibit tolerance to sand.

DISTRIBUTION: Native to Tropical Africa, especially Kalahara, Bechuanaland, and Transvaal.

ECOLOGY: Grows in sands, where average annual rainfall is 6–8 dm with a dry season from April until September or October, at altitudes up to 1,000 m. Ranging from Subtropical Dry through Tropical Very Dry Forest Life Zone, this species has been reported to tolerate an annual precipitation of 6.0–13.2 dm (mean of 3 cases = 11.0 dm), annual temperature of 21.0–27.8°C (mean of 3 cases = 24.4°C), and pH of 5.0–7.4 (mean of 3 cases = 6.5).

CULTIVATION: Mainly collected from native plants. When cultivated, seed are planted in rows like beans.

HARVESTING: Immature pods collected and cooked immediately like other pod-beans. Older pods and beans picked and cooked as vegetables.

YIELDS AND ECONOMICS: Grown and consumed locally in South Africa, mainly by natives.

BIOTIC FACTORS: Following fungi have been reported on this plant: *Colletotrichum* sp., *Meliola pexixigua*, *Microsphaera diffusa*, *Phyllosticta* sp., and *Uromyces jamaicensis*.

REFERENCE:
Engelter, C., and Wehmeyer, A. S., 1970, Fatty acid composition of oils of some edible seeds of wild plants (*Bauhinia esculenta*), *J. Agric. Food Chem.* 18(1):25–26.

CONTRIBUTORS: J. A. Duke, C. F. Reed, J. K. P. Weder.

Caesalpinia coriaria (Jacq.) Willd.

FAMILY: Caesalpiniaceae
COMMON NAME: Divi-divi

USES: Pods of divi-divi contain 40–45% tannin, much more than sumach (*Rhus cotinus* and *Rhus*

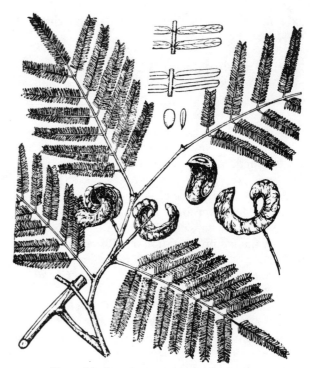

Figure 14. *Caesalpinia coriaria* (Jacq.) Willd.

coriaria) and myrobalans (*Terminalia* spp.). Leather cured with divi-divi is as good as that tanned with oak-bark. In the tanning industry, usually combined with other tanstuffs, as a substitute for gambier (*Uncaria gambier*) and valonia (*Quercus aegilops*). In drum tannage of light leather producing a light yellowish tint. Also in textile industry as a mordant. A powder rich in soluble tannin, (ca. 60%), is obtained from bruised pods, and is used for dyeing black and blue. Pods also yield a good black ink. Wood is reddish-brown, very hard, and produces a red dye. Fallen leaves can be used as fuel or manure.

FOLK MEDICINE: Pods are astringent, and powdered pod is astringent and used as an antiperiodic, tonic, and in treatment of bleeding piles.

CHEMISTRY: This species reported to contain corilagin, gallotannin, tannin, hydrocyanic acid, phellandrene, and shikimic acid. Tannin consists mostly of gallotannin and ellagitannin. Seeds of the genus *Caesalpinia* reported to contain trypsin inhibitors and chymotrypsin inhibitors. Divi-divi liquors have high sugar content (3.2%) and ferment rapidly, especially at tropical temperatures, resulting in an increase of acidity, sedimentation, loss of tannin and development of an objectionable red color. Fermentation can be minimized by use of antiseptics. Leathers tanned with divi-divi liquors are affected by climatic conditions; leather is soft and spongy in wet weather, and lacks pliability in dry weather.

DESCRIPTION: Large bush or poorly formed small spreading tree, 10 m or more tall; stem unarmed, with short bole sometimes 40 cm in diameter; foliage mimosalike; leaves even-pinnate, up to 23 cm long, with 7–8 pairs of pinnae, these 4–8 mm long, glabrous, oblong, green above, pale beneath, main axis and pinnae-axis hairy; racemes or panicles short and dense; calyx 5 mm long; petals pale yellow; filaments reddish; pods thin, oblong or curved, sometimes curled or S-shaped, 2–6 cm long, 1–2 cm broad, indehiscent, not prickly.

GERMPLASM: Assigned to the Middle American Center of Diversity, divi-divi or cvs thereof is reported to exhibit tolerance to insects, poor soil, sand, slope, and waterlogging. ($2n = 24$.)

DISTRIBUTION: Native to tropical American and the West Indies. Introduced and cultivated in India, Pakistan, Burma, Ceylon, Java, and tropical East Africa.

ECOLOGY: Common in open, semiarid country, especially on the dry edges of the tide belt along coasts of southern Mexico, Central America, northern South America and in the West Indies. Thrives best on black cotton soil but can be grown on poor sandy soil. Will live in marshy situations. Grows luxuriantly in clayish calcareous soils, but very slowly on red soils. Grows freely on waste land. Some forms do best on light soil, others on heavy clay. Trees are said to grow best at elevation of 50 m. Hot winds and frost are detrimental to seedlings. Ranging from Warm Temperate Dry through Tropical Dry to Wet Forest Life Zone, divi-divi has been reported to tolerate annual precipitation of 5.9–42.9 dm (mean of 14 cases = 15.1 dm), annual mean temperature of 14.7°–27.5°C (mean of 14 cases = 25.0°C), and pH of 4.5–8.7 (mean of 7 cases = 6.7).

CULTIVATION: Propagated by seeds. In northern India, seeds sown in nurseries in May or June, before beginning of the rains, are transplanted when tall enough (90 cm) to endure the weather. When 9–15 months old, seedlings may be transplanted, spaced 7–8.5 m each, to field during the rainy season. Watering necessary during dry periods for 1–2 years. Mature trees require no care. Forage crops can be raised between the trees.

HARVESTING: Trees begin to bear in 5 years and attain full bearing capacity in 20 years. In India, trees flower and fruit twice a year (Jan.–Feb. and June–July); yield is lower in June–July. In wet areas trees may bear in 3 years, and produce pods freely twice every year. Fallen pods are collected daily and dried before storage. Pods gathered wet, or exposed to moisture, are subject to damage. A yellowish powdery substance surrounds the few seeds and contains up to 50% of superlative tannin easily extracted as 26.5°–32.5°C.

YIELDS AND ECONOMICS: An important source of tannin in tropical America and India, trees yield about 45–135 kg pods per year. Most divi-divi used in the United States comes from Colombia and Venezuela. Venezuela harvests about 10,000 MT of pods annually. In tropical countries, considerable quantities are used locally for tanning and making black dye. Much of the divi-divi from India is imported by England and France. World supply comes from Colombia, Venezuela, Java, and India. Most important importer, was Germany before World War II, and the United States since; United Kingdom also imports. It is the cheapest of South Indian tanning materials; Madras exported 400 MT per year at one time.

BIOTIC FACTORS: Following fungi are known to attack divi-divi: *Fomes lucidus*, *Micropeltis domingensis*, and *Zignoella caesalpiniae*.

CONTRIBUTORS: J. A. Duke, C. F. Reed, J. K. P. Weder.

Caesalpinia echinata Lam.

FAMILY: Caesalpiniaceae
COMMON NAMES: Brazilwood, Poa Brazil
SYNONYM: *Guilandina echinata* (Lam.) Spreng.

USES: Source of important dye used for tinting paper, calico, and other materials. By use of mord-

Caesalpinia echinata Lam.

Figure 15. *Caesalpinia echinata* Lam.

ants, various colors, red, brown, violet, and black, are produced. Straight-grained, fine-textured pieces of the wood are used for making violin bows. Heartwood is bright orange, turning deep red or reddish-brown upon exposure; it is lustrous, and rather uniform except for an occasional darker striping; sapwood is thin, white or yellowish, very hard, heavy, strong and resilient, very durable, and takes good polish.

FOLK MEDICINE: Said to be astringent and used for diarrhea.

CHEMISTRY: No data available.

DESCRIPTION: Forest trees up to 35 m or more, tall with slender, symmetrical bole up to 1 m in diameter, clear of branches for 17–20 m, covered with a thin rough bark; branches brown, armed with short spines; leaves alternate, doubly compound, with many small leaflets, similar to those of honey-locust (*Gleditsia*), ovate and obtuse; flowers singly in simple clusters, streaked with yellow or red; fruits oblong, brown, bristly on outside with small points; seeds smooth and red-brown.

GERMPLASM: Assigned to the South American Center of Diversity, Brazilwood or cvs thereof is reported to exhibit tolerance to low pH and slope.

DISTRIBUTION: Native to the coastal forests of tropical Eastern Brazil from Bahia southward; most abundant in Pernambuco and adjacent states. Now most abundant from Niteroi, State of Rio, and in areas south of Bahia and in Alagoas.

ECOLOGY: Thrives in hot tropical forests along the coast of Brazil. Ranging from Tropical Moist to Tropical Wet Forest Life Zones, Brazilwood is reported to tolerate annual precipitation of 11.2–27.8 dm (mean of 3 cases = 20.6 dm), annual mean temperature of 24.4°–26.6°C (mean of 3 cases = 25.7°C), and pH of 4.5–7.1 (mean of 3 cases = 5.5).

CULTIVATION: Reproduces naturally in the forests of Brazil. Not known to be formally cultivated. Propagates by seed.

HARVESTING: Early demand for Brazilwood was so great that the business was made a royal monopoly. Methods of exploitation were wasteful and natural stands were almost exterminated by the late 1920s. Presently, harvesters must go inland to get the wood, as the tree rarely occurs in the second growth. Trees are cut down, the bark removed and the logs taken to the factories, where the dye is extracted.

YIELDS AND ECONOMICS: Entire trees are used. Demand for Brazilwood has been materially reduced by competition of coal-tar dyes, but there is some trade in Brazilwood from Bahia.

BIOTIC FACTORS: No serious pests and disease are reported.

REFERENCES:
Jose de Souza, B., 1939, Opoa-brazil na historia nacional, *Brasiliana Ser. V* **162**:1–267.
Jose de Souza, B., 1941, Sao Paulo, *Trop. Woods* **67**:38–39.

CONTRIBUTORS: J. A. Duke, C. F. Reed.

Caesalpinia sappan L.

FAMILY: Caesalpiniaceae

COMMON NAMES: Sappanwood, Indian redwood, False sandalwood, Indian brazilwood

USES: Wood yields a valuable red dye for cotton, silk, and wool fabrics. Tree is the source of redwood or Brazilwood of commerce. Commercially valuable parts are the wood and pods. Wood known

Caesalpinia sappan L.

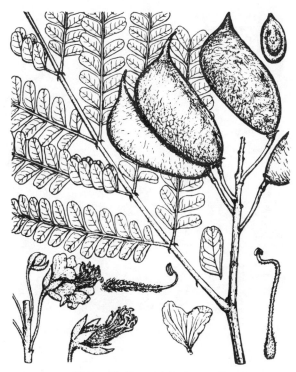

Figure 16. *Caesalpinia sappan* L.

DESCRIPTION: Small- to middle-size tree, thorny, shrubby; branches and inflorescence axes beset with few retrorse prickles; leaves alternate, 20–45 cm long, 10–20 cm broad, pinnate; petioles 4.5–6 cm long, lightly tomentose; rachis 10–32 cm long; stipules in shape of prickles; pinnae in 8–12 pairs, papery, oblong, oblique at base, glabrous above, downy beneath, 5–15 cm long, 2–3 cm broad, the rachilla 4–11 cm long with minute retrorse prickles; pinnules 10–36, oblong-obovate, 1.5–1.8 cm long, about 0.9 cm broad, gland-dotted beneath; racemes or panicles rusty pubescent, the primary peduncles 30–40 cm long, the flowering 9–15 cm long; bracts ovate-acuminate, about 6 mm long; flowers fragrant, 2–3 cm long on pedicels 1.5–1.6 cm long; calyx-tubes about 3 mm long; corollas yellow, the uppermost lobes cuneate, other obovate, all clawed and gland-punctate; filaments densely tomentose at bases and middles; pods glabrous, thick, flattened, obliquely oblong, prominently beaked, wood, polished-brown, 7.5–10 cm long, 2–3-seeded. Fl. August in Burma.

GERMPLASM: Assigned to the Indochina–Indonesian Center of Diversity, sappanwood or cvs thereof is reported to exhibit tolerance to sand and slope. ($2n = 24$.)

DISTRIBUTION: Native to south and central India, Ceylon, Burma, and Malaya, Pegu, and Tenasserim. Generally cultivated in central India, but runs wild in many places.

ECOLOGY: Requires a tropical climate, thriving on sandy riverbanks. Ranging from Subtropical Moist through Tropical Very Dry to Tropical Wet Forest Life Zones, sappanwood is reported to tolerate annual precipitation of 7.0–42.9 dm (mean of 5 cases = 23.3 dm), annual mean temperature of 24.2–27.5°C (mean of 5 cases = 26.1°C), and pH of 5.0–7.5 (mean of 3 cases = 6.5).

CULTIVATION: Propagated easily from seed. Usually cultivated in the forest, where it is native. Usually abundant in areas of adaptation.

HARVESTING: Pods are gathered, pounded and put into cold water. After 2–3 hr the mixture is rubbed and mixed with a solution of iron sulfate. Wood is either cut into pieces or pounded and then boiled in water for 5–8 hr. Dye is extracted from bark by

as 'bakam' or 'sappam' wood; takes a fine polish, does not warp or crack, weighs 83–98 kg/hl, and is useful for inlaying work and to a limited extent for cabinet making and for walking sticks. Dye contains brazilin, which is soluble in water and alcohol. Small quantity is sufficient to dye several meters of cotton fabric. Fast on silk but not on cotton, and used extensively for dyeing wood, and mats and for printing calico. Bakam gives bright red and violet shades, and with garcine produces a chocolate tint. Also used to make a red powder known as 'abir' and 'gulai'; a red water extract is used for holy festivals in India. Bark and pods yield similar dyes. Pods contain ca. 40% tannin used for production of light leather goods and in mixed chrome tannages. Roots give a yellow dye. Pod-cases and bark contain tannin used to produce black shades in dyeing.

FOLK MEDICINE: The wood decoction is a powerful emmenagogue and, because of its tannic and gallic acids, is an astringent used to relieve mild cases of dysentry and diarrhea; given internally for certain skin ailments.

CHEMISTRY: Seeds are reported to contain trypsin inhibitors and chymotrypsin inhibitors.

Caesalpinia sappan L.

boiling water to the desired consistency and tint. In some areas wood is cut into chips or rasped into a powder and extracted twice in hot water. Chips of wood steeped in water yield a red dye that is intensified by alkalies. Addition of turmeric and sulfate of iron gives a liver-colored dye; indigo gives purple. Sappan dye is not permanent; as mordants tannin and alum are used for cotton and a mixture of alum and cream of tartar for wool. In Pegu, used to give red tint to silk; in Madras, to dye straw plants for hat-making. The deep orange extract from wood is fermented before use as a dye, so that brazilin is converted to brazilein.

YIELDS AND ECONOMICS: The amount of wood chips used varies with amount of dye, so data are unavailable. An important dye-making wood, pod and bark in India, Burma and Malaya. Trees are plentiful in areas of adaptation and most are used locally.

BIOTIC FACTORS: Fungi known to attack this plant include the following: *Auricularia auricula-judae* and *Meliola caesalpiniae*.

REFERENCE:
Khin Khin Thi, 1971, The Burmese Caesalpiniaceae, *Union Burma J. Life Sci.* 4:405.

CONTRIBUTORS: J. A. Duke, C. F. Reed, J. K. P. Weder.

Caesalpinia spinosa (Mol.) Ktz.

FAMILY: Caesalpiniaceae
COMMON NAMES: Tara, Huarango, Guaranga
SYNONYMS: *Caesalpinia pectinata* Cav.
Caesalpinia tinctoria (H.B.K.) Benth.

USES: Tara pods contain about twice as much tannin as sumac *(Rhus)*. Tannin used for dyeing and tanning leather from sheep, goat and kid skins, and for making ink. Sometimes grown as a shrub for hedges in Peru for keeping out cattle, pigs, goats, and human beings. Larger forms provide a good, durable wood.

FOLK MEDICINE: Powder within the pods used as an eyewash.

CHEMISTRY: Pods contain nearly 50% tannin.

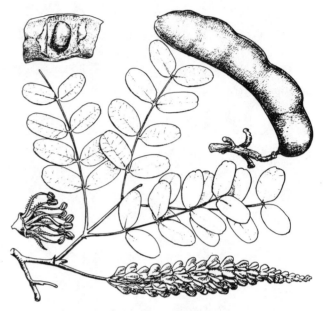

Figure 17. *Caesalpinia spinosa* (Mol.) Ktze.

DESCRIPTION: Shrub or tree, with spreading spinose gray-barked densely leafy branches; leaves smooth or with sparse, short prickles, with 2–3 pairs of pinnae; pinnae often 1 dm long, with about 8 pairs of subsessile, firm, reticulate-veined, oblong-elliptic, glabrous leaflets, oblique at base, rounded at apex, about 2.5 cm long, 1 cm broad; flowers reddish-yellow, in narrow racemes 8–12 cm long; pedicels puberulent, about 5 mm long, articulate below the short calyx tube; larger calyx segments serrulate, about 6 mm long, the petals less than twice as long, about as long as the stamens; pods flat, about 10 cm long, 2.5 cm broad, containing 4–7 large round seeds; seeds black at maturity.

GERMPLASM: Assigned to the South American Center of Diversity, tara is readily grown on forest slopes. ($2n = 24$.)

DISTRIBUTION: Native to the Cordillera Region of Bolivia, Peru, and northern Chile; also occurring in Ecuador, Colombia, Venezuela, and Cuba. Cultivated in Peru, Bolivia, Ecuador, Colombia, Venezuela, and Cuba. Planted in Peru for source of tannin and dye. Introduced to and long cultivated in North Africa, notably Morocco; also in Tropical East Africa.

ECOLOGY: In South America tara grows in the forests and semidesert areas of the Interandine region, as along the higher cooler inner slopes of both Cordilleras of Ecuador. Similar localities in North Africa and elsewhere are preferred. Ranging from Warm Temperate Dry through Tropical Very Dry to Tropical Wet Forest Life Zones, tara is reported to tolerate annual precipitation of 6.6–17.3 dm (mean of 4 cases = 10.5 dm), annual mean temperature of 14.7°–27.5°C (mean of 4 cases = 23.0°C), and pH of 6.8–7.5 (mean of 3 cases = 7.1).

CULTIVATION: Propagation by seeds. Usually started in nurseries and transplanted out when seedlings ca. 10–15 cm tall.

HARVESTING: Pods collected when mature. Tannin easily extracted from pods by steaming in water at 50°–60°C for hours at three different times. Pods contain about 48% tannin. Tanning solution has little color, producing nearly colorless leather; pH 3.3 is favorable for fixation. Sheep and goat skins tanned with tara are white, instead of red as with quebracho, and do not discolor under extended exposure. Insolubles of tara remain in the vats and agitators, which facilitates processing.

YIELDS AND ECONOMICS: Many pods are collected from wild trees, and are a regular article of trade in markets in Lima, Peru. The tannin is used extensively in South America and Morocco for tanning sheep and goat skins, producing an excellent light-colored leather.

BIOTIC FACTORS: No serious pests or diseases have been reported for this plant.

REFERENCE:
Sprague, T. A., 1931, The botanical name of tara, *Kew Bull. Misc. Inform.* **2**:91–96; *Trop. Woods* **26**:25.

CONTRIBUTORS: J. A. Duke, C. F. Reed.

Cajanus cajan (L.) Millsp.

FAMILY: Fabaceae
COMMON NAMES: Pigeon pea, Dhal, Gandul, Red gram, Congo pea, Gungo pea, No-eye pea, Pois d'Angole
SYNONYM: *Cajanus indicus* Spreng.

USES: Pigeon peas are popular food in developing tropical countries. Nutritious and wholesome, the green seeds (and pods) serve as a vegetable. Ripe seeds are a source of flour, used split (dhal) in soups or eaten with rice. Dhal contains as much as 22% protein, depending on cv and location. Tender leaves are rarely used as a potherb. Ripe seeds may be germinated and eaten as sprouts. Plants produce forage quickly and can be used as a perennial forage crop or green manure. Often grown as a shade crop for tree crops or vanilla, a cover crop, or occasionally as a windbreak hedge. In Thailand and N. Bengal, pigeon-pea serves as host for the scale insect that produces lac or sticklac. In Malagasy, the leaves are used as food for silkworms. Dried stalks serve for fuel, thatch, and basketry.

FOLK MEDICINE: Morton (1976) lists many folk medicinal uses for pigeon pea. In India and Java, the young leaves are applied to sores. Indochinese claim that powdered leaves help expel bladder-stones. Salted leaf juice is taken for jaundice. In Argentina the leaf decoction is prized for genital and other skin irritations, especially in females. Floral decoctions are used for bronchitis, coughs, and pneumonia. Chinese shops sell dried roots as

Figure 18. *Cajanus cajan* (L.) Millsp.

Cajanus cajan (L.) Millsp.

an alexeritic, anthelminthic, expectorant, sedative, and vulnery. Leaves are also used for toothache, mouthwash, sore gums, child delivery, dysentery. Scorched seed, added to coffee, are said to alleviate headache and vertigo. Fresh seed are said to help incontinence of urine in males; immature fruits are believed of use in liver and kidney ailments.

CHEMISTRY: Analysis of dhal (without husk) gave the following values: moisture, 15.2; protein, 22.3; fat (ether extract), 1.7; mineral matter, 3.6; carbohydrate, 57.2; Ca, 9.1; and P, 0.26%; carotene evaluated as vitamin A, 220 IU, and vitamin B_1, 150 IU per 100 g. Sun-dried seeds of *Cajanus cajan* are reported to contain per 100 g: 345 calories, 9.9 g moisture, 19.5 g protein, 1.3 g fat, 65.5 g carbohydrate, 1.3 g fiber, 3.8 g ash, 161 mg Ca, 285 mg P, 15.0 mg Fe, 55 μg β-carotene equivalent, 0.72 mg thiamine, 0.14 mg riboflavin, and 2.9 mg niacin. Immature seeds of *Cajanus cajan* are reported to contain per 100 g, 117 calories, 69.5% moisture, 7.2 g protein, 0.6 g fat, 21.3 g total carbohydrate, 3.3 g fiber, 1.4 g ash, 29 mg Ca, 135 mg P, 1.3 mg Fe, 5 mg Na, 563 mg K, 145 μg β-carotene equivalent, 0.40 mg thiamine, 0.25 mg riboflavin, 2.4 mg niacin, and 26 mg ascorbic acid/100 g. Of total amino acids, 6.7% is arginine, 1.2% cystine, 3.4% histidine, 3.8% isoleucine, 7.6% leucine, 7.0% lysine, 1.5% methionine, 8.7% phenylalanine, 3.4% threonine, 2.2% tyrosine, 5.0% valine, 9.8 aspartic acid, 19.2% glutamic acid, 6.4% alanine, 3.6% glycine, 4.4% proline, 5.0% serine with 0 values for canavanine, citrulline, and homoserine. Methionine, cystine, and tryptophane are the main limiting amino acids. However, in combination with cereals, as always eaten, pigeon pea contributes to a nutritionally balanced human food. Oil of the seeds contains 5.7% linolenic acid, 51.4% linoleic, 6.3% oleic, and 36.6% saturated fatty acids. Seeds are reported to contain trypsin inhibitors and chymotrypsin inhibitors. Fresh green forage contains 70.4% moisture, 7.1 crude protein, 10.7 crude fiber, 7.9 N-free extract, 1.6 fat, 2.3 ash. The whole plant, dried and ground contains 11.2% moisture, 14.8 crude protein, 28.9 crude fiber, 39.9 N-free extract, 1.7 fat, and 3.5 ash.

DESCRIPTION: Perennial woody shrub, mostly grown as an annual for the seed; stems strong, woody, to 4 m tall, freely branching; root system deep and extensive, to about 2 m, with a taproot. Leaves alternate, pinnately trifoliolate, stipulate; stipels small, subulate; leaflets lanceolate to elliptic, entire, acute apically and basally, penninerved, resinous on lower surface and pubescent, to 15 cm long and 6 cm wide. Inflorescence in terminal or axillary racemes in the upper branches of the bush. Flowers multicolored with yellow predominant, red, purple, orange occur in streaks or fully cover the dorsal side of the flag, zygomorphic. Pods compressed, 2–9-seeded, not shattering in the field. Seeds lenticular to ovoid, to 8 mm in diameter, about 10 seeds/g separated from each other in the pod by slight depressions. Germination cryptocotylar.

GERMPLASM: Many cvs differ in height, habit of growth, color of flower, time of maturity, color, and shape of pods, and color, size, and shape of seed. Perennial types assume a treelike appearance, yield well the first year but poorer in later years; suitable for forage, cover purposes, shade and for hedge plants. Annual (weak perennial) types are small plants grown as field crops, mainly cultivated for seed, with very good quality white-seeded cvs ('Gujerat' in Ceylon) and red-seeded cvs (common in areas south of Bombay). Also in India and Ceylon the cvs 'Tur 5' and 'Tenkasi' are extensively grown. High-yield, short-duration Indian cvs include 'Co-1,' 'Kanke 3,' 'Kanke 9,' 'Makta,' 'Pusa ageta,' 'Sharda,' 'T-21,' and 'UPAS-120.' In Florida day-neutral 'Amarillo' can be sown and harvested at different times throughout the year. Other good cvs are 'Morgan Congo,' 'Cuban Congo,' and 'No-eye Pea.' Of the better-yielding cvs in trials in Uganda 'CIVE1,' 'UC948,' 'UC2288,' 'UC3035,' and 'UC16' are "spray types," with secondary branches almost as long as the main stem, and there are few tertiaries; 'UC1377' and 'UC959' are "bush types." Assigned to the Hindustani and African Centers of Diversity, pigeon pea or cvs thereof is reported to exhibit tolerance to disease, drought, frost, high and low pH, laterite, nematodes, photoperiod, salt, sand, virus, waterlogging, weed, wilt, and wind. ($n = 11$), ($2n = 22, 44, 66$.)

DISTRIBUTION: Probably native to India, pigeon pea was brought millennia ago to Africa, where different strains developed. These were brought to the new world in post-Columbian times. Truly wild *Cajanus* has never been found; they exist mostly as remnants of cultivations. In several places *Ca-*

janus persists in the forest. The closest wild relative, *Atylosia cajanifolia* Haines, has been found in some localities in East India. Most other *Atylosias* are found scattered throughout India; a group of endemic *Atylosia* species grows in North Australia. In Africa *Cajanus kerstingii* grows in the drier belts of Senegal, Ghana, Togo, and Nigeria. Pigeon peas occur throughout tropical and subtropical regions and in the warmer temperate regions (as North Carolina) from 30°N to 30°S.

ECOLOGY: Pigeon pea is remarkably drought resistant, tolerating dry areas with less than 65 cm annual rainfall, even producing seed profusely under dry zone conditions, as the crop matures early and pest damage is low. Pigeon pea is more or less photoperiod-sensitive; short days decrease time to flowering. Under humid conditions pigeon pea tends to produce luxuriant vegetative growth, rain during the time of flowering causes defective fertilization and permits attack by pod-caterpillars. Annual precipitation of 6–10 dm is most suitable, with moist conditions for the first 2 growing months, drier conditions for flowering and harvest. Grows best at 18°–29°C, some cvs tolerate 10°C under dry conditions and 35°C under moister conditions. Some cvs are sensitive to waterlogging and frost, other more tolerant. It will grow in all types of soils, varying from sand to heavy clay loams, well-drained medium heavy loams being best. Some cvs tolerate 6–12 mmhos/cm salinity. Ranging from Warm Temperate Moist to Wet through Tropical Desert to Wet Forest Life Zones, pigeon pea has been reported to tolerate annual precipitation of 5.3–40.3 dm (mean of 60 cases = 14.5 dm), annual mean temperature of 15.8°–27.8°C (mean of 60 cases = 24.4°C), and pH of 4.5–8.4 (mean of 44 cases = 6.4).

CULTIVATION: Seeds are sown where desired, in pure strands at about 9–22 kg/ha for rows, but sometimes are broadcast and germinate in about 2 weeks. In India pigeon pea may be grown mixed with other crops or in alternate rows (from 1–3 of a kind) with 3–10 rows of sorghum, groundnuts, sesame, cotton, pineapples, millets, or maize. For pure crops pigeon pea should be sown 2.5–5 cm deep in rows 40–120 cm by 30–60 cm. For mixed crops, it should be sown in rows spaced 1.2–2.1 m depending on the associated crop. About 3–4 seeds may be planted in each hill, and later thinned to 2 plants. Plants show little response to fertilizers, e.g., mixed plantings with millet in India showed negative response to N. For 1 month, pigeon pea shares the intercultivation of the main crop. In the tropics, 20–100 kg ha phosphoric acid is recommended. S, with or without P, can significantly increase seed yield and nitrogen fixation. Early cvs start podding in 12 wk and mature in 5–6 mos. Late cvs require 9–12 mos. The crop may be ratooned for forage or let persist for 3–5 yr. Seed yields drop considerably after the first year, and disease build-up may reduce stand.

HARVESTING: In India pigeon peas are sown in June–July. In North India annual, medium, and late cvs flower in January and yield a first crop in March–April. In Central and South India early and medium cvs flower in October–November, yielding in December–January. Very early cvs have not been widely accepted. In East Africa harvests are taken in June–July. In the Caribbean areas, green pods are harvested for home consumption or canning. Caribbeans have developed dwarf cvs with more uniform pod maturity that are mowed and threshed with a combine. In many countries occasional and frequent harvests are taken. Depending on the cv, the location and time of sowing, flowering can occur from 100–430 days. In harvesting a first crop, pods may be picked by hand. Mature crops are harvested by cutting the whole plant with a sickle. Cut plants, often still with green leaves, are dried in the field. Threshing by wooden flails or trampling is carried out on threshing floors. Grain is then cleaned by winnowing. Mechanical threshing and seed cleaning are possible. Commercial dhal is prepared in two ways: *Dry Method:* Grains are sun-dried for 3–4 days, partially split with a stone mill, then treated with vegetable oil and stored. Sesame oil is used for storage of less than 1 month (coconut oil has been used successfully in Sri Lanka). For longer storage, castor oil is used. The oil is absorbed into the seed coat and facilitates the final splitting of the dhal. Split pigeon peas are separated and cleaned from the seed coats by sieving and winnowing. Dhal prepared by the dry method has perfect half-globular shape, cooks soft, and commands a high price. From 40 kg of seeds, 18 kg of clean good quality dhal are extracted. *Wet Method:* Pigeon peas are soaked in water for about 6 hr, drained, and mixed with fine well-sieved soil at a rate of 2–40 kg of seeds. The mixture is heaped during the nights and dried in the sun for a couple

Cajanus cajan (L.) Millsp.

of days. Impurities are then removed and the seed split into halves on a stone mill. About 31 kg of dhal is milled from 40 kg of seed. Oil is not used for wet preparation of the dhal. Dhal prepared by either method is sold without added oil, but at home 1 kg oil per 40 kg dhal may be used. Dhal as such is not kept for longer than 2–3 months.

YIELDS AND ECONOMICS: Green-pod yields vary from 1,000 to 9,000 kg/ha. Dried seed yields may reach 2,500 kg/ha in pure stands, but more often average about 600 kg/ha. Of seven promising cvs in Uganda, 'C1VE1' yielded 889 kg seed/ha with a grain/straw ratio of 0.318, '16' had the highest seed yield of 1225 kg/ha and a grain/straw ratio of only 0.224 (Khan and Rachie, 1972). India's pigeon pea production, 1,818,000 MT from 2,540,000 ha (1975), exceeds that of any other country. Pigeon pea is cultivated commercially (for canning) in the Dominican Republic, Trinidad, Puerto Rico, and Hawaii and mostly for home consumption in Africa, Kenya, Malawi, Tanzania, and Uganda. Elsewhere in the tropics it usually is a crop of kitchen gardens and hedges. As such, in virtually every tropical country, it contributes valuable vegetable protein to the diet, even though not reported in statistics. In 1975, Asia led world production, 1,845,000 MT, averaging 706 kg/ha; Africa produced 70,000 MT, averaging 406 kg/ha; North America produced 41,-000 MT, averaging 1415 kg/ha; South America produced 4000 MT, averaging 449 kg/ha. India was the leading producer country, with 1,818,000 MT, but averaged only 716 kg/ha, as compared with the Dominican Republic where yields were reported at 2,194 kg/ha (FAO, 1975). However, the Indian yields are dry seed while the Dominican yields are fresh seeds or pods.

BIOTIC FACTORS: Many fungal diseases (31), involving 45 pathogens, are known; the most serious is wilt disease *(Fusarium udum)*, favored by soil temperatures of 17°–20°C. The fungus enters the plant through the roots and may persist in soil-borne stubble for a long time. The only effective control measure is development of resistant cvs (e.g., 'C-11,' 'C-36,' 'NP-15,' 'NP-38,' and 'T-17'). Rotation with tobacco and intercropping with sorghum is said to decrease the wilt problem. Other fungi include *Cercospora* spp., *Colletotrichum cajanae, Corticium solani, Diploidia cajani, Leveillula taurica, Macrophomina phaseoli, Phaeolus manihotis, Phoma cajani, Phyllosticta cajani, Phytophthora* sp., *Rhizoctonia bataticola, Rosellinia* sp. *Sclerotium rolfsii,* and *Uredo cajani* (rust). So far, economic damages by these have been small or negligible, but rust is locally of some importance. Pigeon pea is also attacked by the bacterium *Xanthomonas cajani* and the sterility mosaic and yellow mosaic viruses. Sterility mosaic is being recognized as a serious economic threat. Of minor importance are the nematodes isolated from pigeon pea. They include: *Helicotylenchus cavevessi, H. dihiptera, H. microcephalus, H. pseudorobustus, Heterodera* spp., *H. trifolii, Hoplolaimus indicus, Meloidogyne hapla, M. incognita acrita, M. javanica, J. javanica bauruensi, Pratylenchus* spp., *Radopholus similis, Rotylenchulus reniformis, Scutellonema bradys, Scutellonema clathricaudatum, Trichodorus mirzai, Tylenchorhynchus brassicae, T. indicus, Xiphinema campinense,* and *X. ifacolum.* Damage caused by insect pests is a major constraint on yield in most areas. Few of the more than 100 species of insects recorded as damaging the crop in India can be regarded as major pests. The podborer *Heliothis armigera* is regarded as the key pest throughout Africa and Asia. It is particularly damaging on early formed pods. In many parts of India the podfly, *Melanagromyza obtusa,* takes over as the dominant pest later in the season. In some areas, a newly recognized hymenopteran pest, *Taraostigmodes,* can also cause extensive pod damage late in the season. Pests that can be locally or seasonally important are plume moth *(Exelastis atomosa)*, blue butterfly *(Euchrysops cnejus)*, leaf tier *(Eucosma critica)*, bud weevil *(Ceuthorrhynchus aspurulus)*, spotted podborer *(Maruca testulalis)*, pea podborer *(Etiella zinckiella)*, and bugs *(Clavigralla* spp.). A blister beetle *(Mylabris pustulata)* which destroys flowers can be a spectacular but localized pest. Thrips *(Frankliniella sulphurea, Taeniothrips nigricernis)* may cause premature flower drop. In general, lepidopterous borers damage the determinate (clustering) plants, while podfly damages the later indeterminate cvs. The indeterminate cvs have a greater compensatory potential and, where pests are not controlled, commonly yield more than do the clustering types. In the West Indies, the leafhopper *Empoasca fabilis* is combatted with malathion while podborers *Elasmopalpus rubedinellus, Ancylostomia stercorea,* and *Heliothis virescens* are combatted with DDT, Dipterex and Gardona. In Trinidad the black aphis *Aphis craceivora* may develop heavy infestations. Bruchids *(Callosobruchus* spp.) attack the crop in the fields and then build up in stored pods or seeds. The use of insecticides is feasible but, as yet, uneconomic. A few farmers use insecticides; DDT is still the most effective and least expensive. During its first 60 days, pigeon pea requires weed control. Preemergence chloramben, though effective, may slightly damage the crop. In the West Indies: (1) premergence prometryne with postemergence paraquat spray, (2) alachlor plus linuron, and (3) terbutryne up to 9 wk after application, have proved useful in weed control.

REFERENCES:

Cross, L. A., and Thomas, S. M., 1968, *Pigeonpea 1965–1967*, Texaco Food Crops Demonstration Farm Bulletin No. 5.

De, D. N., 1974, Pigeon pea, in: *Evolutionary Studies in World Crops, Diversity and Change in the Indian Subcontinent* (Sir J. Hutchinson, ed.), pp. 79–87, Cambridge University Press.

Gooding, H. J., 1962, The agronomic aspects of pigeon pea, *Field Crop Abstr.* **15**:1–5.

Hammerton, J. L., 1971, A spacing/planting date trial with *Cajanus cajan* (L.) Millsp., *Trop. Agric.* **48**(4):341–350.

Henderson, T. H., 1965, Some aspects of pigeon pea farming in Trinidad, *Occ. Ser. Dep. Agric. Econ. Farm Mgmt.*, University of West Indies, No. 3.

Khan, T. N., and Rachie, K. O., 1972, Preliminary evaluation and utilization of pigeon pea germplasm in Uganda, *East Afr. Agric. For. J.* **38**(1):78–82.

Krauss, F. G., 1921, The pigeon pea *(Cajanus indicus)*: Its culture and utilization in Hawaii, *Hawaii Agric. Exp. Stn., Bull.* **46**:1–23, Washington, D.C.

Morton, J. F., 1976, The pigeon pea *(Cajanus cajan* Millsp.), a high protein tropical bush legume, *Hortscience* **11**(1):11–19.

Mukherjee, D., 1960, Studies on spacing of *Cajanus cajan* (L.) Millsp., *Indian J. Agric. Sci.* **30**:177–184.

Riolliano, A., Perez, A., and Ramos, C., 1962, Effects of planting date, variety, and plant population on the flowering and yield of pigeon pea *(Cajanus cajan* L.), *J. Agric. Univ. P.R.* **46**:127–134.

Thevasagayam, E. E., and Canagasingham, L. S. C., 1960, Some observations on the insect pests of dhal *(Cajanus cajan)* and their control, *Trop. Agric.* **116**(4):287–298.

CONTRIBUTORS: J. A. Duke, L. J. G. van der Maesen, H. Pollard, C. F. Reed, J. K. P. Weder.

Calopogonium mucunoides Desv.

FAMILY: Fabaceae

COMMON NAMES: Calopo, Frisolilla

USES: The mass of intertwining stems rooting at nodes, thick foliage, and the mat of dead leaves

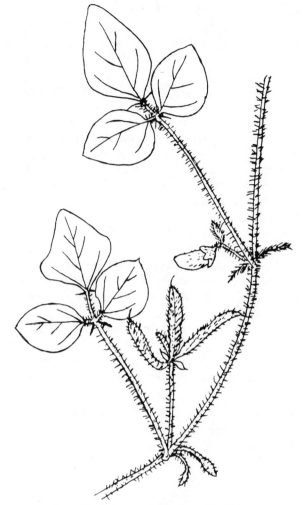

Figure 19. *Calopogonium mucunoides* Desv.

below make this a valuable crop for erosion control and soil humification. Good cover crop for coconuts and young rubber plants. In India and Burma, used as cover crop in rubber and other plantations; sometimes less satisfactory than other legumes. In Java, Malaya and Ceylon used for new clearings to improve the soil. Used for forage in some countries but not relished by cattle.

FOLK MEDICINE: Leaves reportedly used for filariasis in Tonga.

CHEMISTRY: Seed contains 10.3% fat and 5.8% total N. The fat contains arachidic, palmitic, oleic, and linoleic acids.

DESCRIPTION: Annual or perennial vine; stems twining and trailing, densely pilose; 1 m or more

Calopogonium mucunoides Desv.

long; leaves trifoliolate, leaflets ovate to rhomboid-ovate or elliptic, 4–10 cm long, 2–5 cm broad, obtuse or subacute and apiculate at apex, rounded at base, lateral leaflets oblique, appressed pilose or pubescent on both surfaces, often more so beneath; petiole 2–6 cm long, pilose; rachis 0.5–2.2 cm long; stipules ovate-lanceolate, 4 mm long; racemes axillary, up to 20 cm long, on peduncles 0–17 cm long, ferrugineous pilose; fascicles up to 6-flowered, often well-separated, or sometimes single and subsessile; pedicels very short; bracts and bracteoles lanceolate, 4 mm long; calyx campanulate, 6–8 mm long, pilose, teeth subulate, the upper 2 about as long as tube, lower 3 longer, teeth 6–8 mm long, standard blue, 0.7-cm long, 5–6 mm broad, emarginate; corolla mauve to blue, a little longer than calyx, glabrous; pods linear-oblong, straight or falcate, several together, densely brownish-hirsute, 5–8-seeded, transversely impressed between seeds, fruiting pedicels 2–3 mm long; seeds dark brown or yellow-brown, oblong to squarish, 2.5–4.0 mm long, 2.5–3 mm broad, 1.5–2 mm thick; hilum small, round, central, with a short impressed groove at one end. Fl. Dec.; Fr. Dec.–Jan.

GERMPLASM: Assigned to the Middle American Center of Diversity, calopo or cvs thereof is reported to exhibit tolerance to laterite, low pH, and shade. ($2n = 36$.)

DISTRIBUTION: Native to tropical America, probably Guiana; now very widely cultivated and often escaped in Old World tropics, as well as tropical South America, Central America, and West Indies.

ECOLOGY: Thrives in humid tropics. Because it produces shallow roots with many nodules, it is unsuitable for drier areas where rainfall is less than 87 cm; it does best in closed forest situations with rainfall 125 cm or more, where it remains green throughout dry season. Lives from low elevations up to 2,000 m (in East Africa), but does not thrive under shade. Plants wither in dry season and are susceptible to fire. Ranging from Subtropical Moist to Wet through Tropical Very Dry to Wet Forest Life Zones, calopo is reported to tolerate annual precipitation of 8.7–42.9 dm (mean of 21 cases = 22.4 dm), annual mean temperature of 18.7°–27.4°C (mean of 21 cases = 25.0°C), and pH of 4.3–8.0 (mean of 16 cases = 5.7).

CULTIVATION: Seed usually broadcast rather thickly as many seeds may get washed away by heavy rains. With proper care, crop soon becomes established and covers soil with mass of hairy creeping stems that root at nodes. They form dense cover 30–60 cm thick on surface of soil with masses of leaves that are shed and form layer up to 2.5 cm thick, thus smothering out weeds and retaining moisture in soil. Leaves remain green in dry season. Because the stems root at the nodes, old stems may also be used for propagation. In some areas plants are propagated from softwood cuttings in wet weather. In dry areas or in dry weather, crop is best propagated by seed. Crop often intercropped with citrus, rubber, coconuts, and other permanent plantings, as plants tend to hold the soil and add humus. Green manure can be dug in later to enrich the soil still further.

HARVESTING: Crop makes complete ground cover 60 cm thick in 5 months. Fruits and seeds are relatively small and difficult to harvest. However, in wet areas, crop reproduces itself by self-sown seed and produces year after year on the same land. Under Zanzibar conditions plant produces abundance of seeds and may be readily regenerated by hoeing the land where the old crop has died. In wet–dry climates, it is a self-regenerating annual.

YIELDS AND ECONOMICS: Yield after about 6 months is about 60 MT green manure/ha, equivalent to 1,255 kg ammonium sulfate fertilizer. Extensively cultivated as a green manure in tropical areas throughout the world, especially in Old World tropics. Used for grazing seeded with grass cvs in Australia and Brazil. Gowda (1974) reports N fixation of nearly 500 kg/ha in pot experiments, nearly 250 in field experiments.

BIOTIC FACTORS: Following fungi have been reported on this crop: *Cephalosporium eichorniae, C. zonatum, Cercospora borinquensis, Corticium solani, Didymosphaeria calopogonii, Gloeosporium calopogonii, Leptosphaeria calopogonii, Meliola bicornis* var. *calopogonii, Monotospora dominicana, Mycosphaerella calopogonii, Phakospora pachyrhizi, Sclerotium rolfsii, Sporodesmium bakeri, Septoria calopogonii*. Nematodes isolated from this crop include: *Helicoylenchus* sp., *Meloidogyne arenaria, M. incognita* and var *acrita, Pratylenchus brachyurus, P. coffaea, Radopholus similis, Roty-*

lenchulus reninformis, Scutellonema bradys and *Xiphinema* spp.

REFERENCE:
Gowda, H. S. G., 1974, Forage legume inoculation and grass-legume association studies, *Mysore Agric. J.* **9**:515–516.

CONTRIBUTORS: J. A. Duke, A. E. Kretschmer, Jr., C. F. Reed.

Canavalia ensiformis (L.) DC.

FAMILY: Fabaceae

COMMON NAMES: Jackbean, Horsebean, Gotani bean, Overlook bean

SYNONYM: *Dolichos ensiformis* L.

USES: Jackbean is mainly cultivated for green manure, as a soil cover for erosion control, and for forage. As a green manure, it is intercropped with cacao, coffee, and sugarcane. Young pods and immature seeds are used as a vegetable. Flowers and young leaves are steamed as a condiment in Indonesia. Roasted beans sometimes used as a substitute or adulterant of coffee. Ripe dry beans may be eaten after long cooking, but contain a mild poison. Seed coat should be avoided. In Indonesia, the seed may be boiled twice, peeled, left in running water for 2 days, fermented for 3–4 more days and finally cooked again. Said to be unpalatable and indigestible as cattle feed.

FOLK MEDICINE: No data available.

CHEMISTRY: Jackbean meal contains a thermolabile toxin, causing haemorrhage of mucosa of stomach (in rats), but apparently harmless when mixed with wheat meal up to 30%. Jackbean contains several globulins including canavalin and concanavalin A and B and, in the meal, a crystalline diamino acid, canavanine. Immature pods and beans contain per 100 g: 82 calories, 78.5% moisture, 6.9% protein, 0.5% fat, 13.3% carbohydrate, 3.3% fiber, and 0.8% ash. Pods contain (zero moisture basis): 4.5% crude protein, 1.5% fat; 42.1% N-free extract, 48.1% crude fiber, 3.8% ash, 0.3% Ca, and 0.01% P. Dried seeds reported to contain per 100 g, 347 calories, 10.7% moisture, 24.5 g protein, 2.6 g fat, 59.0 g carbohydrate, 7.4 g fiber, 3.2 g ash, 158 mg Ca, 298 mg P, 7.0 mg Fe, 0.77 mg thiamine, 0.15 mg riboflavin, and 1.8 mg niacin. Kay (1978) reports 11–15.5% moisture, 23.8–27.6% protein, 2.3–3.9% fat, 45.2–56.9% carbohydrate, 4.9–8.0% fiber, 2.7–4.2% ash, 30–158 mg Ca/100 g; 54–298 mg P/100 g, 141 mg K/100 g, 19 mg Mg/100 g, 7 mg Fe/100 g, 2 mg niacin/100 g, 0.4 mg pantothenic acid/100 g, 0.4 mg riboflavin/100 g, and 8.5 mg thiamin/100 g. The amino acid composition is (mg/gN): glutamic acid 644, threonine 275, serine 316, alanine 275, glycine 241, valine 288, methionine 85, isoleucine 250, leucine 453, tyrosine 219, phenylalanine 322, lysine 344, histidine 169, arginine 294, tryptophane 75. Green seeds contain per 100 g: 82 calories, 78.5% moisture, 6.9 g protein, 0.5 g fat, 13.3 g carbohydrate, 3.3 g fiber, 0.8 g ash, 33 mg Ca, 66 mg protein, 1.2 mg Fe, 15 µg vitamin A, 0.22 mg thiamine, 0.10 mg riboflavin, 2.0 mg niacin, and 32 mg ascorbic acid. The genus *Canavalia* is reported to contain the following toxins: choline, hydrocyanic acid, trigonelline. Seeds are reported to contain trypsin and chymotrypsin inhibitors. Bogdan (1977) cites data indicating that dry herbage contains 13.8–16.0% crude protein, 2.1–2.9% ether extractives, 26.5–35.7% crude fiber, and 41.2–43.5% N-free extractive. Digestibility of these nutrients was 56–59%, 57–69%, 38–61%, and 70–72%, respectively.

DESCRIPTION: Plant usually grown as an annual, but may become a perennial climber; stems erect or semierect, or climbing, bushy, 0.6–1.6 m long, glabrous or adpressed-pubescent; leaves trifoliolate, leaflets elliptic to ovate-elliptic or oblong, 5.7–20 cm long, 3.2–11.5 cm broad, obtuse, subacute, or acuminate, more or less cuneate, glabrescent or sparsely pubescent on both surfaces, venation raised and reticulate on both surfaces; petioles

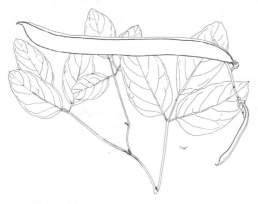

Figure 20. *Canavalia ensiformis* (L.) DC.

Canavalia ensiformis (L.) DC.

2.3–11 cm long; rachis 1–3.5 cm long; petiolules 1–11 mm long, densely pubescent; stipules soon deciduous; racemes axillary, pendulous, 5–12 cm long, on peduncle 10–34 cm long; pedicels 2–5 mm long; bracteoles obtuse, ca. 2 mm long; calyx 1.5 cm long, sparsely pubescent, tube 6–7 mm long, upper lip 5 mm long and truncate; standard rose to purple, rounded, 2.5–3 cm long, emarginate; pods linear-oblong, flat, sword-shaped, 15–35 cm long, 3–3.5 cm broad, containing 12–20 seeds, each valve with a sutural rib and an additional one just below it; seeds white or ivory with brownish mark near grayish hilum, oblong, compressed, 1.45–2.1 cm the longest dimension, 1–1.5 the shortest, 0.7–1 cm thick; hilum 5.5–9 mm long, grayish; germination phanerocotylar. Fl. April–June, depending on locality. About 750 seeds/kg.

GERMPLASM: Both climbing and dwarf (var. *nana*) bushy cvs are grown. Jackbean is assigned to the Hindustani, South American, and China–Japanese Centers of Diversity. It may have originated from *Canavalia plagiosperma* Piper. Jackbean or cvs thereof is reported to tolerate disease, drought, fungus, insects, low pH, salt, sand, shade, slope, virus, and waterlogging. ($2n = 22$.)

DISTRIBUTION: A prehistoric American Indian domesticate found in southwest United States; native from Mexico south to Brazil and Peru, and in the West Indies; now cultivated throughout the tropics as a cover crop or as a green manure crop. Cultivated to a limited extent in India, Indonesia, Taiwan, Tanzania, Kenya, and Hawaii. Cultivated in east tropical Africa up to 1,800 m, and sometimes seen as an escape.

ECOLOGY: Jackbean is relatively drought-resistant and more resistant to waterlogging and salinity than many other legumes. Responds as a short-day plant. Fares best with evenly distributed annual rainfall of 9–12 dm, and temperature of 15°–30°C. Once established, withstands very dry conditions due to deep root system. Grows on a wide range of soil types but does better with pH of 5–6. Tolerates shade. Ranging from Warm Temperate Moist through Tropical Very Dry to Wet Forest Life Zones, jackbean is reported to tolerate annual precipitation of 6.4–42.9 dm (mean of 20 cases = 17.1 dm), annual mean temperature of 14.4°–27.8°C (mean of 20 cases = 23.4°C), and pH of 4.5–8.0 (mean of 16 cases = 6.1).

CULTIVATION: Propagated by seeds, sown in rows 75 cm apart, with 45–60 cm between plants in the row, or 30–40 cm on the square. Preparation before sowing may be limited to a light hoeing, followed by a dressing of a complete fertilizer that should be raked into top layer of soil before seeds are sown. Application of N is said to depress yield. For green manure, 225 kg/ha superphosphate is recommended. When planted in rows, seed rate is 25–30 kg/ha when planted as a green manure crop, seed are broadcast at rate of 40–60 kg/ha. Seed covered to 2 cm. When grown for cover crop, sown at during rainy season. When allowed to climb, plants require strong, durable supports.

HARVESTING: Pods used as vegetable when three-quarters grown. Seed harvested as pods become ripe. When overripe, pods shatter. Green pods produced in 3–4 months, but ripe seed require 180–300 days. As green manure, climbing cvs give more foliage.

YIELDS AND ECONOMICS: Yield of green matter is 20–60 MT/ha; dried beans, 1.5 MT/ha. Yields of seed range from 700 to 5,400 kg/ha. In Brazil, seed yields range from 800 to 1,200 kg/ha; in Puerto Rico, from 4,300 to 5,400. Widely cultivated throughout the tropics, and particularly valuable as a green manure crop between rows of sugarcane, coffee, tobacco, rubber, and sisal, and as cover crop for cacao, coconut, citrus, and pineapple. Large-scale plantings are reported from Congo and Angola.

BIOTIC FACTORS: Under natural conditions flowers are visited by bees and cross-pollination is about 20% or more. However, bagged flowers do set pods and seeds. Crop is attacked by few insect pests, as the fall armyworm, *Spodoptera frugiperda,* and the pod weevil *Sternechus tuberculatus.* Seed are rather resistant to infestation during storage. Most serious fungus disease is caused by *Colletotrichum lindemuthianum,* a root disease. However, other fungi attacking this plant include: *Alternaria brassicae* var *phaseoli, Ascochyta phaseolorum, Cercospora canavaliae, C. canavaliicola, C. cruenta, C. ternatea, Cercosporella canavaliae, Colletotri-*

chum canavaliicola, C. lindemuthianum, Corticium rolfsii, C. solani, Dendryphium canavaliae, Dietelia canavaliae, Elsinoe canavaliae, Heterosporium lagunense, Fusicoccum canavaliae, Gloeosporium canavaliae, Mycosporella venezuelensis, Pellicularia filamentosa, Phyllosticta canavaliae, Physalospora guignardiodes, Rosellinia bunodes, Pythium artotrogus, P. debaryanum, P. irregulare, P. rostratum, P. splendens, Sclerotium rolfsii, Septoria canavaliae, Dimeriellopsis costaricensis. Virus diseases isolated from this crop include: Abutilon mosaic, asparagus bean mosaic, Brazilian tobacco streak, euphorbia mosaic, green or yellow-green virus, *Marmor vignae* var *catjang*. Nematodes isolated from jackbean include: *Heterodera glycines, Meloidogyne incognita acrita, Pratylenchus brachyurus,* and *Rotylenchulus reniformis.*

REFERENCES:
Dey, Dipali, 1970, Cytological studies on *Canavalia ensiformis* and *Dolichos lablab, Indian Biol.,* **2**(2):54–59, illus.
Piper, C. V., 1925, The American species of *Canavalia* and *Wenderothia, Contrib. U.S. Nat. Herb.* **20**:555–585.
Sauer, J. J., 1964, Revision of *Canavalia, Brittonia* **16**:106–181.

CONTRIBUTORS: J. A. Duke, C. F. Reed.

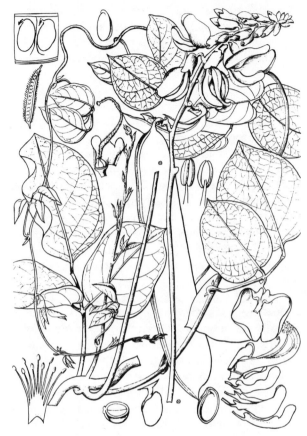

Figure 21. *Canavalia gladiata* (Jacq.) DC.

Canavalia gladiata (Jacq.) DC.

FAMILY: Fabaceae
COMMON NAME: Swordbean

USES: Swordbean used as cover crop, forage, and vegetable. Young green pods are used extensively as vegetable in India, Burma, Ceylon, and East Asian countries. Fully grown seeds of white-seeded races, while still in pods may be cooked and eaten as substitute for broadbeans. They should be thoroughly boiled in salt water with two changes of water. A few should be eaten in a test meal. If there are no harmful effects, as diarrhea or headaches, larger quantities may be eaten in subsequent meals. In Asia, they may be soaked overnight, boiled until soft in water with sodium bicarbonate, then rinsed, boiled in new water, and finally pounded for use in curries or like mashed potatoes. Roasted seed have been used as coffee substitutes.

FOLK MEDICINE: Pink seeds are employed in traditional Chinese medicine.

CHEMISTRY: Swordbeans have higher protein and salt contents than do French beans. The dried seeds are reported to contain per 100 g: 318 calories, 14.9% moisture, 27.1 g protein, 0.6 g fat, 53.8 g total carbohydrate, 11.6 g fiber, and 3.6 g ash. Fresh beans contain: moisture, 88.6; protein, 2.7; fat, 0.2; mineral matter, 0.6; carbohydrates, 6.4; and fiber, 1.5%; carotene, evaluated as vitamin A, 40 IU/100 g. Raw green pods contain per 100 g: 34 calories, 89.2 g moisture, 2.8 g protein, 0.2 g fat, 7.3 g total carbohydrate, 1.5 g fiber, and 0.5 g ash. The beans are said to contain 0.00972% hydrocyanic acid and a toxic saponin.

DESCRIPTION: Plant usually grown as an annual, but may become a perennial climber; stems woody, climbing with some semierect cvs, up to 10 m long; leaves large, trifoliolate; leaflets ovate, 7.5–20 cm

Canavalia gladiata (Jacq.) DC.

long, 5–14 cm broad, acuminate, rounded or cuneate, sparsely pubescent on both surfaces; petioles 5–12 cm long; rachis 2.5–5 cm long, petiolules 4–7 mm long; stipules ca. 2 mm long and thick; racemes axillary, 7–12 cm long, on peduncles 4–20 cm long, bearing several flowers in succession; flowers inverted with standard on bottom, not on top; pedicels 2 mm long; bracteoles 1 mm long, obtuse; calyx glabrous or nearly so, tube ca. 1 cm long, upper lip 5–7 mm long, standard white, 3–5 cm long; pods large, linear-oblong, slightly compressed, 20–40 cm long, 2.5–5 cm broad, widest near apex, curved with strongly developed ridges; seeds 8–12, 2.5–3.5 cm long, red, pink, reddish-brown to almost black, or white, oblong-ellipsoid, strongly compressed; hilum dark brown, 1.5–2.5 cm long, nearly as long as seed. Germination phanerocotylar.

GERMPLASM: Cultivars vary, particularly in length of pods and number of seed. *C. g.* var *alba* (Makino) Hisauchi, is an endemic Japanese cultigen. In India, three cvs are recognized; those with flowers and seeds red; flowers white but seeds red; and those with flowers and seeds white (considered most wholesome, but not the only one grown for human consumption). Assigned to the Indochina–Indonesian and China–Japanese Centers of Diversity, swordbean or cvs thereof is reported to exhibit tolerance to drought, low pH, salt, sand, shade, slope, virus, and waterlogging. ($2n = 22$, 44.)

DISTRIBUTION: Swordbean is of Old World origin and is probably derived from *C. virosa* Wight & Arn., which grows wild in tropical Asia and Africa. Most material referred to this species from Africa belongs to *C. virosa*. However, swordbean is widely cultivated in Far East, especially India, and now widely spread throughout the tropics. (Africa, America, Asia, Australia) and naturalized in some areas.

ECOLOGY: Swordbean is relatively drought-resistant and more resistant to saline soils than many other legumes. Requires a tropical climate, growing from sea level up to 1,000 m. Fares best with evenly distributed annual rainfall of 9–15 dm and fairly high temperatures (15°–30°C). Due to deep penetrating root system, once established, withstands very dry conditions. Grows on a wide range of soil types, but some cvs are susceptible to waterlogging. Tolerates some shade. Ranging from Warm Temperature Moist through Tropical Moist to Tropical Wet Forest Life Zones, swordbean is reported to tolerate annual precipitation of 6.4–26.2 dm (mean of 7 cases = 14.2 dm), annual mean temperature of 14.8°–27.4°C (mean of 7 cases = 22.4°C), and pH of 5.0–7.1 (mean of 6 cases = 6.0).

CULTIVATION: Propagated by seeds, sown at a depth of 5–7.5 cm. Plants are usually spaced 45–60 cm apart in rows 75–90 cm apart. In Hong Kong, seeds sown singly in pots in February, and seedlings transplanted or sown in March to May. In India, seeds sown from mid-April to late June, set out singly, 5–7.5 cm deep, in rows 60 cm apart along a strong high fence. Seed rate, from 25 to 40 kg/ha. Stems attacked by beetle grubs that kill young shoots. Infected branches should be cut off and burned. At season's end, crop should be dug up and burned. Plants should not be kept for more than 2 yr. Preparation of soil before sowing seed may be limited to a light hoeing, followed by a dressing of a complete fertilizer, raked into top layer of soil. When grown as vegetable, usually grown near dwellings, growing over walls and trees. Otherwise, grown on fences.

HARVESTING: Seedlings bear pods from mid-June to early December; older plants begin bearing a month later. Pods should be harvested when 10–15 cm long and before they swell and become tough. Seeds harvested as pods become ripe. When overripe, pods shatter. Produces green seeds in 3–4 months, mature seed in 5–10 months. Plants are good green manure, but less used as such than *C. ensiformis*.

YIELDS AND ECONOMICS: Yields of green matter for forage are 40–50 MT/ha; dried beans, 700–1,500 kg/ha. Seed yields are 700–900 kg/ha. Most extensively cultivated and used in the Far East, especially in India, Burma, Ceylon, and Indochina. Occasionally grown in other tropical regions, especially for forage.

BIOTIC FACTORS: Under natural conditions, flowers are visited by bees; cross-pollination is about 20% or more. However, bagged flowers set pods and seeds. Crop is attacked by few insect pests, as the fall armyworm, *Spodoptera frugiperda,* and beetle grubs that bore into the stems. Seed are rather resistant to infestation during storage. Most serious fungus disease is the root disease, *Colletotrichum lindemuthianum*. Other fungi attacking this plant include: *Acrothecium canavaliae, Cercospora canavaliae, Cerotelium canavaliae, Cladosporium herbarum, Corticium rolfsii, Dietelia canavaliae, Elsinoe canavaliae, Gloeosporium canavaliae, Mycosphaerella canavaliae, Nectria confluens, Nematospora coryli, Physalospora guignardioides, Sclerotium rolfsii*. It is also attacked by the bacterium, *Xanthomonas phaseoli*, asparagus bean mosaic virus and the nematodes, *Meloidogyne* spp., including *M. incognita acrita*.

CONTRIBUTORS: J. A. Duke, C. F. Reed.

Canavalia plagiosperma Piper

FAMILY: Fabaceae

COMMON NAME: Oblique-seeded jackbean

USES: An ancient Indian crop plant, found at various archaeological sites dating from about 2500 B.C., along coastal Peru. Now cultivated as green manure and forage plant in West Indies and northern South America.

FOLK MEDICINE: No data available.

CHEMISTRY: No data available.

DESCRIPTION: Annual herb; stems climbing, terete, 1 to several m long, sparsely strigillose; petioles as long as leaflets; leaves trifoliolate, leaflets membranous, chartaceous or somewhat coriaceous, obscurely reticulate, broadly ovate, obtuse or rounded at base, acute at apex, 10–15 cm long; pubescence short and white, very sparse on both surfaces, petiolules pubescent; inflorescence about 10-flowered, bracteoles 1.5 mm long, acute; pedicel 1 mm long; calyx green spotted with black, strigillose, about 13 mm long, upper lip shorter than tube, upper edge constricted behind nonapiculate tip, lowest tooth 2 mm long, acute, slightly exceeding acute laterals, lower lip with 3 broad deltoid acute subequal teeth; standard 2.25 mm long; corolla

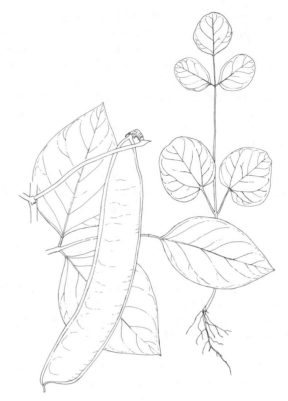

Figure 22. *Canavalia plagiosperma* Piper

purple, with wings as long as the blunt keel; pods linear, up to 25 cm long, 4 cm broad, compressed, spirally dehiscent, tipped with recurved beak, the intermediate ridge 5 mm from the sutural one; seeds ca. 10/pod, oblong and oblique, much compressed, somewhat shiny, abruptly narrowed at the micropylar end, 26–27 mm long, 17–18 mm broad, 10 mm thick, ochraceous-salmon, light brown or white with brown blotches; hilum about 11 mm long, wholly on the oblique micropylar end of seed, black encircled by a narrow brown band.

GERMPLASM: A form of this species found in Trinidad crossed with *C. ensiformis*. Testa of the seeds segregated in F_2 generation with 9 cinnamon, 3 cinnamon, and white, 3 white-speckled cinnamon and 1 white. Oblique-seeded jackbean is assigned to the South American Center of Diversity. ($2n = 22$.)

DISTRIBUTION: An ancient Indian crop plant along coastal Peru; elsewhere in tropical South America; introduced and cultivated in tropical America, West Indies, and Mauritius.

Canavalia plagiosperma Piper

ECOLOGY: All specimens seem to come from cultivated areas, along roadsides or in other artificial habitats. Requires a tropical climate. Ranging from Tropical Very Dry to Tropical Wet Forest Life Zone, oblique-seeded jackbean is reported to tolerate annual precipitation of 7.0–13.2 dm (mean of 2 cases = 10.1 dm), annual mean temperature of 21.0°–27.5°C (mean of 2 cases = 24.2°C) and pH of 5.0–7.5 (mean of 2 cases = 6.2).

CULTIVATION: Propagation by seeds. No details available on cultivation. As a green manure plant, it furnishes lush green foliage that falls to the soil, adding humus.

HARVESTING: No data available.

YIELDS AND ECONOMICS: Locally cultivated as green manure in West Indies, northern South America and in Nicaragua; some preliminary testing in Mississippi and Florida.

BIOTIC FACTORS: The fungus, *Cercospora canavaliae,* has been reported on this species.

CONTRIBUTORS: J. A. Duke, C. F. Reed.

Figure 23. *Cassia alata* L.

Cassia alata L.

FAMILY: Caesalpiniaceae
COMMON NAMES: Ringworm bush, Emperor's candlestick, King-of-the-forest
SYNONYM: *Herpetica alata* (L.) Raf.

USES: Probably best known as a medicinal herb. Bark contains a tanning material. Roots (juice) used in West Africa for tattooing or tribal markings. Plant is highly decorative. Bees are attracted to the flowers. May be poisonous to stock, poultry, and fish.

FOLK MEDICINE: Leaves used medicinally, taken internally as aperient, astringent, expectorant, purgative, taenifuge, and tonic. Leaves plus lime juice given as anthelmintic. Used as an antiparasitic, the leaves contain chrysophanic acid (2.2%). An ointment of leaves in vaseline is recognized remedy for ringworm and other parasitic skin diseases; the effect is enhanced in mixture with lime juice or common salt. Poultice of leaves used to hasten suppuration and on foul ulcers. Flowers taken internally as tonic and used for skin diseases. Seed also taken internally for skin diseases.

CHEMISTRY: Seeds of the genus *Cassia* are reported to contain trypsin and chymotrypsin inhibitors. The genus *Cassia* is reported to contain chrysarobin, cinnamaldehyde, decanal, hydrocyanic acid, isochaksine, ricinoleic acid, and saponin. Allen and Allen's survey (1976) suggested that there are 10 non-nodulated species of *Cassia,* like *C. alata,* for each nodulated species in the subgenera Fistula and Senna.

DESCRIPTION: Glabrous, short-lived herb or shrub, usually dying back to the ground (as in Texas), 3–5 m tall, with few thick branches; young stems and immature leaves puberulous with minute spreading hairs, eglandular; leaves spirally arranged, mostly 30–75 cm long, the petiole and rachis eglandular; leaflets 6–12 pairs, broadly oblong or obovate, 6–17 cm long, 3–7 cm broad, the terminal pair largest, obliquely rounded or truncate at base, obtuse or somewhat retuse at apex; stipules lanceolate to obliquely triangular, 0.6–2 cm long, often greenish-orange or yellowish; flowers in dense elongate spikelike simple or few-branched racemes 1–3 dm long, borne terminally or in the upper axils; bracts petallike, ovate-orbicular, imbricate, 15–25 mm long, deciduous, orange; pedicels very short; sepals straw-colored or light yellow, about 1 cm long; petals nearly equal, obovate, clawed, 1.5–2 cm long, bright yellow drying with conspicuous venation, 1.7–2.2 cm long, 0.9–1.2 cm broad; stamens normally 10, the 3 uppermost much-reduced, the 3 lowest with large anthers; ovary pubescent; pod 12–17 cm long, 1.3–2 cm broad, to 3 cm across the wings, dehiscent, transversely septate, with a crenate-margined wing running longitudinally along middle of each vavle; seeds brown, flattened at right angles to axis of pod, deltoid-rhombic, 30–40 per pod, acuminate to hilum, with a longitudinal ridge along each face, 6–8 mm long, 4–5 mm broad, 1.5–2.5 mm thick. Fl. Aug.–Oct. (in Texas); Fl. and Fr. Oct.–April (in Jamaica).

GERMPLASM: Assigned to the Middle American and Hindustani Centers of Diversity, ringworm bush or cvs thereof is reported to exhibit tolerance to high pH, insects, laterite, smog, and waterlogging. ($2n = 12, 24, 28$.)

DISTRIBUTION: Common in Bangladesh and many parts of India and Pakistan, especially the western peninsula; widespread in tropical countries, as Zanzibar, Pemba, Tanzania, Ghana, Senegal to Cameroons, Trinidad, Tobago, and Jamaica. Cultivated as a perennial herb in south and central Texas, becoming a shrub in lower Rio Grande Valley. Native to tropical America.

ECOLOGY: A tropical to subtropical plant, found from sea level to ca. 1,000 m altitude. Often escaped from cultivation and locally common in swampy places. Also common in village clearings, as in Ghana. Not particular as to soil, as long as it is moisture-holding. Ranging from Warm Temperate Dry to Moist through Tropical Very Dry to Wet Forest Life Zones, ringworm bush is reported to tolerate annual precipitation of 6.4–42.9 dm (mean of 22 cases = 19.5 dm), annual mean temperature of 14.7°–29.9°C (mean of 22 cases = 23.3°C) and pH of 4.3–8.0 (mean of 15 cases = 6.1).

CULTIVATION: Propagated from seeds, usually scattered about in clearings, or planted in rows at edge of forest.

HARVESTING: Leaves are harvested as needed, since plants are perennial and evergreen almost throughout the year in the tropics. Leaves gathered and dried, eventually placed in containers until needed for local medicinal use.

YIELDS AND ECONOMICS: Data unavailable on yields; leaves rarely collected all at once. Plants produce new leaves as others are harvested. Locally an important medicinal plant, especially in Africa, India, Pakistan, and other tropical areas.

BIOTIC FACTORS: Flower attractive to bees and probably used for honey. Fungi reported on this plant include the following species: *Asterina elaeocarpi, Cercospora chamaecristae, C. simulata,* and *Mycosphaerella cassiae.*

REFERENCE:
Allen, E. K., and Allen, O. N., 1976, The nodulation profile of the genus Cassia, in: *Symbiotic Nitrogen Fixation in Plants* (P. S. Nutman, ed.), pp. 113–121, International Biology Program No. 7.

CONTRIBUTORS: J. A. Duke, C. F. Reed, J. K. P. Weder.

Cassia auriculata L.

FAMILY: Caesalpiniaceae
COMMON NAMES: Avaram, Matara tea, Tanner's cassia, Tarwar
SYNONYM: *Cassia densistipulata* Taub.

USES: Shrub best known for its astringent bark, used for tanning heavy hides where color is of minor importance. It dyes leather to a buff color; also used to modify other dyes. Bark contains excellent fiber that can be made into rope. A large species of silkworm feeds on the leaves. In India, shrub is usually browsed by goats and cattle. Branches used as chewing sticks and toothbrushes. Roots used in tempering iron and steel. Infusion of leaves slightly aromatic and used as substitute for tea in Ceylon, where it is called 'matar.' In East Indies and some parts of India an alcoholic beverage

Cassia auriculata L.

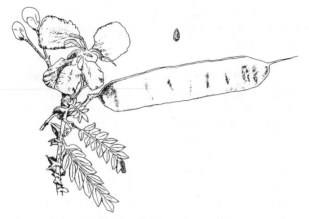

Figure 24. *Cassia auriculata* L.

is prepared by adding the bruised bark to a solution of molasses and allowing the mixture to ferment. In India, tender pods, leaves, and flowers used as food by poor people, but not cultivated. Plants also used for revegetating barren tracts and as a green manure crop; also has ornamental flowers.

FOLK MEDICINE: Plant is considered diuretic. Leaves, fresh or dried, used in infusion for a very cooling drink. Bark is highly astringent and used in place of tannic acid or oak bark for gargles, enemas, and as an alterative. Flowers and flower-buds used in a decoction for diabetes. Seeds reported as refrigerant and attenuant, and used in diabetes, chylous urine, ophthalmia, and conjunctivitis. Roots are astringent, and a decoction given as an alterative and used for skin diseases. Leaves and fruits used as anthelmintic.

CHEMISTRY: Bark contains 15–22% tannin.

DESCRIPTION: Evergreen shrub or small tree, up to 7.5 m tall; branchlets pubescent; bark thin, cinnamon-brown, with numerous brownish spots; leaves compound, petioles eglandular, rachis with a narrow subulate or fusiform gland between each of the 6–13 pairs of leaflets; stipules persistent, leafy, broadly reniform, 7–22 mm broad, one side produced into a subulate point; leaflets oblong-elliptic to obovate-elliptic, 1–3.7 cm long, 0.4–1.2 cm broad, rounded and mucronate at apex, puberulous or pubescent; racemes corymbose, 2–8-flowered, aggregated into rounded terminal panicles; sepals rounded at apex; petals yellow, large, 1.7–3 cm long, clawed, crisped on margins; stamens 10, the 3 lower with large anthers and long filaments, 4 with medium anthers and short filaments and 3 upper ones with reduced anthers; pods flat, strap-shaped, oblong-linear, straight, 6–18 cm long, 1.2–2.3 cm broad, indehiscent, with papery valves transversely undulate between seeds, shortly pubescent; distinct areole on each face 3–3.5 by 0.5–0.75 mm.

GERMPLASM: Assigned to the Hindustani Center of Diversity, avaram or cvs thereof is reported to exhibit tolerance to grazing and weeds. ($2n = 14, 16, 28$.)

DISTRIBUTION: Native to India, Ceylon and Burma; cultivated in Ceylon, India, Ghana, and Tanzania, especially in dry regions.

ECOLOGY: Plants are not exacting in soil and climatic requirements; hence easy and cheap to raise. Grows wild in woodland and wooded grassland, often in dry regions, from 150 to 610 m altitudes in East Africa. Needs full sun. Thrives best on fairly rich soil that drains well, should be moist, but not wet. Light porous soils are preferable. Ranging from Warm Temperate Dry through Tropical Very Dry to Wet Forest Life Zones, avaram is reported to tolerate annual precipitation of 3.8–42.9 dm (mean of 11 cases = 15.6 dm), annual mean temperature of 16.2°–27.5°C (mean of 11 cases = 23.1°C), and pH of 5.0–7.7 (mean of 7 cases = 6.5).

CULTIVATION: Propagated from seeds or stem cuttings. Seeds may be sown in furrows, cross-plowed and hoed at right angles to make another furrow or broadcast. Planted 5–12.5 cm apart in rows, one furrow is closed by plowing the next. Seed planted 10–15 cm deep, thinned during the first season. Weeding and cultivation unnecessary, but stimulate growth. Vigorous plants attain a height of 3 m and girth of 11 cm in 2 years; 5.3 m and 21 cm in 4 years. When 2–3 years old, the branches and twigs are cut, twig bark is stripped and dried in small cornets. The coppiced bushes send out many shoots and these may be harvested after 1 year.

HARVESTING: Bushes are ready to harvest in 3 or 4 years. Stripped bark constitutes 20–23% of the twig. Tannin content increases with age, but after 3 years the increase is not considerable. Size of twigs is more important than age. Twigs (thicker than a lead pencil in size) that have not developed a corky bark seem best. Plants grown on lime-rich soils are richer in tannin than those grown on red loam or gravelly soil. Extracts for tanning leather are as effective from unstripped as from stripped bark. Avaram tannin penetrates the hide quickly and produces a light colored leather with elastic grain and good tensile strength. The leather develops an objectionable black-red color on exposure to light, but this is obviated by subjecting the avaram-tanned leather to a final bath of myrobalan (*Terminalia chebula* Retz.). For local use, bark is dried and marketed in large hessian bags, each containing 100–120 kg.

YIELDS AND ECONOMICS: Yield of bark averaged about 1540 kg green bark/ha, with about 9,030 trees/ha, in 4-year-old plantation. Mainly produced in India, especially in Madras, mainly from native plants. This forms about one-third of India's requirements, the rest (ca. 28,000 MT/year), is obtained from Mysore and Hyderabad.

BIOTIC FACTORS: No serious pests or diseases reported.

CONTRIBUTORS: J. A. Duke, C. F. Reed.

Cassia occidentalis L.

FAMILY: Caesalpiniaceae

COMMON NAMES: Coffee senna, Bricho, Stypticweed, Stinkweed

SYNONYM: *Ditremexa occidentalis* (L.) Britt. & Rose

USES: Primarily medicinal; all parts have some purgative effect; also a restorative febrifuge, with reputed diuretic and sudorific properties. In Ceylon, tender pods, leaves, and flowers are eaten by the poorer people. This plant increases soil fertility, especially in exhausted peanut fields. Said to be richer in potash than *Cassia tora* or *C. sophera*. Said to be mildly toxic to various stock animals, which ordinarily avoid it. Seeds or leaves, ground up or mixed with charcoal, are used to blacken blackboards. Plant parts, e.g., roots and leaves, may be used as substitute for quinine. In Africa and Mauritius seeds used as substitute for coffee, but

Figure 25. *Cassia occidentalis* L.

contain no caffeine. Undried seeds are poisonous, because of presence of phytotoxins, but roasting destroys their purgative properties. Alcoholic infusions are said to be slightly insecticidal.

FOLK MEDICINE: Leaves used as poultices for toothache in the Dutch Indies and for headache in Malaya. Used for many and various diseases and ailments in Africa. Plant is diuretic (hence the common name piss-a-bed), febrifuge, purgative and tonic; leaves, roots, and seeds are purgative; seeds and leaves used externally in skin diseases and as an antiperiodic; root used for snakebite and in infusion used by American Indians as an antidote for various poisons.

CHEMISTRY: Chrysarobin, cinnamaldehyde, decanal, emodin hydrocyanic acid, isochaksine, ricinoleic acid, saponin have been reported from the genus (for relative toxicity, see Appendix). Seeds contain tannic acid, mucilage (36%), fatty oils (2.6%), emodin, and a toxalbumin. According to WOI, seed oil contains 19.7% saturated fatty acids, 31.4% linoleic acid, 30.7% oleic acid, 6.3% linolenic acid. Seeds

Cassia occidentalis L.

analyzing 20.6% protein and 2.7% oil had per 16 g N: 6.2 g lysine, 1.9 g methionine, 7.8 g arginine, 4.2 g glycine, 2.3 g histidine, 3.9 g isoleucine, 6.9 g leucine, 5.1 g phenylalanine, 3.1 g tyrosine, 3.9 g threonine, 5.1 g valine, 4.6 g alanine, 10.2 g aspartic acid, 18.0 glutamic acid, 0.0 g hydroxyproline, 3.7 proline, 5.3 g serine. Chrysarobin (0.25%) has been isolated from the benzene extract of the seeds. Seeds are reported to contain chymotrypsin inhibitors.

DESCRIPTION: Erect, glabrous, malodorous annual or subshrubby perennials lasting 2–3 years; stems 1–2 m tall, sulcate, usually with few to several strongly ascending branches spirally arranged, up to 25 cm long, compound; petioles with a large sessile hemispherical to ovoid or subglobose gland shortly above base; rachis eglandular; leaflets 4–6 pairs, lanceolate to ovate, acute or acuminate, the base only slightly asymmetric, terminal pair longest, 3–6 cm long, 1–2.5 cm broad, somewhat glaucous beneath, closing at night; stipules linear-lanceolate, acuminate, 4–6 mm long, caducous; flowers few together in short bracteate axillary racemes, almost umbellate; peduncles 3–5 mm long; bracts acute; sepals oblong, obtuse, 6–9 mm long, usually glabrous outside; petals yellow or yellow-orange, broadly ovate or elliptic, wilting by mid-day, 10–18 mm long; stamens 10, the 3 uppermost much-reduced, the 3 lowest with elongate filaments, 4 smaller ones on shorter filaments; ovary pubescent on faces; pod linear, slightly curved upwards or sometimes nearly straight, 8–12.5 cm long, 0.5–1.0 cm broad, indehiscent or tardily dehiscent, compressed, at first flat, maturing turgid, medially dark brown and elevated over the seeds, the margins paler; seeds 15–30 per pod, ovoid, compressed, dull brown, 4–5 mm long, 3.5–4.5 mm broad; testa minutely pimpled, with an elliptic areole 2.6 by 1.5 mm on each face. Fl. most of year in Jamaica and other tropical areas; Feb.–Mar. in Rhodesia; Aug.–Nov. in Texas; Aug.–Sept. in other parts of United States; Feb.–Mar. and July in Ghana.

GERMPLASM: Assigned to the Indochina–Indonesian Center of Diversity, coffee senna or cvs thereof is reported to exhibit tolerance to high pH, laterite, virus, and weeds. ($2n = 26, 28$.)

DISTRIBUTION: Pantropical, possibly originating in tropical America, in many places a weed.

ECOLOGY: Well-adapted to many ecological areas of the tropics from sea level up to 1,740 m altitude. Weed of cultivated fields, waste places, coastal areas, roadsides, waste ground near villages and buildings, grasslands, lake shores, pastures, and disturbed places. Tolerates subtropical and warm temperate conditions; in United States grows from Florida and Texas north to Virginia, Illinois, Iowa, and eastern Kansas. Often troublesome in sugar plantations. Ranging from Warm Temperate Dry to Moist through Tropical Desert to Wet Forest Life Zone, coffee senna is reported to tolerate annual precipitation of 6.4–42.9 dm (mean of 32 cases = 15.7 dm), annual mean temperature of 12.5°–27.8°C (mean of 32 cases = 22.8°C), and pH of 4.5–8.4 (mean of 26 cases = 6.3).

CULTIVATION: Plants propagated by seed, planted in gardens to a limited extent; rarely cultivated as a crop. Because of weedy nature, plant is copiously available for native use.

HARVESTING: Plant parts collected from natural sources by natives. Seeds, leaves, roots, or flowers collected by hand in quantities as needed. Rarely appears as a commercial product.

YIELDS AND ECONOMICS: Important plant for local medical uses in many tropical areas, especially in Africa, Philippine Islands, Hawaii, West Indies, and East Indies.

BIOTIC FACTORS: Following fungi have been reported on this plant: *Alternaria tenuis, Cercospora nigricans, C. occidentalis, C. pinnulaecola, Colletotrichum capsici, Pericomia byssoides, Phyllosticta cassiae-occidentalis, Phytophthora parasitica, Ramularia cassiaecola, Sarcinella cassiae, Sphaerotheca cassae*. Nematodes isolated from this legume include the following species: *Heterodera glycines, H. marioni, Meloidogyne arenaria, M. incognita acrita, Pratylenchus pratensis, Rotylenchulus reniformis*. Cultivation in fields is supposed in Senegal to be inimical to the root parasite, *Striga senegalensis*.

REFERENCE:
Niranjan, G. S., and Gupta, P. C., 1973, Chemical constituents of the flowers of *Cassia occidentalis, Planta* 23(3):298–300.

CONTRIBUTORS: J. A. Duke, C. F. Reed, J. K. P. Weder.

Cassia senna L.

FAMILY: Caesalpiniaceae
COMMON NAME: Alexandrian senna
SYNONYMS: *Cassia acutifolia* Del.
Cassia lanceolata Collad.

USES: Grown or harvested as a drug plant.

FOLK MEDICINE: Leaves and pods are source of Alexandrian senna of commerce, a drug generally preferred over East Indian senna, as it is milder, but has the same action. Used as a laxative and cathartic, generally combined with aromatics and stimulants to modify its griping effects. Dried, pulverized leaves are applied to wounds and burns. Entire plant used as a febrifuge.

CHEMISTRY: Purgative qualities are due largely to anthraquinone derivatives. The plant is reported to contain also rhein, aloe-emodin, kaempferol, isorhamnetin, chrysophanic acid, sennacrol, sennapicrin, and cathartomannite. The seeds reportedly contain trypsin inhibitors. The valve of the pods is reported to contain trypsin and chymotrypsin inhibitors. Dried seeds of a species of *Cassia* called *senna* contained per 100 g: 346 calories, 10.3% water, 18.0% protein, 1.3% fat, 66.9 g total carbohydrate, 6.5 g fiber, 3.5 g ash, 205 mg Ca, 184 mg P, and 1.04 mg thiamine.

DESCRIPTION: Low, bushy shrub, 40–60 cm high, leaves paripinnate, 3–9 pairs of grayish-green, lanceolate leaflets, 1.5–3.5 cm long, 0.5–1.2 cm wide, unequal at the base, broadest below the midrib, pubescent to glabrous, subsessile, the stipules subulate, spreading or reflexed, 2–4 mm long; racemes axillary, erect, many-flowered; flowers yellow; pods flattened, 4–6 cm long, 1.8–2.5 cm wide, green, taste sweetish but sickly; odor characteristic, somewhat like tea; valves chartaceous, thinly puberulous, faintly transversely veined; seeds obovate-cuneate, compressed.

GERMPLASM: Apparently no breeding or selection has been undertaken. The plant is cross pollinated by insects. Assigned to the African Center of Diversity, Alexandrian senna or cvs thereof is reported to exhibit tolerance to disease, drought, high pH, insects, poor soil, and sand.

DISTRIBUTION: Originally described from North Africa; now distributed from Egypt along Red Sea to Sudan and desert parts of Nubia. Grown in Egypt and frequently in southern Sudan.

ECOLOGY: Grows on poor sandy soils, survives as a perennial with occasional irrigation. In best-growing areas avg annual precipitation is about 7.5 dm, mean ann temp ca. 24°C, but the plant ranges from Warm Temperate Moist to Tropical Dry Forest Life Zone, tolerating annual precipitation of 6.4–19.4 dm (mean of 7 cases = 13.3 dm), annual mean temperature of 15.8°–27.1°C (mean of 7 cases = 22.6°C), and pH of 5.0–7.1 (mean of 5 cases = 5.8).

CULTIVATION: Senna is propagated by seed. In India planted in nursery beds, usually in December, transplanted in February, when 15 cm high, spaced 90 cm each way, on raised beds or well-prepared patches of open ground. Soil should be thoroughly preworked and free of weeds; vegetable compost or manure should be incorporated. Vigorous leafy plants are obtained by watering during the hottest days of April and May, but shade is not needed.

HARVESTING: Leaves and pods picked from wild and cultivated plants. In Nubia, two crops are harvested annually—one in September after the rains, the other in April. In Sudan leaves are

Figure 26. *Cassia senna* L.

Cassia senna L.

gathered during the winter. In India, where the plant is stripped of its leaves 3 times per season, picking begins in May. Leaves are immediately spread out to sun dry. Senna leaves are graded carefully, and all foreign materials including twigs and leaves are removed.

YIELDS AND ECONOMICS: Up to 1,120 kg/ha of cured leaves may be harvested 70 days after sowing. Sudan exports about 450,000 kg senna leaves annually to the United States and pods to the United Kingdom. Southeast India also produces senna commercially. Alexandrian senna leaves by the barrel commanded ca. $0.75/kg in 1972 and $1.50 in 1977. In bales, the pods commanded ca. $1.00/kg in 1977.

BIOTIC FACTORS: Few important insects, fungi or nematodes are reported to infect this plant. Said to be cross-pollinated by insects.

REFERENCES:
Anonymous, 1972, Senna leaves, *Chem. Market Res.* **July 3, 1972**:22.
Ayoub, A. T., 1975, Sodium and cation accumulation by senna (*Cassia acutifolia*), *J. Exp. Bot.* **26**:891–896.
Ayoub, A. T., 1977, Some primary factors of salt tolerance in senna (*Cassia acutifolia*), *J. Exp. Bot.* **28**:484–492.
Fairbairn, J. W., and Shrestha, A. B., 1967, The taxonomic validity of *Cassia acutifolia* and *C. angustifolia*, *Lloydia* **30**(1):67–72.

CONTRIBUTORS: J. A. Duke, C. F. Reed, J. K. P. Weder.

Ceratonia siliqua L.

FAMILY: Caesalpiniaceae

COMMON NAMES: Carob, Algarrobo, Carob bean, John's bread, Locust, St. John's bread

USES: Carob is primarily cultivated for its fruit (pod) and seeds, both high in sugar and calcium, low in protein and fats. It is used in "health foods" as a chocolate substitute and in livestock feeds, especially for cattle. From the ground endosperm, gum is made, known as locust bean gum, gum tragon, tragosol, carubin, and carob flour. The gum has great water-absorbing qualities and makes an excellent stabilizer, used in many food products, in

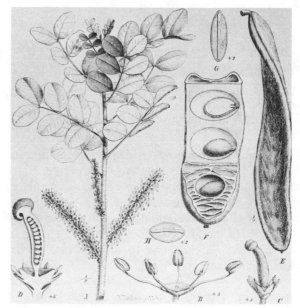

Figure 27. *Ceratonia siliqua* L.

the manufacture of paper, in cosmetics and drugs, and in the chemical industry. The pods, richer in sugar than the seeds, are popularly used in foodstuffs, juices, and flour. The sugar can be extracted from the pods by alcohol and the by-product, molasses, can be fermented to ethanol. Roast seeds have been used as a coffee substitute. Alcoholic beverages have been made from infusions of the pods. Carob is also used in textile printing, synthetic resins, insecticides, and fungicides. American imports are mostly for flavoring tobacco, and in cosmetics. The hard and heavy wood is used locally to make furniture and wheels. Bark contains ca. 50% tannin.

FOLK MEDICINE: Pod is reported to have been used as an anticatarrhal, demulcent, and resolvent. The leaf is astringent. Various portions of the plant are said to be used as antitussive, astringent, pectoral, and purgative.

CHEMISTRY: Carob flour has been reported to contain per 100 g: 11.2 g water, 180 calories, 4.5 g protein, 1.4 g fat, 80.7 g carbohydrate, 7.7 g fiber, 2.2 g ash, 352 mg Ca, and 81 mg P. Germ meal contains: 326 calories, 9.8% moisture, 47.3 g protein, 2.8 g fat, 34.1 g carbohydrate, 3.8 g fiber, and 6.0 g ash. The seeds contain: water, 8.9–13.6; protein, 14.4–19.7; fat, 1.8–3.1; N-free extract, 55.7–62.5; fiber, 6.9–8.3; and ash 2.3–3.6%. The pods contain: water, 3.7–24.7; protein, 2.1–7.2; fat, 1.2–4.0; N-free extract, 24.5–48.4; reducing sugars, 3.0–20.5; sucrose, 7.0–43.6; fiber, 3.1–15.3; and ash, 1.8–3.9%. Van Etten *et al.* reported 47% protein (dry weight) for the seed and classified the amino acids (grams per 16 g N) as follows: 5.6 lysine, 1.0 methionine, 11.8 arginine, 5.3 glycine, 2.5 histidine, 3.5 isoleucine, 6.5 leucine, 3.2 phenylalanine, 3.5 tyrosine, 3.6 threonine, 4.4 valine, 4.1 alanine, 9.0 aspartic acid, 28.0 glutamic acid, 0.1 hydroxyproline, 4.0 proline, and 5.0 serine with 92% of the N as amino acids. The genus *Ceratonia* has been reported to contain the following chemicals: isobutyric acid, lactose, maltose, saponins, and tannins. Vitamin E is also reported, along with A-amino-pimelic acid, ceratose, galactomannans (lacking in embryo, abundant in endosperm; Ahmet and Vardar, 1975), leucoanthocyanins, D-pinitol, primverose, and an unnamed alcohol. Carob groats (bruised grain) are reported to contain 10.6% water, 5.0% protein, 0.3% fat, 68.6% N-free extract (including 26.1% sucrose, 14.3% invert sugar, and 0.4% dextrin, 12.4% fiber), 3.1% ash including 0.17% phosphoric acid and 0.49% lime. Carob fruits are reported to consist of 90% pulp, 8% kernels, and 2% other materials. The kernels consist of 30–33% husk, 23–25% germ, and 42–46% endosperm. Carob bean gum is reported to contain 88% D-galacto-D-mannoglycan (ratio of galactose to mannose about 1 : 4), 4% pentosan, 6% protein, 1% cellulose, and 1% ash. Carob beans are reported to contain the following saccharides: glucose, fructose, xylose, sucrose, and two unestimated reducing disaccharides, one of which contains xylose and glucose. The nutritive value of carob nearly equals that of barley; 147 kg carob corresponding to ca. 129 kg barley, 100 kg wheat, or 70 kg beans. Seeds and commercial bean meals are reported to contain trypsin inhibitors and chymoptrypsin inhibitors.

DESCRIPTION: Long-lived (up to 100 years) evergreen tree, to 13 m tall, 3.6 m in diameter; trunk branching at about 3 m into several spreading branches; leaves rather large, pinnate, with 2 to 6 pairs of dark green, glossy, leathery, oval leaflets; flowers red, appearing in late fall or early winter; fruit thick, sweet, mucilaginous, fleshy, green, becoming somewhat hard, brown and brittle; seeds small, hard flattened, brown, 10–18 per pod; roots strong, deep penetrating. Trees mostly dioecious, with some male or female only, thus necessitating planting some male trees among the others, or grafting male branches on the others to insure good fruit production.

GERMPLASM: Popular "varieties" are *tylliria*, the common tree with the largest pods; *agria*, the wild type; *apostolika*, an unbudded type; *koundaurka*, tree with weeping habit; *koumbota*, a large tree. Great variability in this species affords genetic material for developings cvs. Assigned to the Mediterranean Center of Diversity, carob or cvs thereof is reported to exhibit tolerance to drought, high pH, heat, insects, limestone, poor soil, sand, slope, and smog. ($2n = 24$).

DISTRIBUTION: Common in the eastern Mediterranean region, particularly in Cyprus, Israel, and Lebanon and occurring wild in elevated parts of Yemen, and wild or feral in Algeria and Morocco. Cultivated from antiquity in the Mediterranean Sea area at 0–500 m in latitudes 27°–42°. Introduced to California, southern Mexico, South Africa, Australia, and India.

ECOLOGY: Hot dry summers and cold wet winters of the Mediterranean climate are ideal, with a mean temperature of 10°C in January, 27°C in July, with a range of 5°–37°C. Carob is damaged at −3°C, and dies at −4°– −5°C. In native habitats the carob grows in shallow, rocky soils on hillsides or in deep chalky soils; heavy, poorly drained soils are unsuitable. Ideal soil is a clayey loam containing lime. Tree requires rainless autumn for maturing. Ranging from Warm Temperate Dry through Subtropical Dry to Moist Forest Life Zone, carob has been reported to tolerate annual precipitation of 3.1–40.3 dm (mean of 13 cases = 10.5 dm), annual mean temperature of 12.7–26.5°C (mean of 13 cases = 20.3°C), and pH of 5.0–8.3 (mean of 10 cases = 7.3).

CULTIVATION: Carob may be propagated by seed, grafting or cuttings. Seeds are removed from the pods (soon after maturity, before hardening), mixed with coarse sand and kept moist in greenhouse or between layers of sterilized sacking or burlap. When they show signs of swelling and growth, they are planted in a propagating bed, composed of clay rolled and packed hard, upon which the seeds are placed and covered lightly with sand and good soil

Ceratonia siliqua L.

to the depth of 1.2 cm. When the seedlings show two sets of leaves they are transferred to 6.5-cm pots containing clay soil; transplants allowed to stand 24 hr without water, then watered freely. At 12–15 cm high, they are transferred to tins or boxes 15 × 20 × 20 cm with a good potting soil. Plants 1–1.8 m tall are outplanted 7 m each way in poor, rocky soil; 12–14 m each way in good fertile soil. On hillsides the trees can best be set in terraces constructed on the contour. Since seedlings are very slow to bear and sex is indeterminable for a long time, budding or grafting to a good cv is advisable. Cuttings may be, but seldom are, grown with bottom heat and careful treatment. In Israel and Cyprus, high yielding cvs are maintained by grafting in the third year to the healthiest plants. In South Africa grafting is done from mid-May to end of June, care being taken that the bark opens easily. The best method is to graft on the branches, not the stem, leaving the smaller branches to utilize the winter deposits of sap. These branches may be cut off the following year. Seedlings may be budded the second year in the field when the stalk is about 0.8 cm in diameter. Usually grown as a dryland crop, the tree produces better under irrigation. Cultivation disturbs the root system and is not recommended. Oil or chemical herbicides may be used.

HARVESTING: Carob bears in 5–7 years. After flowering, the fruit requires 6–8 months to mature. Flowers appear in the fall or early winter, fruit develops slowly during the late winter and spring, and matures late the following summer or in the early autumn. From August on, the pods are knocked down with long poles. They are dried for a month, then transported to a factory for crushing.

YIELDS AND ECONOMICS: Yield varies greatly. Spontaneous trees 25–30 years old yield ca. 200–270 kg of pods annually. Wild stock after grafting has yielded 400–450 kg/tree; cultivated stands may produce 10 MT/ha with 50 trees/ha, but average yields rarely exceed 2.5 MT/ha. It is estimated that 1,000 kg pod yield 100 kg seed yielding 35 kg gum. Distilled pods are said to yield 18 percent alcohol. Cyprus, the leading producer of carob, has encouraged production by subsidizing costs in nonirrigated areas, hoping to increase by 200,000 trees annually. Carob is also produced in Crete, Greece, Sicily, Sardinia, Majorca, Spain, Portugal, Palestine, Tunisia, and Algeria. Britain and Italy are the main importers. In 1956, the United States imported 1,100 MT pods, 6,700 MT gum. Labor accounts for 30% of the price.

BIOTIC FACTORS: Carob trees are pollinated by wind and insects; apiaries are recommended in the orchards. Pests include the carob moth (*Ectomyelois ceratoniae*), and midge (*Asphondylia gennadii*), the navel orangeworm (*Amyelois transitella*), boring beetle (*Cerambyx velutinus*), and oleander scale (*Aspidiotus nerii*). *Meloidogyne* nematodes have been isolated. Rodents are fond of the roots. For the 25 fungi found on carob, economic importance is unknown; those most likely to cause damage are: *Alternaria tenius, Amphichaeta ceratoniae, Cercospora ceratoniae, Diplodia natalensis, Fusarium oxysporum, F. solani, Glomerella cingulata, Oidium ceratoniae, Phytophthora cactorum, Sphaerella cuprea,* and *Verticillium alboatrum*; many of the others are basidiomycetes attacking the bark.

REFERENCES:

Ahmet, M., and Vardar, Y., 1975, Chemical composition of carob seed, *Phyton* 33(1):63–64.
Binder, R. J., *et al.*, 1959, Carob varieties and composition, *Food Technol.* 3:213–216.
Charalambous, J. (ed.), 1966, The composition and uses of carob bean, *Cyprus Agric. Res. Inst.,* Nicosia.
Coit, J. E., 1962, Carob varieties, *Fruit Var. Hort. J.* 15:75–77.
Coit, J. E., 1951, Carob or St. John's Bread, *Econ. Bot.* 5:82–96.
Condit, I. J., The carob in California, *Calif. Exp. Stn., Bull.* 309.
Davies, W. N. L., 1970, The carob tree and its importance in the economy of Cyprus, *Econ. Bot.* 24:460–470.
Orphanos, P. I., and Papaconstantinou, J., 1969, The carob varieties of Cyprus, *Cyprus Agric. Res. Inst., Tech. Bull.,* No. 5, Nicosia, Government of Palestine, Sept.

CONTRIBUTORS: J. A. Duke, R. Hegnauer, C. F. Reed, J. K. P. Weder.

Cicer arietinum L.

FAMILY: Fabaceae
COMMON NAMES: Chickpea, Bengal gram, Dhal, Garbanzo, Gram, Pois chiche

USES: Chickpea is valued for its nutritive seeds with protein content of 14–30%; roasted chickpea without husks may contain 22.5%. Seeds are eaten

Figure 28. *Cicer arietinum* L.

fresh or as dry pulse, parched, boiled, fried, or in various dishes. Sprouted seeds are eaten as a vegetable or salad. Young plants and green pods are eaten like spinach. Dhal is the split chickpea without its seedcoat, dried and cooked into a thick soup or ground into flour for snacks and sweetmeats. Chickpeas yield 21% starch suitable for textile sizing, giving a light finish to silk, wool, and cotton cloth. An adhesive may also be prepared; although not water-resistant, it is suitable for plywood. Gram, gram husks, and green or dried stems and leaves are used for stock feed; whole seeds may be milled directly for feed for horses and cattle. Leaves are said to yield an indigolike dye. Acid exudates from the leaves can be applied medicinally or used as vinegar. In Chile, a cooked chickpea–milk (4:1) mixture was good for feeding infants, effectively controlling diarrhea.

FOLK MEDICINE: Glandular secretion of the leaves, stems, and pods consists of malic and oxalic acids, giving a sour taste. In India these acids used to be harvested by spreading thin muslin over the crop during the night. In the morning the soaked cloth is wrung out, and the acids are collected in bottles. Medicinal applications include use for aphrodisiac, bronchitis, catarrh, cutamenia, cholera, constipation, diarrhea, dysentery, dyspepsia, flatulence, snakebite, sunstroke, and warts. Acids are supposed to lower the blood cholesterol levels. Seeds are considered antibilious. Pliny is said to have recommended chickpea as diuretic, galactagogue, and as a prophylaxis against skin disease. Seeds are said to be soporific. In the 16th century, chickpeas were believed to be aphrodisiac: "Curiously enough, lentils were considered to have the opposite effect, and this was probably the reason why the lentil was included in the diet in monasteries on meatless days" (Van der Maesen, 1972).

CHEMISTRY: Constituents vary among samples and methods of analysis, but the seeds contain more oil (4–6%) than do most pulses. Raw whole seeds contain per 100 g: 357 calories, 4.5–15.6% moisture, 14.9–24.6 g protein, 0.8–6.4 g fat, 63.5 g total carbohydrate, 2.1–11.7 g fiber, 2.0–4.8 g ash, 140–440 mg Ca, 190–382 mg P, 5.0–23.9 mg Fe, 0–225 μg β-carotene equivalent, 0.21–1.1 mg thiamine, 0.12–0.33 mg riboflavin, and 1.3–2.9 mg niacin. Ground seeds contain 376 calories, 6.4% moisture, 25.9 g protein, 5.3 g fat, 59.3 g total carbohydrate, 1.2 g fiber, 3.1 g ash, 10.6 mg Fe, 0.38 mg thiamine, 0.14 mg riboflavin, and 1.8 mg niacin. Ripe sun-dried chickpeas have been reported to contain (per 100 g): 362 calories, 11.0% water, 19.4% protein, 5.6 g fat, 60.9 g carbohydrate, 2.5 g fiber, 3.1 g ash, 114 mg Ca, 387 mg P, 2.2 mg Fe, 0.46 mg thiamine, 0.20 mg riboflavin, and 1.2 mg niacin. Ascorbic acid ranges from 4.3–13.8 mg (low with yellow, high with green cotyledons). Boiled and roasted seeds contain similar amounts. Sprouting is said to increase the proportionate amounts of ascorbic acid, niacin, available iron, choline, tocopherol, pantothenic acid, biotin, pyridoxine, inositol, and vitamin K (Hulse, 1975). Various specific and nonspecific chemicals—choline, saponins, biochanines, flavonoids—have been found in chickpea. Malic and oxalic acid exudation of the plant may soil trousers and shoes. Wild species often have similar glandular secretions. The amino acid composition of seeds with 19.4% protein, 5.5% oil is (per 16 g N): 7.2 g lysine, 1.4 g methionine, 8.8 g arginine, 4.0 g glycine, 2.3 g histidine, 4.4 g isoleucine, 7.6 g leucine, 6.6 g phenylalanine, 3.3 g

Cicer arietinum L.

tyrosine, 3.5 g threonine, 4.6 g valine, 4.1 g alanine, 11.7 g aspartic acid, 16.0 g glutamic acid, 0.0 g hydroxyproline, 4.3 g proline, and 5.2 g serine. Percent fatty acid compositions are: 'Deshi': oleic 52.1, linoleic 38.0, myristic 2.74, palmitic 5.11, and stearic 2.05; 'Kabuli': oleic 50.3%, linoleic 40.0, myristic 2.28, palmitic 5.74, stearic 1.61, and arachidic 0.07%. Whole grain also contains adenine, choline, fructose, glucose, inositol, saccharose, and starch. Seeds are reported to contain trypsin inhibitors and chymotrypsin inhibitors. The leaves contain 4–8% protein. Young shoots contain 60.6 moisture, 8.2 protein, 0.5 fat, 27.2 carbohydrate, 3.5 ash, 0.31 Ca, and 0.21% P. On a dry weight basis, the residues contain 12.9 crude protein, 36.3 fiber, 38.1 N-free extract, 1.5 fat, 11.2 ash, 2.2 Ca, 0.5 P, 0.5 Mg, 0.3 Na, and 3.0% K.

DESCRIPTION: Annual herb, erect or spreading, much branched, 0.2–1 m tall, glandular pubescent, olive, middle, dark green or bluish green in color. Root system to 2 m deep, but major portion up to 60 cm. Leaves imparipinnate, glandular-pubescent with 3–8 pairs of leaflets and a top leaflet (rhachis ending in a leaflet); leaflets ovate to elliptic, 0.6–2.0 cm long, 0.3–1.4 cm wide; margin serrate, apex acuminate to aristate, base cuneate; stipules 2–5-toothed, stipels absent. Flowers solitary, sometimes 2–3 per inflorescence, axillary; peduncles 0.6–3 cm long, pedicels 0.5–1.3 cm long, bracts triangular or tripartite, up to 2 mm long; calyx 7–10 mm long; corolla white, pink, purplish (fading to blue), or blue, 0.8–1.2 cm long. Pod rhomboid-ellipsoid, 1–2 (–4)-seeded, 1.4–3.5 cm long, 0.8–2 cm wide, inflated, glandular-pubescent. Seed color cream, yellow, brown, black, or green, rounded to angular obovoid, 0.4–1.4 by 0.4–1 cm; seed coat smooth or wrinkled, or tuberculate, laterally compressed with a median groove around two-thirds of the seed, anterior beaked; germination cryptocotylar. Fl. summer (Meditt., Middle East), winter (India); Fr. summer–fall (Meditt., Middle East), March–April (India).

GERMPLASM: Many "varieties" have been described, but are not widely recognized. Improved cvs have been developed, often for local adaptation. Flower and seed color and size, growth duration, yield, and disease resistances vary. Cream-seeded or 'Kabuli' chickpeas (Mediterranean and Middle Eastern origin) have the largest seeds, and grow well under irrigation. 'Desi' chickpeas (Indian distribution) have smaller seeds, and yield better in India and often elsewhere. Hybrids between Kabuli and Desi have produced strains with medium-size seeds and fair yields. The bulk of chickpeas grown in India is from unselected land races. Improved cvs have so far failed to affect yield at the farmers' level. Some released cvs with improved yield are: 'G-86' and 'Chafa' (large golden yellow seeds); 'Type 87' (large brown seeds for parching); 'Dharwar No. 18-12' (wilt-resistant); 'Ayelet', 'Ofra' (cream seeds, blight-resistant, Israel); 'L-550' (cream seeded); 'JG-62' and 'Annigeri' (early maturity, brown seeds); 'Radhey' and 'T 3' (Uttar Pradesh); 'BEG-482' (Andhra Pradesh); 'Plovdiv 19' (blight-resistant, Bulgaria); 'Kaka,' 'Kourosh,' and 'Pyrouz' (Iran); 'C-214' and 'G-130' (frost tolerant, India); 'Zimistoni' (cold resistant, Uzbekistan). Assigned to the Hindustani, Central Asian, and Near Eastern Centers of Diversity, chickpea or cvs thereof is reported to exhibit tolerance to disease, drought, high pH, photoperiod, and virus. ($2n = 16[14,24]$.)

DISTRIBUTION: The closely related annual species, *C. reticulatum* Ladizinsky and *C. echinospermum* P. H. Davis, occur in SE Turkey. Wild *C. reticulatum* is interfertile with the cultivated pulse and morphologically closely resembles cultivated *C. arietinum*. It should be regarded as the wild progenitor. Botanical and archeological evidence show that chickpeas were first domesticated in the Middle East. Widely cultivated in India, Mediterranean area, the Middle East, and Ethiopia since antiquity. Brought to the new world, it is now important in Mexico, Argentina, Chile, and Peru. On trial in Australia. Wild species are most abundant in Turkey, Iran, Afghanistan, and Central Asia.

ECOLOGY: Grown usually as a rainfed cold-weather crop or as a dry climate crop. Thrives on a sunny site in a cool, dry climate on well-drained soils (clayey, but not too heavy) of pH 5.5–8.6. Grows well on poor soils, but generally grown on heavy black or red soils. Frost, hailstones, and excessive rains damage the crop. Though sensitive to cold, some cvs can tolerate temperatures as low as −9.5°C in early stages or under snow cover. Daily temperature fluctuations are desired with cold nights with dewfall. Optimum conditions include 18°–26°C day and 21°–29°C night temperatures and annual rainfall of 6–10 dm. Relative humidity of 21–41 is optimum for seed set. Occasionally irrigated, some chickpea cvs tolerate higher salinity (conductivity of 1.2 mmhos/cm) than most grain legumes. Protein content of the seeds is greatly influenced by soil and locality. Rhizobium data are on the whole inconclusive. High pH inhibits nodulation. Low pH (4.6) encourages fusarium wilt. In virgin or for the first planting in heavier soils inoculation is said to increase yield by 10–62%. Although spoken of as "day-neutral," chickpea is a quantitative long-day plant, but flowers in every photoperiod. In Tanzania at 2°S latitude, a few Mediterranean cvs did not flower, however. Russian cultivars, adapted to high latitude, are relatively insensitive to photoperiod. Ranging from Warm Temperate Dry to Moist through Tropical Desert to Moist Forest Life Zones, chickpea is reported to tolerate annual precipitation of 2.8–15.0 dm (mean of 41 cases = 6.7), annual mean temperature of 6.3°–27.5°C (mean of 41 cases = 17.0°C), and pH of 4.6–8.3 (mean of 35 cases = 7.2).

CULTIVATION: Chickpeas are propagated from seeds. Soil is worked into a rough tilth, clods broken and field leveled; a fine tilth is not necessary. Seed is broadcast or (more often) drilled in rows 25–60 cm apart, spaced at 10 cm between seeds, at a depth of 2–12 cm with soil well pressed down. Seed is sowed in spring (late March–mid-April in Turkey, United States; February–March–April around the Mediterranean) when the ground has warmed or when the rains recede (mid-September to November, rarely later in India and Pakistan; September-January or April, Ethiopia) depending on the region. Seeding rates vary from 25–40 kg/ha to 80–120 kg/ha, the higher for large-seeded cvs such as 'Kabuli.' Chickpea may be cultivated as a sole crop, or mixed with barley, lathyrus, linseed, mustard, peas, safflower, sorghum, or wheat. In rotation it often follows wheat, barley, rice, or teff. In India, chickpeas are also grown as a catch crop in sugarcane fields and as a second crop after rice. Although usually considered a dry farm crop, chickpeas develop well on rice lands. In most areas, chickpeas are intercultivated once about 3–4 weeks after sowing; thereafter, crop develops enough shade to smother weeds. In other areas light weedings are recommended. Fertilizers or manure have often failed to increase yields substantially. Luxuriant growth on rich soils may result in poor seed set. In India, 27 kg P_2O_5 proved best fertilizer regime, improving yields from 481 kg to 677 kg/ha. Molybdenum applications have increased yields by 16%. Often, however, phosphate application does not increase yields. On poor soils, manure or compost is beneficial. Seed inoculation improves yield only for crops grown for the first time or after rice, where Rhizobium populations are naturally low or absent. Irrigation at 45 and 75 days after planting is useful. For vigorous branching, the plants may be topped about the time they begin to branch. In India and Pakistan, sheep and goats may graze lightly when the plants are 30 cm tall. The main factor limiting yield is poor stand due to incomplete emergence.

HARVESTING: Chickpeas mature in 3–7 months, leaves turn brown, dry, and fall. Seeds not fully ripe can be eaten as a snack or cooked as fresh peas. For this purpose, plants are usually pulled and sold in the markets in bunches. For dry seeds, the plants are harvested at maturity or slightly earlier (when leaves begin to yellow) by cutting them close to the ground or uprooting. The plants are stacked in the field for a few days to dry; seeds are threshed by trampling or beating with wooden flails. The chaff is separated from the grain by winnowing. Tall cvs are suitable for mechanized harvesting. Chickpeas are usually stored in bags, but are more subject to insect damage than when stored in bulk. Proper cleaning, drying, and aeration are necessary to control seed beetles. A thin coating with vegetable oil can reduce storage damage. Sometimes baskets made of twisted rice straw are used as storage containers. Gram "dhal" is prepared by dehusking and splitting the small or medium sized but plump seeds in mills. The seed are sprinkled with water and heaped overnight to hasten loosening of the seed coat and to facilitate milling. Dhal may be polished. It is used as such or ground to flour ("basin") and may be mixed with wheat and barley flour.

YIELDS AND ECONOMICS: Yields usually average 400–1,600 kg/ha, but can surpass 2,000 kg/ha, and in experiments have attained 5,200 kg/ha. Yields from irrigated crops are 20–180% higher than those from rainfed crops (Egypt: 1,724 kg/ha). Yields for brown-seeded types average higher (1,700 kg/ha) than yields for green-seeded types (1,200 kg/ha). In India unstirred plots yielded 290 kg/ha when unweeded and 749 kg/ha when weeded. In a 3-cv trial in India, DM yields ranged from 9,400 to 12,000 kg/ha. Kay (1978) reported 90 MT/ha green matter per year. In 1975 world seed yield was 590 kg/ha. In India chickpea or gram ranks 5th among grain crops, and is the most important pulse crop (1975: 7,150,000 ha). In India and Pakistan most chickpeas are consumed locally, about 56% retained by growers. Seeds sold in Indian markets are of mixed brown and yellow 'Desi,' cream 'Kabuli,' or green-seeded 'Desi' type, each for a different price. Gram in large quantities is sold as split peas (dhal) or flour. In United States and Europe, chickpeas are marketed dried, canned, or in various vegetable mixtures. In Europe, mashed chickpeas from the Mediterranean are sold canned. Mashed chickpea mixed with oils and spices (hummus) is a popular hors d'oeuvre in the Mediterranean Middle East. In 1975, Asia produced 4,985,000 MT, averaging 574 kg/ha, led by India, which produced 4,055,000 MT, averaging 567 kg/ha; Africa produced 364,000 MT, averaging 649 kg/ha. North America produced 260,000 MT, averaging 1,020 kg/ha; Europe 118,000 MT, averaging 771 kg/ha. Following the leader India, Pakistan produced 535,000 MT, averaging 549 kg/ha; Mexico 260,000 MT, averaging 1020 kg/ha; Ethiopia 236,000 MT, averaging 781 kg/ha (FAO, 1976).

BIOTIC FACTORS: Gram is self-pollinated. Bees visit the flowers but cross pollination is rare. The main fungi that affect chickpea are *Fusarium oxysporum* (= *orthoceras*) f. *ciceri,* causing the plant to wilt from about 1 month after sowing onwards, and *Ascochyta (Mycosphaerella) (Phyllosticta) rabiei* blight, the most serious disease in North India, Pakistan, and Middle East (sometimes causing 100% losses). Blight causes brown spots on leaves, stems, and pods; the plants dry; infected seeds as well as dried infected tissue carry the fungus. *Uromyces ciceris-arientini* may cause a serious rust in humid areas of India and Mexico. Other fungi known to attack gram include *Alternaria* sp., *Ascochyta pisi, Botrytis cinera, Colletotrichum capsici, C. dematium, C. trifolii, Didymella rabiei, Corticum solani, Fusarium solani, F. vasinfectum, Leviellula taurica, Macrophomina phaseoli, Mycosphaerella tulasnei, Neocosmospora vasinfecta, Operculella padwickii, Phytophthora citrophtora, Ph. cryptogaea, Ph. megasperma, Pythium debaryanum, P. ultimum, Rhizoctonia bataticola, Rh. solani, Rh. napi, Sclerotium rolfsii, Sclerotinia sclerotiorum, Verticillium albo-atrum*. Some of these fungi may become of economic importance. Root-knot nematodes *(Meloidogyne arenaria)* are troublesome in parts of India. Viruses isolated from chickpea include alfalfa mosaic, pea enation, pea leaf roll, sweetpea streak, bean yellow mosaic, and cucumber mosaic. Most recently, stunt virus was noted as a part of the wilt complex. Chickpeas may be attacked by pod-boring insects. In India, gram podborer or gram caterpillar (*Heliothis armigera* = *obsoleta*), the most important pest, feeds on leaves and developing seeds. In Iran, it causes as much as 90% crop loss. White ants may be a serious problem in sandy soils of North India. Other pests in India include the semilooper *Plusia orichalcea*, the red gram plume moth. *Exelastis atomosa*, and the cutworms, *Agrotis ipsilon,* and *Ochropleura flammatra*. Cutworms (*Agrotis* spp.) and lesser army worms *(Spodoptera exigua)* can be of local importance. Miner flies, e.g., *Liriomyza cicerini,* can reduce yields by 40%. Carbaryl, DDT, Parathion or Thiofos are reported to control these miners. Many insects *(bruchids)* are pests of stored grain. Chickpeas stored as dhal harbor fewer bruchids, the worst pest of stored grain. *Callosobruchus chinensis* lowers seed viability, reduces thiamin and tryptophane, increases free fatty acid content, and adversely affects flavor. For control of bruchids, dusting with BHC, DDT, derris, lindane, or pyrethrum or fumigation with methyl bromide, have been recommended. In some situations, 'P1091' is rather resistant to bruchid attack owing to its rough testa. Mechanical weeding increased yield fourfold in India. Among herbicides recommended in other countries are alachlor, benfluralin, chlorazine, linuron, noruron, prometryne, simazine, and trifluralin.

REFERENCES:
El Baradi, T. A., 1977, Pulses 2.—Chickpeas, *Abstr. Trop. Agric.* **3**(3):9–18.
Hulse, J. H., 1976, Problems of nutritional quality of pigeonpea and chickpea and prospects of research, in: *ICRISAT, 1976,* International Workshop on Grain Legumes (Proceedings), January 13–16, 1975, pp. 189–207, Hyderabad, India.
Maesen, L. J. G. van der, 1972, *Cicer* L., a monograph of the genus, with special reference to the chickpea (*Cicer arietinum* L.), its ecology and cultivation, *Commun. Agric. Univ. Wageningen* **72-10**:342.

Singh, K. B., and Auckland, A. K., 1975, Chickpea breeding at ICRISAT, in: *ICRISAT, 1975,* International Workshop on Grain Legumes (Proceedings), pp. 3–17, Hyderabad, India.

CONTRIBUTORS: S. Chandra, J. A. Duke, H. Pollard, C. F. Reed, L. J. G. van der Maesen, D. Zohary.

Clitoria laurifolia Poir.

FAMILY: Fabaceae

COMMON NAMES: Laurel-leaved clitoria, Butterfly pea

SYNONYMS: *Clitoria cajanifolia* Benth.
Martiusia laurifolia (Poir.) Britt. & Wils.
Neurocarpum cajanifolium Presl

USES: Cultivated especially in Malay Peninsula and Southeast Asia for fodder and green manure, sometimes for a cover crop. Has stiff, erect stems; throws out suckers; a good contour hedge to hold soil. Planted as an ornamental in many areas.

FOLK MEDICINE: Leaves used for pimples in Batavia.

CHEMISTRY: No data available.

DESCRIPTION: Perennial shrub with deep, woody roots; stems semidecumbent, or more usually several and erect, appressed-pubescent upward, 2–9 dm tall; leaves trifoliolate, 5–8 cm long; stipules triangular-lanceolate, acuminate, 4–6 mm long; petioles short, usually less than 6 mm long; leaflets 3, oblong or linear oblong, 3–10 cm long, rounded, emarginate, or sometimes acute at apex, obtuse or acutish at base, rather thin, glabrous above, sparsely short-pubescent and pale beneath; stipels minute and deciduous; peduncles 0.5–3 cm long, 1–2-flowered; bractlets ovate; calyx pubescent or glabrous, 2–2.8 cm long, its lobes ovate or ovate-lanceolate, about one-third the length of tube; corolla purple to nearly white, 5–6 cm long; pod linear-oblong, 2.5–4 cm long, about 8 mm thick, stipitate, glabrous, valves ridges and convex; seeds 5–7 per pod, globose to ovoid-globose, viscid, ca. 3 mm long.

GERMPLASM: Assigned to the Indochina-Indonesian Center of Diversity, laurel-leaved clitoria is said to tolerate heavy soil, sand, and waterlogging. ($2n = 24$.)

DISTRIBUTION: Native to tropical South America. Introduced widely in tropics, especially in Southeast Asia, mainly by cattle to whose hair the viscid seeds adhere. Introduced to West Africa and sometimes cultivated for forage; also in West Indies, Singapore, Malay Peninsula, Thailand, Borneo, and Java.

ECOLOGY: Widely distributed in thickets, along roadsides, especially in sandy soil or low, wet ground. Also common on open red clay slopes, up to about 33 m elevation. Requires a tropical climate. Often becomes weedy. Ranging from Subtropical Dry through Tropical Moist Forest Life Zone, laurel-leaved clitoria is reported to tolerate annual precipitation of 5.3–27.8 dm (mean of 5 cases = 18.9 dm), annual mean temperature of 21.0°–27.3°C (mean of 5 cases = 24.6°C), pH of 4.5–5.5 (mean of 5 cases = 5.0).

CULTIVATION: Propagated by seeds which have a viscid covering and adhere to coats of animals. Sown by hand in rows when cultivated or more commonly broadcast for forage or cover crop. Requires little care once established.

HARVESTING: Plants cut and used for forage, or allowed to grow as cover crop. In some areas

Figure 29. *Clitoria laurifolia* Poir.

Clitoria laurifolia Poir.

plowed under as green manure just before planting other crop. Plants said to increase the nitrogen content of the soil.

YIELDS AND ECONOMICS: An important legume in several tropical countries in South America, West Indies, West Africa, and Southeast Asia, used for cover crop, forage, green manure, and for contour planting for soil erosion control.

BIOTIC FACTORS: The following fungi have been reported on this plant: *Corticium salmonicolor, Phyllosticta* sp., *Rhizoctonia bataticola, Rh. solani, Uromyces neurocarpi, U. rostratus*. The nematode *Meloidogyne incognita* has been isolated from the plant.

CONTRIBUTORS: J. A. Duke, C. F. Reed.

Clitoria ternatea L.

FAMILY: Fabaceae
COMMON NAMES: Butterfly pea, Blue pea, Nazerion, Cordofan pea, Winged-leaved clitoria
SYNONYMS: *Ternatea vulgaris* H.B.K.
Clitoria caerulea Hort.

USES: Leaves used as feed for sheep and goats. Seeds yield a blue dye. Seeds and root bark contain tannin. Blue dye obtained from the corollas can be substituted for litmus. In the Philippine Islands, pods are sometimes eaten by humans. In Amboina, flowers are used to tint boiled rice a cerulean color. Often planted as an ornamental cover crop. Sometimes grown in temperate greenhouses as an ornamental.

FOLK MEDICINE: Leaves, roots, and seeds are used for a wide range of diseases. Leaves used for fevers, earache and ulcers. Roots are bitter and have powerful purgative and diuretic properties; also used as a demulcent and vermifuge. Roots may promote nausea and vomiting, and many consider them unsafe as medicine. Roots and flowers mixed with root-bark or indrani (*Vitex negundo*) given internally for snake bite, especially cobra bite. Said to relieve cystitis, swellings, and swollen joints and to be used as a refrigerant.

CHEMISTRY: Leaves used as forage contain 11% digestible proteins, about 53% total digestible nu-

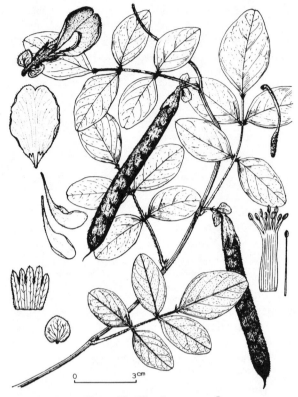

Figure 30. *Clitoria ternatea* L.

trients, and about 39% starch. Raw pods and seeds reported to contain per 100 g: 66 calories, 80.0% moisture, 3.8 g protein, 0.4 g fat, 15.0 g carbohydrate, 4.8 g fiber, 0.8 g ash, 40 mg Ca, 24 mg P, 0.4 mg Fe, 4 mg Na, 309 mg K, 0.04 mg thiamine, 0.18 mg riboflavin, 1.4 mg niacin, 248 mg ascorbic acid. Seed meal contains (in g/16 g N) 6.1 lysine, 1.0 methionine, 3.5 methionine cystine, 3.6 phenylalanine, 6.9 phenylalanine tyrosine, 7.4 arginine, 4.1 glycine, 2.4 histidine, 4.2 isoleucine, 7.4 leucine, 2.2 threonine, 1.2 tryptophane, 4.4 valine, 3.5 alanine, 9.3 aspartic acid, 15.6 glutamic acid, 3.3 proline, and 5.0 g serine (Van Etten *et al*, 1961).

DESCRIPTION: Perennial twiner from a large taproot; stems pubescent, up to 5 m long; leaves pinnate with 5(–9) ovate or elliptic leaflets, these mostly obtuse and shortly emarginate at tip, 2–5 cm long, 1–3 cm broad, short-petioled; flowers solitary on very short axillary pedicels, usually resupinate, bracteoles rounded-oblong, up to 9 mm long; petals commonly deep blue, paler and tinged yellow at base, or light blue or white; standard 3–5 cm long, beautifully marked; wing-petals lightly fused to keel; vexillary stamen free; pods 5–11 cm long, 8–10 mm broad, flattened or turgid, linear-oblong, straight, thin; seeds ellipsoidal-compressed, 5 mm long, nearly black, numerous in a pod (8–10); 100 seeds weigh 4.1 g. Fl. and Fr. all year in Jamaica and most other tropical areas.

GERMPLASM: Variations include narrow-leaved, white-flowered, and double-flowered forms. One Cuban study compares 5 cvs, 'Conchita Clara' (most drought resistant), 'Negra' and 'Jaspeada' from Mexico, and 'Indio Hatuey' and 'Oriente' from Cuba (Matos and Torre, 1971). Assigned to the Indochina–Indonesian Center of Diversity, butterfly pea or cvs thereof is reported to exhibit tolerance to drought, high pH, laterite, photoperiod, slope, virus, and weeds. ($2n = 16$.)

DISTRIBUTION: Pantropical, native to Old World tropics, now widely cultivated. Hardy in central Florida; usually grown as a biennial. Occurs among hedges throughout the tropics.

ECOLOGY: Common in cultivation, but also escaped and naturalized in waste places, field margins, thickets, from sea level to 1,600 m altitude. Grows best in full sun. Needs moisture, but does not tolerate waterlogging. Bogdan (1977) reports that it grows on seasonally waterlogged black clay. Thrives in rich soil, but grows in ordinary well-drained garden soil. In greenhouse, requires no special care, and makes a good pot plant for winter bloom if tips are pinched to form a bushy plant. At Ibadan, flowers 4–6 weeks after sowing and continues throughout the year, perhaps peaking soon after the dry season. Self-pollinating, the stigma does not extrude from the keel if tripped. Ranging from Subtropical Dry to Moist through Tropical Very Dry to Wet Forest Life Zones, butterfly pea is reported to tolerate annual precipitation of 3.8–42.9 dm (mean of 24 cases = 14.3 dm), annual mean temperature of 19.4°–27.9°C (mean of 24 cases = 24.4°C), and pH of 4.5–8.7 (mean of 20 cases = 6.7).

CULTIVATION: In the US, propagated from seed, started indoors in March to bloom in July; seeds germinate in about 7 days at 17°–21°C; sown outdoors in May, seeds germinate in 10–15 days. May also be grown from cuttings. Commercial inoculants and native cowpea rhizobia are effective in nodule formation. Plants usually grown in gardens along with other flowers; rarely grown as commercial plants. Intersown with grasses (e.g., *Brachiaria mutica, Cynodon dactylon, Digitaria decumbens, Hyparrhenia rufa, Panicum maximum*). In Columbia, clitoria comprised 25–35% of stand the first year, declining to less than 5% after a year of grazing or cutting for forage.

HARVESTING: Plant parts collected as needed, and used either fresh or dried depending on use. Covers the ground in 4–6 weeks when sown 25 cm apart in rows 1 m apart.

YIELDS AND ECONOMICS: Of 5 cvs, Mexican 'Conchita Clara' was more drought resistant and produced highest herbage yields—82 MT WM/ha rainfed, 84 MT irrigated—40–64 rainfed, 55–78 MT/ha irrigated. Of economic importance, mainly for native uses, only in tropical countries where adapted. Used, however, by large populations in India, East Indies, and Philippines. Sown in Uttar Pradesh, yielded 24.3 MT/ha fresh weight; dry matter, 21,8%. Yields of 3 MT DM/ha reported from Senegal, 4.4 MT from the Sudan.

BIOTIC FACTORS: Following fungi have been reported: *Ascochyta pisi, Cercospora clitoriae, C. clitoridis, C. cruenta, C. ternateae, Chaetoseptoria wellmanii, Cladosporium herbarum, Colletotrichum capsici, Corticium solani, Erysiphe polygoni, Fusarium semitectum, Macrophomina phaseoli, Macrosporium commune, Melanconiella clitoridis, Phyllosticta clitoridicola, Septoria lablabina*. Nematodes include *Meloidogyne incognita, M. javanica*, and *Pratylenchus brachyurus*.

REFERENCES:
Crowder, L. V., 1974, *Clitoria ternatea* (L.) Due. as a forage and cover crop: A review, *Niger. Agric. J.* **11**(1):61–65.
Matos, E., and de la Torre, R., 1971, Trials with five populations of *Clitoria ternatea* L. *Rev. Cub. Cienc. Agric.* **4**:217–221 (as cited by Bogdan, 1977, not seen).

CONTRIBUTORS: J. A. Duke and C. F. Reed.

Coronilla varia L.

FAMILY: Fabaceae
COMMON NAMES: Crownvetch, Axseed

USES: Crownvetch is primarily cultivated for forage; easily digested and in food value compares favorably with alfalfa. Its extensive root system makes crownvetch useful for erosion control, high-

Coronilla varia L.

Figure 31. *Coronilla varia* L.

way embankments, especially along superhighways, spoil banks in strip-mine areas, and for ground covers.

FOLK MEDICINE: Said to be cardiac, diuretic, purgative, stimulant, tonic, and toxic.

CHEMISTRY: Plant said to contain cytisine, saponin, and tannin; seeds to contain trypsin inhibitors. Crownvetch is said to be a nonbloating legume. Cotyledons of the seeds have a bitter glucoside, coronillin, that imparts a quininelike taste to flour that is made in USSR. Plants are considered somewhat poisonous because of this glucoside.

DESCRIPTION: Hardy, long-lived, perennial legume, spreading by underground creeping roots; stems herbaceous, branching from base of plant, climbing, spreading or ascending, 0.3–1.2 m high, glabrous, hollow and weak; leaves alternate, green, odd-pinnate compound sessile, with 11–25 pairs of leaflets, these entire, oblong or elliptical, blunt and mucronate, 0.6–2 cm long, 0.3–1.2 cm wide; flowers pealike, in dense umbels of 10–20 flowers, on long axillary peduncles; calyx tube (sepals 5) broadly campanulate with very short teeth; corolla (petals 5) 10–15 mm long, variegated, white shading to rose, lavender or violet; legume 2–6 cm long, 4-angled, erect, 3–7-seeded, indehiscent but articulated into joints 4–6 mm long with a terminal beak; seeds tightly enclosed in pod. Fl. May–Aug., Fr. autumn.

GERMPLASM: Present cvs have been developed primarily for erosion control, soil conservation, hay, or forage. 'Chemung,' 'Emerald,' and 'Penngift' are 3 cvs used extensively for soil conservation in the United States. Strains and cvs range widely in vigor, plant type, number, and fineness of stems, time of maturity and percentage of tannin (range: 7–11%). Breeding objectives include enhancement of seedling vigor, recovery after cutting, productivity of plant and seeds, and improved palatability as forage and pasturage. Assigned to the Eurosiberian Center of Diversity, crownvetch or cvs thereof is reported to exhibit tolerance to drought, frost, grazing, insects, low pH, mine spoils, mycobacteria, poor soil, slope, smog, and weeds. ($2n = 24$.)

DISTRIBUTION: Native to central and southern Europe, now common throughout the Mediterranean region east to south and central Russia, Crimea, Causasus, western Syria, Iran, and Transcaspia. Has been introduced into the United States and other temperate regions and often becomes a noxious weed.

ECOLOGY: Crownvetch thrives on a wide range of soils and in a wide range of climatic conditions. Neutral to alkaline soils are preferred; very acid soils should be limed; well-drained soils are essential. Plants are drought-tolerant and frost and winter hardy, remaining evergreen year-round unless there is a heavy frost. They turn brown when killed, but may green up again when the temperature goes above 10°C. Ranging from Cool Temperate Moist to Wet through Warm Temperate Dry to Moist Forest Life zones, crownvetch was reported to tolerate annual precipitation of 5.2–13.6 dm (mean of 14 cases = 8.6), annual temperature of 7.0°–22.5°C (mean of 14 cases = 11.5), and pH of 4.8–7.8 (mean of 13 cases = 6.5).

CULTIVATION: Seeds are sown at 5–10 kg/ha, 0.6–1.3 cm deep, in well-prepared, firmed and properly fertilized seedbed or areas. Crown divisions are planted 60 cm apart. Plants spread and cover the area in 18–24 months. Closer spacing insures quicker coverage. Before planting, seed should be inoculated with a specific strain of bacteria. Seed is hard and should be scarified. Germination is very slow, many seedlings fail; 2 years or more may be required to produce satisfactory stands. Crop may be planted any time except when the ground is very dry or frozen. Once seedlings are well established and growing vigorously, little care is required, except for occasional watering for the first few weeks.

HARVESTING: When cut for hay at the full bloom stage, regrowth is usually slow, resulting in only 1 or 2 cuttings per year. Earlier first cutting may allow sufficient regrowth for a good second cutting without injuring stands. Studies are in progress to determine the effects of number of cuttings and time of cuttings on yields and longevity of stands. Studies of grazing methods and stocking rates are needed. The slow recovery when closely cut for hay indicates that plants should not be completely defoliated by grazing animals. Continuous moderate to light grazing may be the most suitable means to maintain a productive state.

YIELDS AND ECONOMICS: Crownvetch gives forage yields roughly comparable with those of alfalfa, ca. 12.5 MT/ha. Used commercially as a soil binder and for eroded banks by highway departments throughout the United States, including Hawaii. Cost from nurseries is about $.15 per crown. Seed yields average only ca. 112 kg/ha, but yields up to 895 kg/ha have been reported. At State College, Pa., mean yield of hand-harvested seed was 257 kg/ha in 1970, a year of abnormally high precipitation and lush foliage growth, and 530 kg/ha in 1971, a drier-than-normal year. Field losses during commercial harvest were 61% and 58% respectively (Al-Tikrity *et al.*, 1974). Very abundant in southern Pa., southeastern Ohio, and western Md; almost as abundant along highways as tall fescue; one of the most important soil binders in northeastern United States.

BIOTIC FACTORS: Crownvetch requires cross-pollination for setting good crops of seed. Bumblebees seem to be more effective than honeybees; 7–10 colonies per hectare are recommended. Crop is resistant to most root-knot nematodes, but *Meloidogyne* sp. and *Pratylenchus penetrans* have been isolated from it. Resists most insects and rodents, but several fungi, none very serious, attack it: *Ascospora melaena, Aserina melaena, Cercospora coronillae-variae, C. ratuensis, Cercosporella varia, Erysiphe pisi, E. polygoni, Leptosphaeria niessleana, Macrosporium phomoides, Mycosphaerella coronillae-variae, Omphalospora melaena, Ophiobolus cestianus, Peronospora coronillae, P. trifoliorum, Phoma coronillae-variae, Pyrenochaeta erysimi, Ramularia coronillae, Rhizoctonia crocorum, Sphaerella melaena, Stagonoporopsis coronillae, Trichopeziza sulphurea,* and *Uromyces anthyllides*.

REFERENCES:
Al-Tikrity, W., McKee, G. W., Clarke, W. W., Peiffer, R. A., and Risius, M. L., 1974, Seed yield of Coronilla varia L., *Agron. J.* **66**:467–468.
Anderson, E. J., 1959, Pollination of Crownvetch, *Glean. Bee Cult.* **87**:590–593.
Hawk, V.B., 1962, 'Emerald' crownvetch—A new legume fills many needs for conservation plans, *Iowa Soil Water* **6**(3):10.
Hawk, V. B., and Scholl, J. M., 1961, Seed treatment aids crownvetch seedings, *Crops Soils* **14**(6):19.
Kenson, P. R., 1963, Crownvetch—A soil conserving legume and a potential pasture and hay plant, *U.S. Dep. Agric., Agric. Res. Serv.* **34–53**:1–9.
McKee, G. W., 1962, 'Penngift' crownvetch (Reg. No. 2), *Crop Sci.* **2**:356.
Uhrova, A., 1935, Revision der Gattung *Coronilla* L. Bot., *Centralbl. Beihefte* **53**:174.

CONTRIBUTORS: J. A. Duke, C. F. Reed.

Crotalaria brevidens Benth.

FAMILY: Fabaceae
COMMON NAME: Slenderleaf crotalaria
SYNONYMS: *Crotalaria brevidens* var *intermedia* (Kotschy) Polhill
Crotalaria intermedia Kotschy
Crotalaria purpureo-lineata Bak. f.

USES: Plant cultivated for green forage in tropical Africa; very useful for soil erosion control. Mucilage Gum is obtained by dry milling of seeds.

FOLK MEDICINE: No data available.

CHEMISTRY: The genus *Crotalaria* has been reported to contain the following "toxins": hydrocyanic acid, hydroxysenkirkine, indican, jacobine, longilobine, monocrotaline, retronecine, retrorsine, riddelline, senecionine, and seniciphylline. Raw leaves contain 74.5 moisture, 8.8 g protein, 1.6 g

Crotalaria brevidens Benth.

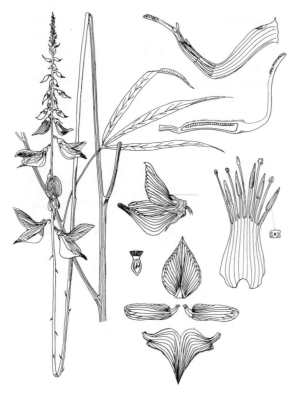

Figure 32. *Crotalaria brevidens* Benth.

ash, 222 mg Ca, and 0.8 mg Fe/100 g. Material planted at Chiromo, Kenya, germinated in 3–5 days, had 4 edible leaves per plant with 27.5% protein (ovendry basis) 36 days after planting, 5.2 leaves with 30.9% protein in 45 days, 6.7 leaves with 33.0% protein in 53 days, 14.4 leaves with 29.3% protein in 61 days, 17.0 leaves with 28.3% protein in 67 days, 22.8 leaves with 28.6% protein in 70 days, and 29.0 leaves with 26.7% protein in 86 days. (Imbamba, 1973). Seeds contain 22.8% mucilage (N-free), 3% oil, and 32.2% protein.

DESCRIPTION: Erect or decumbent annual or short-lived perennial, 4–20 dm tall, with numerous ascending branches from the base upwards; stems covered with short appressed or somewhat spreading hairs; leaves 3-foliolate, the leaflets linear or lanceolate to narrowly elliptic-oblong or elliptic, 4–10 cm long or longer (14 cm), 33 broad, the upper ones proportionately longer and narrower, glabrous above, appressed puberulous beneath; petiole 20–60 mm long; stipules absent; racemes terminal and leaf-opposed, pedunculate, 10–48 cm long, with numerous rather closely arranged flowers; bracts linear-caudate or subulate, expanded at base, 1–3.5 mm long; bracteoles inserted at base of calyx, or near middle of pedicel, subulate or filiform, 0.5–2 mm long, appressed-puberulous or glabrous, the lobes subulate or acuminate, much shorter to longer than the tube; standard ovate or elliptic, cream to clear yellow, veined reddish-brown, glabrous outside; wings a little shorter to a little longer than the keel, often with a dark mark at base; keel shortly rounded, with a long projecting straight or slightly recurved untwisted beak, 12–24 mm long; pods subsessile, narrowly cylindrical, elongate, often a little curved at ends 35–50 mm long, 5–7 mm broad, puberulous or pubescent, with up to 80 seeds or more; seed oblique-cordiform, variable in size, mostly 2–2.5 mm long, 1.75–2 mm broad, 1.25–1.5 mm thick, medium yellow to strongly orange-yellow, red to dark brown or dark gray-blue, smooth, glossy or not. Seeds 281/g, or 5 g/1,000 seeds.

GERMPLASM: Several botanical varieties are recognized. *C. brevidens* var *intermedia* (Kotschy) Polhill, with a puberulous calyx 5–8 mm long, and keel 17–24 mm long, grows in Sudan, Uganda, Kenya, Tanzania, Rwanda, Ethiopia, and Congo west to Nigeria. *C. brevidens* var *parviflora* (Bak. f.) Polhill, with puberulous calyx 3.5–4.5 mm long, and keel 12–14 mm long, grows only in Kenya. Assigned to the African Center of Diversity, slenderleaf crotalaria is reported to exhibit tolerance to slope and waterlogging. ($2n = 16$.)

DISTRIBUTION: Native to tropical Africa from Sudan, Ethiopia, Kenya, and Tanzania, west to Congo, Nigeria, introduced to Morocco, West Indies (Puerto Rico, Guadelupe, Martinique); Costa Rica. Sometimes cultivated.

ECOLOGY: Grows well in grassland, deciduous woodland and bushland, from 600–2,700 m, rarely on termite mounds, and in seasonal swamp grassland. Persists on cultivated ground, roadsides, in clearings of upland dry evergreen forests, but may be in rather moist situations. Plants are adaptable and thrive on nearly any subtropical or tropical soil. Sensitive to herbicides. Ranging from Warm Temperate Moist through Tropical Dry to Moist Forest Life Zones, slenderleaf crotalaria is reported to tolerate annual precipitation of 11.2–26.7 dm (mean of 4 cases = 16.9 dm), annual mean temperature of 16.0°–26.2°C (mean of 4 cases = 22.2°C), and pH of 6.0–7.1 (mean of 2 cases = 6.6).

CULTIVATION: Propagated by seeds, sown broadcast or in rows like beans. Plants develop rapidly and require little attention after established.

HARVESTING: Plants harvested for green forage or more often unharvested and used for erosion control.

YIELDS AND ECONOMICS: Limited cultivation for green forage, but more extensive for erosion control, especially in West Indies, and tropical Africa.

BIOTIC FACTORS: No serious pests or diseases are reported.

REFERENCES:
Imbamba, S. K., 1973, Leaf protein content of some Kenya vegetables, *East Afr. Agric. For. J.* **38**(3):246–251.
Miller, R. H., 1967, Crotalaria seed morphology, anatomy and identification, *U.S. Dep. Agric., Agric. Res. Serv., Tech. Bull.* **1373**:1–73, illus.
Tookey, H. L., Pfeifer, V. F., and Martin, C. R., 1963, Gums separated from *Crotalaria intermedia* and other leguminous seeds by dry milling, *Agric. Food Chem.* **11**(4):317–321.

CONTRIBUTORS: J. A. Duke, C. F. Reed.

Figure 33. *Crotalaria juncea* L.

Crotalaria juncea L.

FAMILY: Fabaceae

COMMON NAMES: Sunnhemp, Indian hemp, Madras hemp, Brown hemp, Sannhemp

SYNONYMS: *Crotalaria benghalensis* Lam.
Crotalaria cannabinus Royle
Crotalaria sericea Willd.
Crotalaria tenuifolia Roxb.
Crotalaria viminea Wall.

USES: Sunnhemp is cultivated for the strong bast fiber extracted from the bark, which is more durable than jute. Fiber is used in twine, rug yarn, cigarette and tissue papers, fishnets, sacking, canvas, and cordage. Wet sunnhemp fiber is stronger than dry, and is fairly resistant to mildew, moisture, and microorganisms in salt water. One of the oldest known fibers in the Indo-Pakistan subcontinent, as mentioned in ancient Sanskrit literature. Widely grown throughout the tropics as green manure, the dried stalks and hay are used as forage for livestock. Seeds are reported poisonous to livestock, but are fed to horses in the Soviet Union and to pigs in Rhodesia.

FOLK MEDICINE: Seeds sometimes used medicinally, said to purify the blood. Seeds are also used in impetigo and psoriasis and as an emmenagogue.

CHEMISTRY: Raw sunnhemp contains 0.61% ash, 9.6% hygroscopic water, 2.8% aqueous extract, 0.55% fat and wax, 80.0% cellulose, and 6.4% pectin bodies, (CSIR, 1950). Dried as cattle feed, the stalks contain 14.4% moisture, 1.1% ether extract, 11.3% albuminoids, 35.8% carbohydrate, 27.4% woody fiber, and 6.4% soluble mineral matter. Seeds contain 8.6% moisture, 34.6% crude protein, 4.3% fat, 41.1% starch, 8.1% fiber, and 3.3 ash (CSIR, 1950). Seeds are reported to contain trypsin inhibitors, and to be poisonous to cattle. Seeds contain 12.6% oil, with 46.8% linoleic acid, 4.6% linolenic acid, and 28.3% oleic acid, with, by difference 20.3% saturated acids (Zafar *et al.*, 1975).

Crotalaria juncea L.

DESCRIPTION: Tall herbaceous shrubby annual, 1–3 m tall, vegetative parts covered with short downy hairs; taproot long, strong, with many well-developed lateral roots, and numerous much-branched, lobed nodules up to 2.5 cm in diameter; stems to 2 cm in diam.; leaves simple with minute pointed stipules; petiole short, about 5 mm long with pulvinus; blades linear elliptic to oblong, entire, 4–12 cm long, 0.5–3 cm broad, bright green; inflorescence a terminal open raceme to 25 cm long with very small linear bracts; flowers showy, small with 5 hairy sepals, shortly united at base, the lobes pointed, with 3 lower sepals united at tips, separating in fruit; petals deep yellow, the standard erect, about 2.5 cm in diameter, rounded, sometimes streaked purple on dorsal surface, the wings shorter and keel twisted; stamens 10, almost free to base, 5 with short filaments and long narrow anthers and 5 with long filaments and small rounded anthers; fruit an inflated pod about 3 cm long, 1 cm wide, grooved along the upper surface, with a short pointed beak, light brown when ripe, several seeded, softly hairy; seeds numerous, small, flattened, dark-gray to black, loose in the pod at maturity, 33,000 seeds/kg. Fl. July–August in Rhodesia.

GERMPLASM: Of several well-recognized types, the best known in Pakistan are 'Madaripur' and 'Serajgnaj,' the first one considered better, having creamy-white fibers of good strength, relatively free from dirt. There are cvs for different rainy seasons: 'Bhadai san,' May, June–Oct., Nov.; 'Rabi san,' Oct., Nov.–Feb. 'Somerset' is a good cv in Rhodesia. In India kharif and rabi sunnhemps are known. 'T6' is day-neutral. 'Cawnpore 12,' a proved cv of 'Kharif sunn' of India, is superior in yield to all other types, but takes 2–2½ months longer than 'Beldanga Early' to mature. It is resistant to stem-break disease. 'Ullapora,' a rabi variety, is superior in yield and quality. Assigned to the Hindustani Center of Diversity, sunnhemp or cvs thereof is reported to exhibit tolerance to disease, drought, insects, laterite, poor soil, slope, virus, and weeds. ($2n = 16$.)

DISTRIBUTION: Believed to be native to India and Pakistan. Now cultivated throughout India (from the foothills of the Himalayas to Ceylon), Pakistan, in Uganda and Rhodesia, and in the western Hemisphere (e.g., Brazil) where it was introduced early in the 19th century.

ECOLOGY: Sunnhemp is the fastest growing species of the genus and is very effective in smothering weeds. Almost any well-drained soil is suitable for the kharif crop. Sunnhemp grown during the rainy season is used mainly as green manure; fiber is not considered of good quality. For fiber sunnhemp is grown on fairly light well-drained soils that retain sufficient moisture during the growing season. Sunnhemp is a short-day crop, but vegetative growth is favored by long days, although seed set may be poor. Tolerant of drought, but not of salt and frost. Ranging from Cool Temperate Steppe to Tropical Very Dry through Tropical Wet Forest Life Zones, sunnhemp is reported to tolerate annual precipitation of 4.9–42.9 dm (mean of 29 cases = 14.9 dm), annual mean temperature of 8.4°–27.5°C (mean of 29 cases = 22.5°C), and pH of 5.0–8.4 (mean of 24 cases = 6.2).

CULTIVATION: Seed is broadcast on plowed ground by hand or a device consisting of a canvas bag containing the seed with a blower attached. Then the land is cross-plowed. Seed is sown at different rates at different times depending on the use of the crop. In India, for fiber, seed is broadcast at 96 kg/ha under dry growing conditions, and less on irrigated fields. In Pakistan seed is sown at rate of 120–240 kg/ha. The heavy seed rate insures upright erect stems, which help to smother weeds, produces fine fiber and increases yield. Height of stalks in crops varied from 1.15 to 1.75 m (to 2 m), with an average thickness of 1.2 cm about the middle of the stalk. In Africa seed is sown in late Nov.–Dec., even as late as Jan. Cultivation is not required except for weeding, which is usually not necessary if the land has been well prepared. Sunnhemp is a dryland crop. Where irrigated, furrows are opened to separate fields into small plots. If there is no rain after sowing, field is irrigated along these furrows. Crop is irrigated once in 10–15 days, lightly compared to other crops, such as tobacco and chilies. Excess moisture is harmful during the first 2 weeks after germination. Seedlings emerge in about 3 days and soon form a thick cover. No manure is applied. Sunnhemp is often grown as a green manure, in rotation with tobacco, vegetables, dry grains, rice, corn, cotton, sugar cane, pineapples, coffee, and orchard crops.

HARVESTING: In some areas plants are harvested at the flowering stage (100–108 days). In Rhodesia the crop is cut when stems turn yellow along most of their length. If grown for seed, plants are harvested after seed are well set, and before pods are dry, so that no seed is lost during cutting and bundling prior to threshing. In Pakistan crop is harvested when pods are ripe. Harvesting at flower stage gives a good fiber, but precludes profit from the seed crop. There are no significant differences in strength and quality of fiber from plants retted at flowering time and those retted when seeds are fully mature. Harvest for fibers and seeds from the same crop is difficult because by the time pod ripens, cold weather has set in, and hard stems do not ret well. Plants are cut with knives or pulled and allowed to remain in the field for 1–2 days until dried leaves fall off easily. Stalks are then tied into bundles and retted for 4–6 days. However, in Rhodesia, 10–14 days are required during July, 8–9 days in August, and 5–8 days in September–October. The number of days required for retting depends on water temperature, locality, time of year, weather, depth and source of water, thickness of stalks, and quantity of straw in relation to volume of water. Cement tanks are preferred for retting, but earth pits, dams, weirs, streams, and backwater pools of rivers are also used. Shallow water from 1–1.3 m deep is satisfactory. If more than one ret is to be carried out in a pool, flow of water must be sufficient to prevent fouling the water, which discolors the fiber. Four or five men can remove and stack 1 ton of straw per day. Bundles are stood on end (15–20 placed with butts on ground at a sufficient angle to permit air circulation in all directions). Thus the straw dries in 1–2 weeks. By standing each bundle up and fanning out the butts, drying time is reduced to 4 days. By leaning bundles on each side of a rack, drying time is reduced to 3–4 days. Various machines are used to decorticate the straw. A 6-hp engine is minimal for economical and speedy decortication. Fiber is stripped from the stalks by hand, washed and hung over bamboo poles to dry in the sun. Cut straw with a yellowish tinge requires 10 days to 3 weeks to bleach so as to give fiber of a satisfactory color. Stems cut while green bleach when exposed directly to the sun but have to be turned at least twice. For seeds, a crop is allowed to stand until pods are fully ripe. Stems are cut close to the ground and left in the field to wither for a few days, reducing the retting period for fiber extraction. In humid districts in Ceylon this cannot be done since the fiber deteriorates. After all leaves are removed, stems are bundled and stored for additional drying. They are then threshed by beating small bundles held by hand against a plank, placed in a sloping position over a threshing mat. Dried stems are then ready for retting.

YIELDS AND ECONOMICS: Handweeded South Carolina material given ca. 20 : 25 : 50 NPK/ha yielded ca. 12.5 MT DM/ha, with ca. 40 : 55 : 100 NPK/ha yielded ca. 13 MT DM/ha, and with ca. 60 : 80 : 150 NPK/ha yielded ca 14 MT DM/ha. Corresponding unweeded yields were ca. 9, 10, and 11 MT DM/ha (White and Haun, 1965). Yields are 500–1000 kg/ha for seed and 560–900 kg/ha for fiber. In Rhodesia 330 kg fiber/ha; in Pakistan 500–600 kg/ha. Strength of cordage fiber of sunnhemp is 185 kg, as compared with 157 kg for cotton rope, 132 kg for hemp and 102 kg for coir. Fiber elements are easily separated with sodium hydroxide solution or chromic acid. Extracted fiber is about 0.5–1.0 cm long, with an average diameter of 0.03 mm, among the broadest of bast fibers. World production of sunnhemp is 130,000 MT, principally produced by India, Brazil, West Pakistan. India grows about 360,000 ha of sunnhemp annually, producing between 80,000 and 100,000 MT fibers, with ca. 20–30% being exported to the United Kingdom, United States, and Belgium.

BIOTIC FACTORS: Sunnhemp is cross-fertilized by bees. It is attacked by many fungi: *Alternaria crotalaticola, Asperigilla versicolor, Ceratocystis fimbriata, Ceratostomella fimbriata, Cercospora canescens, C. crotalariae, C. demetrioniana, Chaetomium globosum, Cladosporium herbarum, Colletotricum crotalariae-junceae, C. curvatum, Corticium solani, Corynespora cassiicola, Curvularia penniseti, Dactuliophora tarrii, Fusarium acuminatum, F. equiseti, F. lateritium, F. moniliforme, F. oxysporum, F. scirpi, F. undum, F. vasinfectum, Gibberella fujikuroi, Leveillula taurica, Macrophomina phaseoli, Microsphaeria diffusa, Mycosphaerella pinodes, Myrothecium roridum, Nematospora coryli, Penicillium wottmanni, Periconia epiphylla, Phyllosticta crotalariae, Sclerotium rolfsii, Sphaerella crotalariae, Synchytrium phaseoli-radiati, Thielaviopsis basicola, Uromyces decoratus*. The following bacteria also infect sunnhemp: *Bacillus megatharium, Pseudomonas cyamopsicola, Ps. syringae, Ps. viridiflora, Xantho-*

Crotalaria juncea L.

monas patelii, and *X. vignicola*. The following viruses have been isolated: alfalfa mosaic, alsike clover mosaic, bean chlorotic ringspot, bean mosaic, bean necrosis, Brazilian tobacco streak, chlorotic mottle, mosaic (*Marmor vignae* var. *catjang*), mosaic, and witches broom. *Striga asiatica, S. hermonthica,* and *S. lutea* are parasitic on sunnhemp. Black beetles are serious pests in Rhodesia. The two most important and serious insect pests are the sunnhemp moth (*Utethesia pulchella*) and the stem borer (*Enarmonia pseudonectis*). Pod borers lower seed production. Among the nematodes are *Anguina* sp., *Aphelenchoides* sp., *Ditylenchus* sp. *Helicotylenchus canescens, H. cavenessi, H. dihystera, Heterodera glycines, Meloidogyne hapla, M. incognita acrita, M. thamesi, Peltamigratus negeriensis, Pratylenchus brachyurus, P. coffeae, P. vulnus, Rotylenchulus reniformis, Rotylenchus coheni, Scutellonema clathricaudatum, Trichodorus* sp., *Tylenchus* sp., and *Xiphinema longicaudatum*.

REFERENCES:

Chaudhai, S. D., 1950, Sunnhemp in East Pakistan, *Agric. Pak.* **1**:156–160.

Fox, D. H., 1954, Sunnhemp fiber production, *Rhod. Agric. J.* **42**:6–13.

Howard, A., and Howard, G. C., 1910, Studies in Indian fibre plants. I. On two varieties of Sann (*Crotalaria juncea*), *Mem. Dep. Agric. india, Bot. Ser.* **3**(3).

Montgomery, B., 1954, The bast fibers, in: *Matthew's Textile Fibers*, 6th ed. (Herbert R. Mauersberger, ed.), Wiley, New York.

Paul, W. R. C., and Chelvanayagam, A. V., 1936, Sunnhemp in the Jaffna Peninsula, *Trop. Agric.* **86**(1):23–27.

Singh, B. N., and Singh, S. N., 1936, Analysis of *Crotalaria juncea* with special reference to its use in green manure and fiber production, *J. Am. Soc. Agron.* **28**:216–227.

White, G. A., and Haun, J. R., 1965, Growing *Crotalaria juncea*, a multipurpose legume, for paper pulp, *Econ. Bot.* **19**(2):175–183.

CONTRIBUTORS: J. A. Duke, C. F. Reed.

Crotalaria lanceolata E. Mey.

FAMILY: Fabaceae

COMMON NAME: Lanceleaf crotalaria

USES: Cultivated for green pasture and forage. Very useful for soil erosion control.

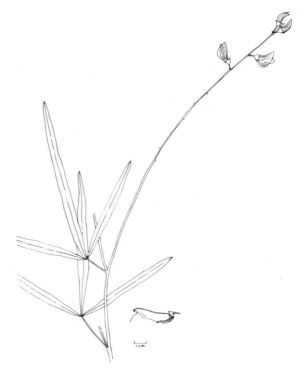

Figure 34. *Crotalaria lanceolata* E. Mey.

FOLK MEDICINE: No data available.

CHEMISTRY: Seeds contain 18.5% mucilage (N-free), 2.5% oil and 31.1% protein.

DESCRIPTION: Erect annual, up to 17 dm tall, laxly branched; stems ribbed, appressed puberulous, or rarely shortly spreading pubescent; leaves 3-foliolate, (rarely some of the upper 1-foliolate); leaflets mostly linear to lanceolate, the lower ones sometimes elliptic, 100 or more mm long, 3–22 mm broad, glabrous, sometimes hairy above, thinly covered with appressed hairs beneath; petioles up to 80 mm long; stipules absent; racemes with many closely arranged flowers, up to 35 mm long; bracts subulate or filiform, expanded at base, 1–3 mm long; bracteoles inserted at base of calyx, narrow, up to 1 mm long, appressed puberulent, usually thinly so; lobes subulate or acuminate, shorter than the tube; standard broadly elliptic or suborbicular, yellow, with reddish-purple veins glabrous outside; wings shorter or longer than the keel; keel rounded about the middle, with an in-curved or projecting untwisted beak 6.5–11 mm long; pods 2.2–4 mm long, 4–6 mm across, black, subsessile, appressed puberulous, 24–50 seeded; seeds oblique-cordiform, 1.5–2.25 mm long, 1.5–1.75 mm broad, 1–1.25 mm thick, strong yellow to brown-orange, or orange-red, lustrous to glossy, smooth; hilum open, the sinus at an acute angle. Seeds 374,850/kg; 381 seeds/g or 2.8 g/1,000 seeds.

GERMPLASM: Two subspecies are recognized in tropical Africa; subsp. *lanceolata* has the keel rounded shallowly, with a short, rather incurved beak 6.5–8.5 mm long and 4–5 mm broad, shorter than the wings and the pods 28–35 mm long; sp. *contigua* Polhill has the keel strongly rounded, with a longer incurved beak 8–9 mm long and 7–8 mm broad, shorter than the wings, and pods 24–32 mm long. Assigned to the African Center of Diversity, lanceleaf crotalaria or cvs thereof is reported to exhibit tolerance to laterite, slope, and waterlogging. ($2n = 16$.)

DISTRIBUTION: Native to Africa from Rhodesia and Tanzania, south to northeast Cape Province and Malawi. Now pantropical; sometimes cultivated.

ECOLOGY: Grows in grassland of valleys and in deciduous woodland, along riverbanks, lake shores, and other damp places, extending up to the margins of upland rain forests; often persisting as weed of cultivated ground and roadsides. Thrives from sea level to 2,150-m elevations. Sensitive to herbicides. Ranging from Cool Temperate Moist through Tropical Moist Forest Life Zone, lanceleaf crotalaria has been reported to tolerate annual precipitation of 8.7–28.2 dm (mean of 10 cases = 14.9 dm), annual mean temperature of 11.5°–26.2°C (mean of 10 cases = 20.3°C), and pH of 5.5–7.1 (mean of 8 cases = 6.1).

CULTIVATION: Propagated by seeds, sown broadcast at seeding rate of 8–11 kg/ha. Mostly broadcast over areas for soil erosion control.

HARVESTING: Plants grow rapidly and provide green pasture in short time.

YIELDS AND ECONOMICS: Wild and cultivated from East tropical Africa to Cape Province; occasionally cultivated elsewhere in the tropics for erosion control.

BIOTIC FACTORS: The fungi *Fusarium vasinifectum* (*F. oxysporum*) and *Uredo harmsiana* have been reported on this plant. Brazilian tobacco streak virus also affects plants. Nematodes isolated from this *Crotalaria* sp. include: *Heterodera glycines*, *Meloidogyne arenaria*, *M. hapla*, *M. incognita acrita*, *M. javanica*, and *M. thamesi*.

CONTRIBUTORS: J. A. Duke, C. F. Reed.

Crotalaria pallida Ait.

FAMILY: Fabaceae
COMMON NAME: Smooth crotalaria
SYNONYMS: *Crotalaria brownei* Bert. in DC.
Crotalaria hookeri Arn.
Crotalaria mucronata Desv.
Crotalaria striata DC.

USES: Widely cultivated throughout the tropics for pasture, forage and as a cover crop in coconut, rubber, and tea plantations. Very useful for erosion control. Seeds sometimes used as substitute for coffee. Seeds when washed and cooked, are said to be nontoxic.

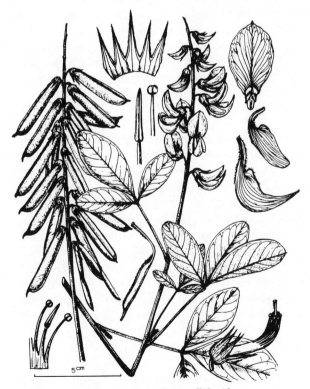

Figure 35. *Crotalaria pallida* Ait.

Crotalaria pallida Ait.

FOLK MEDICINE: No data available.

CHEMISTRY: Leaves contain an alkaloid poisonous to goats; dried leaves are not toxic. Seeds also contain poisonous alkaloid. Seeds contain 13.3% mucilage (N-free), 2.4% oil, and 27.5% protein.

DESCRIPTION: Erect, well-branched annual or short-lived perennial, up to 3 m tall; stem with short appressed hairs, herbaceous to suffruticose; leaves 3–4 foliolate, the leaflets variable, elliptic, obovate-elliptic or obovate, 100 mm long, 25–65 mm broad, glabrous above, thinly appressed puberulous beneath; petiole 20–85 mm long; stipules filiform, to 3 mm long, caducous or absent; racemes shortly pedunculate, to 30 cm long with numerous closely arranged flowers; bracts linear, to 5 mm long, caducous before the flowers open; bracteoles inserted at base of calyx, filiform, to 3 mm long, sometimes small and obscure; calyx becoming basally truncate and deflexed against the pedicel, 6–7.5 mm long, appressed puberulous, the lobes narrow, acuminate, subequal to or longer than the tube; standard elliptic, yellow, often with reddish-brown veins, glabrous outside; wings much shorter than the keel; keel rounded about the middle with a narrow, slightly projecting, untwisted beak; pods linear-oblong, short-stipitate, incurved, 4–5 cm long, pubescent or glabrescent, with 30–40 seeds; seeds oblique-cordiform, ca. 3.5 mm long, smooth or minutely papillose, mottled ochre and dark gray-green or brown, dull or glossy. Seeds 183/g or 5.9 g/1,000 seeds.

GERMPLASM: In the typical variety leaflets are elliptic with the widest point 0.45–0.6 of the distance from base to apex, mostly 6–13 cm long, acute to rounded at apex. *C. pallida* var. *obovata* (G. Don) Polhill (Synonym: *C. falcata* DC., *C. obovata* G. Don), has the leaflets obovate-elliptic to obovate with the widest point 0.6–0.8 of the distance from base to apex. They are ca. 3–7 mm long, rounded, or retuse at the apex. Abundant around shores of the large East African lakes, along major rivers, and common in coastal regions of West Africa. Assigned to the African Center of Diversity, smooth crotalaria or cvs thereof is reported to exhibit tolerance to drought and slope. ($2n = 16$.)

DISTRIBUTION: Pantropical, partly adventive, predominantly in Old World, West to East Africa, India generally, Ceylon, Malaysia, Australia, Hawaii; Mexico to Brazil; West Indies; Florida to Mississippi.

ECOLOGY: Grows along river banks, edges of lakes, extending into woodland and grassland; also persisting on cultivated ground. Typical form generally in rather drier situations than var *obovata*. Sometimes cultivated. Thrives from sea level to 1,500-m elevations. Requires tropical climate, and no frost. Tolerates variable climatic conditions, growing basically as a xerophyte, sometimes in rather dry situations. Ranging from Subtropical Dry to Moist through Tropical Moist Forest Life Zone, smooth crotalaria is reported to tolerate annual precipitation of 8.7–36.1 dm (mean of 9 cases = 18.3 dm), annual mean temperature of 16.0°–26.2°C (mean of 9 cases = 22.2°C), and pH of 4.5–7.1 (mean of 7 cases = 6.0).

CULTIVATION: Propagated by seeds, usually broadcast. Crop may be planted in rows also. Plants develop rapidly and require little care.

HARVESTING: Stands cutting well; crop lasts 1.5–2 years.

YIELDS AND ECONOMICS: For green forage, from either open or shade-grown situations, yields average 10.5–10.7 MT/ha. Cultivated or adventive in many areas of the tropics throughout the world. Mainly grown for forage and pasture; useful for soil-erosion control. Grown for pasture in southeastern United States, Hawaii, and North Africa.

BIOTIC FACTORS: No serious pests and diseases are reported.

CONTRIBUTORS: J. A. Duke, C. F. Reed.

Crotalaria spectabilis Roth

FAMILY: Fabaceae

COMMON NAME: Showy crotalaria

SYNONYMS: *Crotalaria cuneifolia* Schrank
Crotalaria leschenaultii Macfadyen
Crotalaria macrophylla Weinm.
Crotalaria retzii Hitchcock
Crotalaria sericea Retz.

USES: Formerly cultivated in the United States as forage (being replaced by other plants as all parts

Crotalaria spectabilis Roth

Figure 36. *Crotalaria spectabilis* Roth

of this plant contain monocrotaline). Plants are toxic to cattle under field and experimental conditions. Very useful for erosion control. Plants are spectacular ornamentals, flowering for long periods. A fairly strong fiber is extracted from this plant.

FOLK MEDICINE: Plants are used in treatment of scabies and impetigo in India.

CHEMISTRY: Plants are poisonous to livestock: monocrotaline lowers blood pressure in dogs; lethal dose for chickens is 65 mg/kg.

DESCRIPTION: Annual herb; stems erect, usually much-branched, to 2 m tall, angled, sulcate, subglabrous; leaves simple, petiole 2–8 mm long, blade oblanceolate or obovate, 8–14 cm long, 3–8 cm broad, glabrous above, appressed-pubescent beneath; stipules obliquely oblong-ovate, 3–7 mm long, 2–4 mm broad; racemes short peduncled, rather lax, 21–45 cm long, many-flowered; bracts cordate, 12–20 mm long, 6–10 mm broad, acute or acuminate; bracteoles inserted near the middle of pedicle, lanceolate, 1–2 mm long; calyx 11–14 mm long, with the tube protracted on lower side, glabrous; upper lobes broadly triangular or slightly ovate-triangular, longer than the tube; standard suborbicular, yellow, glabrous outside; wings obovate-oblong, longer than the keel; keel rounded about the middle, with slightly in-curved twisted beak, 13–14 mm long; pod short-stipitate, broadly oblong-clavate, 45–50 mm long, 18–20 mm broad, glabrous, 20–24-seeded; seeds oblique-cordiform, with the narrow end strongly in-curved, about 4 mm long, smooth, brown. Seeds about 66,150/kg; weight 77 kg/hl.

GERMPLASM: Assigned to the African Center of Diversity, showy crotalaria or cvs thereof is reported to exhibit tolerance to high pH, heavy soil, nematodes, insects, slope, virus, and weeds. ($2n = 16$.)

DISTRIBUTION: Probably native to tropical Asia and now distributed throughout the tropical region of Asia and America. Introduced to Puerto Rico from the Antilles. Widely cultivated and tending to become naturalized in many areas, e.g., Kenya, Tanzania, Virginia, Georgia, and Texas.

ECOLOGY: Well adapted to heavy and compact soils, fields, gardens, and cultivated ground. Locally naturalized as a weed of cultivated ground in tropical East Africa at altitude from 1,100 to 1,500 m. Sensitive to herbicides, especially 2-4D. Magnesium deficiencies may induce a malady known as "bronzing." Ranging from Cool Temperate Moist through Tropical Dry to Wet Forest Life Zones, showy crotalaria is reported to tolerate annual precipitation of 9.0–28.2 dm (mean of 17 cases = 14.3 dm), annual mean temperature of 11.5°–27.8°C (mean of 17 cases = 21.6°C), and pH of 4.9–8.0 (mean of 14 cases = 6.3).

CULTIVATION: Propagated by seed, which germinate rapidly. Seeding rate varies from 28–34 kg/ha broadcast. Plants develop rapidly; require little attention after establishment.

HARVESTING: Plants produce an abundance of organic material and seeds.

YIELDS AND ECONOMICS: Organic material yields about 50 MT/ha. Abundance of seeds, but no precise data available. Presently plants used mostly

Crotalaria spectabilis Roth

for erosion control. Monocrotaline content makes plants poisonous to livestock. Cultivated mainly in India and southeastern United States.

BIOTIC FACTORS: Following fungi have been reported on this species of Crotalaria: *Ascochyta pisi, Botrytis cinerea, Cercospora demetrioniana, Choanephora cucurbitarum, Colletotrichum crotalariae, Diaporthe crotalariae, Erysiphe polygoni, Fusarium moniliforme* var. *subglutinans, F. oxysporum* f. *tracheiphilum, F. udum* var *crotalariae, F. vasinifectum, Gibberella fujikuroi* var. *subglutinans, Macrophomina phaseoli, Oidium erysiphoides* var. *crotalariae, Pellicularia filamentosa, Phymatotrichum omnivorum, Pyrenochaeta* sp., *Rhizoctonia* sp., *Sclerotium baticola, S. rolfsii, Septoria* sp., and *Vermicularia dematium*. Viruses known to attack this plant include the following: alfalfa mosaic, alsike clover mosaic, bean chlorotic ringspot, bean local chlorosis (6 strains), bean mosaic, bean necrosis, Brazilian tobacco streak, chlorotic mottle, pelargonium leaf curl, subterranean clover stunt, whiteclover mosaic, and crotalaria mosaic. Also attacked by the bacterium, *Pseudomonas syringae*. Nematodes reported from this plant include the following species: *Aphelenchus avenae, Axonochium amphicolle, Helicotylenchus dihystera, H. schachtii, Meloidogyne hapla, M. incognita acrita, M. javanica, M. thamesii, Pratylenchus penetrans, P. vulnus, P. zeae, Radopholus similis, Rotylenchulus reniformis*, and *Tylenchorhynchus claytoni*.

REFERENCE:
Senn, H. A., 1939, North American species of *Crotalaria*, Rhodora **41**:326.

CONTRIBUTORS: J. A. Duke, C. F. Reed.

Cyamopsis tetragonoloba (L.) Taubert

FAMILY: Fabaceae

COMMON NAMES: Guar, Clusterbean, Siambean, Calcutta lucerne

SYNONYM: *Cyamopsis psoralioides* DC.

USES: Guar is grown as a cover crop, green manure, forage for cattle, and protein supplement in cattle feed; as shade plant for young shoots of ginger in India, and as food for human consumption. Young

Figure 37. *Cyamopsis tetragonoloba* (L.) Taubert

pods are eaten like stringbeans, or may be dried, salted, or fried in oil until crisp; mature pods cooked as a vegetable; mature seeds used as a substitute for lentils; young leaves are eaten like spinach in Africa. Dehulling and degerminating the seeds and grinding the remaining endosperm gives guar gum with 5–8 times the thickening power of starch, used as flocculant and filter acid in mining industry, as an appetite depressant; used in cosmetics, hand lotions and creams, and in paper manufacturing. Chemurgically guar is a basic building block for polymers. Seed flour, containing mannogalactan, used to strengthen paper and to thicken and stabilize salad dressings, bakery products, and ice cream. Cake remaining after gum extraction furnishes good high-protein fodder.

FOLK MEDICINE: Medicinally, fruit is laxative, used in biliousness. Leaves are eaten to cure night-blindness; boiled seeds used for poultices for head swellings, enlarged liver, plague, and swelling caused by broken bones. Ashes of burnt guar, mixed with oil, are applied to boils on animals.

CHEMISTRY: Green pods contain per 100 g: 82.5 g water, 3.7 g protein, 0.2 g fat, 2.3 g fiber, 9.9 g carbohydrate, 1.4 g ash, 0.13 g Ca, 0.05 g P, 5.8 mg Fe, 330 IU vitamin A and 49 mg vitamin C. The hay contains 25.21% crude protein, 13.82% crude fiber, 43.59% N-free extract, 0.87% ether extract, and 16.51% total ash. Green forage contains 6.63% protein, 40.73 carbohydrate, and 0.67 ether extract. Another analysis suggests 80.8% moisture, 3.1 crude protein, 0.4 ether extract, 4.4 crude fiber, 8.0 N-free extract, 3.3 ash, 0.61 Ca, 0.07 P. The seeds contain (digestible) 33.23% crude protein, 39.93% carbohydrate, and 2.96% ether extract. Seeds are reported to consist of 14–17% hulls, 43–47% germ, and 35–42% endosperm. The chemical compositions of hulls, germ, and endosperm are as follows: moisture 10%; protein 5.0%, 55.3%, and 5.0%; fat 0.3%, 5.2%, and 0.6%; crude fiber 36.0%, 18.0%, and 1.5%; ash 4.0%, 4.6%, and 0.6%, respectively. Commercial guar gum is reported to contain 10–15% moisture, 5–6% protein, 2.5% crude fiber, and 0.5–0.8% ash. The carbohydrate moiety is a nonionic polysaccharide, called guaran (a mannogalactan), composed of D-mannose and D-galactose in molar ratio 2 : 1, with a linear chain of poly (1 → 4) β-D-mannopyranose, and α-D-galactopyranosyl units attached by (1 → 6) linkages to every other D-mannopyranosyl unit. Seeds are reported to contain trypsin inhibitors but commercial guar flour is reported to contain no inhibitors. The seed meal contains per 16 g N 4.0 g lysine, 1.4 g methionine, 2.0 g methionine-cystine, 3.7 g phenylalanine, 7.0 g phenylalanine tyrosine, 12.5 g arginine, 5.1 g glycine, 2.5 g histidine, 3.2 g isoleucine, 5.9 g leucine, 2.8 g threonine, 1.9 g tryptophane, 4.2 g valine, 4.2 g alanine, 10.2 g aspartic acid, 20.1 g glutamic acid, 3.1 g proline, and 4.9 g serine.

DESCRIPTION: Bushy branching annual to 3 m tall; well-developed lateral root system, with large light-colored nodules, branches angular, grooved, pubescent, some varieties glaucous; leaves alternate, trifoliolate, leaflets 5–10 cm long, ovate, serrate; flowers small, borne in dense axillary racemes, pink to white, zygomorphic; fruit a compressed oblong pod to 12 cm long, beaked; seeds 5–12 per pod, oval, white, gray or black, about 5 mm long, very hard, with a rough gritty surface. Pods form from just above the ground to the top of the plant, in dense clusters, hence the name clusterbean.

GERMPLASM: Tolerances to bacteria, disease, drought, fungi, heat, heavy soils, high pH, salt, and viruses have been reported in the guar gene pool. Best cvs grown for commercial production of guar in the United States are: 'Texsel,' 'Mesa,' 'Groehler Hall,' 'Brooks,' and 'Mills.' 'Brooks' is the first improved cv resistant to major diseases, Alternaria leafspot and bacterial blight. Many cvs grown in India and other countries. India recognizes three main types: (1) *Deshi*-mostly a rainfed seed crop 1.2–1.5 m tall; (2) *Pardeshi*, mostly a vegetable (the pods) crop, 1.5–1.8 m tall; and (3) *Sotiaguvar*, grown mostly for fodder or green manure, 2.5–3.0 m tall. Plants vary from 1-leader branch type to heavily branched, maturing in 125–135 days to 160–175 days; some are prolific seed producers, while others make excellent soil-improving crops. 'Brooks,' 'Mills,' and 'Pusa Sadabaha' are day-neutral, but most cvs are photosensitive. ($2n = 14$.)

DISTRIBUTION: Not surely known in the wild state, guar was once considered indigenous to India and Pakistan. However, Hymowitz suggests that guar arose by transdomestication of the drought-tolerant African native *C. senegalensis*. Exhibiting a Hindustani Center of Diversity, guar is cultivated in drier tropical areas, such as Brazil, Palestine, southwestern United States, and Australia. In India, Pakistan, and southwestern United States, plant is cultivated for gum production.

ECOLOGY: A drought-tolerant summer legume, guar requires only 4–5 dm annual rainfall; growing best in the United States with 9 dm. When moisture is short, growth stops until moisture becomes available; crop responds well to irrigation. Adequate available moisture ensures maximum production of forage and seed. Excessive rain after maturity causes seed to turn black and shrivel, lowering the yield and quality. Areas with high rainfall and humidity are best for green manure. Sensitive to frost and shade, guar requires a minimum frost-free period of 110–130 days and does best with a soil temperature of 25°–30°C. Guar grows well in a wide range of soils, thriving on alluvial and sandy loams, with good well-drained subsoil and pH 7.5–8.0. Should be grown in rotation, for the benefit it imparts to succeeding crops. Rotates well with cotton, grain sorghum, corn, and vegetables. Is a good summer cover on small grain land, and produces good yields of beans when planted soon after grain is harvested and moisture is available. Ranging from Warm Temperate Dry to Moist through Tropical Thorn Woodland to Moist Forest Life Zone, this species has been reported from stations with an annual precipitation of 3.8–24.1 dm (mean of 16 cases = 7.5 dm), annual temperature of 7.8°–27.5°C (mean of 16 cases = 22.0), with 2–12 consecutive frost free months, each with at least 60 mm rainfall, and pH of 5.3–8.3 (mean of 14 cases = 7.2).

CULTIVATION: Seedbed prepared the same as for corn, grain sorghum, or cotton; soil should be firm, free of weed, with row surface slightly above general ground level. Seeded in 110–120 cm rows generally, in 30–60 cm rows if moisture is adequate. Good-quality, certified seed of recommended cvs, resistant to disease should be inoculated before planting with a special guar inoculant or cowpea (Group "E") inoculant. The inoculated cv fixes atmospheric nitrogen for guar in amounts similar to those for cowpeas or other legumes. S ± P increases nodulation significantly. Spraying with Mo and tryptophane also increases nodulation, and Zn is sometimes required. Seed planted 2.5–3 cm deep when soil temperature is about 21°C, requires higher temperature than cotton. Planted in March and August, depending upon adaptation to the region (in S. Texas, April–May 31; CW Texas, May 15–July 1), at rate of 4–6 kg/ha in single rows, 6–8 kg/ha in double rows, and 10–15 kg/ha broadcast, so that germination assures 5 plants per 30 cm of row. Should not be broadcast where moisture is insufficient to support greater plant populations. Guar usually is planted with grain sorghum equipment, with straight instead of beveled or tapered holes, or with equipment designed for seeding vegetable or oil seed crops. It forms well without fertilization. Guar requires high level of phosphorus, 200–250 kg/ha of superphosphates, which also increases the yield of succeeding crops.

HARVESTING: Green pods start forming 45–55 days after sowing, peak in 75–80 days; seeds ripen 110–160 days after sowing. Guar seed do not shatter. For best quality seed, crop should be harvested soon after maturity when pods are brown and dry or when moisture content is less than 14%. In the developing nations, guar is hand-harvested and threshed, in developed areas grain combines are often used with few adjustments. When harvested for hay, leaves shatter readily, so that crop is cut when lower pods turn brown. Other crops are better suited for hay production. For maximum tonnage of green manure, crop should be turned under when lower pods begin to turn brown (usually ca. 90 days). In some regions guar is grazed, usually after frost, to reduce bloat problems. Makes good winter dry forage; cattle and sheep do well when grazed on dry guar stubble after harvest.

YIELDS AND ECONOMICS: Grown as a manure, guar adds 50–150 kg N/ha. Rainfed guar yields ca. 10,000 kg/ha green fodder and 700 kg/ha seed. Yields may be doubled with irrigation, with seed yields up to 1,750 kg/ha. 'Brook' yields 1,280 kg/ha, other cvs 1,900 kg/ha. The guar bean market is based on price of splits (endosperm portion of seed with hull and germ removed). Grade established by purchasers depending on the moisture, foreign material and weight per bushel. In India and Pakistan guar farming is a valuable source of foreign revenue, providing perhaps 90% of the U.S. requirement for guar. India exports gum to Greece, Italy, Netherlands, Spain, Switzerland, and the United States. Gum shipped to Europe is usually refined there before being shipped to the United States. In 1963, more than 5 million kg gum were exported from India; about 30% was sent directly to the United States, at ca. $0.18/kg, valued at $1 million. Around 1975, guar seed commanded about $0.12 per kg. Average production costs were about $39.93/ha and return ca. $55.00/ha. Edible guar gum commanded nearly $1.00/kg, f.o.b. New York. In 1976 trade sources estimated world annual production of guar at 100 million tons.

BIOTIC FACTORS: Seed yields may be greatly reduced by weed competition, especially with johnsongrass. Early preparation of field and mechanical cultivation during period reduce weed infestation. Preplanting applications of trifluralin, and preemergence chlorthal, EPTA, linuron and naptalam are recommended as herbicides, whereas 2,4DES, noruron, and prometryne are said to injure the guar. Postemergence weed control is reported with 4-(2,4-dichlorophenoxy) butyric acid. Vernolate is showing promise in India. Principal fungal and bacterial diseases are: *Alternaria brassicae* (leaf spot), *Fusarium caeruleum* (root rot), *Oidiopsis taurica* (powdery mildew), *Phymatotrichum omnivorum* (Texas root rot), *Rhizoctonia bataticola* (*Macrophomina phaseoli*), (root rot), *Sclerotium rolfsii* (root rot), *Xanthomonas cyamopagus,* and *X. cyamopsidis* (bacterial blight). Applications of organic manure are said to reduce disease incidence. Other fungi known to attack guar are: *Alternaria cyamopsidis, Cercospora psoraleae, Colletotrichum capsici, Fusarium conglutinans, F. moniliforme, Macrophomina phaseoli, Memnoniella echinata, Myrothecium roridum, Oidium* sp., and *Rhizoctonia solani*. Viruses infecting guar are: Abutilon mosaic, Brazilian tobacco streak, Dolichos enation mosaic, sunnhemp mosaic, curly top (transmitted by beet leafhopper, *Eutettix tenella*), and tobacco necrosis. Girdle is stem damage caused by alfalfa leafhopper *(Stictocephla festuca)*, especially found when guar fields are adjacent to alfalfa fields. The clusterbean pod gall midge *(Aspondylia cyamopsii)* reduces pod yields in India. Nematodes *Caconema radicicola* and *Heterodera marioni* (root knot) infest the crop.

REFERENCES:

Hodges, R. J., *et al.,* 1970, Keys to profitable guar production. Fact sheet, *Texas A. & M. Univ. Agric. Ext. Serv.* **L-907**:1–4, College Station, Texas.

Hymowitz, T., 1972, The transdomestication concept as applied to guar, *Econ. Bot.* **26**:49–60.

Rowland, B. W., 1945, The use of guar in paper manufacture, *Chemurgic Dig.* **4**(23):369, 372–376, illus.

Shelton, G., 1956, Guar, a double purpose legume, *Soil Water* **June:**14–15.

CONTRIBUTORS: J. A. Duke, T. Hymowitz, C. F. Reed, J. K. P. Weder.

Derris elliptica (Roxb.) Benth.

FAMILY: Fabaceae

COMMON NAME: Derris

USES: Powdered root is used as effective insecticide for agricultural crops, and also to stupefy fish, poison arrows, and destroy vermin and parasites. Rotenone used in sprays, aerosols, and powdered dusts. For dusting, the root is ground to fine powder and diluted with talc or clay to desired concentration. Leaves are said to be poisonous enough to kill cattle.

FOLK MEDICINE: Except for action on external parasites, this poisonous plant appears to have few folk medical uses.

CHEMISTRY: The genus Derris has been reported to contain rotenone. Other toxic substances occur

Figure 38. *Derris elliptica* (Roxb.) Benth.

in the root, e.g., *dl*-toxicarol, tephrosin, and deguelen. Roots contain 6.4% moisture, 5.1% crude rotenone, 17.5% ether extract. A phenolic substance, starch, sucrose, fats, waxes, saponins, resins, and tannins are also present (WOI).

DESCRIPTION: Climbing or trailing perennial shrubs with short stems that send out numerous slender branches up to 17 m long, forming thick cover and often rooting at each node; often climbing over trees and other vegetation; leaves alternate, leaflets opposite, the odd one distant; flowers white to purple, in racemes; pods indehiscent, 1- to several-seeded. Rarely sets fruit because roots are dug before flowering.

GERMPLASM: Several variations: two introduced to Puerto Rico are 'Sarawak Creeping' and 'Changi No. 3,' they are commercially superior to any other selections. Assigned to the Indochina–Indonesian

Derris elliptica (Roxb.) Benth.

Center of Diversity, derris or cvs thereof is reported to exhibit tolerance to heavy soil, insects, poor soil, shade, slope, and some waterlogging. ($2n = 22, 20, 24$.)

DISTRIBUTION: Native from Malaysia to New Guinea, Java, and East India; widely spread to Old World tropics, and then to Central and South American tropics, including West Indies.

ECOLOGY: Grows well at low altitudes, but has been reported up to 450 m in Puerto Rico and up to 1500 m in Java; growth is noticeably slower above 800 m. Fares better with mean annual rainfall of 20 dm or more; with less precipitation, irrigation may be necessary to develop roots of maximum quality. Distribution of moisture over year important; dry spells longer than 4 months are unfavorable. Fine sandy loams to sandy clay loams are preferable, because roots can be dug easily and soil particles fall away readily. Soils should be well drained, but derris grows nearly as well on poorly drained land. Land should be flat or slightly sloping. Erosion probably is most serious during first few months after planting and again just after harvesting. Land planted to derris should be deep and free of stones. If roots are harvested toward close of rainy season and field immediately replanted, erosion is minimized. Ranging from Subtropical Moist through Tropical Dry to Wet Forest Life Zone, derris has been reported to tolerate annual precipitation of 13.5–41.0 dm (mean of 6 cases = 24.7), annual temperature of 23.3°–27.4°C (mean of 6 cases = 25.7), and pH of 4.3–8.0 (mean of 5 cases = 6.2).

CULTIVATION: Propagated readily by stem cuttings, about 8 mm in diameter and 45 cm long, with 2 or more nodes or buds. Cuttings may be planted directly at field spacing, or be rooted in shaded or unshaded propagation beds, in bamboo pots or baskets. Cuttings root in about 3 weeks and are ready to set out in 6 weeks. If cuttings are exposed to sun, leaves should be removed to prevent stems from drying from excessive evaporation. Under light shade, cuttings root easily, but unshaded nurseries are practical for rooting cuttings on large scale in some areas. Use of rooted cuttings reduces amount of weeding required and assures uniform stand of plants in field. Chinese market gardeners plant long cuttings, which have been twisted into a circle, spacing them about 2 m apart. Plants are allowed to ramble over ground. During growth, pig manure is applied to soil. By time derris cuttings are well rooted, many plants have developed vines several decimeters long. Because vines are often intertwined, they are cut back with machete to stump about 15 cm long to facilitate transplanting. If rooted cuttings are not to be planted for a day or so after digging, they are tied in bunches and put in "mud bath" to prevent excessive drying, kept in shade and protected from sunlight and wind. Field to be planted to derris should be plowed and disked. Furrows 10 cm deep are spaced 1 m apart. Rooted cuttings are placed flat on bottom of furrow, and if no fertilizer is to be applied, furrow is then filled with soil, leaving 5 cm of cutting projecting above surface. If fertilizer is to be added, about 5 cm of soil is firmed over rooted cutting, fertilizer scattered on soil and covered with remaining loose soil. Fertilizer should be added in this manner only when cuttings are planted in rainy season. When cuttings are planted in dry season, furrows are filled with soil and fertilizer added after the rains begin; at this time fertilizer is scattered along rows and covered lightly. After plants have become well established, they form dense mat, which permits root development in superficial layers of soil. Derris does not respond to heavy applications of ammonium sulfate; many nitrogen-fixing nodules form on roots. The K and P requirements of derris are high, compared to N. Trellising, common on small farms, is rarely employed on large plantings (where it is often used as catch crop between young rubber trees). Slender 3-m poles are placed near plants and are stabilized by trellising 2 rows simultaneously. Derris growing on trellises produces 3 times as many usable cuttings as plants allowed to trail over ground.

HARVESTING: Yield increases with age, but quality gradually decreases after second year. Growth is best at about 26 months. In many areas entire crop is not harvested at one time, but roots are lifted as required. Local climatic and soil conditions, field spacing and efficiency of harvesting labor affect yield. At harvest, tops are cleared from fields, roots removed from soil, and nurseries are prepared and planted simultaneously. Tops of vines are removed by cutting with machetes a mat into strips 3 rows wide and 6–7 m long, and rolled for removal from field. Thickest stems are removed for cutting material and treated as described above. Roots are grubbed out of soil to depths of 1 m with special hoes. Ordinarily on large plantings, all roots are removed and the area replanted. In small plantings where plants are usually trellised, tops are left in place and all roots are removed, except for those directly under the plants; this root-pruning practice allows for several harvests from same plant. It is highly desirable to clean field before harvesting roots, as pieces of stems and foreign matter add to labor of cleaning roots for market. Roots are cut with plow to depth of 40 cm (81% of roots occur in this area), and are removed from loosened soil with potato forks. A small tractor has sufficient power to open the 40 cm furrow in a single operation; use of oxen requires 2 cuts in same vertical plane, the first cut about 20 cm deep and the second about 40 cm. Roots should be removed from loosened soil of first cut before second cut is made.

YIELDS AND ECONOMICS: In 1963 more than 1,500 MT crude derris roots were imported in the United States with an additional 500 MT rotenone extract. Market quotations are based on roots having 5% rotenone and 8–12% moisture; roots with less than 3% rotenone are usually not accepted. Labor for a nontrellised crop in Puerto Rico: nurseries, 10%; weeding, 19%; harvesting, 64%; and miscellaneous, 7%. Yields of roots are directly correlated with total light plants receive during growth. Plants harvested at 25–27 months of age yield 800–1,700 kg/ha air-dry roots, when trailing system of culture is used, and 3,000 kg/ha when plants are trellised. Roots of same diameter taken from same set of plants may vary more than 100% in rotenone content. Roots with intact bark can be dried in direct sunlight without destruction of rotenone and should be spread out until the thickest are sun dry. Drying out of doors requires 7–15 days, in well-aerated sheds, about 3 weeks. In Far East, growers prefer to cut fresh root into pieces 3.5–5 cm long, and when dry to ship them in bags as 'chips.' In Far East plantings at different field spacing (0.6 × 0.6–1 m × 1 m) give different yields; however, wider spacings decrease root yields. Closer spacings permit complete ground cover more quickly and reduce cost of weeding. Cuttings of different diameters give different yields of root and all in 31 months, cuttings 5–8 mm yield 1,000 kg air-dry roots/ha, with 5.3% rotenone; 8–15 mm, 1050 kg/ha with 5.3% rotenone; and 15–35 mm, 1100 kg/ha, with 4.5% rotenone.

BIOTIC FACTORS: Following fungi have been reported on derris: *Asterina* sp., *Colletotrichum derridis, Corticium salmonicolor, Diphragmium koodersii, Fomes lignosus, Hapalophragmium phaseoli, Phyllosticta derridis, Phyllachora yapensis. Pratylenchus coffeae* is one nematode isolated from derris. Insect pests include: beetles attacking roots (*Dinoderus bifoveolatus* and *D. minutus*); leafwebber (*Hedylepta indicata*); and leaf-tiers (*Proteides mercurius pedro, Polygonus leo leo*) which damage leaves.

REFERENCES:
Jones, H. A., and Smith, C. M., 1936, Derris and cube, approximate chemical evaluation of their toxicity, *Soap* June:113–115.
Moore, R. H., 1943, Derris culture in Puerto Rico, *P.R. Ext. Stn.* **24**:1–17, illus. Mayaguez, P.R.
Moreau, R. E., 1944, Derris agronomy: An annotated bibliography and a critical review, *East Afr. Agric. J.* **10**(2):75–82.
Roark, R. C., 1932, A digest of the literature of Derris (*Deguelia*) species used in insecticides, 1747–1931, *U.S. Dep. Agric., Misc. Publ.* **120**:1–86.
Sievers, A. F., 1940, *The Production and Marketing of Derris Root*, U.S. Department of Agriculture, Division of Drugs and Related Plants, Washington, D.C., 23 pp.
White, D. B., 1945, Propagating Derris by cuttings, *Agric. Am.* **5**(8):154–156, illus.

CONTRIBUTORS: J. A. Duke, C. F. Reed.

Derris malaccensis (Benth.) Prain

FAMILY: Fabaceae

COMMON NAMES: Tuba root, Tuba merah, Tuba rabut, Sarawak erect

USES: Roots a source of rotenone, used as insecticide, larvicide, and a fish poison. Roots have nitrogen-fixing bacterial nodules, and plants used for soil improver; planted in mixed cultivation with coconuts, kapok, rubber, and cocoa.

FOLK MEDICINE: No data available.

CHEMISTRY: Root contains toxicarol, malaccol, an isoflavone, a phenolic resin and a resin containing rotenone, deguelin, and elliptone. Roots richer in total toxic substances than are those of *D. elliptica*, but contain less rotenone. Roots contain up to 9% rotenone.

Derris malaccensis (Benth.) Prain

Figure 39. *Derris malaccensis* (Benth.) Prain

DESCRIPTION: Erect climbing shrub; stems do not trail nor form cover; leaves alternate, compound, light green; young foliage bright red-brown; petioles and lower surface smooth; leaflets opposite, the odd one distant; flowers white to purple; pods indehiscent, 1- to several-seeded, develop rarely.

GERMPLASM: In Malaya the following cvs are recognized, along with their rotenone content: 'Kinta,' 0.5% percent rotenone and 19% ether extract; 'Tuba merah,' 1% and 19.5%; and 'Sarawak erect,' 3% and 20.4%. *D. malaccensis* var *sarawakensis* is the most important commercial type. Assigned to the Indochina–Indonesian Center of Diversity, tuba root or cvs thereof is reported to exhibit tolerance to insects, poor soil, and slope. ($2n = 22, 24$.)

DISTRIBUTION: Indigenous to Malaya, cultivated on small scale in India, Borneo and West Indies. Not cultivated as frequently as *D. elliptica*.

ECOLOGY: Grows on most soils, especially on clay loams with some sand; soils should be light; heavy clay soils make harvesting difficult. With temperature of 26°–27°C and relative humidity of 80–90%, it tolerates any amount of sunshine. In Java and Malaysia crop grows up to 1,580 m altitude. Ranging from Subtropical Moist through Tropical Dry Forest Life Zone, tuba root is reported to tolerate annual precipitation of 15–25 dm and annual mean temperature of 25°–27°C.

CULTIVATION: Plants do not produce seed in 2½ years, when root crop is usually harvested. Vegetative propagation by stem cuttings assures uniform quality with respect to toxicity. Cuttings, about 45 cm long, are readily rooted in nursery beds with light shade. In about 6 weeks they are ready to be planted out. Often long cuttings are twisted into a circle and planted about 2 m apart. Plants are also grown on ridges spaced 1 m apart in the rows giving ca. 12,000 plants/ha. In Japan cuttings taken nearest the roots are thought to be best. Plants 2 years old should provide several hundred 2-node cuttings. Cuttings show fewer failures from this species than from *D. elliptica*. During growth, pig manure is applied to soil. If planted close enough, plants grow rather erect, but occasionally are allowed to ramble over ground.

HARVESTING: Root maturity varies according to cultural conditions; the average period of growth is 2 years. Toxicity of root is highest 19–27 months after planting the cuttings; in some areas 15 months is sufficient. Fertilizing with manure, rice bran, or human excrement does not significantly increase toxicity of roots. In Taiwan, fish meal added to soil gives good results. In Borneo, roots reported to be most toxic from June to August; in Japan, rotenone highest July–Oct., according to location. Roots ca. 0.6 cm in diameter give highest yield of rotenone; larger ones are definitely less toxic. Roots are lifted as required in some areas.

YIELDS AND ECONOMICS: Weight of fresh roots from single plant is 450–900 g, according to spacing, with an average weight loss of 40% during drying.

Japan is main importer of tuba root from Malaysia, and reexports it as 'derris soap.' Borneo, Philippine Islands, and Malaysia are the main exporters of tuba root. Main importers are Japan, United States, Great Britain, France, and Germany.

BIOTIC FACTORS: Several fungal diseases and insect pests cause local damage to plants, but none serious enough to affect crop.

REFERENCES:
Georgi, C. D. V., 1939, Variation in toxic content of roots of *Derris malaccensis* var. *sarawakensis* with increase in age of plant, *Malay. Agric. J.* **27**(4):134–140.
Georgi, C. D. V., and Teik, G. L., 1938, Further selection experiments with *Derris malaccensis, Malay. Agric. J.* **26**:4–17.

CONTRIBUTORS: J. A. Duke, C. F. Reed.

Desmodium intortum (Mill.) Fawc. & Rendle

FAMILY: Fabaceae

COMMON NAMES: Greenleaf, Beggarlice

SYNONYMS: *Desmodium hjalmarsonii* Standl.
Desmodium trigonum (Sw.) DC.
Hedysarum intortum Mill.
Meibomia intorta (Mill.) Blake

USES: Widely recognized as a potentially valuable pasture and forage legume for the wetter tropical and subtropical regions of the world. As both leaves and pods may adhere to clothing, some may regard this species as more trouble than it is worth.

FOLK MEDICINE: Cuna Indians regard greenleaf for love potions, perhaps reflecting the doctrine of signatures. Leaf decoctions of other desmodium species are used for consumption, convulsions, and various types of sores, including venereal sores.

CHEMISTRY: The genus *Desmodium* has been reported to contain bufotenine, dimethyl tryptamine, and donaxine, and the seed may contain trypsin inhibitors and chymotrypsin inhibitors.

DESCRIPTION: Perennial, liana or subshrub, or semiprostrate herb with stems ascending, often trailing

Figure 40. *Desmodium intortum* (Mill.) Fawc. & Rendle

on low bushes, some cvs or vars spreading or repent; stems 1–3 m long, unbranched, slender, trisulcate and lineate, densely uncinate-pubescent, and sometimes rather densely spreading-pilose; petioles 1.8–5.5 cm long, rachis 0.6–1.2 cm long; petiolules 3 mm long; leaflets ovate-acuminate, mucronate, rounded at base, shining and glandular-uncinate-puberulent or -pilose, or both above, paler and pilose beneath; terminal leaflet 3.5–7.5 cm long, 1.5–4.5 cm broad, lateral leaflets 2.2–6 cm long, 1–3.3 cm broad; stipules up to ca. 7 mm long; stipels ca. 3–5 mm long; inflorescences in indeterminant racemes, axillary or terminal; rachis densely uncinate-pubescent; primary bracts up to 8 mm long, secondary bracts not always present; pedicels uncinulate-puberulent to glabrescent in age, 6–8 mm long; calyx puberulent and somewhat pilose on teeth, central tooth of lower lobe 5 mm long, lateral teeth 4 mm long, upper bifid lobe 4 mm long; standard 9 mm long, 6 mm broad, wings subelliptic,

Desmodium intortum (Mill.) Fawc. & Rendle

8 mm long, 3 mm broad, keel-petals subfalcate, 8 mm long, 3 mm broad; flowers violet to purplish, lavender-pink, pink or rarely cream; loment stipitate, up to 9-articulate, the articles semirhomboid to semiorbicular, 4 mm long, 2 mm broad. Seed size 1.2–1.6 g/1000 for *D. aparines*, 1.8–2.0 g/1,000 for *D. uncinatum*. Fl. Feb.–May; Fr. by frost. (North of equator.)

GERMPLASM: Rotar has gathered strong evidence suggesting that much that has been called greenleaf desmodium is *Desmodium aparines* (link) DC; not *D. intortum*. Rotar regards *D. aparines* as having less acutely pointed leaves than *D. intortum,* and darker green leaflets with two distinct "watermarks" along their upper surface. *D. aparines* seed he regards as smaller, 1.3 to 1.5 g/1,000, cf. 1.9 g/1,000 for *D. intortum*. *D. intortum* has more triangular stems and longer pedicels, and the standard petal differs. Quite possibly this writeup involves data derived from at least these two species. Several early-flowering, large-seeded, and vigorous segregates have been obtained from hybrids between this species and *D. sandwicense,* a species with which it hybridizes freely and produces great variability. The confusion between *D. aparines* and *D. intortum* is probably compounded by hybridization with other species. Other variations occur in growth, persistence, flowering time, seed yield, seed size, leafiness, nodulation, disease resistance, and salt tolerance. The best known cultivar is 'Greenleaf' from Australia, an improved agrotype, not as high in tannins. Although tannins have been shown to reduce protein digestibility, there is no evidence that they have any effect on the grazing ruminant. Other cvs include: 'HES-4331' (a rather uniform Hawaiian parent line of greenleaf and the recent release 'Kuiaha'), 'HES-4247' (a mixture including the entire range of types), 'C-50' from Taiwan, and 'Medellin' from Colombia. Assigned to the South American Center of Diversity, greenleaf or cvs thereof is reported to exhibit tolerance to disease, grazing, low pH, salt, and shade. ($2n = 22$.)

DISTRIBUTION: Native to tropical America, usually above 500 m, but now widely distributed throughout the tropics and subtropics.

ECOLOGY: Common in thickets and forests, roadsides, in grasslands, along coasts, from sea level up to ca. 2,400-m elevation (mostly 500–2,000 m). Thrives in moist to wet land, sometimes in thickets and grassy land. Often a weed in coffee plantations. Frequent in rocky places along streams and river banks. In pine, oak or mixed forests up to 2,400-m elevations. Tolerating no frost (or light frosts, e.g., in Queensland), it is one of the fastest growing tropical legumes during cool seasons. The seed crop in Queensland is often lost to frosts; although the flowers are destroyed, the plants survive. Although reported from some tropical stations, it seems better adapted to subtropical situations. Tolerates higher pH limestone soils in the Bahamas; also tolerates sandy spodosols. Adaptability to saline soils varies; some cvs are suitable for such soils. Ranging from Warm Temperate Moist through Tropical Very Dry to Wet Forest Life Zones, greenleaf has been reported to tolerate annual precipitation of 5.3–40.3 dm (mean of 20 cases = 14.4), annual temperature of 7.3°–27.1°C (mean of 20 cases = 21.3), and pH of 4.5–7.1 (mean of 17 cases = 5.9).

CULTIVATION: Propagated by seed, sown on a well-prepared seedbed of alluvial, limestone, or sandy soil. Seed should be inoculated at planting with *Rhizobium* CB-627, to produce a well-nodulated stand. Commercial cowpea inoculant does well. Crop grows well in association with *Panicum maximum* and *Setaria anceps,* but not with *Chloris gayana*. On sandy spodosols, greenleaf desmodium mixes well with *Digitaria decumbens* and *Paspalum notatum*. Grass/legume pasture mixtures in the tropics have proved beneficial in increased animal production, their value being greatest during the dry season. It is being successfully grazed with *Hemarthria altissima* in the Bahamas. Seeding rates vary from 1.12 to 5.6 kg/ha, depending on the accompanying grass. If seed are not available, cuttings may be used and root readily. In Hawaii, it is propagated by cuttings 15–30 cm long, spread with a manure spreader, and disked in. Plants flower under short days (less than 12 hours). Rotar speaks of *D. aparines* as a "long–short"-day plant. Planting late in summer affects flowering in the first season. Time of cutting in the second season or later summers also affects seed production. If certain cvs are cut in late September in Hawaii they may remain vegetative during the following short-day period (winter). Plants often form a dense mass of flowers and foliage over other plants. Most cvs do not flower the first year, but all flower the second season and set some seed before frost. In warmer climates, they flower over a longer season. *Desmodium aparines* responds well to P fertilization.

HARVESTING: In Rhodesia crop withstands grazing for 6 years if adequate P is supplied. In Uganda, grazing for at least 3 years has been reported. Crop mostly grazed by animals, or cut by hand for forage.

YIELDS AND ECONOMICS: A valuable pasture and forage legume, usually grown with a grass, for wetter tropical and subtropical areas. Cultivated for pasture and forage in Central and South America, tropical East Africa, Australia, and other areas of adaptation. Total yield of legume increases with longer cutting intervals of 4–12 weeks; 4.23 MT DM/ha/yr; 8 weeks, 6.54 MT DM/ha/yr; 12 weeks, 8.03 MT DM/ha/yr. Best cut at height of 5–10 cm; shorter cuttings will destroy the plant in short order. Mixed with pangola and bahia, greenleaf DM yields are ca. 12.5 MT/ha with 11.5–12.4% crude protein. Crude protein yields are ca. 1.5 MT/ha/yr. In Hawaii desmodium-pangola yields approach 27 MT/ha supporting beef production of 450–675 kg beef/ha/yr. Seed yields are highest from plants with large number of inflorescences with flowers which are readily self-tripped. In Florida seed yields are reported to be low in absence of large bee populations, but in Hawaii, self-tripping *D. aparines* sets seed well without bees. Yields also affected by flower density and seed size.

BIOTIC FACTORS: Fungi reported on this plant include: *Parodiella perisporioides*, *Phakopsora meibomiae*, *Synchytrium citrinum*, and *Uromyces hedysaripaniculati*. Plants are also susceptible to legume little-leaf disease, which has prevented their success in Brazil and probably other South American countries. A good stand of bees may improve seed set, but apparently is not needed for *D. aparines*.

REFERENCES:
Hutton, E. M. and Coote, J. N., 1972, Genetic variation in nodulating ability in "Greenleaf" *Desmodium*, *J. Aust. Inst. Agric. Sci.* **38**:68–69.
Imrie, B. C., 1973, Variation in *Desmodium intortum*, a preliminary study, *Trop. Grassl.* **7**(3):305–311.
Jones, R. J., 1973, The effect of frequency and severity of cutting on yield and persistence of *Desmodium intortum* cultivar "Greenleaf" in a subtropical environment, *Aust. J. Exp. Agric. Anim. Husb.* **13**(61):171–177.
Mills, P. F., 1968, Kuru vine (*Desmodium intortum*), *Rhodes. Agric. J.* **65**:59.
Rotar, P. P., and Chow, K. H., 1971, Morphological variation and interspecific hybridizing among *Desmodium intortum*, *D. sandwicense* and *D. uncinatum*, *Hawaii Agric. Exp. Stn., Tech. Bull.* **82**:28 pp.
Vallis, I., and Jones, R. J., 1973, Net mineralization of nitrogen in leaves and leaf litter of *Desmodium intortum* and *Phaseolus atropurpureus* mixed with soil, *Soil Biol. Biochem.* **5**:391–398.
Younge, O. R., Plunknett, D. L., and Rotar, P. P., 1974, Culture and yield performance of *Desmodium intortum* and *D. canum* in Hawaii, *Hawaii Agric. Exp. Stn., Tech. Bull.* **59**.

CONTRIBUTORS: J. A. Duke, A. E. Kretschmer, Jr., H. L. Pollard, C. F. Reed, P. P. Rotar.

Dipteryx odorata (Aubl.) Willd.

FAMILY: Fabaceae

COMMON NAMES: Tonka bean, Tonga, Cumaru, Tonquin

SYNONYMS: *Coumarouna odora* Aubl.
Coumarouna punctata Blake

USES: Tonka beans are cultivated for the seed which yield coumarin, used to give a pleasant fragrance to tobacco, a delicate scent to toilet

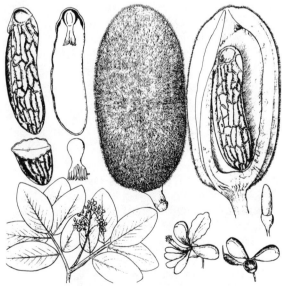

Figure 41. *Dipteryx odorata* (Aubl.) Willd.

Dipteryx odorata (Aubl.) Willd.

soaps, and a piquant taste to liqueurs. Extract is used also in foodstuffs, e.g., cakes, candies, and preserves, and as a substitute for vanilla, as a fixing agent in manufacturing coloring materials, in snuffs and in the perfume industry. Most important in the United States for flavoring tobacco. The timber is said to be resistant to marine borers, perhaps because it contains 0.01% silicon dioxide.

FOLK MEDICINE: Said to be used for cachexia, fumigation, narcotic, nausea, whooping cough, and tonic. In the People's Republic of China, seed extracts are used rectally to treat schistosomiasis.

CHEMISTRY: The active constituent of the seed is coumarin (1–3%). According to notes from the FDA: "Dietary feeding of coumarin to rats and dogs causes extensive liver damage, growth retardation, and testicular atrophy." Five isoflavones have been isolated from the heartwood: retusin, retusin 8-methyl ether, 3′-hydroxyretusin 8-methyl ether, odoratin, and dipteryxin (Hayashi and Thomson, 1974). A sample of Malaysian Tonka bean oil showed palmitic acid (6.1%), stearic (5.7%), oleic (59.6%), linoleic (15.4%), and C_{26}-C_{24} acids (13.2%). The neutral fraction of the bark of *Dipteryx odorata* of Venezuelan origin, yielded unspecified quantities of methyl palmitate, methyl margarate, methyl stearate, methyl oleate, methyl linolenate, and methyl linoleate. The major flavoring principle of Tonka beans is coumarin. In 19 samples of beans from Venezuela and Trinidad (uncured) coumarin content was 2.1–3.5% (dry weight), and moisture 6.9–8.4%; in 1 sample of Brazilian beans moisture was 10.2% and coumarin 2.7%. Surinam beans contained 2.4% coumarin and 29.2–33.7% alcoholic extractive; Trinidad beans, 1.3% coumarin and 49.5% alcohol extractive; and Imburana beans 2.6% coumarin.

DESCRIPTION: Compact tree 25–30 (–40) m tall, with trunk to 1 m in diameter; bark smooth, gray, leaves alternate, pinnate with 3–6 leaflets, leathery, glossy, dark green, the winged rachis projecting beyond leaflets; flowers rose-colored, corolla 10–12 mm long; fruit an indehiscent pod 7–10 cm long, 3–6 cm broad, yellow-brown to mahogany at maturity, with one seed; seed dark-colored, 3–5 cm long, 1–2 cm wide, weighing ca. 2.9 g each. Seedlings have a taproot which develops lateral anchor roots that penetrate the soil, forming a dense mass of feeding roots. Fl. Mar.–May; Fr. June–July.

GERMPLASM: 'Sarapia' is a very large fruit and seed type; 'Angasture,' a Venezuelan type; 'Para,' a Brazilian type. Dwarf forms are known and could be used as rootstocks. Assigned to the South American Center of Diversity, Tonka bean or cvs thereof is reported to exhibit tolerance to insects, poor soil, shade, slope, and waterlogging. ($2n = 32$.)

DISTRIBUTION: Tonka bean is abundant along tributaries of the Orinoco River in Venezuela, and common in the Brazilian States of Amazonas, Para, and Mato Grosso. It has been introduced in the West Indies, Trinidad, and Dominica.

ECOLOGY: Thrives in shady, damp situations, especially along river banks, where the rainfall is 15–28 dm annually. A calcifuge, tolerates poor, well-drained soils, but grows best on more fertile soils rich in humus. In Trinidad it grows up to 330 m in altitude. Ranging from Tropical Dry to Tropical Wet Forest Life Zone, Tonka bean is reported to tolerate annual precipitation of 13.5–40.3 dm (mean of 4 cases = 25.9), annual mean temperature of 21.3°–26.6°C (mean of 4 cases = 25.0), and pH of 5.0–8.0 (mean of 4 cases = 6.2).

CULTIVATION: Usually propagated by seed, but can be propagated by budding, cuttings, and marcottage. Seeds lose viability soon after ripening. Whole seeds germinate in 4 weeks, endocarpless seeds in 1–2 weeks. No particular cultivation is required in native regions. Bulk of crop is still produced from wild trees.

HARVESTING: Fallen pods are harvested in Jan.–March, fresh pods in June–July. Hard outer shell is removed and the kernels (beans) spread out for 2–3 days to dry. They are then bagged and shipped by boat to nearby towns. There, beans are soaked in rum or alcohol (45–67%) for several days. When the rum is drained off, beans are again dried, becoming coated with a white crystalline deposit of coumarin, and ready for export. Beans are ground and soaked in rum for about 3 months. Resulting liquid, rich in coumarin and highly aromatic, is decanted and sprayed over tobacco, giving it a distinctive fragrance.

YIELDS AND ECONOMICS: Yields of bean per tree are 1–3.5 kg/yr. Plenty are available from wild trees so few are cultivated. Major producer is Venezuela followed by Brazil and Colombia. Formerly Trinidad was the leading extract producer but Venezuela is now the leader. The United States is the principal importer, especially for the tobacco industry. Competition from synthetic coumarin and vanillin has reduced demand.

BIOTIC FACTORS: Tonka beans are pollinated by insects. Bats eat the pulpy flesh of the fruit and are the worst pests. Among the fungi known to attack Tonka beans are: *Anthostomella abdita, Diatrype ruficarnis, Macrophoma clavuligera,* and *Myiocopron cubense.*

REFERENCES:
Anonymous, 1947, Tonka beans, *Econ. Bot.* **1**:175 [repr.: 1946, *Braz. Bull.* **3**(58):1.]
Anonymous, 1977, Combined oral furapromidium and rectal *Dipteryx* in *Schistosomiasis japonica, Chin. Med. J.* **3**:103.
Ciferri, R., 1927, Qualche notizia sulla "Fava Tonka" (*Coumarouna punctata* Blake), *Inst. Agric. Colon. Ital.,* Florence, Italy, 16 pp. illus.
Hayashi, T., and Thomson, R. H., 1974, Isoflavones from *Dipteryx odorata, Phytochemistry* **13**:1943–1946.
Pound, F. J., 1938, History and cultivation of the tonka bean (*Dipteryx odorata*) with analysis of Trinidad, Venezuela and Brazilian samples, *Trop. Agric. (Trinidad)* **15**:4–9, 29–32.

CONTRIBUTORS: J. A. Duke, N. R. Farnsworth, J. Langenheim, C. F. Reed.

Genista tinctoria L.

FAMILY: Fabaceae

COMMON NAMES: Dyer's greenwood, Woodwaxen, Dyer's greenweed

SYNONYMS: *Genista depressa* Bieb.
Genista hungarica A. Kerner
Genista marginata Besser
Genista mayeri Janka
Genista tanaitica Smirnov.
Genista ovata Waldst.
Genista patula Bieb.
Genista polygalaefolia Hort., non DC.
Genista sibirica Hort.
Genista tetragona Besser

USES: All parts of the plant, especially the flowering tops, yield a yellow or green dye once used especially for wool (largely superseded by *Reseda luteola* with wood from *Isatis tinctoria*). Extraction of flowers with petroleum ether produced 0.16% of a concrete that yielded 53% of alcohol-soluble absolute. On steam distillation, the concrete gave 2.3% of yellowish volatile oil having a "heavy" and "green" odor. Seeds have been suggested as a coffee substitute. Buds are picked and used in sauces as a caper substitute. Cows sometimes eat the leaves and produce bitter milk.

FOLK MEDICINE: Both flower tops and seeds have been used medicinally; the oil having cathartic, diuretic, and emetic properties. Said to be used for abscesses, albuminuria, aperient, dropsy, gout, gravel, hydrophobia, hypertensive, laxative, narcotic, purgative, rheumatism, sciatica, scrofula, sudorific, stimulant, stones, tumors, and vasoconstrictor.

Figure 42. *Genista tinctoria* L.

Genista tinctoria L.

CHEMISTRY: Active principle is scoparine, starry yellow crystals soluble in water and alcohol. Dyer's greenwood also contains anagyrine (formerly used in treating edema), cytisine (formerly used like sparteine), genisteine, isosparteine, lupanine, *N*-methylcytisine, pachycarpine, retamine, sparteine (the sulfate of which has been used as an antiarrhythmic and in early stages of labor—said to have narcotic properties), and thermopsine.

DESCRIPTION: Procumbent to erect shrub, 10–200 cm tall; branches striped, glabrous or slightly pubescent, not thorny; leaves alternate, 9–50 mm long, 2.5–15 mm broad, simple, highly variable in shape, mostly oblong-elliptic to oblong-lanceolate, almost glabrous, ciliate, smooth on edges; leaves, calyx and legume glabrous to densely sericeous; flowers borne singly in the axil of each bract in short racemes toward the ends of branches, or in long, simple or compound racemes, panicled at ends of branches; bracts foliaceous; bracteoles ca. 1 mm long; pedicel 1–2 mm long; calyx 3–7 mm long; corolla glabrous, bright yellow, the standard 8–15 mm long, broadly ovate; pod or legume narrow-oblong, 2.5–35 cm long, glabrous or slightly pubescent; 6–10 seeded, brown when ripe. Fl. June–Aug. in United States; April–Aug. in Europe; Fr. early autumn.

GERMPLASM: Species is highly variable in habit, leaf shape, and degree of hairiness. Populations show different combinations of these characters and have been variously referred to as distinct species or subspecies. Most of the plants fall into one of the following groups. Group I (*G. tinctoria* L. sensu stricto, *G. anxantica* Ten., *G. campestris* Janka, *G. elata* Wenderoth, *G. tenuifolia* Loisel., *G. virgata* Willd.) with plants erect or ascending, 20–200 cm tall, leaves 15–35 mm long, 2.5–9 mm broad, oblong or lanceolate, conspicuous lateral veins, glabrous, margins and midrib ciliate, flowers numerous in simple or branched racemes, and the calyx and legume glabrous. Group II (*G. alpestris* Bertol., *G. tinctoria* subsp. *littoralis* (Corb.) Rothm.) with plants procumbent, less than 20 cm tall, leaves 9–12 mm long, 3–4 mm broad, ovate-elliptic to elliptic-oblong glabrous, ciliate along margins and midrib, usually with conspicuous lateral veins, flowers few, and the calyx and legume usually glabrous. Group III (*G. hungarica* A. Kerner, *G. lasiocarpa* Spach., *G. mantica* Pollini, *G. mayeri* Janka, *G. ovata* Waldst. & Kit., *G. perreymondii* Loisel.) has erect or ascending plants, 20–200 cm tall, leaves 20–50 mm long, 6–15 mm broad, ovate to elliptical, usually pubescent, with conspicuous lateral veins, flowers numerous in simple or branched racemes, and the calyx and legume usually pubescent. Group IV (*G. csikii* Kummerle & Jav., *G. depressa* Bieb., *G. friwaldskyi* Boiss. *G. tetragona* Besser) has plants procumbent, not more than 20 cm tall, leaves 10–20 mm long, 3–5 mm broad, lanceolate to oblanceolate, usually pubescent, lateral veins not conspicuous, flowers few and calyx and legume glabrous or pubescent. In addition, var *plena* Hort. has double flowers; var *virgata* Mert. & Koch is a vigorous plant to 2 m tall and pods 3–6-seeded, native to southeastern Europe, and var *humilior* Schneid, a dwarf and compact form, more pubescent, and silky-villous pods, native to Italy. A dwarf form, grown in tufts in English meadows, is said to enrich poor soil. Assigned to the Eurosiberian Center of Diversity, dyer's greenwood or cvs thereof is reported to exhibit tolerance to limestone, low pH, mycobacteria, poor soil, shade, slope, smog, and weeds. ($2n = 45$.)

DISTRIBUTION: Native to central Europe, Atlantic Europe, Scandinavia, western Asia, and Siberia, not uncommon in England, but rare in Scotland. Introduced in North America and somewhat naturalized from southern Maine to Massachusetts, and locally to Washington, D.C. and Michigan.

ECOLOGY: Ranging north to Zone 5 in the United States, dyer's greenwood thrives in dry woods, wood margins and scrub hillsides, chiefly on calcareous and sandy soils. Suitable for temperate areas, and adapted for covering dry, sandy banks and rocky slopes. Also useful for borders and rockeries in the garden. In England grows in meadows, pastures, heaths, and on the borders of fields. Requires a well-drained soil. Ranging from Cool Temperate Moist to Wet through Warm Temperate Dry Forest Life Zone, dyer's greenwood has been reported to tolerate annual precipitation of 5.2–11.6 dm (mean of 12 cases = 8.0 dm), annual temperature of 7.0°–14.7°C (mean of 12 cases = 9.3°C), and pH of 4.5–7.4 (mean of 11 cases = 6.2).

CULTIVATION: Propagated by seeds, layering and greenwood cuttings rooted under glass. For optimum propagation, seeds should be sown as soon as ripe or in early spring. Seeds are sown in pots of light, loamy soil, in a greenhouse frame or prepared seedbed. Seedlings, when large enough to handle, are planted singly in a nursery border; after 2 yr, may be planted in permanent sites. They thrive in any good garden soil, especially a loamy light soil. Plants should be cut back a few times when young to induce a bushy habit and to induce more flowering tops. Some cvs are not long-lived. When signs of deterioration appear, replacements should be started.

HARVESTING: When used for dye, flowering tops are gathered during the summer and immediately sent to the extracting center.

YIELDS AND ECONOMICS: In the past used extensively as a source of a dye, especially for wool. Now of minor importance but used a little in Europe and Asia. Useful as an ornamental plant both in Europe and in North America.

BIOTIC FACTORS: Following fungi have been reported on this plant: *Coniothyrium genistae, C. genisticola, Erysiphe communis, E. genistae, E. martii, E. polygoni, Eutypa ludibunda, Knyaria garciniae, Phyllosticta genistae, Uromyces cytisi,* and *U. genistae-tinctoriae*. No serious insect pests have been reported.

CONTRIBUTORS: J. A. Duke, C. F. Reed

Glycine max (L.) Merr.

FAMILY: Fabaceae
COMMON NAMES: Soybean, Soya
SYNONYMS: *Dolichos soja* L.
Phaseolus max L.
Soja hispida Moench

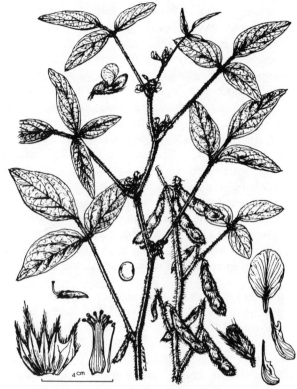

Figure 43. *Glycine max* (L.) Merr.

USES: Seeds are among of the world's most important sources of oil and protein. Unripe seeds are eaten as vegetable and dried seeds eaten whole, split or sprouted. Processed to give soy milk, a valuable protein supplement in infant feeding, and curds and cheese. Soy sauce is made from the mature fermented beans, and soy is an ingredient in other sauces. Roasted seeds used as a coffee substitute. The highly nutritious sprouts are readily consumed in Asia. Seeds yield an edible, semidrying oil, used as salad oil and for manufacture of margarine and shortening. Oil used industrially in manufacture of paints, linoleum, oilcloth, printing inks, soap, insecticides, and disinfectants. Lecithin phospholipids, a by-product of the oil industry, are used as a wetting and stabilizing agent in food, cosmetic, pharmaceutical, leather, paint, plastic, soap, and detergent industries. Demand is increasing for soy meal, a high-protein livestock feed. Meal and soybean protein used in manufacture of synthetic fiber, adhesives, textile sizing, waterproofing, fire-fighting foam and many other uses. Whole beans yield a full-fat flour with about 20% oil; mechanically expressed meal gives low-fat flour with 5–6% oil; solvent-extracted meal gives defatted flour with about 1% oil. The flour is used in bakery and other food products; and as additives and extenders to cereal flour and meat products, and in health foods. The vegetative portions of plant used for silage, hay, pasture, or fodder, or may be plowed under as a green manure. The straw can be used to make paper, stiffer than that made from wheat straw.

Glycine max (L.) Merr.

FOLK MEDICINE: Old Chinese herbals suggest that the soybean was a specific remedy for the proper functioning of the bowels, heart, kidney, liver, and stomach. A decoction of the root is said to be astringent. The meal and flour are used to prepare diabetic foods due to the small amount of starch contained therein. Soybean diets are valued for acidosis. Soybean oil, with a high proportion of unsaturated fatty acid, is recommended to combat hypercholesteremia. Commercial grades of natural lecithin, often derived from soybean, are reported to contain a potent vasodepressor. Medicinally lecithin is indicated as a lipotropic agent. Soybean is listed as a major starting material for stigmasterol, once known as an antistiffness factor. Sitosterol, also a soy byproduct, has been used to replace diosgenin in some antihypertensive drugs.

CHEMISTRY: Raw seeds of *Glycine max* have been reported to contain per 100 g: 139 calories, 68.2 g moisture, 13.0 g protein, 5.7 g fat, 11.4 g carbohydrate, 1.9 g fiber, 1.7 g ash, 78 mg Ca, 158 mg P, 3.8 mg Fe, 0.40 mg thiamine, 0.17 mg riboflavin, 1.5 mg niacin, and 27 mg ascorbic acid. Sprouts contain per 100 g (edible portion): 62 calories, 81.5% moisture, 7.7 g protein, 1.8 g fat, 8.0 g total carbohydrate, 0.7 g fiber, 1.0 g ash, 52 mg Ca, 58 mg P, 1.1 mg Fe, 30 mg Na, 279 mg K, 25 mg B-carotene equivalent, 0.19 mg thiamine, 0.15 mg riboflavin, 0.8 mg niacin, and 10 mg ascorbic acid. Dried yellow seeds are reported to contain 400 calories, 10.2 g moisture, 35.1 g protein, 17.7 g fat, 32.0 g carbohydrate, 4.2 g fiber, 5.0 g ash, 226 mg Ca, 546 mg P, 8.5 mg riboflavin, and 2.2 mg niacin. The Wealth of India indicated the mineral composition (percentage on fresh-weight basis): 2.09 K, 0.38 Na, 0.22 Ca, 0.0081 Fe, 0.0012 Cu, 0.24 Mg, 0.59 P, 0.02 Cl, 0.0032 Mn, 0.406 S, 0.0022 Zn, and 0.0007 Al; also I, Mo, B, Ni, and Si. Green feed of soy contains 12.56% fiber, 23.7 fiber, 52.1 N-free extract, 2.2 ether extract, 1.9 CaO, 0.57 P_2O_5, 1.4 MgO, and 2.4% K_2O. Hay contains 15.0% crude protein, 29.1 fiber, 42.6 N-free extract, 1.3 either extract, 12.0 total ash, 2.9 CaO, 0.60 P_2O_5, 1.2 MgO, 0.3 Na_2O, and 2.0% K_2O. Soybean straw contains 16.0% moisture, 7.4% protein, 2.0% ether extract, 28.3% N-free extract, 26.1% fiber, and 10.2% fiber. Nutritional analyses of dozens of soybean products appear in the Food Composition Table for Use in East Asia. The sprouts, now popular among health faddists, contain 86.3% water, 6.2 protein, 1.4 fat, 5.3 carbohydrates, 0.8 ash, and, per 100 g sprout, 48 mg Ca, 67 mg P, 1 mg Fe, 180 IU vitamin A, 0.23 mg thiamin, 0.20 mg riboflavin, 0.8 mg niacin, and 33.8 mg vitamin C. Soybean lecithin contains 11.7% palmitic, 4.0 stearic, 8.6 palmitic, 9.8% oleic, 55.0 linoleic, 4.0 linolenic, and 5.5% C22–C20 acids (including arachidonic). A globulin (glycinine) accounts for 80–90% of the total nitrogen protein of the seed. Glycinine contains 1.1% cystine, 1.8 methionine, 5.4 lysine, 1.7 tryptophane, 2.1 threonine, 9.2 leucine, 2.4 isoleucine, 4.3 phenylalanine, 3.9 tyrosine, 2.2 histidine, 1.6 valine, 8.3 arginine, 0.7 glycine, 1.7 alanine, 5.7 aspartic acid, 19.0 glutamic acid, and 4.3% proline. *Glycine* has been reported to contain betaine, choline, guanidine, hydrocyanic acid, isovaleraldehyde, maltose, oxalic acid, saponin, trigonelline, and tryptophane.

DESCRIPTION: Bushy, rather coarse annual herb; stems up to 1.8 m tall, sometimes vinelike, terete toward the base, more or less angled and sulcate to subquadrangular above, gray-brownish or tawny, hirsute to pilose with pale hairs; leaves pinnately trifoliolate, their petioles 2–20 cm long, from subterete and sparsely pilose or glabrescent to strongly angled, sulcate and hirsute, the rachis 0.5–3 cm long; stipules broadly ovate, abruptly acuminate, 3–7 mm long, conspicuously several-nerved, more or less strigose; leaflets membranous, broadly ovate, suborbicular, oval or elliptic-lanceolate, 3–14 cm long, 2.5–10 cm broad, the terminal seldom appreciably larger than the lateral which is usually more or less inequilateral, generally acute, but frequently obtuse and mucronulate, occasionally deltoid-acuminate, tapering to rounded or subtruncate at base, usually sparsely silky-strigose on both surfaces or glabrate above, occasionally rather densely strigose-velutinous below, their petiolules 1.5–4 mm long, usually densely hirsute, stipels narrowly lanceolate to setaceous, 1–3.5 mm long, bracts from broadly to narrowly lanceolate 4.5–5.5 mm long, several-nerved, strigose; racemes axillary, irregular, often leafy, very short, 10–35 mm long, usually rather compactly few (5–8) flowered, the peduncle and pedicels often reduced and concealed by a densely hirsute vesture, the flowers sometimes single or paired in the lower axils; bractlets from broadly to narrowly lanceolate, 2–3 mm long, strigose, caducous; flowers on usually densely hirsute to glabrescent pedicels 0.25–3 mm long;

calyx 5–7 mm long, setose to appressed-hirsute or strigose, the teeth subequal, lanceolate to lanceolate-attenuate, the upper pair generally united to above the middle, the bracteoles setaceous, appressed, setose, 2.5–3.25 mm long; corolla white, pink, greenish blue, violet or purple, 4.5–7 mm long, the standard suborbicular-obovate to subreniform, emarginate, somewhat longer than the narrowly oblong wings which much exceed the keel, porrect or somewhat upturned near the apex; pod oblong, subfalcate, pendent, 25–75 mm long, 8–15 mm broad, coarsely hirsute or setose, the bristly hairs up to 2.5 mm long, yellowish-brown; seeds 2–3 per pod, ovoid to subspherical or irregularly rhomboidal, 6–11 by 5–8 mm, greenish cream or grayish-olive to reddish black, smooth, the caruncle scalelike, membranous, erect or appressed, about one-third to one-half the width of the hilum. Fl. summer; Fr. fall; varying as to locality.

GERMPLASM: Plants extremely variable and many varieties and cultivars named or developed, bred for resistance to diseases, for flowering time control, climatic and edaphic conditions and oil or protein content. Agricultural agents should be consulted for the best local cv. Thirty-six trivial variants have been described as subspecies and varieties, and many horticultural cvs have been developed. Several germplasm collections of soybean are described in Hill (1976). Assigned to the China-Japanese Center of Diversity, soybean or cvs thereof is reported to exhibit tolerance to aluminum, bacteria, disease, frost, fungi, hydrogen flouride, high pH, heavy soil, insects, laterites, limestone, low pH, mycobacteria, nematodes, photoperiod, pesticides, smog, smut, and viruses. ($2n = 40$.)

DISTRIBUTION: Widely cultivated, not known in the wild state. Believed to be a cultigen from *Glycine ussuriensis,* reported to grow in China, Japan, Korea, Russia, and Taiwan.

ECOLOGY: Subtropical plant, but cultivation extends from the tropics to 52°N. In the United States, production is greatest in the corn belt. I observed it as one of the more frequent cultivars at 47°N in Nan Char, People's Republic of China. Does not withstand excessive heat or severe winters. A short-day plant. Requires 5 dm water for good crop. Grows best on fertile, well-drained soils, but tolerates a wide range of soil conditions; pH 6.0–6.5 preferred. Soils must contain the proper nitrogen-fixing bacteria. When grown on the same land for 2–3 successive years, yields increase. Crop suited to a dry zone, to a low or mid-country wet zone or under irrigation. Soybeans grow better than many other crops on soils that are low in fertility, droughty or poorly drained. Many high-latitude cvs do very poorly in low latitude. Ranging from Cool Temperate Moist to Wet through Tropical Very Dry to Wet Forest Life Zones, soybean has been reported to tolerate annual precipitation of 3.1–41.0 dm (mean of 108 cases = 12.8), annual mean temperature of 5.9°–27.8°C (mean of 108 cases = 18.2), and pH of 4.3–8.4 (mean of 98 cases = 6.2).

CULTIVATION: Propagated by seed. Seedbed preparation for soybeans is similar to that for corn or cotton; very thorough cultivation to provide a deep loose seedbed. Most growers prefer fall or early spring plowing to plowing immediately before planting. Weeds should be destroyed by light disking, thorough harrowing or by use of cultivators, immediately before planting, such that weeds do not get ahead of the soybeans. Soil temperatures and day length determine the best time to plant (at or after corn-planting time in most areas). Full-season cvs, which take most of the growing season to mature, produce highest yields when planted with or soon after corn. Rate of seeding varies with area: in northern United States, narrow rows 46–68 cm wide produce the highest yields; in southern United States there is little advantage in rows closer than 90 cm for spring plantings. Soybeans are often planted with planters designed for other crops, but are adapted for soybeans by special plates. They are sometimes planted with a drill in which all the feed cups are covered, except for those needed for row planting. A row planter provides uniform depth of seed (should be 2.5–5 cm). Seeding rate depends on cv or size of seed, width of row and germination of seed. A good rate is 1 seed per 2.5 cm of row. Close spacing encourages rapid growth of soybeans and helps control weeds, but spacings closer than 2.5 cm may seriously increase lodging. Excessive lodging causes difficulty in combining and reduces yields. Seeds are often treated, any time before planting, even in the preceeding fall at harvest, to protect them from soil-borne diseases. Soybeans need inoculation with a commercial culture of nitrogen-fixing bacteria unless the bacteria, which persist for a number of years, are known to be in

Glycine max (L.) Merr.

the soil. Some farmers do not inoculate if nodulated soybeans have been grown there in the past 4–5 yr. Inoculants from other legumes are not effective on soybeans. In the absence of nodulation, soybeans require nitrogen fertilizer for maximum yields. Soybeans fit well into many rotations, with corn, small grains, or other legumes; or in rotation with cotton, corn, or rice; or planted after early potatoes and vegetables, or after winter grain; or planted when grass, clover, or row crops have failed. Fertilizer needs vary with the soil and the cropping system. Soil tests indicate specific needs. Fertilizer often is applied to other crops in the rotation and soybeans may not need additional fertilizer. On soils of low fertility, fertilizers increase yields. If plants are nodulated properly, nitrogen fertilizer is not needed. Fertilizer containing potash is injurious to germination when in direct contact with the seed. Fertilizer may be applied in bands 5–7.5 cm to the side and 5 cm below the seed; or soil and fertilizer may be mixed if 2.5 cm are left between fertilizer and seed. Broadcast fertilizer should be plowed under or disked in. Soybeans are more acid tolerant than other legumes but respond to lime application on acid soils. Weed competition is serious, and may reduce yields by 50%. Early cultivation prevents weeds from becoming established ahead of the soybeans. Both rowed and drilled soybeans can be cultivated effectively with a rotary hoe, drag harrow, or weeder even before soybeans emerge, but plants are easily injured just before and during emergence. After emergence danger of breakage is less if cultivation is done during the hot part of the day. For final cultivation, row cultivating equipment is used. Cultivation should be only deep enough to destroy weeds, and should be discontinued when it damages the plants. In Taiwan, handweeding doubled the yield as compared to the weedy control, (1,600 versus 800 kg/ha). Alachlor (2 kg/ha), chloramben (2), linuron (0.25), and nitrolen (3 kg/ha) also doubled the yield; the first two controlled weed grasses most effectively. Soybeans are usually not grown under irrigation, at least in the United States. A good crop usually requires about 50 cm of water. In most areas where soybeans are grown, moisture is adequate. Soybeans tolerate dry soil conditions before they bloom, but drought during the pod-filling stage seriously reduced yields and seed quality. During this stage, supplemental irrigation is successful. Contour of the land largely determines the type of irrigation. Row or flood irrigation may be used on land that has been leveled and prepared for it. Heavy, infrequent irrigations usually give better results and require less labor than frequent light irrigations. Time between irrigations depends on the type of soil and the weather. Double cropping of soybeans, usually by alternating with a small grain crop such as barley or wheat, has increased in warmer regions of the United States. Typically, the small grain is seeded in fall and harvested in spring. Then soybeans are seeded for the summer growing season. Both conventional plowing and "no till" planting in small grain stubble are widely used. Improvements in planting equipment, herbicides, and development of early-maturing cultivars of small grains have contributed to the increased use of double cropping.

HARVESTING: All seeds on a soybean plant mature at essentially the same time. Then the leaves drop rapidly and stems dry. Combines should be adjusted frequently during the day so as to reduce losses due to splits and mechanical damage to the seeds, which can amount to 10–20% of the crop. Combining seed for planting requires special care to prevent mechanical damage. As seed moisture drops below 12%, germination damage because of mechanical injury increases. The best combine cylinder speed threshes properly but does not crack seed. Soybeans should be stored in clean, dry bins. The most important function of good storage is to control moisture content. Moisture content of every load should not exceed 11% for 1-yr storage period or 10% for 5-yr storage period. Excessive moisture may cause molding, heating, and spoiling. Dry beans do not deteriorate appreciably in quality during 1 year or even longer. Viability deteriorates rapidly if seed is stored beyond the first planting season following harvest. Soybeans make a versatile emergency hay crop because they are adapted to a wide range of planting dates. They should supplement and not substitute for alfalfa, clover or other hay crops. Soybean hay is difficult to cure, and loss of leaves and spoilage during curing may reduce quality. There are, however, many good hay or forage types of soybeans, which usually have fine stems and small, dark-colored seed. When drilled 5–7.5 bu/ha (136–204 kg/ha), seed cvs can equal forage cvs both in quantity and quality of hay production. Time to cut soybean hay ranges from the time the pods begin to form to the time when the seeds reach full size. A widely used guide in harvesting soybean hay is to cut during the first

Glycine max (L.) Merr.

favorable weather after the seeds are half developed. Soybean hay commonly is cured in the swath 1–2 days, then raked into small windrows. Unless drying conditions are good, the windrows may need turning once or twice before the hay is ready to be baled. A roller-crusher attachment on the mower hastens the curing process because crushed stems lose moisture more rapidly than intact stems. Soybeans require 75–200 days to mature seed, depending on cv and region.

YIELDS AND ECONOMICS: Soybean yields average 5 MT/ha for hay, about 1,700 kg/ha (60 bu/ha) for beans. High-yielding cvs, adapted to the locality and grown under proper culture and favorable conditions yield more than twice the average yield. Some farmers have produced yields of more than 125 bu/ha (3,400 kg/ha). In Taiwan, TK-51 irrigated with 200 MT H20/ha yielded 3.1 MT/ha beans and 7.6 MT DM/ha compared with the control with 2.8 MT beans and 7.1 MT DM. World production of soybeans in 1970 was 46,521,000 MT grown on 35,019,000 ha, yielding 1330 kg/ha. North America produced about 31 million MT; Mainland China, 11.5 million MT; Latin America, 1.9 million MT; South America, 1.1 million MT; Asia, 1.2 million MT. Economically, this is the most important grain legume crop in the world. World production in 1975 was 68,356,000 MT on 46,463,000 ha, averaging 1471 kg/ha. North America led with 42,317,000 MT, of which the US, averaging 1909 kg/ha, produced 41,406,000 MT. Asia excluding Russia produced 13,727,000 MT, averaging 828 kg/ha. South America produced 11,109,000 MT, averaging 1759 kg/ha. Russia produced 600,000 MT, averaging 750 kg/ha. Europe produced 442,000 MT, averaging 1369 kg/ha, Africa 96,000 MT, averaging 482 kg/ha, Oceania 64,000 MT, averaging 1404 kg/ha. China was estimated to be second to the United States, producing 12,062,000 MT, Brazil third with 10,200,000 MT, Indonesia fourth with 560,000 MT, and Mexico fifth with 545 MT. New Zealand's reported yields were highest at 3,000 kg/ha, Canada 2,322 kg/ha, Paraguay 2,160 kg/ha, Turkey 2,000 kg/ha, and Colombia 1,913 kg/ha. In 1975 the effective demand for oilcake was growing faster than that for fats and oils. Soybean oil production was up 19% in 1970 over 1969, from 5.7 to 6.8 million MT. Also prices were up as much as 46%. On a protein cost-per kilogram basis, soybeans are a cheap source of protein. Recent development in the utilization of soybean protein in the form of concentrates, isolates, and textured protein for human consumption offers a possible solution to the world's protein needs.

BIOTIC FACTORS: Insects known to attack soybeans include corn earworms, Mexican bean beetles, bean leaf beetles, velvetbean caterpillars, lesser cornstalk borers, stink bugs, and other insects. Occurrence, prevalence and rate of reproduction of soybean insects vary greatly from one part of the country to another. All insects can be controlled by timely dusting or spraying with the proper insecticide. Local agricultural agents should be consulted for advice. Mexican bean beetles are said to ignore soybeans when snapbeans are planted nearby. The more important fungal diseases of soybeans are: *Alternaria* sp. (leaf spot), *Cephalosporium gregatum* (brown stem rot), *Cercospora kikuchii* (purple seed stain), *Cercospora sojina* (frogeye leaf spot), *Corynespora cassiicola* (target spot), *Diaporthe phaseolorum* var *caulivora* (stem canker), *Diaporthe phaseolorum* var *sojae* (pod and stem blight), *Erysiphe polygoni* and *Microsphaera diffusa* (powdery mildews), *Fusarium orthoceras* (root rot), *Glomerella glycines* and *Colletotrichum truncatum* (anthracnose), *Macrophomina phaseoli* (charcoal rot), *Melanopsichium missouriense* (soybean smut), *Nematospora coryli* (yeast spot), *Peronospora manshurica* (downy mildew), *Phakopsora pachyrhizi* (soybean rust), *Phyllosticta sojicola* (Phyllosticta leaf spot), *Phymatotrichum omnivorum* (root rot), *Phytophthora megasperma* (Phytophthora rot), *Pythium ultimum* and *P. debaryanum*, *Rhizoctonia leguminicola* (black patch), *Rhizoctonia solani*, *Sclerotinia sclerotiorum* (stem rot), *Sclerotium rolfsii* (blight), and *Septoria glycines* (brown spot). Virus diseases include: soybean mosaic, bud blight, and yellow mosaic. Insect vectors are known for all but one of the important soybean viruses, aphids being the most important vector. Types of nematodes attacking soybeans include: sting (*Belonolaimus longicaudatus*), ring (*Criconemoides*), spiral (*Helicotylenchus*), lance (*Hoplolaimus*), pin (*Paratylenchus*), root lesion (*Pratylenchus*), stubby root (*Trichodorus*), and stunt (*Tylenchorrhynchus*). In Taiwan, Thailand, and eastern Australia, soybean rust is the most important disease. As the soybean spreads, soybean rust could spread to new areas (e.g., the United States). Direct relation was demonstrated between the amount of carbohydrate exuded by germinating soybeans and seed rot caused by *Pythium*. In the United States, yield

Glycine max (L.) Merr.

losses of as much as 10% may be caused by nematodes; rootknot nematodes (*Meloidogyne* spp.) cause 4%, soybean cyst nematodes (*Heterodera glycines*) 4%, and other nematodes ca. 2%. As many as 50 species from 20 genera are reported to feed on soybeans. Disease-resistant cvs of soybeans have been developed and are available in most production areas. The use of disease-resistant cvs is the most effective means of reducing losses from disease. Also available are cvs that resist rootknot and cyst nematodes. The use of resistant host plants may be the most desirable and ecologically sound method for managing plant-feeding insect populations. On soybean cvs with normal pubescence, high populations of the potato leafhopper *Empeasca fabae* do not develop. The major insect pests for which resistance has been found include the velvetbean caterpillar [*Anticarsia gemmatalis* (Hubner)], Mexican bean beetle [*Epilachna varivestis* (Mulsant)], tobacco budworm [*Heliothis virescens* (Fabricus)], corn earworm [*Heliothis zea* (Boddie)], green cloverworm [*Plathypena scabra* (Fabricus)], and the soybean looper [*Pseudoplusia includens* (Walker)].

REFERENCES:
Asian Vegetable Research and Development Center, 1976, *Soybean Report '75*, Shanhua, Taiwan, Republic of China, 68 pp.
Caldwell, B. E., Howell, R. W., Judd, R. W., Johnson, H. W. (eds.), 1973, *Soybeans: Improvement, Production, and Uses*, Argonomy Series No. 16, American Society of Agronomy, Inc., Madison, Wisc., 681 pp.
Hill, L. D. (ed.), 1976, *World Soybean Research,* The Interstate Printers and Publishers, Inc., Danville, Ill., 1073 pp.
Johnson, H. W., *et al.,* 1967, Growing soybeans, *U.S. Dep. Agric., Farmers Bull.* **2129**:1–10, illus.
National Soybean Corp Improvement Council, 1966 (rev. ed.), *Soybean Farming*, 34 pp., illus., Urbana, Ill.
Strand, E. G., 1948, Soybeans in American farming, *U.S. Dep. Agric. Tech. Bull.* **966**:1–66, illus.

CONTRIBUTORS: T. E. Devine, J. A. Duke, C. F. Reed, R. J. Summerfield.

Glycine wightii (R. Grah. ex Wight & Arn.) Verdc.

FAMILY: Fabaceae
COMMON NAME: Perennial soybean
SYNONYMS: *Notonia wightii* Wight & Arn.
Johnia wightii Wight & Arn.
Glycine javanica auctt. mult.

USES: Plant extensively cultivated in tropical Africa, Asia, and Brazil for forage. Grown as a cover, green manure or forage, fed green or as hay. In Malawi, the leaves are cooked as a potherb.

FOLK MEDICINE: No data available.

CHEMISTRY: Hay grown in a soil transitional between gray-hydromorphic and red-yellow podzolic in Brazil had higher crude protein content in early stages than alfalfa with a similar crude fiber content.

Date cut	Stage cut	Crude protein (%)	DM (%)	Ether extract (%)	Crude fiber	Ash	N-Free extract (%)
Feb. 10	Vegetative	19.60	89.12	2.69	31.67	6.99	31.61
Apr. 20	Floral initiation	14.21	91.34	1.77	30.62	7.77	36.97
Jul. 10	Fl. and green Fr.	11.20	90.81	2.81	44.79	6.25	30.76

Indian analyses of dry hay give 17.1% protein, 36.6 crude fiber, 1.4 ether extract, 12.8 ash, with 57.07% total digestible nutrients.

DESCRIPTION: Perennial climber or trailer, often woody at base, 0.6–4.5 m long; rootstock often thick and woody; stems glabrescent to densely velvety with usually ferrugineous hairs: leaves pinnately trifoliolate, their petioles 2–12 cm long, more or less sulcate and reflexed-hirsute, the rachis 2–20

Figure 44. *Glycine wightii* (R. Grah. ex Wight & Arn.) Verdc.

Glycine wightii (R. Grah. ex Wight & Arn.) Verdc.

mm long, hirsute to glabrate, the stipules ovate, 1.2–5 mm long, strongly several-nerved, strigose; leaflets thin, oval to ovate, 3–10 cm long, 1.5–7 cm broad, acute to acuminate or broadly obtuse and mucronate at apex, abruptly tapering to rounded or almost truncate at base, more or less strigose to glabrate above, the margins and veins sometimes densely strigose, generally pilose to silky-strigose and prominently veined beneath, the secondary veins 4–7 pairs, usually hirtellous or strigose, the stipels setaceous, 1–3.5 mm long, bracts persistent, ovate, 3–4 mm long, strigose to glabrate; racemes axillary, very compact, stout, oblong-ovoid, often almost headlike, densely flowered (70–150 fls), and tapering to an acute apex during the early stages of anthesis to somewhat interrupted as the rachis elongates, 2–18 cm long, on peduncles 3.5–5.5 mm long, appressed-pilose; bractlets setaceous, 2.5–3.5 mm long, hirtellous, flowers exceeded by the unusually long fascicle-bracts and setaceous calyx-teeth, on short glabrous to minutely sericeous pedicels, 0.5–1 mm long, somewhat nodose at base; calyx 3.75–5.5 mm long, sparsely strigillose to hirsute, the subequal, setaceous or lanceolate-setaceous teeth about 3 times the length of the tube; approximating the calyx in length, but sometimes even shorter, the standard suborbicular, 4–6 mm long, the claw one-third the length of the blade, auriculate, the wings oblong, 3–5 mm long, auriculate, with blade and claw subequal, the claw of the subequal keel nearly equaling the oval, truncate-based blade; pod oblong-linear, 2–3 cm long, 4–5 mm broad, transversely septate and more or less constricted between the seeds, margined, more or less densely hirsute or strigose, the spreading or appressed hairs reddish-brown; seeds 3–5, short-oblong to subquandrangular, biconvex with flattened sides, 2.25–3.5 mm by 1.5–2.5 mm, reddish brown, minutely granular, the caruncle liguliform, scarious, white, generally recurved.

GERMPLASM: Several subspecies and varieties are recognized, especially in tropical Africa where the main distribution of the species occurs; *G. wightii* subsp. *wightii* var *longicaudia* (Schweinf.) Verdc. (*G. longicaudia* Schweinf., *G. moniliformis* Hochst. ex A. Rich. and *G. micrantha* Hochst. ex A. Rich.) is the most plentiful and widespread form in tropical Africa, differing from typical plants in having the stems often much branched, the racemes more or less interrupted, many (20–100) but rather loosely flowered, 3–25 cm long, on peduncles 2–6 cm long, pods linear, 3–5 mm broad, and seeds 3–7 per pod, greenish to straw-colored to dark purplish-brown, sometimes mottled; *G. wightii* sp. *petitiana* (A. Rich.) Verdc. (*Johnia petitiana* A. Rich.) from Ethiopia; *G. wightii* subsp. *petitiana* var *mearnsii* (DeWild.) Verdc. from Congo, Uganda, and southern Ethiopia; and *G. wightii* sp. *pseudojavanica* (Taub.) Verdc. from Zaire, Uganda, Kenya, Tanzania, and Angola. Assigned to the African Center of Diversity cvs of this species are reported to tolerate grazing, low pH, salt, shade, slope, savannah, and weeds have been reported. Attempts to cross this with soybean have been unsuccessful. ($2n = 22, 44$.)

DISTRIBUTION: Native mainly in Tropical Africa (from Zaire to Uganda, Kenya, Tanzania, and Angola) and Asia (India, Ceylon, Java, and Malaysia). Cultivated in tropical America.

ECOLOGY: Twining or scrambling over shrubs or small trees, sometimes completely covering them, on edges of woods or in thickets, on roadsides, and in grassy or fallow fields. Sometimes found in secondary forests, gallery forests, savannas, or in cultivated fields. Adapted to a tropical climate from sea level up to 2,000 m in altitude. Often on plains in India and Ceylon. Ranging from Subtropical Dry to Moist through Tropical Moist to Wet Forest Life Zone, this perennial soybean has been reported from areas with annual precipitation of 5.3–42.9 dm (mean of 15 cases = 13.5), annual mean temperature of 18.0°–27.4°C (mean of 15 cases = 21.5), and pH 5.0–7.1 (mean of 13 cases = 6.0).

CULTIVATION: Propagated from seeds or rooted cuttings of woody stems. In most areas plants are self-seeding. For new stands, seeds are scattered about the edge of forests or thickets, and plants soon cover the entire area. It may be mixed with such grasses as *Panicum maximum* and *Pennisetum purpureum*.

HARVESTING: Vegetative parts of plant cut for forage throughout the year as needed. Hay from the vegetative and start-of-blooming-stages have higher crude protein content than alfalfa with the same crude fiber content. Digestible crude protein content was highest and nutritive ratio was narrowest for hay cut in the vegetative stage. Hay from

Glycine wightii (R. Grah. ex Wight & Arn.) Verdc.

perennial soybean cut at any stage produced weight gain in all animals fed the hay.

YIELDS AND ECONOMICS: Hay yields of 7 MT/ha are reported. In Florida, mixed with grasses, yields compare favorably with those of siratro, averaging 8–10 MT/ha (A. E. Kretschmer, unpublished data). An important forage plant in the hot tropical areas of Africa, India, and Ceylon, and to a lesser degree in Malaysia and Brazil.

BIOTIC FACTORS: A Kenyan study showed that most of 43 introductions behaved as self-pollinators, while several tetraploids behaved as cross pollinators. Fungi causing diseases in perennial soybean include the following: *Cercospora sojina*, *Cercosporella* sp., *Dactuliophora glycines*, *Ascochyta sojaecola*, *Aecidium glycines*, *Meliola bicornis*, and *Synchytrium dolchi*. It is also attacked by the Alfalfa mosaic virus, Glycine strain. Nematodes isolated from this plant include *Meloidogyne incognita acrita* and *M. javanica*.

REFERENCES:
Hermann, F. J., 1962, A revision of the genus *Glycine* and its immediate allies, *U.S. Dep. Agric. Tech. Bull.* **1268**:1–82, illus.
Hymowitz, T., 1970, On the domestication of the soybean, *Econ. Bot.* **23**:408–421.
Lima, C. R., and Santo, S. M., 1972, Nutritive value of hays from different growth stages of perennial soybean (*Glycine javanica*), *Resq. Agropec. Bras. Ser. Zootec.* **7**:59–62.
Verdcourt, B., 1966, A proposal concerning *Glycine* L., *Taxonomy* **15**:34–36.

CONTRIBUTORS: J. A. Duke, A. E. Kretschmer, Jr., C. F. Reed.

Figure 45. *Glycyrrhiza glabra* L.

Glycyrrhiza glabra L.

FAMILY: Fabaceae
COMMON NAMES: Common licorice, Licorice root, Spanish licorice root

USES: Licorice, grown primarily for its dried rhizome and roots, is used as a condiment and to flavor candies and tobaccos. Roots contain glycyrrhizin, which is 50 times as sweet as sugar. In India, it is chewed with betel. Spent licorice is used in fire-extinguishing agents, to insulate fiberboards, and as a compost for growing mushrooms; also in feed for cattle, horses, and chickens. The bulk of licorice production is used by the tobacco industry and for the preparation of licorice paste, licorice extract, powdered root, and mafeo syrup.

FOLK MEDICINE: It is considered demulcent, expectorant, estrogenic, and laxative. Used in treating Addison's disease, as a tonic and blood purifier, for internal inflammations, topical dressings, thirst, colds, and sore throats. Also said to be used for alexeritic, alterative, appendicitis, asthma, bladder ailments, bronchitis, cough, deodorant (lf), depurative, diabetes, diuretic, dropsy, emollient, eyes, febrifuge, pectoral, refrigerant, scalds (lf), stomach ulcer, sudorific, and tuberculosis. With a long history of use for indigestion and inflamed stomach, licorice provides two derivatives which reduce or cure ulcers.

CHEMISTRY: *Glycyrrhiza* has been reported to contain saponin and tannic acid, and the seeds to contain trypsin inhibitors and chymotrypsin inhibitors. *Glycyrrhiza glabra* contains 2-β glucuronosyl glucuronic acid, glycyrrhizin, and isoliquiritigenin-4-glucoside. One analysis showed 20% moisture, 12–16% glycyrrhizin, 8% reducing sugars, 8% non-reducing sugars, 30% starch and gums, 5% ash, and 13–17% undetermined. *Warning:* Excessive licorice ingestion can lead to cardiac dysfunction and severe hypertension (Lewis and Elvin-Lewis, 1977).

DESCRIPTION: Erect, branching, perennial herb, to 1 m tall, with a branching rhizome; leaves alternate, pinnate, with 9–17 ovate, entire, pubescent leaflets, with minute stipules; flowers axillary, in many-flowered racemes or spikes, zygomorphic, 1.3 cm long, lavender to violet, white or bluish; pods 7–10 cm long, ovate or oblong-linear, compressed, scarcely dehiscent, 1–6-seeded, smooth or prickly; seeds few, reniform. Fl. June–July.

GERMPLASM: So-called *"typica"*—Spanish, Italian, Greek—commands highest price because of lack of bitterness; so-called *"glandulifera"*—Russian, Anatolian; *"violacea"* Iraq; *"typica* X *violacea"*—Syrian. Assigned to the Mediterranean and Eurosiberian Centers of Diversity, common licorice or cvs thereof is reported to exhibit tolerance to drought, grazing, high pH, heavy soil, heat, mycobacteria, salt, slope, waterlogging, and weeds. ($2n = 16$.)

DISTRIBUTION: Native to Eurasia, northern Africa, and western Asia. Widely naturalized in temperate regions. The Oriental type of licorice root is grown in the Near East, Syria, Iraq, Caucasus of Russia, Iran, and northern China; the Spanish type in Spain, Italy, and Greece. Modern plantations have been successful in California and Yorkshire.

ECOLOGY: Licorice grows best in dry, sunny, hot climate, faring best in deep, rich, heavy river-bottom soils. Requires annual rainfall of 50–65 cm, and adequate soil moisture; overflow does not harm crop. A definite dry season seems beneficial. Ranging from Boreal Moist through Subtopical Dry Forest Life Zone, this species has been reported to tolerate annual precipitation of 4.0–11.6 dm (mean of 13 cases = 6.5), annual temperature of 5.7°–25.0°C (mean of 13 cases = 13.9), and pH of 5.5–8.2 (mean of 10 cases = 7.1). Often becomes a weed. Attempts are made to eradicate it so the soil may be used for more productive crops.

CULTIVATION: Licorice may be grown from seed, but usually from cuttings, suckers or crown divisions. In spring, crown divisions are set ca. 45 cm apart in rows wide enough to permit cultivation. New plantings must be watered adequately. The crop is ready to harvest in 3–5 yr when rhizomes and roots have developed an extensive system 2–3.5 m deep and several m wide. Attempts to propagate by seed in Kashmir were unsuccessful. Cuttings take fairly well. Crop should be weeded occasionally and may be grown with a catch crop, such as cabbage, carrot, or potato.

HARVESTING: Roots are harvested by hand, preserving the crown for replanting. In the Near East, root is usually dug during the rainy season (Oct.–Apr.), stacked, dried and baled. In other areas, only the offshoot roots are dug, leaving the taproot to regenerate in 2–3 yr. Roots are sorted, cleaned, and weighed. The long straight roots, trimmed of the rejected small branches, are washed, cut into lengths, then piled in huge stockpiles for curing through the following summer until dry enough to bale. Stacks are covered during rainy periods and frequently turned over. Dried roots are pressed into 125-kg bales and shipped. In the United States roots are stored in warehouses near the source. Different brands of licorice extract are made by blending roots gathered from different regions. Many American manufacturers have established extraction plants where the root is grown and collected (mainly in Iran) and import the concentrate. For licorice extract, roots and rhizomes are ground into pulp and boiled in water, and the extract is concentrated by evaporation.

YIELD AND ECONOMICS: Unprocessed roots range from 2,000 to 4,000 kg/ha. Yields as high as 22,000 kg/ha of air-dry root (50 MT green root) have been reported from Russia. Because the plant may be harvested only once every 3–5 yr, production is not always lucrative. Recent Russian literature suggests that cultivated licorice produces 2–3 times as much root as natural stands. In 1957, the United States imported 133,500 kg of licorice root extract, mostly from Spain and Turkey. In 1958, the United States imported ca. 1,450,000 kg licorice, from

Glycyrrhiza glabra L.

sia, Greece, and Iraq. Within the last decade, whole dried root commanded ca. $0.60/kg, granulated root from $0.60 to $0.90/kg, and powdered root $0.75 to $0.98/kg.

BIOTIC FACTORS: The following fungi attack the licorice plant: *Cercospora cavarae, Diplodia glycyrrhizae, Leveillula leguminosarum, L. taurica, Microsphaeria diffusa,* and *Uromyces glycyrrhizae.*

REFERENCES:
Edmonds, P., 1968, Licorice: Primitive harvesting, modern manufacture, *Tobacco* Dec. 27, 1968.
Houseman, P. A., 1944, Licorice root as a chemurgic crop for America, *Chem Dig.* **3**(6):84–88, illus.
Houseman, P. A., and Lacy, H. T., 1929, The licorice root in industry, *Ind. Eng.* **Oct.**:915–917, illus.
Molyneux, F., 1975, Licorice production and processing, *Food Technol. Aust.* **June**:231–234.
Penick Research Dept., 1958, *Glycyrrhiza and Its Derivatives (A Bibliography)*, 16 pp., S. B. Penick, New York.
Walker, W. W., 1952, Licorice: Dark mystery of industry, *Atlantic Monthly* **Nov.**:23–26.

CONTRIBUTORS: J. A. Duke, C. F. Reed, J. K. P. Weder.

Glycyrrhiza lepidota Pursh

FAMILY: Fabaceae

COMMON NAME: American licorice

USES: Roots have an agreeable sweet flavor and were used, especially by American Indians, as a spice and masticatory. Used much like licorice. A good soil binder, but potentially a noxious weed.

FOLK MEDICINE: Licorice is a favorite cold and cough medicine.

CHEMISTRY: Roots contain ca. 6% glycyrrhizin.

DESCRIPTION: Perennial herb, 6–9 m tall; stems erect, leafy, arising from stout roots, herbage glandular-viscid; leaves alternate, pinnate, with minute stipules; petioles short; leaflets 15–19, oblong-oblanceolate, mucronate-pointed, sprinkled with small scales when young and with corresponding dots when mature; flowers in many-flowered, dense, spikelike racemes, short-peduncled; calyx 2-lipped, the upper lip nearly entire, the 3 lower lobes not coalescent so high; corolla whitish or pale yellow; stamens diadelphous, the uppermost free, 9 filaments fused, alternating large and small; pod dry, indehiscent or slightly dehiscent, laterally compressed, oblong, 12–15 mm long, covered with hooked prickles, with 2–6 seeds. Fl. Apr.–Aug.

Figure 46. *Glycyrrhiza lepidota* Pursh

GERMPLASM: Assigned to the North American Center of Diversity, American licorice or cvs thereof is reported to exhibit tolerance to sand, slope, and weeds. ($2n = 16$.)

DISTRIBUTION: British Columbia to California and Mexico, east to Arizona, Texas, Missouri, and Arkansas, northeast to New York and Ontario.

ECOLOGY: Frequent in alluvial and sandy soils, often in stream beds or damp roadside ditches. Grows under various climatic and soil conditions in temperate North America. Ranging from Cool Temperate Steppe to Tropical Moist Forest Life Zone, American licorice is reported to tolerate annual precipitation of 3.5–5.3 dm (mean of 4 cases

= 4.4), annual mean temperature of 6.9°–8.4°C (mean of 4 cases = 7.5), and pH of 7.3–7.5 (mean of 3 cases = 7.4).

CULTIVATION: Not known to be cultivated.

HARVESTING: Long fleshy roots gathered and eaten by Indians and some settlers.

YIELDS AND ECONOMICS: No yield or economic data avialable.

BIOTIC FACTORS: Following fungi are known to attack American licorice: *Cylindrosporium glycyrrhizae, Erysiphe polygoni, Microsphaera diffusa, Septoria glycyrrhizae,* and *Uromyces glycyrrhizae.*

CONTRIBUTORS: J. A. Duke, C. F. Reed.

Hedysarum coronarium L.

FAMILY: Fabaceae

COMMON NAMES: Sulla, Sulla sweetvetch, Spanish or Italian sainfoin, French honeysuckle

USES: Sulla or sulla sweetvetch is grown extensively in countries bordering the Mediterranean for forage and green manure. It is fed green or as hay, comparable in nutritive value to red clover. Chief value of sulla lies in its use as a soil-improving crop. Introduced into India as an ornamental.

CHEMISTRY: Sulla forage, from plants in full flower, contains 2.26% crude protein, 0.38% crude fat, 5.16% crude fiber, 1.78% ash, and 7.42% N-free extract containing galactans and arabans.

DESCRIPTION: Small, deep-rooting perennial or biennial herb; stems robust, 30–120 cm tall, branched, sparsely appressed-pubescent; leaves pinnately compound, with 3–5 pairs of leaflets, 15–35 mm long, 12–18 mm broad, elliptic to obovate orbicular or oval, glabrous or subglabrous above, pubescent beneath; flowers in crowded racemes of 10–35, ca. 2 cm across; calyx sparsely to densely pubescent, the teeth about as long as the tube; corolla 12–15 mm long, bright red-purple, carmine or rarely white; pods divided transversely

Figure 47. *Hedysarum coronarium* L.

into 2–6 segments which separate at maturity, smooth yellow surface with rough sharp projections, turning brown at maturity; seeds 2–2.5 mm in diameter, discoid. Fl. Apr.–July.

GERMPLASM: Assigned to the Mediterranean center of diversity, sulla or cultivars thereof has been reported to exhibit tolerance to frost, limestone, slope, and virus. ($2n = 16$.)

DISTRIBUTION: Native to north temperate regions of the world, especially in central and western Mediterranean regions; introduced in North America, Australia, and India.

ECOLOGY: Ranging from Boreal Moist through Tropical Dry Forest Life Zone, sulla has been reported to tolerate annual precipitation of 4.6–23.6 dm (means of 12 cases = 10.0), annual temperature of 5.7°–29.9°C (mean of 12 cases = 16.5), and pH of 5.5–8.0 (mean of 7 cases = 6.6). At home in rich

Hedysarum coronarium L.

pastures and cultivated ground, sulla adapts easily to deep, calcareous, well-drained soils. Plants withstand light frost, but usually killed back by heavy freezing. Thrives in Mediterranean climates. Recommended for Tunisia, growing well on marls and clay limestone where the annual precipitation is 4.5–6 dm.

CULTIVATION: Propagated by seeds or by layering. A stand is difficult to establish, partly because of poor seed-bed preparation, hard seed, and lack of inoculation. Hard seed may be scarified with wet sand between two hard surfaces. Boiling seeds for 3–5 min also softens the seed coats sufficiently for prompt germination. Inoculation should be provided. Soil from inoculated plants of sulla is not usually available, but pure cultures usually are available from Agricultural Stations. Crop is still experimental in the United States. Trials should be made to determine the best regional planting dates, cultivation methods, and harvest times. Most agronomic processes are similar to those used for alfalfa.

HARVESTING: With irrigation, hay may be cut three times a season. Under ordinary conditions, however, it is less satisfactory than alfalfa. Seed could be harvested locally from successful stands and could then be flailed out by hand increase.

YIELDS AND ECONOMICS: Average yields of green forage range from 100–125 MT/ha under irrigation, and 15–40 MT/ha rainfed. Major production areas are in the countries bordering the Mediterranean and in parts of Australia.

BIOTIC FACTORS: Following fungi cause diseases in sulla: *Anthostomella lasullae, A. sullae, Cercospora arimimensis, Erysiphe martii, E. polygoni, Leptosphaeria circinans, Phoma hedysari, Placosphaeria onobrychidis, Pleospora herbarum,* and *Uromyces appendiculatus*. It is also attacked by nematodes of the genus *Meloidogyne*.

REFERENCE:

Come, D., and Semodeni, A., 1973, Gases released from seed coats during imbibition. I. Application to the study of the hardness of *Hedysarum coronarium* L. seeds. (Fr), *Physiol. Veg.* **11**:171–177.

CONTRIBUTORS: J. A. Duke, C. F. Reed.

Indigofera arrecta Hochst. ex A. Rich.

FAMILY: Fabaceae

COMMON NAMES: Natal, Bengal, or Java indigo

USES: This species has been a chief source of the blue dye, indigo. The indigotin content of plant is 0.8–1.0%, and leaves contain up to 4% of the flavanol glycoside, kaempferitrin, that on hydrolysis yields rhamnose and kaempferol. It is also grown as a cover crop or intercrop in cocoa, coffee, oil palm, tea, and rubber plantations, and as a green manure crop for rice. Recommended in rotations before cotton. Manurial constituents include N, P, K, and Ca. It is often grown in closed forests as green manure crop. Plants give shade and protection as well as keep down weeds and improve the soil by their plentiful supply of nodules. Plants give denser shade, last longer and endure dry weather better than pigeon-pea.

FOLK MEDICINE: Said to be used for antiseptic, astringent, bladder, bugbites, epilepsy, hysteria, liver, snakebite, spleen, and ulcer.

CHEMISTRY: Leaves contain the following minerals: N, 4.46; P_2O_5, 0.02; K_2O, 1.95; and CaO, 4.48%. The genus *Indigofera* is reported to contain hydrocyanic acid and indican, and other poisonous nitrogenus compounds. In 26 species of *Indigofera*, the Poisonous Plant Research Laboratory of the USDA found toxic organic nitro compounds in many, if not most species.

DESCRIPTION: Stout, rather woody, herb or undershrub, 1–2 or rarely up to 3 m tall; perennial, but

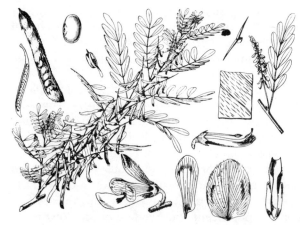

Figure 48. *Indigofera arrecta* Hochst. ex A. Rich.

Indigofera arrecta Hochst. ex A. Rich.

often cultivated as an annual; stems erect, somewhat ridged, rather densely strigulose, the hairs white or brown; stipules subulate-setaceous, 2–9 mm long; rachis strigulose, up to 6 cm long, including petiole of up to 1.5 cm; stipellae subulate, less than 1 mm long; leaves blue-green, imparipinnately compound, 10–12.5 cm long; leaflets 7–8 pairs with one odd terminal leaflet, narrowly elliptic-oblong, up to 2 cm long, 7 mm broad, strigulose beneath, usually glabrous above; racemes many-flowered, axillary, sessile, up to 5 cm long, but usually shorter; bracts lanceolate, ca. 1 mm long, caducous; pedicels about 1 mm long, strongly reflexed in fruit; calyx brown strigulose, ca. 1.5 mm long; calyx-lobes triangular, as long as tube; corolla brown strigulose outside; stamens 3–4 mm long; pod brown, straight, slightly tetragonal, somewhat torulose, but not in the plane of the sutures, 12–17 mm long, about 2.2 mm broad, up to 2.8 mm thick; seeds 4–6 per pod, shortly oblong, rhombic in cross section.

GERMPLASM: Some variations have been noted, namely those plants that grow in lowlying arid places and others that have leaflets strigose above instead of glabrous. Assigned to the Africa Center of Diversity, natal indigo or cultivars thereof has been reported to exhibit tolerance to shade and weeds. ($2n = 16$.)

DISTRIBUTION: Native throughout tropical Africa, South Africa (Transvaal, Natal), southern Arabia, and Madagascar. Introduced into India east through Java. Cultivated extensively in various areas of India, Assam, and Java. Its range has been greatly extended because of its cultivation for indigo.

ECOLOGY: Native in deciduous and upland evergreen bushland, areas of cultivation and secondary growth and along forest margins. Plants require a hot, moist climate, with rainfall not under 175 cm yr, although it is grown in some areas with rainfall from 40–180 cm per annum. In Africa it grows from 300–2,700 m in altitude. Crop cannot tolerate more than 2 months of constant moisture in soil. It grows well on well-drained, light, friable soils. Thorough tillage is necessary. Ranging from Warm Temperate Moist through Tropical Moist Forest Life Zone, natal indigo has been reported to tolerate annual precipitation of 8.7–42.9 dm (mean of 5 cases = 20.8 dm), annual mean temperature of 18.7°–27.4°C (mean of 5 cases = 23.6°C), and pH of 5.0–6.5 (mean of 3 cases = 5.8).

CULTIVATION: Propagation is by seed which are produced in large quantities; germination is usually uniform and good. Seeds have hard seed coats and for good germination, must be treated. Usually scarified in a machine or by treatment with concentrated sulfuric acid. Seed sown in rows on well-plowed, manured, level soil. In India usually sown in mid-October, at rate of 6–8 kg/ha; plants spaced ca. 60 cm apart. Germination takes 3–4 days. Crop matures early and withstands submersion under water; grows vigorously and stands lopping better than *Crotalaria* spp. Production of seed is usually poor in plants that have been cut once or more for leaf; crops must be grown exclusively for seed. For seed purposes, sowing is best done in mid-August, in rows 60 cm apart, and crop is left uncut until seeds are fully mature.

HARVESTING: For the leaf crop, plants are ready for first cut by May-June in India. Plants are cut from about 2.5 cm to 15 cm aboveground. Additional crops may be cut at intervals of 6 weeks or more depending on growing conditions. Plants flower in about 3 months from seed, and are ready to harvest. Cut plants are tied in bundles and carried fresh to the factory. Stumps ratoon and form 2–4 crops on same roots within a year. Plants are best grown in rotation with other crops. For dye, green crop is placed in tanks, weighted down with planks and covered with water for fermentation for 12–16 hr, or until leaves become a pale color. Liquid is run off and constantly stirred for 2–3 hr, then the indigo settles to bottom as a bluish mud. This, after water is drained is put into bags that are hung to dry. Afterward, it is cut into cubes ca. 7.5 cm square and is stamped and further dried for export.

YIELDS AND ECONOMICS: Plants yield heavily; with superphosphate manuring, yields of 22–100 MT green matter have been reported in India. Yield of leaves is higher in this than for any other species of *Indigofera*. Recorded output of indigo cake is 137–325 kg/ha. Seed yields of 675–900 kg/ha have been obtained in Bihar, and up to 1,200 kg/ha in Assam. Major producers are India and Java. With advent of aniline dyes demand for indigo became

Indigofera arrecta Hochst. ex A. Rich.

less. In some areas, however, it is grown for other purposes.

BIOTIC FACTORS: Following fungi have been reported on this species: *Diplodia inocarpi, Gloeosporium inocarpi, Hypochnus centrifugus, Nematosporangium indigoferae, Neocosmospora vasinfecta, Parodiella perisporioides, Phoma inocarpi, Pythium indigoferae, Ravenelia marganguensis, Rhizoctonia solani, Rosellinia bunodes, Uredo maranguensis,* and *Uromyces orientalis.* It is also attacked by the bacterium, *Bacillus solanacearum.* Nematodes isolated from indigo include: *Heterodera marioni, Meloidogyne incognita acrita,* and *M. javanica.*

CONTRIBUTORS: J. A. Duke and C. F. Reed.

Indigofera hirsuta L.

FAMILY: Fabaceae

COMMON NAME: Hairy indigo

USES: Hairy indigo is a valuable cover crop, used extensively for soil improvement. Manurial constituents of leaves are nitrogen, phosphoric acid, potash and lime. In West Africa it is a source of indigo dye. In Florida row-crop fields, it is considered a weed (controlled by 2.2 kg/ha napropamide presowing watermelon), but has been used as a cover crop in citrus, and in mixtures with permanent grasses for grazing.

FOLK MEDICINE: A decoction of the leaves is used in Ghana (Gold Coast), as a lotion for yaws and, in Philippines, for diarrhea and as stomachic.

CHEMISTRY: The leaves contain: 2.14 N, 0.29 P_2O_5, 1.84 K_2O, and 4.25% CaO. Hay from plants cut at the flowering stage shows the following values: 10.68 moisture, 13.65 crude protein, 1.41 crude fat, 46.04 N-free extract, 21.00 fiber, and 7.22% ash; digestibility coefficients: dry matter 62.5; protein, 67.0; fat, 61.0; N-free extract, 67.0; and fiber, 53.5%; digestible nutrients: total digestible nutrients, 53.15; and digestible protein 9.14%; starch value, 40.13%; and nutritive ratio, 4.8 (WOI). The plant, possibly poisonous, can severely irritate hooves of animals that graze for extended periods.

Figure 49. *Indigofera hirsuta* L.

Although the plant is used for grazing, the possibility of toxicity should be studied.

DESCRIPTION: Summer annual or biennial herb; stems coarse, erect, or spreading, 0.6–2.3 m tall, becoming woody with age, with long spreading gray or brownish pubescence; leaves pinnate; stipules threadlike, short-petioled; blades 5–12.5 cm long, with 5–11 large obovate leaflets; leaflets 2.5–5 cm long; flowers very small, red or pinkish, in dense long-peduncled axillary racemes, exceeding the leaves in length; pods narrow, 1.3–2 cm long, crowded together densely along stem, densely brown pubescent, pointing downwards. Fl. summer; Fr. Oct. Seed 441,000/kg; wt. kg/hl = 71.

GERMPLASM: Two distinct cultivars are recognized; one a large and late-maturing; the other, a smaller type that matures a month earlier. Assigned to the Africa Center of Diversity, hairy indigo or cultivars thereof has been reported to exhibit tolerance to disease, insects, low pH, nematode, poor soil, shade, slope, virus, and weeds. ($2n = 16$.)

Indigofera hirsuta L.

DISTRIBUTION: Native to tropical Asia, throughout India in the plains, and up to 1,300 m in Kumaon. Introduced in United States about 1910, and adaptable to Gulf Coast region from Florida to Texas. Also introduced and cultivated in other tropical and subtropical areas, e.g., Africa, Australia, and Venezuela.

ECOLOGY: Hairy indigo by its spreading habit provides ample leafy material. Its recovery from lopping is uncertain. It has a comparatively low lime requirement, grows fairly well on moderately poor sandy soil, but does best on fertile sandy loams. In Florida it requires moderately to well-drained soil conditions and is intolerant of waterlogging. Insect pollinated. Ranging from Warm Temperate Moist through Tropical Dry to Wet Forest Life Zone, hairy indigo has been reported to tolerate annual precipitation of 8.7–26.7 dm (mean of 13 cases = 14.5 dm), annual mean temperature of 15.8°–27.8°C (mean of 13 cases = 23.7°C), and pH of 5.0–8.0 (mean of 10 cases = 6.5).

CULTIVATION: Propagation is by seed. Seed lots contain small amounts of hard seed and much that is of low germination. Seed smaller of the early than of the late strain. Seed sown from early to late spring. Early seeding is preferred, using 3–5 kg/ha when drilled in close drills, and 6–10 kg/ha when broadcast in a well-firmed seedbed. Smaller amounts are recommended for seed production, and larger amounts for forage or green pasture. Superphosphate and potash applications increase growth. Suggested amount of 0-10-10 or 0-14-10 fertilizer is 300–500 kg/ha, or its equivalent of phosphate and potash. Spacing about 90 cm. Hairy indigo is a good legume for soil improvement. Makes good growth and matures sufficient seed in Florida to volunteer a satisfactory crop in corn after the last cultivation. Plants have been inoculated naturally wherever grown. Under Florida grazing, hairy indigo is a self-regenerating annual, faring well after burning or disking, or both.

HARVESTING: Crop can be handled as other common hay crops with ordinary farm equipment, and should be cut early (so that it does not become coarse) when ca. 75–90 cm tall. If cut 20–25 cm high in August before blooming time, a second growth may be expected. The aftermath may then be used for grazing. Grazing should be rotational to prevent severe removal of leaves. Seed may be harvested by cutting seed-branches by hand, producers use one of two other methods. Exceptionally large plants may be cut with a mowing machine, allowed to dry in the swath or windrow, and then run through a stationary grain thresher. Stands that are not too heavy may be combined when seed are mature but before shattering. Seed set abundantly but matures in late fall on the large strain, and 3–4 weeks earlier on the smaller strain.

YIELDS AND ECONOMICS: In the United States green matter yields average about 22 MT/ha yr; in India, ca. 10 MT/ha in coconut groves. In Florida, mixed with pangola or bahiagrass, hairy indigo gives annual yields of ca. 5.5 MT DM/ha with average crude protein of about 10%. Seed yields average 100–300 kg/ha. Widely cultivated in the tropics and subtropics for forage and soil improvement, and economically adaptable to poor sandy loamy soils.

BIOTIC FACTORS: Not affected seriously by any diseases or pests, but the following fungi have been reported: *Colletotrichum dematium, Corticium solani, Diplodia* sp., *Oidium* sp., *Rhizoctonia solani* and *Sclerotium rolfsii*. Highly resistant to root knot but the following nematodes have been isolated: *Aphelenchus avenae, Heterodera marioni, Hoplolaimus tylenchiformis, Meloidogyne javanica, Pratylenchus brachyurus,* and *Radopholus similis*. Pot experiments suggest that hairy indigo is antagonistic to root nematodes of the genus *Meloidogyne* and their galls. In field experiments, yields of cabbage, cucumber, and snapbean were significantly higher following a 3-month cover crop of hairy indigo. This treatment was as effective or more effective than certain nematicidal treatments.

REFERENCES:
Elmstrom, G. W., 1976, Napropamide for weed control in watermelon, *Proc. 28th Ann. Meet. South. Weed Sci. Soc.*, p. 173.
Rhoades, H. L., 1976, Effect of *Indigofera hirsuta* on *Belonolaimus longicaudatus, Meloidogyne incognita,* and *M. javanica* and subsequent crop yields, *Plant Dis. Rep.* **60**(5):384–386.
Wallace, A. T., 1957, Hairy indigo, a summer legume for Florida, *Fla. Agric. Exp. Stn., Circ.* **S-98**.

CONTRIBUTORS: J. A. Duke, A. E. Kretschmer, Jr., C. F. Reed.

Indigofera spicata Forsk.

FAMILY: Fabaceae

COMMON NAMES: Indigo, Trailing indigo, Spicate indigo

SYNONYM: *Indigofera endecaphylla* Jacq.

USES: Widespread in cultivation as a source of indigo, derived from the leaves. Also a valuable cover crop for coffee, tea, and rubber. Because of toxicity, should not be recommended as a forage crop.

FOLK MEDICINE: No data found.

CHEMISTRY: Five grams dry plant material is the lethal dose for 1-week-old chicks. Causes abortion in cattle and reportedly toxic to rabbits and sheep. Toxicity is due to indospicine, a hepatotoxic amino acid that interferes with both the synthesis and utilization of arginine. Indospicine inhibits growth and DNA synthesis in liquid cultures of mouse bone marrow cells and hamster kidney cells. Chemical mutagens were applied to develop toxin-free cvs; a few low-indospicine, but no indospicine-free plants were obtained. Species also synthesizes the toxic 3-nitropropionic acid. Analysis of the fresh plant gave the following values: moisture, 74.7; organic matter, 22.1; ash, 3.2; and nitrogen 0.78%. Analysis of the leaves and stems gave the following values: water, 80.5; protein, 4.1; fat, 0.6; soluble carbohydrates, 7.9; fiber, 4.7; and ash, 2.2%; digestible nutrients: protein, 3.1; fat, 0.4; soluble carbohydrates, 6.4; and fiber, 2.8%; nutritive ratio, 3.3; starch equivalent, 12.9 kg/100 kg.

DESCRIPTION: Prostrate or ascending herb, with a thick perennial rootstock; stems sparingly appressed strigulose, ridged, somewhat flattened, 1–2 m tall; stipules broadly scarious at base, tapering to a subulate tip, appressed strigulose, glabrescent along margins; rachis flattened, up to 3 cm long, including a petiole 1–3 mm long, prolonged up to 3 mm beyond the lateral leaflets; leaves compound, the 5–11 leaflets cuneate-obovate or cuneate-oblong, apiculate, varying greatly in size from ca. 3 mm long in reduced forms to 10 mm in average plants to 30 mm in stout forms, usually strigose on both surfaces, rarely glabrous above; racemes densely many-flowered, the fertile portion 2 or more times as long as the 1–4 cm long peduncle; bracts brownish, lanceolate, caducous; pedicels about 0.5 mm long; calyx 2–3 mm long, divided to base appressed strigose; corolla sparsely strigose outside; stamens 3–4 mm long; style about 1 mm long and bent upwards near base; pods reflexed, straight or slightly downward-curved, slightly tetragonal, apiculate, often torulose when immature, strigulose, 11–18 mm long, about 1.7 mm wide and 2 mm thick; seeds 5–8 per pod, subglobular, yellow, smooth. Fl. summer; Fr. early fall.

GERMPLASM: Plants vary greatly and several segregate species or cvs are mainly distinguished by the shape and indumentum of the leaflets and the growth habit, which may vary from usually stunted to quite luxuriant. Two of the most extreme forms are *I. endecaphylla* var *parvula* Chiov. (*I. parvula* sensu A. Rich., non Del.), grows at high altitudes and has leaflets only about 3 mm long, and var *major* Bak. fil., with leaflets up to 3 cm long. All forms grade into one another. Most introductions into Australia were tetraploid perennials with red stems, stoloniferous habits, and poor seeding abil-

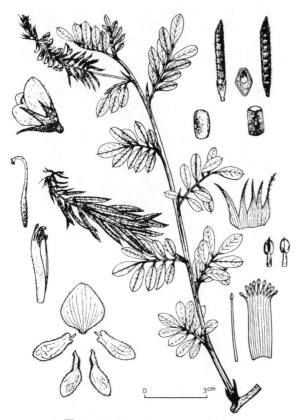

Figure 50. *Indigofera spicata* Forsk.

ity, the remainder green-stemmed annuals or biennials, diploid or tetraploid, nonstoloniferous (or nearly so), freely seeding. Assigned to the Africa Center of Diversity, trailing indigo or cultivars thereof is reported to exhibit tolerance to drought, laterite, and slope. ($2n = 16, 32$.)

DISTRIBUTION: Native to Africa from Senegal to Ethiopia, Congo, southern tropical Africa, South Africa (Transvaal, Natal), and Malagasy, Mascarene Islands, Yemen, India, Ceylon, Southeast Asia, and introduced in America in tropical areas, Hawaii, and Philippine Islands.

ECOLOGY: Does best in clay soils, but gives good cover on sandy soil. Common in disturbed grasslands, cultivated areas and waste places from sea level to about 2,700 m in altitude, but apparently scarce below 700 m. Short-day plant, thrives in areas with 600–1,500 mm annual rainfall. Tolerates some shade and is also drought resistant. Tolerates acid soils and phosphorus deficiencies. Ranging from Warm Temperature Moist through Tropical Moist Forest Life Zone, trailing indigo is reported to tolerate annual precipitation of 8.7–42.9 dm (mean of 7 cases = 20.0 dm), annual mean temperature 16.0°–27.4°C (mean of 7 cases = 22.0°C), and pH of 5.0–7.7 (mean of 4 cases = 6.2).

CULTIVATION: Propagation by seed, sown individually or drilled, in rows about 240 cm apart. Germination is usually poor unless the seed is scarified (e.g., with 40% sulfuric acid). Seed rate 2–4 kg/ha. Can be propagated by cuttings.

HARVESTING: Pods can be hand-harvested, or the whole crop may be cut and threshed late. Seed setting is sometimes poor because of relative humidity at flowering.

YIELDS AND ECONOMICS: Seed yields average about 500 kg/ha. Plants 2 months old can yield 5 MT/ha green matter (ca. 1.5 MT DM) with about 35 kg N. Six-month yields are closer to 25 MT/ha WM (12.5 MT DM) with more than 200 kg N/ha. An important source of the dye indigo in some tropical areas; south tropical Africa, Southeast Asia, Ceylon, India, and Hawaii.

BIOTIC FACTORS: Said not to be affected by any serious diseases or pests.

REFERENCE:
CSIRO, 1972, Tropical pastures, *Annual Report 1971–1972*.

CONTRIBUTORS: J. A. Duke, A. E. Kretschmer, Jr., C. F. Reed, M. C. Williams.

Indigofera tinctoria L.

FAMILY: Fabaceae

COMMON NAMES: Common or Indian indigo

SYNONYM: *Indigofera sumatrana* Gaertn.

USES: Indigo is cultivated for the blue dye, and in southern India as a cover or green-manure crop in coffee plantation and rice fields. Was once a primary source of Indian indigo, but was largely replaced by *Indigofera arrecta*. Leaves are rich in potash and plant is said to be palatable to cattle.

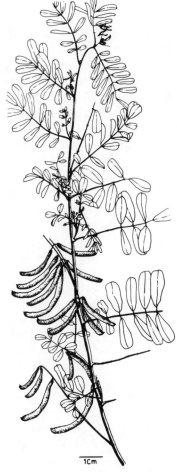

Figure 51. *Indigofera tinctoria* L.

Indigofera tinctoria L.

(Other species of *Indigofera* are quite poisonous.) *Indigofera sumatrana* is grown in India mainly as a green manure, preceeding corn, cotton, and sugarcane.

FOLK MEDICINE: Medicinally, juice of leaves is used as a prophylactic against hydrophobia, and a decoction for blennorrhagia; also an extract of the plant is given for epilepsy, nervous disorders, bronchitis, and as an ointment healing sores, old ulcers, and hemorrhoids. Roots are used for hepatitis, scorpion bites, and urinary complaints. Plant has been used to produce nausea and vomiting.

CHEMISTRY: Leaves contain (dry basis): N, 5.11; P_2O_5, 0.78; K_2O, 1.67; and CaO, 5.35%. A rich source of potash, the ash (4.4%) contains as much as 9.5% of soluble potassium salts. The indigo refuse, after dye extraction, gave on analysis: N, 1.8; P_2O_5, 0.4; and K_2O, 0.3% (WOI). Entire plant contains a glucoside, indican.

DESCRIPTION: Shrubby herbaceous plant, variable in all aspects; 1–2 m tall; stems erect, copiously branched, branches often widely spreading; leaves compound, leaf-rachis including the petiole 2.5–10 cm long; leaflets 5–13, oval, oblong, obovate, 10–27 mm long, 4–17 mm broad; racemes erect, subsessile, about 6 cm long; pedicels 1.5–2 mm long; flowers red, standard orbicular-obovate, 4.2 mm long, on inner side greenish-yellow with radiating purple streaks; pods glabrescent, 15–35 mm long, about 2 mm wide, 25 mm or more long, reflexed, slightly or almost straight; seeds 3–15 (usually 6–12), much longer than broad.

GERMPLASM: Assigned to the Africa Center of Diversity, indigo or cvs thereof is reported to exhibit tolerances to laterite. ($2n = 16$.)

DISTRIBUTION: Native to the Malaysian Archipelago, grows spontaneously in Africa. Introduced and widely cultivated in India, China, Java, Africa, Malagasy, and tropical America, often escapes from cultivation.

ECOLOGY: Widely cultivated in tropical regions, from sea level to 300 m grows well along riverbanks, roadsides and in brushwood and grassy fields. Heavy or continuous rain or too much water may kill plants. Excessive heat and hot winds cause withering. In the US, reportedly restricted to hardiness zone #10. Ranging from Warm Temperate through Tropical Dry to Wet Forest Life Zone, common indigo is reported to tolerate annual precipitation of 6.4–41.0 dm (mean of 13 cases = 15.2 dm), annual mean temperature of 10.5°–27.4°C (mean of 13 cases = 23.1°C), and pH of 4.3–8.7 (mean of 9 cases = 6.5).

CULTIVATION: Depending on the region, seed sown in October or April. In April, cultivation consists of one to two plowings after showers with light harrowing, both before and after broadcast sowing. Crop requires little additional attention and is ripe in June or July. Sowing starts in February in India, with a seed drill. Deep, thorough digging is necessary; then plowing each way, to break up soil and to make a smooth surface; field then rolled with wooden rollers; ground is again plowed 4–5 times; finally field sown and smoothed down. Seeding is at rate of about 20–30 kg/ha. Seeds germinate in 4–5 days. Weeding starts when plants are 5–8 cm tall, with a small cutting tool so as not to disturb seedlings; weeding is continued until the crop is tall enough not to be injured. In some areas fertilization and irrigation may be necessary, but irrigation should be carefully controlled.

HARVESTING: Harvested when the flowers begin to appear, usually in June, earlier or later depending on weather and planting time. Crop should be harvested promptly, as it may be destroyed in a few hours by flooding or heavy rains. Fall sowing usually occupies field for about 8 months, reaped during summer or early fall. Crop is cut with sickle and tied into bundles, that are taken to the factory and packed in steeping vats at 30–35°C for processing. Blue dyestuff is formed by fermentation caused by indican, a yellow amorphous substance with bitter taste, acid reaction, and rather soluble in water, alcohol, or ether. Indigo may be manufactured at any time between early spring and late fall. Dye may best be separated by the alkali and acid process. Indigo quality can be entirely controlled by proper use of acid, sufficient to redissolve the organic and mineral matter precipitated with the dye. Fermented liquor resists decomposition; a second steeping yields 40% more dye.

YIELDS AND ECONOMICS: Yields vary widely according to area, season and cultivation. Yield of plant material is 75–100 bundles/ha (a bundle weighs ca. 130 kg) and yield of dye is 1.6–5.4 kg/ha. Indian yields after 5 months are only ca. 2–½ MT/ha WM, with ca. 100 kg N. After a year, yields are closer to 7 MT with 300 kg N. India produces and exports nearly one-half million kg/yr, from ca. 45,000 ha. Price varies with the region. Synthetic product, indigotine, is manufactured chemically, and used as a substitute.

BIOTIC FACTORS: Following fungi attack indigo plants: *Cercosporella indigoferae, Cladosporium indigoferae, Colletotrichum indigoferae, Nectria cinnabarina, Parodiella perisporioides, Phymatotrichum omnivorum, Ravenelia indigoferae, Ravenelia laevis,* and *Uromyces indigoferae*. Plants are also susceptible to blights, green caterpillars, grasshoppers, locusts, and other insects that feed on the leaves and flowers, especially when plants are young. Plants are attacked by nematode *Heterodera glycines*. High winds, hailstorms, flooding, and heavy rains are destructive to the crop.

REFERENCE:
Ghosh, A. K., 1944, Rise and decay of the indigo industry, *Sci Cult.* 9(11):487–493; 9(12):537, 542.

CONTRIBUTORS: J. A. Duke, C. F. Reed.

Inga edulis Mart.

FAMILY: Mimosaceae

COMMON NAMES: Ice-cream bean, Guava machete, Inga cipo, Guavobejuco, Guava

SYNONYMS: *Inga tropica* Toro
Inga vaga (Vell.) Moore

USES: Plants cultivated for the white and edible pulp of the fruits used in flavoring desserts. Pods consumed by natives. Trees extensively used in Central and South America as shade for coffee, tea, and cacao plants, especially at lower altitudes.

FOLK MEDICINE: No data found.

CHEMISTRY: Seeds of *Inga edulis,* eaten as vegetables, are reported to contain per 100 g, 118 calories, 63.3% moisture, 10.7 g protein, 0.7 g fat, 24.0 g total carbohydrate, 1.6 g fiber, 1.3 g ash.

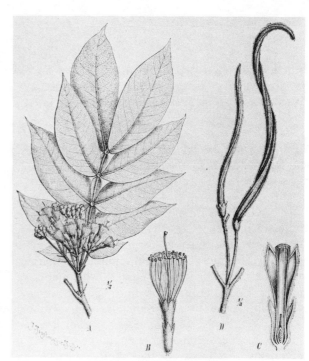

Figure 52. *Inga edulis* Mart.

Pulp of *Inga* spp. contains per 100 g, 60 calories, 83.0 g moisture, 1.0 g protein, 0.1 g fat, 15.5 g total carbohydrate, 1.2 g fiber, 0.4 g ash. Dried seeds of *Inga* spp. contain per 100 g, 339 calories, 12.6 g moisture, 18.9 g protein, 2.1 g fat, 62.9 g total carbohydrate, 3.4 g fiber, 3.5 g ash. Seeds of the genus *Inga* are reported to contain trypsin inhibitors and chymotrypsin inhibitors.

DESCRIPTION: Small trees up to 17 m tall, with broad spreading crown, almost flat; bark gray; trunk usually contorted, cylindrical, 30 cm or more in diameter and branching 1–2 m from base; leaves simply pinnate, 10–30 cm long, with 4–6 pairs of large oval leaflets, each pair separated by a winged rachis, the smallest pair below, the terminal leaflet 1.5 dm long and half as broad, membranous, minutely pubescent on both surfaces; flowers fragrant, sessile, solitary, arranged in crowded heads at tips of stems, peduncles 2–4 cm long, or flowers may be solitary in upper axils and fasciculate and subcorymbose below; calyx puberulent, striate, 5–8 mm long; corolla silky-villous, 14–20 mm long; bractlets oblong-lanceolate, about 5 mm long, caduous by fully anthesis; pods with very thickened sulcate margins, 4-angled, up to 2 m long, 1–3 cm in diameter, flat or twisted, with white sweet pulp

Inga edulis Mart.

surrounding the seeds. Fl. Feb.–May; Fr. several months later.

GERMPLASM: Assigned to the Middle America Center of Diversity, ice cream bean or cvs thereof is reported to exhibit tolerance to waterlogging. ($2n = 26$.)

DISTRIBUTION: Native to Central and South America, from Mexico southward. Introduced to Tanzania, and probably elsewhere in the tropics.

ECOLOGY: Abundant along margins of large rivers; common in thickets usually below the high-water mark, and in wooded swamps. Also in ravines, upland woods at edge of rivers and adjacent rainforests. Requires a tropical climate with plenty of moisture. Found at altitudes up to 600 m. Ranging from Subtropical Dry to Moist through Tropical Dry to Wet Forest Life Zone, ice cream bean is reported to tolerate annual precipitation of 6.4–40.0 dm (mean of 9 cases = 16.9 dm), annual mean temperature of 21.3°–27.3°C (mean of 9 cases = 25.1°C), and pH of 5.0–8.0 (mean of 7 cases = 6.6).

CULTIVATION: Propagates naturally by seeds in the forest. Planted from seedlings grown in nursery and then arranged to give maximum shade to coffee, tea or cacao plants.

HARVESTING: Fruits are harvested when ripe in small amounts as needed. Once established, trees provide shade, for many years over the undercrop of coffee, tea, or cacao.

YIELDS AND ECONOMICS: No data available. Fruits are produced periodically, and nearly continuously. Plants cultivated locally for the fruits in various tropical countries, as in South America, Hawaii, West Indies, and East Africa. Trees also extensively used for shade of coffee, tea, and cacao, especially in Central America.

BIOTIC FACTORS: The following fungi are reported: *Bitzea ingae*, *Catacauma ingae*, *Fusarium semitectum* var *majus*, *Perisporium truncatum*, *Peziotrichum saccardinum*, *Phyllosticta ingae-edulis*, *Ravenelia ingae*, *Rhizoctonia solani*, and *Uredo ingae*. Trees are also affected by a mosaic virus and witches broom.

CONTRIBUTORS: J. A. Duke, C. F. Reed, J. K. P. Weder.

Lablab purpureus (L.) Sweet

FAMILY: Fabaceae

COMMON NAMES: Hyacinth bean, Bonavist, Chicaros, Chink, Egyptian bean, Pharao, Seem, Val.

SYNONYMS: *Dolichos lablab* L.
Dolichos purpureus L.
Lablab niger Medik.
Lablab vulgaris Savi.
Dolichos albus Lour.
Dolichos cultratus Thunb.
Dolichos lablab var *hortensis* Schweinf & Muschler
Lablab leucocarpos Davi
Lablab nankinicus Savi
Lablab perennans DC.
Lablab vulgaris var *niger* DC.

USES: Lablab is grown in tropical and subtropical areas for human food; young pods eaten as snapbeans; beans are cooked and eaten as a vegetable or in salads. Dried seeds also are consumed. In Egypt, pounded seeds are substituted for broad

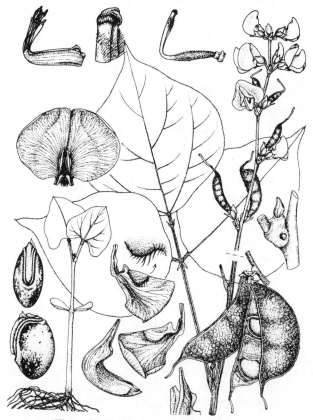

Figure 53. *Lablab purpureus* (L.) Sweet

beans in the fried bean cake "tanniah." Plants are grown as forage for livestock and fed to goats, cattle, hogs, and, mixed with oats, to horses. Also used as green manure. Single plants produce twice the herbage of cowpea; stems are tough and more fibrous, the leaves more succulent. Seeds are sometimes soaked overnight, allowed to sprout, then sundried and stored. Fresh seeds, containing prussic acid, are considered poisonous, and should be thoroughly cooked before eating. As a rule, the darker seeds are more likely to contain cyanide.

FOLK MEDICINE: Seeds are reported to be alexeritic, antispasmodic, aphrodisiac, febrifugal, and stomachic; also used for menopause. The refrigerant leaf infusion is used for colic and gonorrhea. Leaf and flower have been prescribed in menorrhagia and leucorrhea. The leaf is poulticed onto snakebites. For eczema, Malays make a poultice of the leaves with rice flour and turmeric. Juice from the pods is applied to inflamed ears and throats.

CHEMISTRY: Dried seeds contain per 100 g: 334 calories, 12.1% moisture, 21.5 g protein, 1.2 g fat, 61.4 g carbohydrate, 6.8 g fiber, 3.8 g ash, 98 mg Ca, 345 mg P, and 3.9 mg Fe. Raw green pods contain per 100 g edible portion: 30 calories, 87.5% moisture, 3.1 g protein, 0.3 g fat, 8.2 g total carbohydrate, 1.9 g fiber, 0.9 g ash, 75 mg Ca, 50 mg P, 1.2 mg Fe, 2 mg Na, 279 mg K, 160 μg β-carotene equivalent, 0.08 mg thiamine, 0.13 mg riboflavin, 0.60 mg niacin, and 16 mg ascorbic acid. The raw leaves contain per 100 g: 31 calories, 89.1% water, 2.4 g protein, 0.4 g fat, 6.1 g carbohydrate, 6.7 g fiber, 1.4 g ash, 120 mg Ca, 57 mg P, 17 mg Fe, 3145 μg β-carotene equivalent, 0.28 mg thiamine, and 16 mg ascorbic acid. Raw runners contain per 100 g: 33 calories, 86% moisture, 2.8 g protein, 0.2 g fat, 6.8 g total carbohydrate, 1.4 g fiber, 0.6 g ash, 116 mg Ca, 63 mg P, 1.5 mg Fe, and 268 mg K. Dried seeds contain 0.15 mg pyridoxine, 1.2 mg pantothenic acid, 0 μg vitamin B_{12}, and 21.8 μg folic acid. Vitamin C content of raw pods varies from 7.33 to 10.26 mg/100 g, cooked pods 0.77–1.12 mg/100 g (assuming WOI had the values mixed). For seed yielding 23.4 protein and 1.1% oil, the amino acids per 16 g N were: 6.8 g lysine, 0.9 g methionine, 6.6 g arginine, 4.6 g glycine, 3.2 g histidine, 4.4 g isoleucine, 8.5 g leucine, 4.9 g phenylalanine, 3.6 g tyrosine, 4.2 g threonine, 5.2 g valine, 4.5 g alanine, 12.0 g aspartic acid, 15.7 g glutamic acid, 0.6 g hydroxyproline, 4.3 g proline, and 5.4 g serine. Forage composition is: 28.1 fiber, 3.5 fat, 14.2 crude protein, 39.4 carbohydrate, 14.8 ash, 1.98 Ca, and 0.26% P. Dry feed has 28.08% fiber, 3.5% ether extract, 14.8% total ash, 2.77% CaO, 0.6% P_2O_5, 0.97% MgO, 0.55% N_2O, and 3.52% K_2O. Seeds contain trypsin inhibitors and chymotrypsin inhibitors and may contain cyanide and stigmasterol. Said to be a rich source of catechol oxidase. The root is said to be poisonous.

DESCRIPTION: Potential perennial, but cultivated as an annual or biennial (in Brazil), with bushy erect and vining twining forms, extremely variable in all aspects, often behaving as a perennial in the tropics, persisting for 2 or more years; stems twining, hairy, or glabrate, usually 2–3 m, but often to 10 m long, other forms dwarf and bushy; leaves pinnately trifoliolate, leaflets very broad ovate, the lateral ones lopsided, 7.5–15 cm long and nearly as broad, rather abruptly acuminate; flowers purple to pink or white, 2–4 at each node in an elongating raceme, up to 2.5 cm long; pods flat or inflated, 5–20 cm long, 1–5 cm broad, pubescent or smooth, papery, straight, or somewhat curved, green or purple; seeds 3–6, 0.6–1.3 cm long, nearly as wide, flattened, oblong with rounded ends, black or nearly so, but white in white-flowered vars, the hilum and raphe long and prominent; germination phanerocotylar.

GERMPLASM: Many of the natural variations have been named taxonomically. From 39 to 50 varieties are recognized, based on variability of the size, shape, and color of the pods (green, white, purple, or purple margins; fleshy or fibrous), size, shape, or color of the seed (white to yellow to black or reddish-purple), flower characteristics (white, pink, or purple); size of corolla; degree of fragrance; length of peduncle; abundance of flowers; panicles on short peduncles and 10–20-flowered, or on long-stalked peduncles to 30 cm long and 20–30-flowered), color of leaves (green to purple-tinged to purple), and plants glabrous to covered with short, white hairs. 'Darkness,' a black-seeded cv; 'Daylight,' a white-flowered form, grown as ornamentals. 'Highworth' and 'Rongai,' the latter later-flowering, were developed as forage cvs in Australia. Short-day and long-day cvs have been developed, the flowering depending on photoperiod, not temperature. Assigned to the African Center of Diversity, lablab is reported to exhibit tolerance to alu-

Lablab purpureus (L.) Sweet

minum, disease, drought, frost, heat, heavy soil, poor soil, sand, shade, and virus. *Lablab purpureus* ssp. *uncinatus* Verdcourt—More slender inflorescences, legume smaller, about 4 cm long, 1.5 cm wide.

SYNONYMS: *Lablab uncinatus* A. Rich.
　　　　　Dolichos lablab forma *uncinatus* Penzig
　　　　　Dolichos uncinatus Schweinf.
　　　　　Dolichos lablab var *uncinatus* (Schweinf.) Chiov.
　　　　　Lablab niger var *uncinatus* (A. Rich.) Cuf.
　　　　　Lablab niger var *crenatifructus* Cuf.

Lablab purpureus ssp. *bengalensis* (Jacq.) Verdcourt—Pods similar to lima beans and quite dissimilar to those of other bonavist, but races interbred freely.

SYNONYMS: *Dolichos bengalensis* Jacq.
　　　　　Dolichos lablab ssp. *bengalensis* (Jacq.) Rivals
　　　　　Lablab niger ssp. *bengalensis* (Jacq.) Cuf.

Lablab purpureus var *rhomboideus* (Schinz) Verdcourt—A cv with more glabrous leaves, broader bracteoles and the pods minutely puberulous rather than pubescent.

SYNONYMS: *Dolichos lablab* var *rhomboideus* Schinz
　　　　　Dolichos pearsonii Hutch.

($2n = 22, 24.$)

DISTRIBUTION: Some believe lablab is native to India or southeast Asia, others (Zeven and Zhukovsky, 1975) assign it to the African Center of Diversity. It has spread over the entire tropics and is now cultivated throughout the warmer parts of the world. In the United States it is mainly cultivated as an annual ornamental in gardens, from which it frequently escapes and becomes naturalized.

ECOLOGY: Lablab thrives in climates averaging 18°–30°C; high temperatures apparently do not affect its development. Needs rainfall or irrigation during the first 2–3 months after sowing. Later, it is very drought resistant, and continues to grow, producing flowers and seeds for many months, staying green as the weather becomes dry and cool. Plants do not tolerate standing brackish water or waterlogging. Said to thrive on sandy loams (pH 6.5) and heavy clays (pH 5.0). In India lablab is grown as a dry land crop in regions up to 1,830 m in altitude with a rainfall of 6–9 dm. Frost damages the leaves, but light frost does not kill the plants. Root system is very deep and uses residual soil moisture. Ranging from Warm Temperate Dry to Moist through Tropical Desert to Wet Forest Life Zones, the versatile hyacinth bean has been reported from zones with pH of 5.9–7.8, annual precipitation of 200–2,500 mm, and annual temperatures of 9°–27°C. Resistant to many diseases, especially leaf diseases, drought tolerant and has been suggested for allic soils.

CULTIVATION: May be seeded in many ways at 20–70 kg/ha. Sowed alone, and hoed once or twice to eliminate weeds, the crop is used primarily for soil improvement. Best bushy and semibushy varieties are planted in 90-cm rows; growth is similar to that of cowpeas. Hills, 90 cm apart in both directions, may be sown with 6–10 seeds each, thinned to 4 plants after a month, and allowed to climb on poles. They produce a good yield of hay which is easily cured due to the low moisture content. Seed may be sown with corn at rate of 10–20% lablab and fertilized with chicken manure enriched with 10% superphosphate. Lablab grows slowly at first, and does not compete with corn. When corn begins to ripen, lablab starts to vine and grows most vigorously after the corn dries. After the corn is harvested, lablab develops a complete cover, several decimeters thick over the soil. Corn is sown in Oct.–Dec. (in San Paulo) and harvested in March–May, when dry season begins. Lablab begins to flower in May and seed begin to ripen in June; it continues to flower and seed for many months, even in dry, cool weather. The pods do not shatter and the crop may be harvested at any time. Humidity seems to induce disease, so fruit set should coincide with the dry season. Lablab intercropped with corn nearly doubled the corn yield, and beans yielded 909 kg/ha against 424 kg/ha. In Africa lablab precedes cereal crops in rotation. In Kenya seed is planted after the rains start, 20–30 kg/ha, sown at spacing of 0.75 × 1 m. In India, seed sown in June–July, often with finger millet; is often grown as a dry land crop in regions up to 1,830 m with a rainfall of 630–890 mm. When grown with finger millet, lablab is fertilized with 90 kg/ha ammonium sulfate and 40 kg/ha superphosphate of lime after first weeding. Lablab can be mixed with 10% pigeon pea *(Cajanus cajan)* and planted with mechanical seeders on soil improved the preceding year. After 4 months, cattle may graze the mixture. Lablab can be seeded between coffee; after 2 months, no weeding is necessary. In orchards, lablab forms a good organic mulch when cut often. In India 100 kg/ha ammonium sulfate and 250 kg/ha superphosphate are recommended. Grown for grazing in Australia, lablab receives 250 kg/ha molybdenized superphosphate on average soils, 500 kg/ha on poor coastal soils.

HARVESTING: Lablab can be harvested when the pods are dry and yellow (60–300 days). Yield of seeds is relatively poor and harvest is difficult. For seed production, cvs in which the pods retain their form when dry, are most desirable because they are affected little by weather and thresh easily. Vining types give best seed production when grown alone or staked. Earliest cvs ripen in 60 days; others do not bloom in 140 days and are often killed by frost. Plants are pulled by hand and stacked in sun to dry. Ripe pods can be picked alone from a standing crop, giving high-quality seed. Threshing is normally done by flailing plants with sticks, but in India they may be threshed with stone rollers or under feet of oxen. Seed requires thorough winnowing and sieving as weed content is often high. Tending to reseed itself, hyacinth bean is considered a weed by some. In Haiti, e.g., it grows as a waif among other legumes, as the seeds are contaminants.

ECONOMICS AND YIELDS: The nodulating power of lablab is said to be remarkable, even without inoculation. Average seed yields are 450 kg/ha if grown as a mixed crop, 1,460 if grown in monoculture. In India, green pod yields average 2,600–4,500 kg/ha. Fodder yields of 5–10 MT/ha are reported. Lablab has many advantages: high yield of green vegetative mass for improving the soil as a green manure; stays green during droughts; good cover crop; palatable forage (cut, pastured, or as silage); economical seeding methods; easy harvesting of the beans; a high-protein food for humans and farm animals. In Africa and other areas this is becoming an important legume; the flour (thoroughly cooked) is used for bean cakes.

BIOTIC FACTORS: All cvs of lablab are free from severe losses to leaf diseases. Many cvs are subject to both Fusarium root rot and to nematodes. However, many fungi have been found on lablab; the worldwide survey indicates the following: *Alternaria tenuis, A. tenuissima, Ascochyta dolichi, A. phaseolarum, Cercospora canescens, C. columnaris, C. cruenta, C. dolichi* (leaf spot), *Choanephora cucurbitarum* (Dacca, India), *Ch. manshurica, Colletotrichum lindemuthianum* (anthracnose, also occurs on seed in storage), *Corticium solani, Curvularia lunata, C. penniseti, C. spicifera, Cylindrocladium scoparium, Didymella lussoniensis, Discosporella phaeochlorina, Elsinoe dolichi* (China, scab of hyacinth bean), *E. phaseoli, Erysiphe polygoni, Glomerella cingulata, Helminthosporium lablabis, Isariopsis griseola, Leveillula leguminosarum, L. taurica* (powdery mildew), *Macrophomina phaseoli* (Ashy stem blight, controlled by seed dressing), *Myrothecium roridium, Nematospora coryli, Neocosmospora vasinfecta, Oidiopsis macrospora, Parodiella perisporioides, Periconia byssoides, Phakopsora vignae, Phyllosticta dolichi, Phyllachora dolichogens, Phymnatotrichum omnivorum, Phytophthora capsici, Pleospaeropsis dalbergiae, Pyrenochaeta dolichi, Septoria dolichi, S. lablabina, Sphaceloma glycines, Uromyces appendiculatus* (rust), *U. dolicholi, U. vignae*. In Africa (Sudan) *Macrophomina phaseoli* and *Xanthomonas phaseoli* are among the more serious diseases. Spread by rain, these can be controlled by chemical seed treatment. Plants are attacked by the parasitic species of *Striga, S. asiatica, S. gesnerioides,* and *S. hermonthica*. The bacterium, *Bacterium cereus* var *thuringensis* (HB-III) may be used to control pod-boring noctuoid caterpillars *(Adisura atkinsoni)* in India. The gram caterpillar, *Heliothis armigera,* the plume moth, *Exelastis atomosa,* and the spotted pod borer, *Maruca testulalis* are of economic importance. Flowers are destroyed by the *Mylabris* beetles (in Malawi). Cock-shaver larvae (*Schizonycha* sp) damage young seedlings and may be controlled by mercurized seed dressings. In Puerto Rico, the bean leaf beetle, *Cerotoma ruficormis* is a major pest. Aphids and stink bugs *(Coptasoma eribraria)* controlled by insecticides. Among the nematodes that have been isolated from lablab are: *Aphelenchoides bicaudatus, Caconema*

Lablab purpureus (L.) Sweet

radicicola, Heteronema glycines, H. marioni, Hoplolaimus sp., *Meloidogyne incognita acrita, M. javanica, Pratylenchus brachyurus,* and *Rotylenchus reniformis*. The following virus diseases have been isolated from lablab: Mosaic disease (Poona, India, causes chlorotic streaks), Ringspot virus (India), Alfalfa mosaic, Alfalfa yellowing, Alsike clover mosaic, Bean chlorotic ringspot, Brazilian tobacco streak, Enation mosaic, Ringspot mottle, Subterranean clover stunt, White clover mosaic, and Yellow mosaic virus. For weeds, preemergence herbicides have been recommended: chloramben, chlorthal, dinoseb, diphenamid, and trifluralin. Captan and zineb have been recommended for anthracnose; Bordeaux mixture for leaf spot and powdery mildew, and sulfur dust for rust.

REFERENCES:
Draper, W. J., 1967, *Dolichos lablab*—An alternative green manure crop, *Queensl. Bur. Sugar Exp. Stn., Cane Growers Q. Bull.* **30**(4):119.
Patil, G. D., 1958, Anthesis and pollination in Field Wal (*Dolichos lablab* Rox.), *Poona Agric. Mag.* **49**:95–102.
Piper, C. V., and Morse, W. J., 1915, The bonavist, Lablab or Hyacinth Bean, *U.S. Dep. Agric., Bull.* **318**:1–15.
Rangawami Ayyangar, G. N., and Krishman, K. Kuhni, 1935, Studies in *Dolichos lablab* (Roxb.) L. The Indian field and garden bean, II, *Proc. Ind. Acad. Sci.* **2**(1):74–79.

CONTRIBUTORS: J. A. Duke, A. E. Kretschmer, Jr., C. F. Reed, J. K. P. Weder.

Lathyrus hirsutus L.

FAMILY: Fabaceae

COMMON NAMES: Roughpea, Caleypea, Singletary, Wild winter pea

USES: This is an important winter legume for pasture, and hay, and for soil improvement in southern United States; especially valuable for winter and early spring forage.

FOLK MEDICINE: No data available.

CHEMISTRY: Seeds of the genus *Lathyrus* are reported to contain trypsin inhibitors and chymotrypsin inhibitors. The genus *Lathyrus* is reported to contain folic acid, hydrocyanic acid, maltose, pantothenic acid, quercitrin, and saponin.

Figure 54. *Lathyrus hirsutus* L.

DESCRIPTION: Winter annual legume; stems sparsely pubescent, weak, and decumbent, except in thick stands where they are ascending, 20–120 cm long, winged; leaflets 1 pair, 15–80 mm long, 3–20 mm broad, linear or oblong; stipules 10–18 mm long, 1–2 mm broad, linear, semisagittate; racemes 1–3-flowered; calyx-teeth as long as or slightly longer than the tube; corolla 7–20 mm long, red with pale blue wings; pods rough or hirsute, 20–50 mm long, 5–10 mm broad, brown; seeds 5–10 per pod, round, characteristically tuberculate; hilum one-sixth to one-fifth of circumference. Fl. early April. Seeds 33,075/kg; wt. 71 kg/hl. Wt.of seeds about 24 kg/bu.

GERMPLASM: Assigned to the Central Asia and Mediterranean Centers of Diversity, roughpea or cultivars thereof is reported to exhibit tolerance to heavy soil and waterlogging. ($2n = 14$.)

DISTRIBUTION: Native to southern Europe and southwestern Asia. Introduced in southern United States and long established, often as an escape. Cultivated in Alabama, Louisiana, and Mississippi.

ECOLOGY: Well adapted to the cotton-producing areas of southern United States, particularly Mississippi and Alabama. Grows best on well-drained soils, but also grows on soils too wet for clover and small grains. Does better on heavier soils than many other annual legumes. Fares well on lime soils; grows well on acid soils and under wet conditions; however, light sandy soils are best for seed production. Ranging from Boreal Moist to Wet through Warm Temperate Dry to Moist Forest Life Zones, roughpea is reported to tolerate annual precipitation of 3.2–12.9 dm (mean of 12 cases = 8.2), annual mean temperature of 4.3°–23.8°C (mean of 12 cases = 13.6°C), and pH of 5.5–8.2 (mean of 11 cases = 6.6).

CULTIVATION: Plants reseed freely. Percentage of hard seed is high. Seed retain viability for years. Seed usually sown in autumn (United States) as late as Oct.–Nov. (Mediterranean region). In southern Australia, seed sown in October to January, at rate of 70–80 kg/ha broadcast, 60 kg/ha drilled. For new stands, seed are broadcast on a firm seedbed at rate of 20–25 kg/ha of scarified seed and 40–60 kg/ha unscarified when establishing. In the United States seed are broadcast 20–30 kg/ha. If drilled with cereals, half that much seed are required. Liming is advised on acid soils. Plants grow little in winter, but more rapidly in late winter. Plants begin to flower in early spring, depending on location and seasonal conditions. During this period, it is palatable, nutritious pasture, and all livestock types show good gains. Roughpeas, when grazed at maturity, are poisonous to livestock; cattle exhibit a characteristic stiffness or lameness from eating the seed; animals recover rapidly when taken off the roughpea pasture. Grazing should be discontinued when plants begin flowering and forming seed pods. In most areas of the south, roughpeas do not make the large amount of growth sufficiently early in the spring to be a satisfactory soil-improving crop for early planted corn or cotton. High-quality hay can be produced by seeding roughpeas in mixtures with small grains and ryegrass. Roughpeas require an application of phosphorus and potash fertilizers for maximum production.

HARVESTING: Plants mature at or before the wheat harvest. In the United States, it usually precedes earlier planted spring crops such as cotton. Often grown with Abruzzi rye or sorghum for seed production and gives a higher seed yield than when grown alone. Roughpeas are normally cut when the first pods are ripe. In the Mediterranean region, plants are uprooted, carried in bundles to a threshing floor, or allowed to dry before seed is flailed. In the United States and Australia, crop is cut by mower, windrowed to dry and then threshed, or the standing crop can be combined directly. Usually mowed when 90% of pods are brown and vines are wet with dew to decrease shattering. Green seed from the combine heat if stored immediately; must be stacked or spread on threshing floor and aerated for ca. 10 days. In Israel, very late threshing reduces dormancy.

YIELDS AND ECONOMICS: Seed yields average 300–400 kg/ha; in inoculated fields in the United States, yields up to 1,000–2,000 kg/ha have been obtained. Roughpea is an important winter legume over the southern third of the United States and in Europe and Asia. In the United States more than 10,500 MT of seed are produced annually.

BIOTIC FACTORS: Following fungi have been reported on this legume: *Ascochyta* sp., *Mycosphaerella* sp., *Peronospora lathyri-hirsuti*, *Ramularia galegae*, *R. lathyri*, and *Uromyces orobi*. It is also attacked by the Wisconsin pea streak virus, and by the nematodes *Heterodera goettingiana* and *Rotylenchulus reniformis*.

REFERENCE:
Henson, P. E., 1953, Roughpea, *Lathyrus hirsutus*, U.S. Dep. Agric., Soils Agric. Eng. **April**:2 pp. (mimeo).

CONTRIBUTORS: J. A. Duke, C. F. Reed.

Lathyrus sativus L.

FAMILY: Fabaceae

COMMON NAMES: Grasspea, Chickling vetch, Khesari

USES: Primarily cultivated as a cold weather forage crop. Leaves also eaten as a pot-herb and immature pods boiled as a vegetable. Seeds are dehusked and parched before use. Mixed with oil cake and salts, seeds are used as nutritive feed for cattle. Khesari seeds are used in India as an article of diet by the poor in times of famine. They are made into paste balls, put in curry, or boiled and eaten like a pulse.

Lathyrus sativus L.

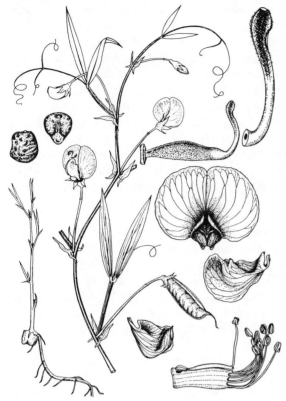

Figure 55. *Lathyrus sativus* L.

Plants are valued for green manure but have weedy tendencies.

FOLK MEDICINE: Oil from the seeds is a powerful and dangerous cathartic that contains a poisonous principle, probably an acid-salt of phytic acid. The seeds are used locally in homeopathic medicine.

CHEMISTRY: Germination of seeds enhances content of vitamins, especially folic acid, biotin and pyridoxine. Normal vitamins are carotene, thiamine, riboflavin, nicotinic acid, biotin, pantothenic acid, folic acid, pyridoxine, inositol, ascorbic acid, and dehydroascorbic acid. Stigmasterol and possibly lycopene are also present. Seeds contain 28.2% protein, 0.6% fat, 58.2% carbohydrate (ca. 35% starch). Another analysis of the seeds reveals per 100 g edible portion: 10 g moisture, 25.0 g protein, 1.0 g fat, 61.0 g total carbohydrate, 15.0 g fiber, 3.0 g ash, 110 mg Ca, 5.6 mg Fe, 70 IU vitamin A, 0.10 mg thiamin, and 0.40 mg riboflavin. The seed contains 34.8% starch of which 30.3% is amylose 69.7% amylopectin. Seeds also contain 1.5% sucrose, 6.8% pentosans, 3.6% phytin, 1.5% lignin, 6.6% albumin, 1.5% prolamine, 13.3% globulin, and 3.8% glutelin. The essential amino acids are (g/16 g N): arginine 7.85, histidine 2.51, leucine 6.57, isoleucine 6.59, lysine 6.94, methionine 0.38, phenylalanine 4.14, threonine 2.34, tryptophane 0.40, and valine 4.68 (very deficient in methionine and tryptophane). People and livestock consuming khesari as the principle article of diet for months develop a paralytic disease known as "lathyrism"; livestock that consume seeds as 30–50% of the diet for 3–6 months develop neurolathyrism, a disease that leads to microgliosis in the anterior horns and lateral cords and partial degeneration of the motor tracts of the spinal cord and anterolateral sclerosis in the dorsolumbar spinal cord, and death in extreme cases. Amino acid derivatives from the seeds of other species of the genus *Lathyrus* and of some species of the genus *Vicia* produced similar effects on experimental animals. Occasional use is harmless. Seeds if soaked in water for 24 hr before cooking are not toxic. Analysis of the leaves gave the following values: moisture, 84.2; crude protein 6.1; fat (ether extr.), 1.0; carbohydrates, 7.6; ash 1.1; Ca, 0.16; and P, 0.1%; Fe 7.3 mg; and carotene (as vitamin A), 6,000 IU/100 g. Analysis of green plant, cut at the flowering stage gave the following values (dry basis): protein, 17.3; fiber, 36.6; fat, 4.47; ash, 6.0; P_2O_5, 0.51; and CaO, 1.08%. Analysis of hay gave the following values: moisture, 14.5; crude protein, 9.9; fat, 1.9; fiber, 36.5; N-free extract, 31.0; and ash, 6.1%; starch equivalent, 12.5 kg/100 g. A sample of hay (ash, 8.0%) contained: P_2O_5, 0.34; and CaO, 0.96%.

DESCRIPTION: Much-branched suberect, straggling or climbing herbaceous winter annual; stems 0.6–9.0 m tall or long; leaves pinnately compound; leaflets usually 2, linear-lanceolate 25–150 mm long, 3–9 mm broad, the upper leaflets modified into tendrils; flowers solitary, axillary, on peduncles 30–60 mm long; corolla 12–24 mm long, flat, reddish-purple, pink, blue or white; pods oblong, 2.5–4.0 cm long, flat, slightly curved, dorsally 2-winged; seeds 3–5 per pod, white, grayish-brown or yellowish, usually spotted or mottled, somewhat smaller than peas, 3–15 mm long; hilum 1/15th to 1/16th circumference of seed. Seeds 11,025/kg; germination cryptocotylar.

GERMPLASM: Plants are classified on basis of color of flowers, markings on pods and size and color of seeds. About 56 types have been distinguished in

India alone (45 *cyaneus*, 10 *roseus*, one *albus*). 'Indore T 2-12' is resistant to rust, with earlier higher yields than other local types in India. Assigned to the Central Asia and Mediterranean Centers of Diversity, khesari or cultivars thereof is reported to exhibit tolerance to drought, heavy soils, high pH, low pH, poor soil, rust, virus, and waterlogging. ($2n = 14$.)

DISTRIBUTION: Origin unknown, but considered native to Southern Europe and Western Asia. Cultivated in central, south, and east Europe, and widely naturalized. Introduced and widely grown in India, Iran, Middle East, and South America. Cultivated in many parts of India, up to 1,300 m in the Himalayas.

ECOLOGY: Grows well on almost all types of soil. A hardy crop suited to dry climates, producing good seed crops on poor soils. Commonly cultivated on heavy clay soils. Black deep retentive soils considered best. Sensitive to acidity, and requires lime on acid soils. Grows on land subject to drought, excessive rain or flood. Grows well on land considered unfit for wheat, rice, cotton, or other more popular pulses. Ranging from Boreal Moist to Wet through Tropical Very Dry Forest Life Zone, khesari is reported to tolerate annual precipitation of 3.2–13.6 dm (mean of 31 cases = 7.4 dm), annual mean temperature of 4.3°–27.5°C (mean of 31 cases = 13.1°C), and pH of 4.5–8.3 (mean of 27 cases = 6.8).

CULTIVATION: Propagated by seed. Some say inoculation is essential before sowing, especially in virgin soil; others say it appears unnecessary. In some temperate regions, sown after rye, or on fallow land. Seeding rate 78–90 kg/ha. Sown in India Sept.–Oct. or even later; their seed rate varies from 12 to 50 kg/ha, according to method of cultivation. Seeds may be sown broadcast or in furrows ca. 3 cm apart. In a well-prepared field, crop comes up as a thick close mass over the entire surface and smothers out weeds. Easy to cultivate. Except for lime on acid soils, other nutrients are rarely needed. Phosphorus application is recommended. Plants may be sown as pure or mixed; often sown with standing rice crop in India; sometimes comes up self-sown; or may be second crop after rice. When grown for green manure, seed rate is about 55–60 kg/ha; said to add about 55 kg/ha N.

HARVESTING: Seeds ripen in 4–6 months and are harvested as soon as leaves begin to turn yellow when pods are not fully ripe; fully ripe pods dehisce and scatter the seeds. May be reaped with sickle or uprooted. Harvested crop is allowed to dry in heaps in field for about a week and then threshed and winnowed. Crop can be cut and fed green, or the standing crop may be pastured; not fit for silage. Can be cured into hay under mild climatic conditions. When fed alone, fresh young plants are reported to be harmful to horses; however, cattle, rabbits, and sheep can consume large amounts without ill effects.

YIELDS AND ECONOMICS: Average yields of seeds, 1,000–1,500 kg/ha; with inoculated seed in the United States, up to 2,000 kg/ha. Yields are proportional to quantity of seed sown. Average crop at seed rate of 40 kg/ha yields about 925 kg/ha of pulse and 3.2 MT/ha of forage in India. At seed rate of about 14 kg/ha in mixed cultivation, yield per ha is about 300 kg of pulse and ½ MT of straw. At a seed rate of 67 kg/ha, grasspea adds ca 62 kg/ha N to the soil. Ukranian experiments with sewage irrigation gave yields of 3120 kg seeds, 34.5 MT WM and 7.7 MT hay/ha. The sewage irrigation slightly decreased the 1,000-seed weight, but markedly increased seed N, P, and K contents and nutritive values of fresh forage and hay. In India this pulse occupies about 4% of the total pulse crop and constitutes about 0.3% of the total pulse production, with about 1.6 million ha, producing about 0.5 million MT of seeds. This pulse is cheap and easy to cultivate and is often cultivated in the tropics, usually as a winter crop.

BIOTIC FACTORS: Plants are self-pollinating. Fungi reported on this plant include the following species: *Ascochyta orobi* and *A. pisi* (both serious), *Erysiphe communis*, *E. polygoni*, *Fusarium orthocerus*, with var *ciceri* and var *lathyri* (wilt, most serious in some parts of India), *Glomerella cingulata*, *Leveillula taurica*, *Macrophomina phaseoli*, *Mycosphaerella ontarioensis*, *Oidium erysiphoides* (mildew), *Peronospora lathyri-palustris* (mildew), *P. viciae*, *Pleosphaerulina hyalospora*, *Pleospora hyalospora*, *Uromyces pisi* (rust), *U. fabae* (rust). Plants are also attacked by the bacterium *Pseudomonas cannabina* and parasitized by *Cuscuta pentagona*. Nematodes attacking this plant include: *Heterodera goettingiana*, *H. schachtii* and *H. trifolii*. Rootknot nematodes are not usually serious

Lathyrus sativus L.

in the United States. In Australia, the red-legged mite, *Halotydens destructor,* attacks grasspea; in India aphids are troublesome.

REFERENCE:
Mozheiko, A. M., Nasonor, Y. F., and Kuzovenko, N. P., 1976, Effect of irrigation with sewage water on yield and quality of peavine [Russ.], *Tr. Khar'k. Skh. Inst.: Ref. Zhur.* **1.55.674:** 1 975 205 75-82 (through CAB Field Crop Abst.).

CONTRIBUTORS: J. A. Duke, C. F. Reed, J. K. P. Weder.

Lens culinaris Medik.

FAMILY: Fabaceae
COMMON NAMES: Lentil, Masurdhal, Tillseed

USES: Lentil is cultivated for its nutritious seed, eaten as dhal and considered the most nutritious of pulses. Split seeds used in soups; flour is used mixed with cereals, in cakes and as a food for invalids and infants. Young pods used as vegetable in India. Husks, dried leaves and stems, and bran fed to livestock. Seeds are a source of commercial starch for textile and printing industries. Green plants make valuable green manure but DM is low. Lentils are used as a meat substitute in many countries.

FOLK MEDICINE: Lentils are supposed to remedy constipation and other intestinal afflictions. In India, lentils are poulticed onto the ulcers that follow smallpox and other slow-healing sores.

CHEMISTRY: Dried seeds contain per 100 g: 340 calories, 12.0% moisture, 20.2 g protein, 0.6 g fat, 65.0 g total carbohydrate, ca. 4 g fiber, 2.1 g ash, 68 mg Ca, 325 mg P, 7.0 mg Fe, 29 mg Na, 780 mg K, 0.46 mg thiamine, 0.33 mg riboflavin, 1.3 mg niacin, and 0 mg ascorbic acid. The starch content is ca. 40.4%, total sugars 2.7%, and reducing sugars 1.8%. Fructose, glucose, stachyose, sucrose, and verbascose have been identified among the sugars. Decorticated seeds contain per 100 g: 344 calories, 9.9% moisture, 25.8 g protein, 1.8 g fat, 58.8 g total carbohydrate, 0.9 g fiber, 3.7 g ash, 24 mg Ca, 271 mg P, 10.6 mg Fe, 0.47 mg thiamine, 0.21 mg riboflavin, and 1.5 mg niacin. Dried seeds contain per 100 g: 0.49 mg pyridoxine, 1.5 mg pantothenic acid, 0 µg vitamin B_{12}, and 110 µg folic acid. Lentils are a good source of B vitamins, containing per 100 g: 0.26 mg thiamine, 0.21 mg riboflavin, 1.7 mg nicotinic acid, 223 mg choline, 107 mg folic acid, 130 mg inositol, 1.6 mg pantothenic acid, 13.2 mg biotin, and 0.49 mg pyridoxine. Mentioned also are 1.6 mg carotene, 4.2 mg ascorbic acid, 0.25 mg vitamin K, and 2.0 mg tocopherol. Vitamins, except folic and pantothenic acid, increase markedly during sprouting. Dry lentil husks contain 11.1% protein (1.3% digestible), 0.7% fat, 47.5% carbohydrate, 25.6% fiber, and 3.1% ash (nutritive ratio: 40). Dry bran contains 20.5% protein, 3.2% fat, 71.7% carbohydrate, 0.43% CaO, 0.79% P_2O_5 with 12.7% digestible protein; lentil hay at 10.2% moisture contains 1.8% fat, 4.4% protein, 50.0% carbohydrate, 21.4% fiber, 10.8% soluble mineral matter, and 1.4 insoluble mineral matter. The seed proteins are 44.0% globulin, 20.6% glutelin, 1.8% prolamine, with 25.9% water-soluble fraction. From seeds assayed at 26.9% protein and 0.8% oil, lysine was 6.7 g/16 N, methionine 0.6 g, arginine 7.8 g, glycine 3.7 g, histidine 2.1 g, isoleucine 3.8 g, leucine 6.6 g, phenylalanine 4.2 g, tyrosine 2.9 g,

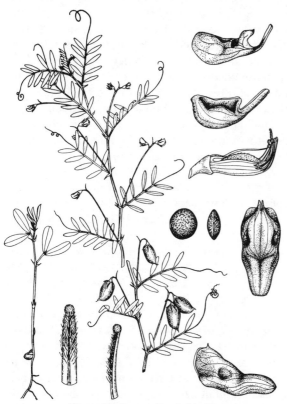

Figure 56. *Lens culinaris* Medik.

Lens culinaris **Medik.**

threonine 3.3 g, valine 4.2 g, alanine 3.5 g, aspartic acid 10.9 g, glutamic acid 14.5 g, hydroxyproline 0.0, proline 3.5, and serine 4.4 g. Seeds are reported to contain trypsin inhibitors and chymotrypsin inhibitors. The seeds also contain amylase, phosphatase, phytase, a saponin (esculenin) and a low-molecular-weight phytohemagglutinin.

DESCRIPTION: Annual bushy herb, erect or suberect, much-branched, softly hairy; stems slender, angular, 25–75 cm tall; leaves compound, pinnate, usually ending in a tendril or bristly; leaflets 4–7 pairs, alternate or opposite, oval, sessile, 1–2 cm long; stipules small, entire; stipels absent; flowers small, pale blue, purple, white or pink, in axillary 1–4-flowered racemes; pods oblong, flattened or compressed, smooth, to 1.3 cm long, 1–2-seeded; seed biconvex, rounded, small, 4–8 mm × 4–8 mm × 2.2–3 mm, lens-shaped, green, greenish-brown or light red speckled with black; seed weight of 100 varies 2–8 g; cotyledons red, orange, yellow, or green, bleaching to yellow, often showing through the testa, influencing its apparent color. Germination cryptocotylar. Fl. July–Sept. (United States); Mar.–Apr. (Near East).

GERMPLASM: Cultivars and varietal differences based on color of flowers and color, shape, and mottling of seeds. Usually placed in two groups of subspecific rank: *Lens culinaris* ssp. *macrosperma* (Baumb.) Baroulina, large-seeded cvs with large flat pods and seeds, flowers large, white or blue. Mediterranean, Africa and Asia Minor; *Lens culinaris* ssp. *microsperma* (Baumb.) Baroulina, small-seeded cvs with small convex pods and small seeds, flowers small, violet-blue to white or pink. Chiefly in southwest and western Asia and Africa. These are reported to give best yields. Lentils may also be classified as summer and winter types. One wild species of the Near East *Lens orientalis* (Boiss.) Hand.-Mazz., shows close morphological similarities and close genetic affinities to *L. culinaris*. It is occasionally interconnected to the cultivated lentil by a series of intermediate types and is interfertile with it. On basis of combined archaeological, botanical, and genetic evidence, Zohary (1973) concludes *L. orientalis* is the wild progenitor of cultivated lentil and that the domestication of this pulse took place (in the millennium B.C.) in the Near East arc (Northern Israel, Syria, South Turkey, North Iraq, Western Iran). Slinkard's careful studies do not support the contention that protein content varies widely. Still in one study of a wide variety of germplasm, protein content ranged from 15 to 30% and was highest in small-seeded cvs. Cystine and methionine vary widely in the land races. The Lentil Gene Bank, University of Saskatchewan, is soliciting seed of *Lens ericoides, L. monbretii, L. nigricans,* and *L. orientalis* and is increasing an array of near-isogenic lines of *Lens culinaris*. Assigned to the Near Eastern Center of Diversity, lentil or cvs thereof is reported to exhibit tolerance to alkali, disease, fungus, drought, high pH, heavy soil, heat, insects, low salinities, poor soil, slope, and virus. Some cvs have slightly higher levels of cold tolerance than others. Photoperiodic responses are more overriding. ($2n = 14$.)

DISTRIBUTION: Lentils probably originated in near East and Mediterranean region; known to ancient Egypt and Greece, where still cultivated. Then they spread northward into Europe as far as the British Isles, east to India and much of China, and south to Ethiopia. Now introduced and cultivated in most subtropical and warm temperate regions of world, and high altitudes of the tropics, as well as Chile and Argentina.

ECOLOGY: Seeds require a minimum of 15°C for germination, with an optimum of 18°–21°C; temperatures above 27°C are harmful; optimum temperatures for yields are around 24°C (opt. temp. range for 'Large Blond' is 19°–29°C, for 'Ancia,' 21°–25°C). Lentils grown as a cold weather or winter crop in the tropics, cultivated from sea level to 3,800 m, but are not suited to humid tropics. They are less damaged by drought than by waterlogging. In Bulgaria small-seeded cvs are more drought-resistant than large-seeded. Thrive on wide range of soils from light loams and alluvial soil to black cotton soils, best on clay soils; tolerate moderate alkalinity. Salt tolerance is higher during germination than during subsequent development. At 20.0 mmhos/cm salinity seed yields are reduced to 50%. Greenhouse studies suggest that tolerance to 3.9 mmhos is more realistic. Lentils are quantitative long-day plants, some cvs tending to be day-neutral. 'Large Blond' requires 14–16 hr to flower while 'Ancia' flowers under 9–16 hr; day-lengths of 15–16 hr promote early flowering. Ranging from Cool Temperate Steppe to Wet through Subtropical Dry to Moist Forest Life Zones, lentil is reported

Lens culinaris **Medik.**

to tolerate annual precipitation of 2.8–24.3 dm (mean of 35 cases = 7.9), annual mean temperature of 6.3–27.3°C (mean of 35 cases = 14.5), and pH of 4.5–8.2 (mean of 32 cases = 6.8).

CULTIVATION: Propagated by seed. Field should be plowed and harrowed to a fine texture. Seed may be either broadcast, or sown in drills at rate of 25–90 kg/ha, in rows 25–30 cm apart. In the United States, lentils are often planted with small grain equipment 17.5–30 cm apart in rows 60–90 cm apart. Lentils are a cool season species and as such are grown in the summer as a summer annual in temperate climates and in the winter as a winter annual in subtropical climates. In India, planting starts late in November spaced at 22.5 × 30 cm when sown monoculture. Two intercultivations are usually sufficient for weed control during establishment. A 2–3–1 NPK mixture is applied at planting, followed by a top-dressing of ammonium sulfate at flowering. Crop, usually rainfed, yields much better under irrigation but does not tolerate poor drainage. In India, lentils may be intercropped with barley, castor, mustard, and rice; in other areas rotated with other crops, especially wheat.

HARVESTING: Crop flowers in 6–7 weeks after planting with early cvs ready to harvest in 80–110 days, late cvs in 125–130 days. Plants cut to ground level and dried for 7–10 days, then combined or threshed. Low moisture is desirable at harvest. In the United States, lentils are harvested with modified cereal harvesters. In Russia problems with mechanical harvesting were overcome by intercropping with a fiber plant *Camelina* that served as a support. Lentils are usually stored in bulk bins or elevators.

YIELDS AND ECONOMICS: Seed yields range from 450–675 kg/ha in dry cultivations, may increase to 1688 kg/ha with irrigation, and have reached 3,000 kg/ha. In India, lentils sown in October at 35–44 kg/ha yielded 4.3 MT green matter in April. The straw-to-seed ratio in one cultivar is ca. 1.2 : 1. In studies of 28 cvs in New Delhi, pulse yields ranged from 558 to 1,750 kg/ha, while DM yields from 2,667 to 3,550 kg/ha. From 20 cvs in Uruguay, yields ranged from 190 to 1,500 kg/ha. Major producer of lentils in the world is India, with about 600,000 ha producing 270,000 MT/yr. World production is about 1.4 million hectares. Other important producers are Pakistan, Ethiopia, Syria, Turkey, and Spain. In 1959 U.S. production was about 18,000 ha, about 95% in the Northwest. In 1975, world production was 1,207,000 MT with an average 640 kg/ha. Asia, excluding Russia, produced 832,000 MT (597 kg/ha); Africa 161,000 MT (594 kg/ha); Europe, 88,000 MT (924 kg/ha); the USSR, 50,000 MT (1200 kg/ha), South America produced 29,000 MT (834 kg/ha). Yields were highest in France (1,695 kg/ha), Argentina (1588 kg/ha), Jordan (1,454 kg/ha), Egypt (1,307 kg/ha), United States (1,303 kg/ha). India produced 500,000 MT, Turkey 125,000 MT, Ethiopia 80,000 MT, Syria 67,000 MT.

BIOTIC FACTORS: Plants are usually self-pollinated, rarely cross-pollinated. Several fungal diseases are recorded on lentil, but lentils are not disease prone. They are seriously affected by fewer diseases than *Phaseolus* and *Pisum*. One of the worst diseases in Asia is the wilt, *Fusarium oxysporum* f. sp. *lentis*, which is favored by light and dry soils (soil moisture ca. 25%). Rust *Uromyces viciae-fabae*, is favored by high humidity and moderate temperatures (17°–25°C). Other fungal diseases recorded on lentil are: *Alternaria tenuis, Ascochyta ervicola, A. pinodella, A. pisi, Botrytis cinerea, Corticium rolfsii, C. solani, Erysiphe communis, E. polygoni, Fusarium avenaceum, F. culmorum, F. orthoceras* var *lentis* (wilt), *F. scirpi, Leveillula leguminosarum, Mycosphaerella tulasnei, Peronospora lentis, P. viciae, Phyllosticta phaseolina, Pythium debaryanum, Rhizoctonia* sp., *Sclerotinia fuckeliana, S. sclerotiorum, Uromyces ervi, U. fabae* (rust), *U. viciaecraccae,* and *Verticillium albo-atrum*. Lentils also attacked by bacterium, *Mycobacterium insidiosum;* by parasitic flowering plants, *Cuscuta* sp. and *Orobanche speciosa;* by viruses, pea enation mosaic, abutilon mosaic, bean yellow mosaic, cucumber mosaic, and euphorbia mosaic; and by nematodes, *Heterodera goettingiana, H. marioni,* and *H. schactii*. The only insect of economic importance in the United States is the pea leaf weevil *(Sitona lineatus)*. Aphids (*Aphis* spp.) and the vetch bruchid, *Bruchus brachialis,* are occasionally troublesome. In SE Asia, the main insect pests are the gram caterpillar, *Heliothis obsoleta;* white ants. *Clotermes* sp.; the gram cutworm, *Ochropleura flammatra;* and the weevil *Callosobruchus analis*. In the United States weeds are controlled by pre-planting application of diallate followed by postemergence application of dinoseb. In Bulgaria, preemergence prometryne followed by postemergence barban, is suggested, as well as linuron, metobromuron, and methoprotryne.

REFERENCES:
Ayoub, A. T., 1976, Salt tolerance of lentil *(Lens esculenta)*, J. Hortic. Sci. **52**(1):163–168.
Evans, L. E., and Slinkard, A. E., 1975, Production of pulse crops in Canada, in: *Oilseed and Pulse Crops in Canada* (J. T. Harapiok, ed), Chap. 12, W. Coop, Fertilizers, Calgary, Alberta.
Slinkard, A. E. (ed.), 1975–1977, *LENS (Lentil Experimental News Service),* Vols. II–IV (annual newsletter).
Williams, J. T., Sanchez, A. M. C., and Carasco, J. F., 1975, Studies on lentils and their variation. II. Protein assessment for breeding programmes and genetic conservation, *Sabrao J.* **7**(1):27–36.
Zohary, D., 1973, The wild progenitor and the place of origin of the cultivated lentil: *Lens culinaris, Econ. Bot.* **26**(4):326–332.

CONTRIBUTORS: J. A. Duke, H. L. Pollard, C. F. Reed, A. E. Slinkard, J. K. P. Weder, D. Zohary.

Lespedeza cuneata (Dum.) G. Don

FAMILY: Fabaceae
COMMON NAMES: Sericea, Perennial lespedeza
SYNONYM: *Lespedeza sericea* Miq., not Benth.

USES: Sericea is grown in the United States for pasture, hay, and seed; also used for soil improvement and erosion control. Forage is not much relished by stock, although seeds are eaten by wild birds. It is better than Korean lespedeza for soil improvement, especially on badly depleted soils, on which it is difficult to establish other legumes. Sericea is a good hay plant that cures easily; quality is adequate for maintenance of beef brood cows.

FOLK MEDICINE: No data available.

CHEMISTRY: Feeding trials show that sericea hay is inferior in nutritive value to Korean or Japanese lespedeza. Based on more than 70 analyses of sericea hay, Miller (1958) reports on a moisture-free basis: 8.1–21.2% crude protein (avg. 15.2), 0.9–5.4% fat (avg. 3.0), 16.8–43.5% crude fiber (avg. 27.5), 3.4–10.0% ash (avg. 6.1) and 40.3–54.1% N-free extract (avg. 48.2). Based on 15 analyses of green forage, he reports (moisture free

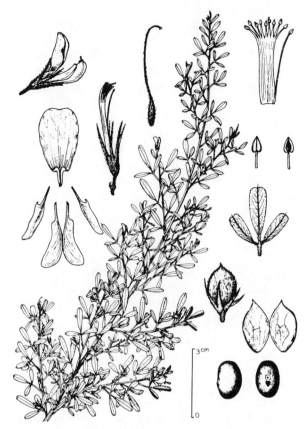

Figure 57. *Lespedeza cuneata* (Dum.) G. Don

basis): 13.1–21.7% crude protein (avg. 18.0), 3.1–4.8% fat (avg. 3.8), 19.8–25.7% crude fiber (avg. 22.7), 5.5–7.0% ash (avg. 6.2), and 46.9–53.0% N-free extract (avg. 49.3). The hay contained 0.8–2.5% Ca (avg. 1.5), 0.1–0.4% P (avg. 0.2), 0.6–1.4% K (avg. 1.0) 0.2–0.3% Mg (avg. 0.2) 0.01–0.06% Fe (avg. 0.03), 77–247 ppm Mn (avg. 125 ppm) and ca. 9.6 ppm riboflavin, and 39.4 ppm carotene. Whole plants contain 5.1–8% tannins; leaves contain 7.5–18% tannin; tannin content is higher if plant is not cut early. In high tannin leaves at the hay stage, tannin content ranged from 6.3 to 8.8%, DDM (digestible dry matter) ranged from 41 to 56% and crude protein 18.6 to 24.2%. In low-tannin leaves, tannin was 2.6–3.5, DDM 55–76 and crude protein 23.1–26.9. Stems have lower values. Seeds yield a semidrying oil with iodine value of 153.8. The genus *Lespedeza* is reported to contain bufotenine, dimethyl tryptamine, shikimic acid, and tannic acid.

DESCRIPTION: Perennial herb, with stiffly erect, tufted or simple, much-branched stems, 60–100 cm

Lespedeza cuneata (Dum.) G. Don

tall, sparsely white appressed-pilose, densely leafy; stipules 3-veined, lanceolate; leaves trifoliolate, with petioles 2–5 mm long; leaflets ascending or erect, linear-cuneate, 10–25 mm long, 2–4 mm broad, truncate and mucronate at apex, glabrous above, sericeous beneath, with slightly in-curved margins; flowers mostly solitary, or few fasciculate in axils of upper leaves; petaliferous flowers white marked with pink or purple, 6–7 mm long; calyx sericeous, 3–4 mm long with narrow lanceolate or lance-subulate lobes; pods ovate, about 3 mm long; apetalous flowers few to many, fasciculate in leaf axils, calyx 1.5–2 mm long, pods near-orbicular, about 3 mm long, longer than calyx. Fl. Aug.–Oct. Seed 771,750/kg; wt. kg/hl = 77.

GERMPLASM: A botanical variety, var *serpens* (Nakai) Ohwi (*L. serpens* Nakai) has prostrate, thick, spreading-pilose stems, thicker, broader leaflets, and flowers with a purplish standard. It is common in grassy places, abandoned lawns and especially common near the sea in Japan and China. Cvs include the following strains developed in the United States: 'Arlington,' developed in Virginia, is productive and widely adapted in the lespedeza region of the United States; 'Interstate,' developed in Alabama, is a multipurpose cv, used on highway right-of-ways, and is shorter growing, more profusely branched, with finer stems and grows more uniformly than other cvs, giving promise for grazing and hay in many regions; and 'Serala,' developed in Alabama, is high-yielding cv, with relatively fine, soft stems, producing numerous stems per plant, and is more palatable than other cvs because of the fine stems. 'Caricea,' developed in North Carolina, resembles 'Serala,' but has a broader crown with more stems that are somewhat less upright in habit. Of improved cvs only 'Serala' and 'Interstate' are available to farmers. Assigned to the China-Japan Center of Diversity, sericea or cultivars thereof exhibit tolerance to grazing, poor soil, slope, and weeds. Lowering the tannin content by breeding increases both the percentage and digestibility of crude protein. Much improvement may be expected from breeding for low tannin and high leaf percentage. ($2n = 20, 18$.)

DISTRIBUTION: Native along Himalayas from Hazara and Kashmir to Bengal, and up to 2,400 m in Nilgiris and 2,100 m in Palnis. Widely introduced and now cultivated in various areas of Orient and in the United States.

ECOLOGY: Plants are adapted to tropical, subtropical and warm temperate areas. Heat resistant, it grows best from May until August in temperate regions. Crop is quite tolerant to low fertility and generally gives a profitable response to lime and fertilizer only on badly depleted, acid soils. Ranging from Warm Temperate Dry to Moist through Subtropical Moist Forest Life Zone, sericea is reported to tolerate annual precipitation of 6.1–23.4 dm (mean of 14 cases = 12.7), annual mean temperature of 9.9–26.2°C (mean of 14 cases = 16.1), and pH of 4.9–7.1 (mean of 12 cases = 6.1).

CULTIVATION: Propagation is by seed; however, percentage of hard seed is high and germination may not exceed 25% unless seed is scarified. Crops are established with difficulty, but then grow well. A firm seedbed, shallow covering of seed, and a herbicide, are essential for good seedling establishment. Seed planted in early spring, about March or April, at 34 kg/ha or more of scarified seed, usually with no production the first year unless a herbicide is used. In some areas broadcast at 11–19 kg/ha. Plant is dormant in winter. In spring new growth comes from crown buds below ground level that were produced the preceding fall. Growth is profuse during rainy season. Lime should be supplied as needed for clover. Seed should be inoculated. Perennial lespedeza is usually sown alone but some authors say it may be seeded with either a winter or spring grain. Recommended fertilizer is as 3–12–12 or 4–12–8 with winter grain, or 0–12–12 or 0–14–7 with spring grain, used at grain seeding time at 300–400 kg/ha. Nitrogen is not recommended when sericea is sown alone.

HARVESTING: Sericea lespedeza contributes little to hay yields the first year in intercropped systems. Too short to be cut with clover or grass, it increases the nitrogen content of the soil, indirectly affecting grass yield. Contributes good pasture in late summer and fall. Meadows that are maintained beyond 1 year become largely grass-lespedeza stands, in which the grass supplies the early growth and the lespedeza the late growth; the combination providing a good distribution of production for several years. For midseason grazing, sericea may be established in permanent sod by the renovation or trash mulch method. Old vegetation must be reduced to a level that does not compete with the sericea seedlings. With use of an herbicide, cutting or grazing may be permitted in midsummer the first year. Sericea must be cut early to produce good hay. Two or three (rarely four) cuttings can be made between May and October in the lower south, but only one cutting in some areas, such as Ohio. Quality of hay is good from plants cut at 25–45 cm in height; however, the tannin content is too high for ideal livestock feed, either as hay or pasture. Thick stands with finer stems yield better hay. Hay cures rapidly. Unless handled properly, high leaf loss results in poor quality, stemmy, low-protein hay. Crop should not be cut too low, as recovery growth develops from axillary buds, not crowns. For good stands in succeeding years, no more than two hay cuttings or one hay cutting followed by a seed crop should be harvested annually. For grazing, pastures taller than 25 cm should be clipped to promote young growth. Pasture is best when plants are tender and growing vigorously. Mowing or grazing in late summer prevents storage in the roots of food reserves for the next spring. Sericea should be rotationally grazed; otherwise stand and productivity decline. Sericea is less palatable than annual lespedeza, and is less digestible than common lespedeza because of the high tannin content of the leaves, and the relatively high lignin content, coarseness, and woodiness of the stems. Cvs, e.g., 'Serala,' have thinner stems and lower tannin content. From stalks left after harvest of seed, paper pulp (with yield of about 50%) is obtained by various chemical treatments; it is a soft-fibered material, with high gloss and easily bleached, suitable for cheap paper. Bleached pulp can be blended with long-fibered pulps and used for speciality papers. Seed is harvested by binder, mower, or combine, but stems should remain on or be returned to land.

YIELDS AND ECONOMICS: Quality of sericea hay best when thick, immature stands are cut at 25–30 cm tall, on land of medium fertility; yield is 3,360–7,620 kg/ha from two cuttings per season. Annual forage yields exceeded 9,000 kg/ha from 2 or 3 cuttings at 9-week intervals. Yield was reduced by 41% at 6-week and by 53% at 3-week intervals. Seed yields average 335–1,000 kg/ha and are largest when entire season's growth is allowed to accumulate and mature. Considered a valuable plant for soil improvement and soil erosion control and, to a lesser extent, for forage. Lespedeza belt in the United States extends from southern Pennsylvania across central Ohio, Indiana, and Iowa, southward to Florida and eastern Texas.

BIOTIC FACTORS: Up to 80% of natural crossing takes place in chasmogamous flowers on selected plants. The occurrence of outcrossed and selfed seed on the same plant is an aid in genetic and breeding studies. Mature seeds are of two types, offering a convenient way to determine the effects of inbreeding and outbreeding. Following fungi have been reported on sericea lespedeza: *Cylindrosporium scoparium, Erysiphe pisi, E. polygoni, Ischnochaeta pisi, Olpidium trifolii, O. viciae, Pleosphaerulina briosiana, Uromyces lespedezae-procumbentis,* and *U. lespedezae-sericeae.* Nematodes damaging sericea include: *Helicotylenchus dihystera, Heterodera glycines, H. marioni, H. trifolii, Meloidogyne arenaria, M. hapla, M. incognita acrita, M. javanica, Paratylenchus projectus, Trichodorus christiei,* and *Tylenchorhynchus claytoni.* Slight damage may be done to plants by American grasshoppers (*Schistocera americana*), armyworms (*Pseudaletia unipuncta*), lespedeza webworms (*Tetralopha scortealis*), three-cornered alfalfa hopper (*Spissistilus festinus*), and other insects.

REFERENCES:

Bailey, R. Y., 1951, Sericea in conservation farming, *U.S. Dep. Agric., Farmers Bull.* **2033.**

Dodd, D. R., Thatcher, L. E., and Willard, C. J., 1948, The lespedezas in Ohio agriculture, *Bull. Agric. Ext. Serv., Ohio State Univ.* **300:**1–8, illus.

Donnelly, E. D., and Anthony, W. B., 1973, Relationship of sericea lespedeza leaf and stem tannin to forage quality, *Agron. J.* **65:**933–994.

Hadlaway, C. W., et al., 1936, Korean lespedeza and sericea lespedeza hays for producing milk, *Va. Agric. Exp. Stn., Bull.* **305.**

Hawkins, G. E., 1955, Composition and digestibility of lespedeza sericea hay and alfalfa hay plus gallotannin, *J. Dairy Sci.* **38:**237–243.

Henson, P. R., 1957, The lespedezas, *Adv. Agron.* **9:**113–157.

Hoveland, C. S., and Anthony, W. B., 1974, Cutting management of sericea lespedeza for forage and seed, *Agron. J.* **66**(2):189–191.

Minton, N. A., and Donnelly, E. D., 1972, Nematode-resistant sericea being developed, *Lespedeza cuneata, Ga. Agric. Res.* **14**(3):7–9.

CONTRIBUTORS: W. A. Cope, J. A. Duke, C. S. Hoveland, C. F. Reed.

Lespedeza stipulacea Maxim.

FAMILY: Fabaceae
COMMON NAME: Korean lespedeza
SYNONYMS: *Kummerowia stipulacea* (Maxim.) Makino
Microlespedeza stipulacea (Maxim.) Makino

USES: Widely grown for forage, primarily for pasture from June until frost. Also considered valuable for hay and green manure and for soil improvement and erosion control. Used on fertile bottomlands as well as on depleted or eroded upland fields or abandoned slopes, in rotation with various crops including cereal grains and flax, or sown in perennial grass pastures or on depleted native pastures. May be used for orchard cover crop or for roadside plantings.

FOLK MEDICINE: No data available.

CHEMISTRY: On the basis of 28 analyses, Miller (1958) reported for Korean lespedeza hay (moisture-free basis): 9.1–19.8% crude protein, (avg. 14.7), 2.4–6.4% fat (avg. 3.5), 25.0–39.0% crude fiber (avg. 30.6), 5.2–10.1% ash (avg. 6.6) and 41.5–46.9% N-free extract (avg. 44.6). On the basis of 100 analyses, he reported that the green forage contained (moisture-free basis): 11.3–25.0% crude protein (avg. 16.5), 1.5–4.7% fat (avg. 2.7), 18.5–46.6% crude fiber (avg. 30.1), 5.9–14.8% ash (avg. 9.3), and 29.6–49.1% N-free extract (avg. 41.4). The hay contained: 0.6–2.0% Ca (avg. 1.0), 0.07–0.38% P (avg. 0.22), 0.6–1.7% K (avg. 1.0), 0.19–0.49% Mg, (avg. 0.31), 0.01–0.10% Fe (avg. 0.02), and 60-211 ppm Mn (avg. 84). Green forage contained: 0.7–1.6% Ca (avg. 1.1), 0.04–0.90% P (avg. 0.34), and 1.36–1.88% K (avg. 1.58).

DESCRIPTION: Annual herb, diffusely branched; stems 1–6 dm long, sparsely pubescent with upwardly appressed hairs; stipules ovate-lanceolate, brown, scarious, persistent, 4–6 mm long; leaves trifoliolate, on petioles 2–10 mm long, lower ones spreading, upper ones nearly erect or appressed; leaflets broadly ovate; flowers in dense leafy spikes 1–2 cm long, in upper axils of leaves, 6–7 mm long, either apetalous or petaliferous, and then petals pink or purple, 6–7 mm long; calyx-tube about 1 mm long, the lobes ovate, equaling the tube; pod oval or obovate, about 3 mm long, 1-seeded, strongly reticulate, rounded at apex. Fl. Aug.–Oct. Seeds 496,125/kg; weight unhulled kg/hl = 51.

GERMPLASM: Several selections and cvs of Korean lespedeza have been developed and are widely cultivated in middle and southern United States. 'Harbin' and 'F.C. 19604' are extra early cvs, especially useful in more northern areas. 'Climax' is a late-maturing cv, selected from introductions from China and is resistant to bacterial wilt of lespedeza and has field resistance to tar spot; matures earlier than 'Kobe' and 10 days later than 'Rowan.' 'Yadkin,' developed in North Carolina, has light-colored flowers, is resistant to tar spot

Figure 58. *Lespedeza stipulacea* Maxim.

and moderately resistant to common rootknot nematodes, and produces more forage and seed than 'Climax.' 'Summit,' developed in Arkansas and Missouri, is a selection of hybrid Climax × Harbin, and is slightly later in maturity than 'Korean,' high yielding, except where rootknot nematodes are problem, and more resistant to bacterial wilt than most cvs; 'Iowa 6,' developed in Iowa, is an early-maturing cv, and carries some resistance to lespedeza wilt. 'Rowan,' developed in North Carolina, is intermediate in maturity between 'Iowa 6' and 'Climax,' and is moderately resistant to the two common nematodes and powdery mildew and is field resistant to tar spot, and is decidedly more productive than other Korean cvs. Assigned to the China–Japan Center of Diversity, Korean lespedeza or cultivars thereof is reported to exhibit tolerance to bacteria, disease, drought, fungus, grazing, limestone, nematodes, slope, and weeds. ($2n = 20$.)

DISTRIBUTION: Native to eastern Asia, North China, Korea, and Manchuria; introduced to the United States from Korea and China.

ECOLOGY: A warm temperate plant, adapted to a wide variety of soils, responding well to rich soils, but making some growth even on poor or acid soils. Much more resistant to high soil acidity than alfalfa, sweetclover, or red clover. Benefits from lime on extremely acid soils. Requires proper crop rotation and lime and fertilizers to maintain soil fertility. Liming to pH 6.0–6.5 and application of 11–20 kg P/ha annually are recommended. Withstands considerable drought once established, but is not truly drought-resistant. For maximum yield requires fairly humid conditions, considerable soil moisture and good fertile soils. Does not grow well on wet, poorly drained soils. Ranging from Cool Temperate Steppe to Wet through Subtropical Moist Forest Life Zone, Korean lespedeza is reported to tolerate annual precipitation of 4.9–16.0 dm (mean of 12 cases = 11.7), annual mean temperature of 8.4°–26.2°C (mean of 12 cases = 16.1), and pH of 5.5–7.3 (mean of 11 cases = 6.3).

CULTIVATION: Propagation by seeds. Scarification of seed not necessary. Korean often requires no seedbed preparation. Method of seeding depends on the use to be made of crop. It is said to be cheaper and easier to sow than alfalfa. Seed sown at rate of 11–17 kg unhulled seed/ha in late February or early March, later further north. Seed germinate in spring, sometimes as early as mid-March. Seedlings apparently tolerate light frost. Seed may also be drilled or broadcast into a cereal grain during late winter or very early spring. In some areas growers prefer to sow seed as thick as 25–30 kg/ha. Plants grow rapidly in warm weather after mid-July and help maintain pasture production during that period. Annual lespedeza is seldom planted alone; if so it may be drilled in March or early April on a firm seedbed, and not very deeply, at rate of 20–30 kg/ha. Simple rotation pasture seed mixture for spring seeding is 15 kg/ha of Korean and 6 kg/ha of orchardgrass or tall fescue. A crop of grass hay may be cut in early part of season, if desired. Also a small grain may be sown each year and the lespedeza only once as it volunteers. Each of the 2 crops may be grazed, cut for hay or allowed to mature seed according to needs. Inoculation of seed with common cultures of rhizobia is more helpful in plantings north of the Ohio River than further south.

HARVESTING: Grazing may start after crop is 5 cm or more tall and continue until frost. Even under close grazing, this crop produces enough seed to volunteer a crop the following season. Heavy grazing usually is not started before mid-July. When grown with small grains, Korean is ready for grazing a short time after grain harvest. Best quality lespedeza hay should be cut during or just before first bloom. If allowed to remain uncut until seed begins to form, hay is coarser and of lower quality. Early cutting stage ensures sufficient time after haying to permit formation of seed for volunteer seeding for following year. Lespedeza contains less moisture than does alfalfa and cures more rapidly in the field. May be windrowed immediately after mowing, and in dry weather may be ready for stacking 24 hr after cutting. Seed shatters only slightly when standing undisturbed, so crop can be allowed to ripen completely and be combined after frost. Harvesting by mowing, cocking and threshing as soon as seed is ripe should be done in mornings when wet with dew or after a light shower so as to reduce shattering. After drying a few days, cocks are threshed in grain thresher adjusted for threshing oats, or in a clover huller adjusted correctly.

YIELDS AND ECONOMICS: Yields of hay vary greatly with soil fertility, soil moisture supply and length

Lespedeza stipulacea Maxim.

of growing season. Yields average 2.5 MT/ha; but under favorable conditions reach 5–7.5 MT/ha. Good quality lespedeza hay is nearly equal to alfalfa in feeding value. Under average conditions yields of unhulled seed are 300–400 kg/ha and on good soils in favorable seasons have reached 800–1000 kg/ha. In extremely dry years or on soils of low fertility, yields may be as low as 100 kg/ha. Extensively cultivated for pasture, forage and for soil erosion and for soil improvement, especially in central and southern United States.

BIOTIC FACTORS: Following fungi have been reported on Korean lespedeza: *Catosphaeropsis caulivora, Cercospora latens, C. lespedezae, Erysiphe polygoni, Microsphaera diffusa, Pythium* sp., *Rhizoctonia bataticola,* and *Sclerotium rolfsii.* It is also attacked by the bacterium, *Phytomonas lespedezae,* and is parasitized by *Cuscuta pentagona,* which can be a serious weed. Nematodes isolated from and for which strains are bred resistant include: *Heterodera glycines, H. lespedezae, Meloidogyne arenaria, M. hapla, M. incognita acrita, M. javanica,* and *Tylenchorhynchus claytoni.* Other nematodes include *Belonolaimus gracilis, Paratylenchus projectus,* and *Trichodorus christiei.* Insect pests include: American grasshopper *(Schistocerca americana)*, armyworm *(Spodoptera frugiperda)*, threecornered alfalfa hopper *(Spissistilus festinus)*, lespedeza webworm *(Tetralopha scortealis)*, leaf tier *(Archips obsoletana)*, whitefringed beetles *(Graphognathus* spp.), and Japanese beetles *(Popillia japonica)*. Local agents should be contacted for control of these pests.

REFERENCES:
Anderson, K. L., 1949, Lespedeza in Kansas, (Contrib. No. 396, Dep. Agron.), *Agric. Exp. Stn., Kansas State Coll. Agric. Appl. Sci.,* **251**:1–19, illus., Manhattan.
Henson, P. R., and Cope, W. A., 1969, Annual lespedezas: Culture and use, *U.S. Dep. Agric., Farmers Bull.* **2113**:1–16, illus.

CONTRIBUTORS: J. A. Duke, C. F. Reed.

Lespedeza striata (Thunb. ex Murr.) Hook. & Arn.

FAMILY: Fabaceae

COMMON NAMES: Japanese or Annual lespedeza, Common lespedeza

SYNONYMS: *Kummerowia striata* (Thunb.) Schindl.
Microlespedeza striata (Thunb.) Makino

USES: Commonly grown in southern United States for pasture and forage; extensively used for erosion control; also used for green manure. Particularly valuable for permanent pastures, as it reseeds itself.

FOLK MEDICINE: No data available.

CHEMISTRY: On the basis of more than 100 analyses, Miller (1958) reports that common lespedeza hay (89.6% DM) contains (moisture-free basis): 8.3–16.5% crude protein (avg. 13.7), 1.4–4.8% fat (avg. 2.7), 22.7–45.8% crude fiber (avg. 32.0), 3.0–7.5% ash (avg. 5.5), and 39.4–49.5% N-free extract (avg. 46.1). On the basis of 12 analyses, the green forage (27.6% DM) contains (moisture-free basis): 12.6–18.4% crude protein (avg. 15.3), 1.7–2.5% fat (avg. 2.1), 29.0–45.3% crude fiber (avg. 36.1), 7.2–13.8% ash (avg. 11.2), and 31.6–35.3% N-free extract. The hay contained

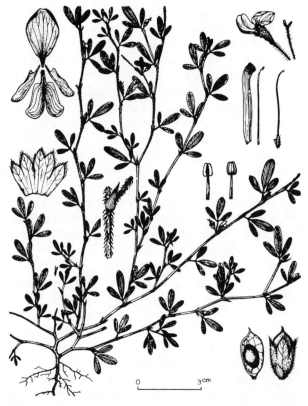

Figure 59. *Lespedeza striata* (Thunb. ex Murr.) Hook. & Arn.

0.8–1.5% Ca (avg. 1.0), 0.12–0.40% P (avg. 0.20), 0.81–1.49% K (avg. 1.05), 0.14–0.45% Mg (avg. 0.29), 0.023–0.068% Fe (avg. 0.031), and 114–217 ppm Mn (avg. 187). The green forage contained 1.00–1.21% Ca (avg. 1.13), 0.19–0.28% P (avg. 0.27), 0.55–1.28% K (avg. 1.16), 0.24–0.41% Mg (avg. 0.27), 0.020–0.045% Fe (avg. 0.032), and 79–251 ppm Mn (avg. 178). Moldy hay is liable to produce a hemorrhagic disease similar to that produced by sweetclover. Seeds are eaten by wild birds and yield a semidrying oil with iodine value of 149.7.

DESCRIPTION: Annual herb, with erect, much-branched, slender stems, 1–4 dm tall; stems with downwardly curved, short, white hairs; stipules narrowly ovate, few-veined, 5–8 mm long, erect, persistent, brown, scarious; leaves trifoliolate, on petioles 1–2 mm long; leaflets oblong-obovate, thin, 10–15 mm long, 5–8 mm broad; flowers 1–3 in upper axils, sessile or on short pedicels up to 4 mm long, pink or purple, the petaliferous ones 6–7 mm long, mingled with apetalous flowers; calyx-tube 1.5–2 mm long, the lobes about equal, oblong, reticulate, about as long as the tube; pods obovate, acute, 3–4 mm long, scarcely reticulate, with short appressed scattered hairs. Fl. Aug.–Oct. Seeds 418,830/kg unhulled; weight unhulled kg/hl = 32.

GERMPLASM: Improved cvs and strains are widely used in southern United States. 'Common,' which is prostrate except in dense stands, is well-adapted in the deep South and is sometimes preferred in permanent pastures. 'Kobe' is more erect form, larger, and coarser than 'Common'; and more productive for hay and pasture. Assigned to the China–Japan Center of Diversity, common lespedeza or cvs thereof is reported to exhibit tolerance to high pH, limestone, poor soil, slope, virus, and weeds. ($2n = 22$.)

DISTRIBUTION: Native to Manchuria, China, Korea, Ryukyus, and Taiwan; introduced and naturalized in North America.

ECOLOGY: Common in waste grounds and roadsides where native. Grows on almost any type of soil, doing well on sandy loam soils of Coastal Plain, on clay soils of Piedmont and on limestone soils in Virginia, Tennessee, and Kentucky. Grows on soils too acid for clover. Suitable for poor or worn out lands. Crop does better on good land and makes best growth on fertile bottomland. On poor soils it responds to both lime and fertilizers. Phosphate increases yields and should be used generally on all poorer soils. Liming to pH 6.0–6.5 is recommended; sometimes an application of 11–20 kg P/ha annually is recommended. Adapted to the climatic conditions of central and lower southern United States. Ranging from Warm Temperate Dry to Moist through Subtropical Moist Forest Life Zone, common lespedeza is reported to tolerate annual precipitation of 4.9–16.3 dm (mean of 14 cases = 11.4), annual mean temperature of 8.4°–26.2°C (mean of 14 cases = 16.4), and pH of 4.9–7.3 (mean of 13 cases = 6.2).

CULTIVATION: Propagated by seed. Freshly prepared land or loose seedbeds should be rolled or otherwise firmed; firm seedbed is essential in establishing a good stand. Should be planted in late February or early March in the Deep South, and in March or early April further north. Sown broadcast or drilled alone or on winter grain. If broadcast seeding is too late (or if ground is too hard for freezing and thawing to work seed under surface), field should be lightly harrowed after seeding. Often grown mixed with grasses or sweetclover, but for hay is grown pure. Inoculation with common cultures of rhizobia is helpful. It is slower to start growth in spring than Korean cvs, but late summer yields are usually good. Since 1945, slightly more than 20% of total lespedeza seed harvest has been 'Kobe.' On the Coastal Plain ca 200–400 kg/ha 0-14-14 fertilizer is recommended. Amount of fertilizer depends to some extent on amount applied to crop preceding the lespedeza in rotation. Where lespedeza is grown each year with winter grains and harvested for hay, the available supply of phosphate and potash may be exhausted.

HARVESTING: Hay is best cutting at or just prior to first bloom; in North Carolina, this is about mid-August. When left until considerable seed is ripe, hay is of poor quality. Lespedeza contains less moisture than alfalfa or red clover and is more quickly cured. Field-cured hay contains somewhat more dry matter than do field-cured alfalfa or clover hay. Common lespedeza cut when about 25 cm tall should be windrowed soon after cutting and in good weather may be hauled to barn in 24 hours. If cut in early morning, it may be stacked late the same

Lespedeza striata (Thunb. ex Murr.) Hook. & Arn.

day. Taller plants should be left longer to dry before moving. To assure a volunteer crop, it should be cut early enough to permit second growth to produce sufficient seed, or left long enough to permit shattering when cut initially. For temporary and permanent pastures, the best growth occurs in summer and provides excellent grazing during this period when ladino and other white clover pastures are low in production. Grazing begins in June or July, depending on location, and may continue until frost. Seed is harvested throughout the region where lespedeza is grown. Seed of striate cvs, 'Common' and 'Kobe,' shatter more readily than seed of 'Korean.' For high yields of seed, it should be harvested, ordinarily by combine, soon after maturing. As soon as seed is mature, plants are mowed and windrowed. After curing in windrow, it is threshed from the combine. Seed from combine must be recleaned to take out weed seeds.

YIELDS AND ECONOMICS: Yields average 2.5–7.5 MT/ha of hay and 100–250 kg/ha of seed. Annual lespedezas are valuable primarily for pastures, and secondarily for hay and soil erosion control and improvement. They increase soil fertility, adding nitrogen and organic matter, grow slowly in spring, make good soil cover in summer, and provide good stubble and debris in winter. Extensively grown in the southern half of the United States.

BIOTIC FACTORS: Following fungi have been reported on striate lespedeza: *Corticium solani, Erysiphe polygoni, Microsphaera diffusa, Thielavia basicola, Uromyces lespedezae,* and *U. lespedezae-procumbentis.* It is also attacked by the bacterium, *Phytomonas lespedezae;* and is parasitized by dodder, *Cuscuta* sp., which can be a serious weed. Nematodes causing damage to striate lespedeza include: *Helicotylenchus dihystera Heterodera glycines, H. lespedezae, H. trifolii, Hoplolaimus galeatus, Meloidogyne arenaria, M. hapla, M. incognita acrita, M. javanica, Paratylenchus projectus, Pratylenchus brachyurus, Rotylenchus,* sp., *Trichodorus christiei,* and *Tylenchorhynchus claytoni.* Insects pests vary with the locality, but those most often causing damage in southern United States are the following: American grasshopper *(Schistocerca americana),* fall armyworm *(Spodoptera frugiperda),* threecornered alfalfa hopper *(Spissistilus festinus),* lespedeza webworm *(Tetralopha scortealis),* leaf tier *(Choristoneura obsoletana),* whitefringed beetles *(Graphognathus* spp.), Japanese beetles *(Popillia japonica).* Local agents should be consulted for control.

REFERENCES:
Anderson, K. L., 1949, Lespedeza in Kansas, (Contrib. No. 396, Dep. Agron.), *Agric. Exp. Stn., Kansas State Coll. Agric. Appl. Sci., Circ.* 251:1–19, illus., Manhattan.
Henson, P. R., and Cope, W. A., 1969, Annual lespedeza: Culture and uses, *U.S. Dep. Agric., Farmers Bull.* 2113:1–16, illus.

CONTRIBUTORS: J. A. Duke, C. F. Reed.

Leucaena leucocephala (Lam.) de Wit.

FAMILY: Mimosaceae

COMMON NAMES: Leucaena, Koa haole, Ekoa, Hediondilla, Zarcilla, Tanta, Jumbie bean, Ipil-Ipil

SYNONYMS: *Leucaena glauca* (L.) Benth.
Leucaena latisiliqua (L.) Gillis

USES: Leucaena is valued as an excellent protein source for cattle fodder, consumed browsed or harvested, mature or immature, green or dry. The nutritive value (discounting toxicities) is equal to or superior to alfalfa. Leucaena also is used in land reclamation, erosion control, water conservation, and reforestation and soil improvement programs, and is a good cover and green manure crop. Use of the leaves as a mulch is said to significantly increase yields of other crops. Seeds yield ca. 25% gum worthy of commercial investigation. Seeds are softened and strung as beans into items of jewelry for tourists in the West Indies. Young pods and shoots are eaten fresh in Mexico and as a cooked vegetable in the Philippines. Ripe seeds are used as a substitute for coffee. Wood is hard and heavy (sp. gr. 0.7), the sapwood light yellow, the heartwood yellow-brown to dark brown. Leucaena is widely used for fuel or charcoal, burning with high heat and low ash. Plants are used in some countries as shade for black pepper, coffee, cocoa, quinine, and vanilla and for hedges. In some places, renegade seedlings have been regarded as weeds. Common strains seed freely and may invade denuded areas, occasionally in dense stands. The depilatory chemical mimosine has been used, experimentally at least, to shear sheep.

Leucaena leucocephala (Lam.) de Wit.

Figure 60. *Leucaena leucocephala* (Lam.) de Wit.

FOLK MEDICINE: Medicinally, the bark, pods, and seeds are eaten for internal pain. A decoction of the root and bark is taken as a contraceptive, ecbolic, depilatory, or emmenagogue in Latin America. In cattle (ruminant), leucaena has no effect on conception. In rats, fertility was reduced.

CHEMISTRY: Mimosine content of seeds is about 10%, youngest leaves up to 10% and edible forage 4–5% (3.5%, on a dry weight basis of the protein). High mimosine intake causes loss of hair in nonruminants, especially horses, mules, donkeys, and hogs. In ruminants, mimosine is immediately converted in the rumen to dihydroxypyridine, a goitrogen that reduces blood thyroxine levels. More than 40% Leucaena in the diet causes loss of hair and weight and cracked hooves. Heating the leaves or adding ferrous sulfate reduces the mimosine or its toxicity. Leaves also contain 0.08% of the glucoside quercitrin. Per gram of N, there are 294 mg arginine, 88 cystine, 125 histidine, 563 isoleucine, 469 leucine, 313 lysine, 100 methionine, 188 methione + cystine, 294 phenylalanine, 231 threonine, 263 tryosine, and 338 mg valine. Raw young leaves are reported to contain per 100 g edible portion: 68 calories, 79.5% moisture, 2.9 g protein, 0.8 g fat, 15.3 g total carbohydrate, 1.8 g fiber, 1.5 g ash, 553 mg Ca, and 51 mg P. Raw tender tops and pods contain per 100 g edible portion: 59 calories, 80.7% moisture, 8.4 g protein, 0.9 g fat, 8.8 g total carbohydrate, 3.8 g fiber, 1.2 g ash, 137 mg Ca, 11 mg P, 9.2 mg Fe, 4730 μg β-carotene equivalent, 0.09 mg riboflavin, 5.4 mg niacin, and 8 mg ascorbic acid. On a moisture-free basis, the immature forage contains 2.8% Ca, 0.12% P, 1.3% K, and 0.22% Mg, 14.3% protein, 2.1% fat, 33.6% crude fiber, 5.7% ash, and 44.3% N-free extract. The genus *Leucaena* is reported to contain hydrocyanic acid, leucaenine, quercitrin, and tannic acid.

DESCRIPTION: Nonspiny arborescent deciduous small tree or shrub, to 20 m tall, fast-growing; trunk 10–25 cm in diameter, forming dense stands; where crowded, slender trunks are formed with short bushy tuft at crown, spreading if singly grown; leaves evergreen (when moisture is not limiting), alternate, 10–25 cm long, malodorous when crushed, bipinnate with 3–10 pairs of pinnae, each of these with 10–20 pairs of sessile narrowly oblong to lanceolate, gray-green leaflets 1–2 cm long, less than 0.3 cm wide; flowers numerous, axillary on long stalks, white, in dense global heads 1–2 cm across; fruit a pod with raised border, flat, thin, becoming dark brown and hard, 10–20 cm long, 1.6–2.5 cm wide, dehiscent at both sutures; seeds copiously produced, 15–30 per pod, oval, flattish, shining brown, 18,000–24,000/kg; taproot long, strong, well-developed. Fl. and Fr. nearly throughout the year.

GERMPLASM: More than 100 cvs and botanical varieties, and several closely related or synonymous species contribute to the leucaena gene pool. The commoner 'Hawaiian' type, native to coastal Mexico, is versatile in adaptation and now a widespread tropical weed. The 'Salvador' type, is tall and tree like, producing twice the biomass of the Hawaiian types. The 'Peru' type, treelike also, branches low on the trunk and several cvs produce abundant forage. A new Australian cv 'Cunningham' (from 'Guatemala' × 'Peru') gives much more edible forage than 'Peru.' Mimosine content varies and cvs lower in mimosine are needed. Colombian cvs and *Leucaena pulverulenta* have less mimosine. Assigned to the Middle America Center of Diversity

Leucaena leucocephala (Lam.) de Wit.

(Yucatan Peninsula), leucaena or cvs thereof is reported to exhibit tolerance to aluminum, disease, drought, fire, high pH, heavy soil, insects, laterite, light frost, limestone, low pH, salt, slopes, weeds, and wind. In some locations, some promise for tolerance to frost and waterlogging is evident. ($2n = 104$).

DISTRIBUTION: Probably native to Mexico and northern Central America, naturalized throughout the West Indies from Bahamas and Cuba to Trinidad and Tobago, and from Texas and Florida to Brazil and Chile. Also naturalized in Hawaii, Pacific Islands, and the Old World tropics.

ECOLOGY: Requires long, warm growing seasons, doing best under full sun. In Indonesia it is grown to 1,350 m, in Africa and Latin America to 1,200 m. Most natural stands are below 500 m in areas of 6–17 dm rainfall; growth rate is slower at higher altitudes. About one dm per month is required for good growth. The plant is known for its drought tolerance. Leucaena grows on a wide range of soils, but thrives on deep clay soils that are fertile, and moist. Tolerates aluminum and soils low in iron and phosphorus. Grows best on neutral or alkaline soils and poorly on acidic latosols unless Mo, Ca, S, and P are added. Intolerant of waterlogging, its deep root system permits it to tolerate many soil types, from heavy soils to porous coral. Ranging from Warm Temperate Dry to Moist through Tropical Very Dry to Wet Forest Life Zone, leucaena is reported to tolerate annual precipitation of 1.8–41.0 dm (mean of 30 cases = 14.9), annual mean temperature of 14.7°–27.4°C (mean of 30 cases = 24.0°C), and pH of 4.3–8.7 (mean of 21 cases = 6.1).

CULTIVATION: Trees are usually propagated by seed. Cuttings do not take well but trees coppice well. Some trees less than 1 yr old produce viable seed. Seeds remain viable from several months to several years. The hard waxy seedcoat must be scarified, usually in water at 80°C for 3–4 min. For forage, seed should be sown 1.0–5.0 cm deep at onset of wet season. Leucaena responds favorably to fertilizer and lime. Tolerates infertile acid soils if the seed is coated with an efficient Rhizobium and lime pelleted and fertilizer containing Ca, Cu, Mo, P, S, and Zn is applied. Minor elements can be sprayed onto the growing plants as a dilute solution. Irrigation and cultivation may be necessary. The crop soon produces a dense stand.

HARVESTING: The crop can be cut at any stage for fodder for cattle; green forage yields are maximum on 10–12 week cycles. Can be cut ca. once a month, depending on water available. Can be rotationally grazed, as the branches are very pliable, and tolerate extensive damage from grazing. The dry leaf meal is becoming an important high-protein component of chicken and pig rations.

YIELDS AND ECONOMICS: Leucaena produced 56 MT/ha/yr green forage in Hawaii near sea level. With adequate moisture green matter yields have been 80 MT/ha. Two-year-old trees have yielded 4.5–7 kg pods per tree. Annual DM yields are ca. 2–20 MT/ha, equivalent to up to 4,300 kg protein/ha, nearly double the yields of alfalfa. With its rhizobium, leucaena can fix up to 500 kg N/ha. On 3- to 8-yr-old trees, annual wood increments vary from 24 to over 100 m^3/ha, averaging 30–40 m^3. Dry leucaena wood has 39% the calorific value of fuel oil (10,000 cal/kg), leucaena charcoal 72.5%.

BIOTIC FACTORS: Leucaena is relatively resistant to the pests and diseases prevalent in Hawaii and elsewhere, but extensive plantation culture may invite the breakdown of this apparent resistance. Twig borers, seed weevils, and termites, as well as damping off may hinder the plant. Herbivorous mammals may be fond of the seedlings. Bananas may do better in the shade of leucaena than in full sunlight due to reduced damage by Sigatoka disease.

REFERENCES:
National Academy of Science, 1977, Leucaena, promising forage and tree crop for the tropics, Washington, DC., 115 pp.
Oakes, A. J., 1968, *Leucaena leucocephala:* Description, culture, utilization, *Adv. Front. Plant Sci.* **20**:1–114.

CONTRIBUTORS: J. C. Brewbaker, J. A. Duke, E. M. Hutton, C. F. Reed, E. C. Williams.

Lonchocarpus nicou (Aubl.) DC.

FAMILY: Fabaceae

COMMON NAMES: Barbasco, Cube, Timbo, White haiari

Lonchocarpus nicou (Aubl.) DC.

SYNONYMS: *Derris nicou* (Aubl.) Macbride
Lonchocarpus floribundus (Benth.) Willd.
Robinia nicou Aubl.
Robinia scandens Willd.

USES: Roots are a source of rotenone, an insecticide for controlling both chewing and sucking insects. Rotenone leaves no harmful residue and thus can be used with safety on garden plants and crops. Roots are also used by natives as fish poisons.

FOLK MEDICINE: Said to be used for insecticide.

CHEMISTRY: Roots contain 0.75–1% rotenone; also deguelin, tephrosin, and toxicarol.

DESCRIPTION: Erect shrub or liana, sometimes reaching the tree tops; leaves compound, petiole up to 2 dm long, with petiolules 4–9 mm long; leaflets 2–9, usually 7 in cultivation, oblong or elliptic to elliptic-lanceolate, 12–35 cm long, 4–12 cm broad, acute at base, more or less gradually caudate-acuminate, appressed pilose beneath, with 6–10 arcuate lateral veins; inflorescence 1–2 dm long, reddish silky tomentose, the peduncles 4–6 mm long, the slender pedicels about 3 mm long; calyx cylindric-campanulate, 4–5 mm long, the acute lobes 2 mm long; petals reddish-violet, silky outside, the suborbicular banner 12 mm broad and nearly as long, the wings nearly straight; pods broadly ovate to oblong-ovate, 4–9 cm long, 2.5–3 cm broad, rounded or subactue at apex, more or less acute at base, strongly compressed, minutely sericeous, 1–2-seeded.

GERMPLASM: In studying the variations in *L. nicou* in the Amazon, some students have placed several varieties or forms under this species, as var *languidus* Hermann, which has large papyraceous leaflets with very long tips averaging 3.5 cm long; var *urucu* (Killip & Smith) Hermann, with thick coriaceous leaflets, and var *utilis* (A. C. Smith) Hermann, with the secondary veins curved rather than straight and ascending as in typical *L. nicou*. Assigned to the South America Center of Diversity, cube or cultivars thereof is reported to exhibit tolerance to insects and shade.

DISTRIBUTION: Widely distributed in the Amazon Basin of Brazil, north to Guyana, Surinam, and west to Peru and Ecuador.

ECOLOGY: Found from sea level to ca. 1,340 m elevations, often in large clumps several ha in area. Fares better in black soil soil with good humus layer. Usually confined to well-drained low areas occupied by tall forests. Requires tropical conditions, annual rainfall about 20 dm, well distributed, mean maximum temperature of 38°C and a mean minimum of 18°C; plants in coastal regions cannot withstand 10°C. Plants benefit from some shade when small, but later make more vigorous growth in full sun. Plants grow best in fairly open well-drained soil, and do not withstand waterlogging. They grow well on acid humus. Root development and gathering better on permeable soil. Ranging from Subtropical Moist to Wet through Tropical Moist to Wet Forest Life Zones, cube is reported to tolerate annual precipitation of 20–31 dm, annual mean temperature of 22°–25°C, and pH of 6.0–7.0.

CULTIVATION: Present practices were developed through trial and error by the Amazon Indians and

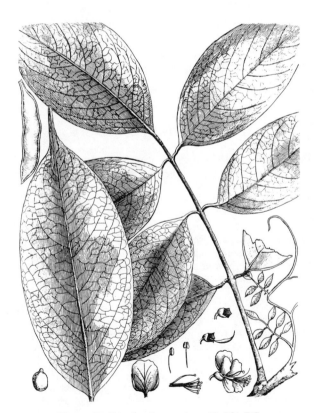

Figure 61. *Lonchocarpus nicou* (Aubl.) DC.

Lonchocarpus nicou (Aubl.) DC.

riverbank settlers to insure an adequate supply of fish poison. Plants are usually established on previously wooded land, as it is considered easier to clear land than to reclaim old cultivated fields. Land is usually cleared during the drier months of the year, so fallen trees dry sufficiently to burn. No further seedbed preparation is practiced. The farmer plants the crop during the next rainy season. Planters propagate with leafless stem cuttings 25–45 cm long and 2–5 cm in diameter. Of good cuttings, kept moist and out of the sun until planted, 80% root. Cuttings, usually with 3–6 axial buds, are inclined at 15°–60° angles in holes made by digging sticks, hoes or machetes. Apical portion, with 1–2 buds, is left exposed; soil is firmed around the rest. One or two cuttings are placed in each hole; 2 or more holes may be dug side by side, or in a circle, square or other design. The tangle of fallen logs makes precise spacing impossible, but spacings of 1–4 m are reported. Spacing is close when only 1–2 cuttings are used at each location, where cuttings may not survive. Lonchocarpus matures in 2.5–3 yr. During the first year it can be conveniently interplanted with food crops such as beans, cassava, corn, bananas, plantains, or okra. During the second and third years, the plants grow 2–3 m tall, and the planting broadens into a bushy thicket. At this stage interplanting becomes impractical, although mats of bananas, plaintains, or pineapples that have become established are left undisturbed. The average planter has several small lonchocarpus fields at various stages of maturity. Fields may be weeded until shade inhibits weed growth.

HARVESTING: Usually during the dry season (May–Aug. in Peru), the roots, which generally spread out laterally but may grow downward, are dug or pulled. Roots of a 2.5-yr-old plant may weigh 0.46–2.2 kg. The collector salvages as much root system as possible. He usually cuts the trunk off ca. 45 cm above the ground with a machete; then pries under the trunk with a stout sharp-pointed pole and lifts the crown enough to locate the roots. Then he cuts the roots on one side of plant and pulls and digs them loose. Finally, he pulls the trunk backward and rips the other roots from the soil. He may harvest ca. 10 kg/hr working 5–6 hr/day. Roots are tied with long pieces of liana in bundles purchased by a local dealer who sells to exporters in the principal shipping centers. Roots for export are stored for several months, until moisture content is reduced from 60% to ca. 20%, then baled and wrapped in burlap or unbleached muslin to prevent further moisture loss. At local grinding industries (Iquitos, Peru and Belem, Brazil), roots are air-dried, chopped, oven-dried, and then ground. The final product (200-mesh or finer) is bagged for shipment.

YIELDS AND ECONOMICS: Yields vary with locality and strain of cube. On virgin land, 2.5–3-yr-old plants normally yield ca. 13.5 MT/ha of green root or 7.5 MT/ha of air-dry root. In Brazil, 3.5-yr-old plants yield 17.5 MT/ha. Attempts have been made to select high rotenone-yielding strains. *L. utilis* contains more rotenone, but has a smaller root system than does *L. urucu,* which has more roots but less rotenone content. The latter is the source of most of the rotenone exported commercially from Brazil, largely from the basins of the Amazon and Solimoes Rivers, shipped mainly from Para and Manaos. The U.S. imports about 1,500 MT crude roots and ca. 1,000 MT rotenone extract. Average shipments of commercial root contain 4–6% rotenone. Cube has been introduced by agricultural stations in several Caribbean and Central American countries, the Philippine Islands and Malaysia.

BIOTIC FACTORS: The following fungi have been reported on this plant: *Dicheirinia archeri* and *D. guianensis.*

REFERENCES:
Cultivation of *Lonchocarpus,* 1938, *Bull. Imperial Inst., London* **36**(2):179–185.
Hermann, F. J., 1947, The Amazonian varieties of *Lonchocarpus nicou* a rotenone-yielding plant, *J. Wash. Acad. Sci.* **37**(4):111–113, illus.
Higbee, E. C., 1948, Lonchocarpus, Derris and Pyrethrum: Cultivation and sources of supply, *U.S. Dep. Agric., Misc. Publ.* **650**:1–16, illus.
Krukoff, B. A., and Smith, A. C., 1937, Rotenone-yielding plants of South America, *Am. J. Bot.* **24**:576–587.
Legros, J., 1939, Some *Lonchocarpus* species, rotenone-yielding plants of South America, *Int. Rev. Agric.* Jan.–Feb.: 11T–61T.
Roark, R. C., 1936, *Lonchocarpus* species (Barbasco, Cube, Haiari, Nekoe, and Timbo) used as insecticides, *U.S. Dep. Agric., Bur. Ent. Pl. Q. Publ.* **E-367**:1–133.
Roark, R. C., 1938, *Lonchocarpus* (Barbasco, Cube and Timbo)—A review of recent literature, *U.S. Dep. Agric., Bur. Ent. Pl. Q. Publ.* **E-453**:1–174.
Wylie, K. H., 1948, World production, supply and price of rotenone, *U.S. Dep. Agric., For. Agric. Rep.* **28**:1–5.

CONTRIBUTORS: J. A. Duke, C. F. Reed.

Lonchocarpus urucu Killip & Smith

FAMILY: Fabaceae

COMMON NAMES: Barbasco, Cube, Timbo urucu

SYNONYMS: *Derris urucu* (Killip & Smith) Macbride
Lonchocarpus nicou var *urucu* (Killip & Smith) Hermann

USES: Roots are used as a source of rotenone, an insecticide widely employed in agriculture for controlling both chewing and sucking insects. Leaving no harmful residue, rotenone can be used safely on garden plants and crops.

FOLK MEDICINE: Said to be used for insecticide (rt).

CHEMISTRY: Rotenone content averages 4.4%.

DESCRIPTION: Erect shrub, often attaining the tree tops; leaves compound, petioles 7–15 cm long, the petiolules 4–9 mm long; leaflets 3–4 pairs, opposite, with a terminal leaflet, spreading pilose beneath, the trichomes more or less golden, thin coriaceous, dark green and lustrous above, dull beneath, obovate-oblong, especially the terminal one, or elliptical, 10–31 cm long, 6–16 cm broad, rounded or obtuse at base, abruptly acuminate at apex, the upper surface including the midvein glabrous; lateral veins 7–12, prominent beneath; inflorescence 1–2 dm long, reddish silky tomentose, the peduncles 4–5 mm long, the slender pedicels ca. 3 mm long; calyx cylindric-campanulate, 4–5 mm long, nearly as broad, the subacute lobes ca. 2 mm long; petals reddish-violet, silky outside, the suborbicular banner 12 mm broad, nearly as long, retuse, the wings nearly straight; ovary minutely sericeous, with 4 ovules; pods broadly ovate to oblong-ovate, 4–9 cm long, 2.5–3 cm broad, rounded or subacute at apex, acute at base, strongly compressed, minutely sericeous, 1(–2)-seeded.

GERMPLASM: Assigned to the South America Center of Diversity, timbo urucu or cultivars thereof is reported to exhibit tolerance to insects and shade.

DISTRIBUTION: Native to the rain forest area of Brazil, where widely distributed throughout the Amazonian region west to Peru and Colombia.

ECOLOGY: Ranging from Subtropical Moist to Wet through Tropical Moist to Wet Forest Life Zones, urucu timbo is reported to tolerate annual precipitation of 20–31 dm, annual mean temperature of 22°–25°C and pH of 6.0–7.0.

BIOTIC FACTORS: The fungus *Dicheirinia archeri* has been reported on this plant. Although these roots contain a natural insecticidal principle, they are subject to infestation by insects unaffected by the rotenone content.

CONTRIBUTORS: J. A. Duke, C. F. Reed.

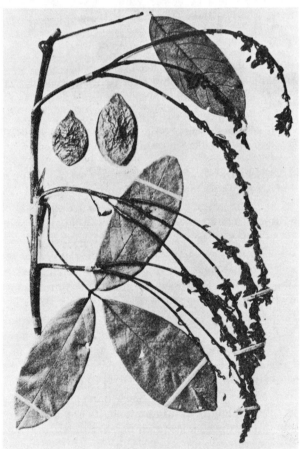

Figure 62. *Lonchocarpus urucu* Killip & Smith

Lotus corniculatus L.

FAMILY: Fabaceae

COMMON NAME: Birdsfoot trefoil

SYNONYMS: *Lotus ambiguus* Besser ex Sprengel
Lotus caucasicua Kuprian

USES: Birdsfoot trefoil is an excellent nonbloating forage. Compared with alfalfa and clover hay, it

Lotus corniculatus L.

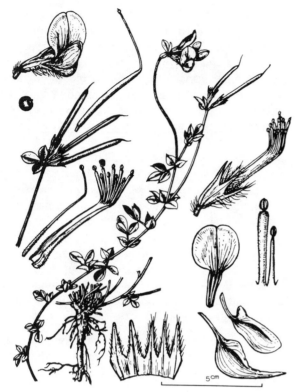

Figure 63. *Lotus corniculatus* L.

has softer stems, lower cellulose content and more carbohydrate. It is a good pasture, hay, and silage crop for horses and cattle, and is of major importance on less fertile, poorly drained soils; also used on fertile soils in permanent pastures and in pastures hard to cultivate. Makes a good drought-resistant ground cover.

FOLK MEDICINE: No data available.

CHEMISTRY: Because flowers of some cvs contain traces of prussic acid, some plants are suspected to be mildly toxic in flowering stage, but they are quite innocuous in form of hay or silage. The species is polymorphic for cyanogenic glucosides. No cases of livestock poisoning reported. Forage is nonbloating because of tannins (0–3% dry weight of proanthocyanidins). Tannin content varies, with season and cv. 'Empire,' 'Winnar,' and 'Leo' are low in tannins, 'Maitland,' 'Franco,' and 'Viking' are high. In composition and nutritive value, it compares favorably with alfalfa and white clover. Analysis of pasture and hay from the United States gave the following values: pasture—total moisture, 80.0; protein, 5.6; fat, 1.0; N-free extr., 9.3; fiber, 2.6; and ash, 1.5%: digestible nutrients—digestible protein, 4.6; total digestible nutrients, 15.0%; nutritive ratio, 2.3; hay—total moisture, 8.8; protein, 14.2; fat, 2.1; N-free extr., 41.9; fiber, 27.0; mineral matter 6.0; Ca, 1.6; P, 0.20; and K, 1.66%; digestible nutrients—digestible protein, 9.8; total digestible nutrients, 55.0%; nutritive ratio, 4.6. The hay contains 7.3 mg/100 g carotene. Milk from cows fed on trefoil hay contains more vitamins A and E than that from cows given alfalfa hay. Currier and Strobel (1977) reported a new glycoprotein ("trefoil chemotactin") which, at micromolar concentrations, attracts six strains of rhizobia. Miller (1958) reported that the hay (90.4% DM) contained (moisture-free basis) 13.4–28.8% crude protein, (avg. 17.8), 2.0–5.3% fat (avg. 3.0), 12.6–32.9% crude fiber extract (avg. 44.3). The green forage (23.3% DM) contained on a moisture-free basis: 14.8–29.1% crude protein (avg. 21.6), 1.2–5.0% fat (avg. 2.5), 13.0–28.4% crude fiber (avg. 22.2), 5.6–10.5% ash (avg. 8.1), and 43.8–49.9% N-free extract (avg. 45.6). The hay contained 1.04–1.92% Ca (avg. 1.6), 0.16–0.49% P (avg. 0.24), 1.47–1.97% K (avg. 1.75), 0.31–0.78% Mg (avg. 0.63), 0.013–0.015 Fe (avg. 0.014), 0.6% Mn, 6.2–7.3 ppm thiamin (avg. 6.8), 14.5–17.6 ppm riboflavin (avg. 16.1), and 43.6–500.0 ppm carotene (avg. 143.9). The genus *Lotus* also contains cytisine, malonic acid, and saponin. Seeds are reported to contain trypsin inhibitors.

DESCRIPTION: Perennial herb, with well-developed long, branching taproot, reaching to depth of 1–1.1 m, with numerous lateral roots; stems glabrous or sparsely white-hairy, arising from a single crown, erect, ascending or decumbent, well-branched, 60–90 cm long, slender, moderately leafy; leaves with 5 broad leaflets, the lower 2 leaflets stipulelike, obliquely ovate, mostly acuminate, 7–10 mm long, 6–10 mm broad, the upper 3 leaflets obovate, rounded at apex, 10–15 mm long, 6–10 mm broad; leaflets of upper leaves lanceolate, acuminate; flowers 4–8 attached in umbels by short pedicels on a long peduncle 5–10 cm long; flowers yellow, the standard sometimes orange, 10–15 mm long; calyx 5–6 mm long, glabrous or finely hairy, teeth as long as tube; standard 10–15 mm long, its broadly rounded limb abruptly passing into a cuneate claw; wings 10 mm long, about equaling the keel, the obovate limb 4 times as long as the claw; keel incurved at a right angle; pods long, cylindrical, brown to almost black, about 2.5 cm long, 3 mm wide, 15–20 seeded; seeds olive-green, brown to almost black, or mottled with black spots, 1.4 mm long, 1.2 mm wide, oval to spherical, attached to ventral suture and at maturity pods splitting along both sutures and twisting spirally to release seeds. Seeds 826,875/kg; wt. kg/hl = 77. Fl. May–Sept.

GERMPLASM: Plants vary greatly as to size, shape, color and pubescence of stems and leaves as well as growth habit. There are many variants in Europe; those that are sparsely to densely pubescent with calyx-teeth shorter than tube, found mainly in northern Europe; those with glabrous or sparse pubescence, smaller leaflets, calyx-teeth about half as long as tube, found mainly along coasts of western and northern Europe; those with a villous or dense pubescence, calyx-teeth slightly longer than tube, found in central and southern Europe; and dwarf forms found in the mountains. In Kenya, var *eremanthus* Chiov. is cultivated and ranges from Sudan and Eritrea to Ethiopia and Tanzania. Cvs are numerous, and are sometimes divided into two types: Empire or New York type, a naturalized ecotype from Albany, New York, which has finer stems, is more prostrate in growth habit, 10–14 days later in flowering, and is more indeterminant in growth and flowering habit (include cv 'Carroll' and 'Dawn'); and European cvs developed from various European ecotypes. Erect-growing cvs include: 'Cascade,' 'Douglas,' 'Granger,' 'Leo,' 'Maitland,' 'Mansfield,' 'Tana,' and 'Viking'; these recommended for hay or rotational pasture. Semi-erect or spreading cvs include: 'Empire' and Empire strains, and are recommended for permanent and close-grazed pasture. Most cvs are winter-hardy. 'San Gabriel' types from South America are winter active. Assigned to the Eurosiberian Center of Diversity, birdsfoot trefoil or cultivars thereof is reported to exhibit tolerance to disease, drought, frost, fungus, grazing, herbicides, high pH, heavy soil, insects, low pH, mine spoils, salt, poor soil, sand, slope, virus, weeds, and waterlogging. ($2n = 24, 12, 26$.)

DISTRIBUTION: Native almost throughout Europe and parts of Asia, at least to Crimea, Caucasus, Central Asia, Iran, and India. Introduced in many areas; Australia, North America (Canada to Dakotas, south to West Virginia, Nebraska, Arkansas, and North Carolina), South America (Argentina, Brazil, Chile, and Uruguay) and elsewhere.

ECOLOGY: Naturally grows in pastures and used as feed for horses and cattle. Adapted to temperate climate of northern hemisphere (U.S. Hardiness Zones 3–10), and in cooler parts of tropics and subtropics (tolerates frosts of −6°C with no damage). Suitable to many types of soil varying from clays to sandy loams; will grow on poorly drained, droughty, infertile, acid, or mildly alkaline soils. Does better on acid soils with liming and is most productive on fertile, moderately well-drained soils (however, it tolerates waterlogging when actively growing); ph 4.5–7.0, with best nodulation from pH 6.0–6.5. A long-day plant requiring 16-hr day-length for full flowering. Thrives with rainfall of 5.5–9 dm/yr. Moderately salt tolerant, showing definite yield reductions only at 6–12 mmhos. Ranging from Boreal Moist to Wet through Subtropical Moist Forest Life Zone, birdsfoot trefoil is reported to tolerate annual precipitation of 2.1–19.1 dm (mean of 59 cases = 8.4 dm), annual mean temperature of 5.7°–23.7°C (mean of 59 cases = 11.1°C), and pH of 4.5–8.2 (mean of 56 cases = 6.4).

CULTIVATION: For agricultural use, the species is usually propagated by seed. May be propagated by stem cuttings; roots develop from callous tissue along internodes on stems and both shoots and roots develop from axillary buds at the nodes. However cuttings are weak and impractical for routine propagation. Seed should be plump with high germination value. Hard seeds amount to about 90% when hand-harvested, and 40% when combined. These seed do not germinate unless scarified to allow rapid imbibition of water. For vigorous seedlings, a fine firm seedbed is required as seedlings lack vigor and do not compete with nurse crops, weeds or grasses. Plowing should be well in advance of seeding to permit rains or irrigation to settle seedbed. Seedbed may then be disked, harrowed and rolled before seeding. Amount of seedbed preparation needed depends on kind of soil, moisture content and season. Seed rate 6–15 kg/ha (broadcast, band-seeded, or drilled). To band seed, drill fertilizer in. Lime is needed on acid soils and usually phosphate and potassium are needed on less fertile soils. Row spacing 20–40 cm. Owing to low seedling vigor and poor competitive ability, cultivation may require herbicidal treatment (pre-plant applications of EPTC or postemergence 2,4DB or dalapon [no grasses] may help). Inoculation is necessary for successful stands in areas where trefoils have not been grown before. Bacteria used to inoculate alfalfa, red clover, and other common legumes are not effective on trefoils. Strains of bacteria that inoculate birdsfoot and narrowleaf trefoil may not be effective on big trefoil. A specific inoculum must be used for proper effects.

HARVESTING: Properly cured hay is leafy and nutritious, comparable in quality to other good legume hays. The erect-growing cvs of European origin are used for hay. Hay is usually cut as plants come into bloom; yields are maximum at about the one-tenth flower stage. Trefoil makes good silage, especially from a second cutting with 18% dry matter. Increases production of permanent pastures and promotes animal gains, and is nonbloating. The feeding value of hay has equaled that of alfalfa and other legumes when fed to sheep and dairy cows. Hay can be stockpiled for later feeding. Primary grazing can be delayed until July 1 without sacrificing high yield of good quality forage. When initial growth has been removed as hay or silage between June 1 and 5, the crop may be grazed anytime up to September 1. If cut for hay, the stubble must contain sufficient green leaves for regrowth. All cvs set seed over an extended period of time, with both ripe pods and new flowers on plant at same time. Some pods dehisce before flowering and pod set are completed. Seed should be harvested when 70–80% of pods are mature. Usually crop is mowed, when most pods are brown, windrowed to dry before combining. Crop may also be harvested and stored like hay, and then threshed, but loss of seed is excessive.

YIELDS AND ECONOMICS: Seed yields vary from 200–520 kg/ha, but average 55–160 kg/ha. Year to year differences in seed set and difficulty in harvest often cause wide variation in average annual seed yields and prices. Beef production from 'Empire' birdsfoot trefoil-bluegrass pasture under rotational grazing was 57% greater than that from a similar pasture without trefoil. Total acreage in the United States increased from about 260,000 ha in 1957 to slightly over 800,000 ha in 1967, and is still increasing. In addition to use as a major pasture legume, either alone or with various grasses, it is used extensively on new highway slopes for improving the soil and controlling erosion.

BIOTIC FACTORS: Birdsfoot trefoil is mainly self-incompatible, and requires insects, usually Hymenoptera, to fertilize the flowers. Following fungi have been reported on birdsfoot trefoil: *Erysiphe communis, E. martii, E. polygoni, Fusarium avenaceum, F. roseum, F. solani, Mycoleptodiscus sphaericus, M. terrestris, Ovularia exigua, Peronospora lotorum, Phomopsis loti* (Phomopsis blight), *Pleospora herbarum, Ramularia schulzeri, Rhizoctonia solani* (crown and root rot), *Sclerotinia sclerotiorum* var *minor, S. trifoliorum, Sclerotium rolfsii* (southern blight), *Stemphylium loti* (leafspot and stem canker), *Thanatephorus cucumeris, Uromyces euphorbiae-corniculati, U. loti, U. striatus,* and *Verticillium* spp. Curly top and tobacco ringspot viruses attack plants, but are not serious. Nematodes isolated from birdsfoot trefoil include: *Helicotylenchus* sp., *Heterodera glycines, H. trifolii, Meloidogyne arenaria, M. hapla, M. incognita, M. incognita acrita, M. javanica, Paratylenchus* sp., *Pratylenchus penetrans. P. pratensis, P. vulvus, Trichodorus christiei,* and *Tylenchorhynchus* sp. Several insects attack the foliage and seeds; these include: meadow spittlebug *(Philaenus spumarius)*, alfalfa plant bug *(Adelphocoris lineolatus)*, potato leafhopper *(Empoasca fabae)*, and clover seed chalcid *(Bruchophagus platypterus)*. White grubs, *Phyllophaga hirticula,* attack root systems and may destroy virtually all roots especially in newly plowed grass sods (demonstrated by C. R. Drake, former ARS plant pathologist at Blacksburg).

REFERENCES:

Currier, W. W., and Strobel, G. A., 1977, Chemotaxis of *Rhizobium* spp. to a glycoprotein produced by birdsfoot trefoil roots, *Science* **196**:434–435.

Henson, P. R., and Schoth, H. A., 1962, The Trefoils—Adaptation and culture, *U.S. Dep. Agric., Agric. Handb.* **223**:1–16, illus.

MacDonald, H. A., 1946, *Birdsfoot Trefoil (Lotus corniculatus L.). Its Characteristics and Potentials as a Forage Legume,* Memoir 261, Cornell University Agricultural Experiment Station, Ithaca, N.Y.

Seaney, R. R., and Henson, P. R., 1970, Birdsfoot Trefoil, *Adv. Agron.* **22**:119–157.

U.S. Department of Agriculture, Agricultural Research Service,

1967, Trefoil production for pasture and hay, *Farmers Bull.* **2191**:1–16, illus.

CONTRIBUTORS: J. A. Duke, M. B. Forde, J. D. Miller, C. F. Reed, J. K. P. Weder.

Lotus tenuis Wald. & Kit. ex Willd.

FAMILY: Fabaceae

COMMON NAME: Narrowleaf trefoil

SYNONYM: *Lotus tenuifolius* (L.) Reichenb. *non-* Burm. *fil.*

USES: Narrowleaf trefoil is grown for pasture and hay in areas with dry soils difficult to drain after irrigation.

FOLK MEDICINE: No data available.

Figure 64. *Lotus tenuis* Wald. & Kit. ex Willd.

CHEMISTRY: Sometimes high in cyanogenic glucosides. No tannins in forage. Procyanidin in roots.

DESCRIPTION: Perennial herb, with shallow root system, and many stems from a single crown; stems slender, weak, more or less decumbent, 20–90 cm long, glabrous or sparingly pubescent; leaves compound, leaflets narrow, linear, or linear-lanceolate, tapering, 5–15 mm long, 1–4 mm broad, the lower ones stipulelike and lanceolate, the upper ones lanceolate to lanceolate-linear and acuminate; flowers 1–4 together, on long peduncles; calyx glabrous, 4–5 mm long, the teeth equal to usually shorter than the tube; corolla 6–12 mm long, yellow drying blue, standard 7–9 mm long, 5–7 mm broad, with rounded-reniform limb and cuneate claw; wings obovate-oblong; keel, 6–7 mm long; pods linear, cylindric, 15–30 mm long, 2–3 mm wide. Seeds 882,000/kg; wt kg/hl = 77. Fl. June–July; Fr. summer.

GERMPLASM: Several cvs have been developed, but all have fine stems, narrow leaves and grow close to ground. Stocks usually are named as to the state of origin, as California, New York, or Oregon narrowleaf. 'Los Banos' was developed in California and is grown extensively throughout the state. 'New York' is somewhat more winter-hardy than those forms of the West. Assigned to the Eurosiberian Center of Diversity, narrowleaf trefoil or cultivars thereof is reported to exhibit tolerance to alkali, high pH, Hg, salt, weeds, and waterlogging. ($2n = 12, 24$.)

DISTRIBUTION: Native throughout most of Europe except northeast and extreme south eastward to Crimea, Caucasus, and Central Asia. Introduced to the United States, especially in New York's Hudson River Valley region. Naturalized at two locations in the Shenandoah Valley of Virginia (*Lotus Newsletter,* **1976**(7):19).

ECOLOGY: Native in wet meadows and sands, but adaptive to dry soils that do not drain well after being flooded or irrigated. Narrowleaf trefoil may be nearly as hardy as birdsfoot. Old stands occur near Watkins Glen, N.Y., which is anything but mild climatically. It is also found in the Hudson River Valley near Albany, N.Y. In Oregon it grows well with relatively mild winters. It is well suited to heavy, poorly drained soils in northern United

Lotus tenuis Wald. & Kit. ex Wild.

States and does well on heavy-textured clay soils. In California, it grows well on saline and alkaline soils. Does well at pH of 4.5–7.9, with best nodulation at 6.0–6.5. Ranging from Cool Temperate Steppe to Wet through Warm Temperate Moist Forest Life Zone, narrowleaf trefoil is reported to tolerate annual precipitation of 4.4–11.6 dm (mean of 10 cases = 7.0 dm), annual mean temperature of 7.0°–16.9°C (mean of 10 cases = 10.2°C), and pH of 5.6–8.0 (mean of 9 cases = 6.7).

CULTIVATION: Plants can be propagated by stem cuttings, but crops are usually propagated by seed. A fine, firm seedbed is required to produce vigorous trefoil seedlings, as seedling plants lack vigor and do not compete with nurse crops, weeds, or grasses. Plowing should be well in advance of seeding to permit rains or irrigation to settle seedbed. Seedbed may be disked, harrowed and rolled before seeding. Seed rate 6–8 kg/ha, banked, drilled, or broadcast with cyclone seeder. On less fertile soils, P and K may be needed: 50–60 kg/ha P_2O_5; 68–80 kg/ha K_2O. Row spacing 30–40 cm. Inoculation is necessary for successful stands in areas where trefoils have not been grown before. Bacteria used to inoculate alfalfa, red clover, and other common legumes are not effective on trefoils. Strains of bacteria that inoculate big trefoil may not be effective on narrowleaf trefoil. A specific inoculum must be used for proper effects. The same Rhizobium strain that is used with birdsfoot trefoil is effective on narrowleaf.

HARVESTING: Properly cured hay is leafy and nutritious, comparable in quality to other good legume hays. Narrowleaf trefoil is an excellent pasture legume in the Hudson Valley Region of New York and in the Pacific Northwest. When initial growth has been removed as hay or silage during early June, the aftergrowth may be grazed up to early September. If cut for hay, the stubble must contain sufficient green leaves for regrowth. As seeds ripen unevenly, it is best reaped or mowed when most pods are brown.

YIELDS AND ECONOMICS: Yields are generally lower for narrowleaf trefoil than for birdsfoot trefoil. However, the crop can be planted on solonetz soils, where it thrives better than birdsfoot. Seed yields average 100–200 kg/ha, with fertilizer 150–250 kg/ha. An important pasture and hay crop in the area of adaptation; in the United States this is Upper New York State and the Pacific Northwest from Oregon south throughout California.

BIOTIC FACTORS: Trefoils are mainly self-incompatible, and require insects, especially Hymenoptera, to fertilize the flowers. Following fungi have been reported on narrowleaf trefoil: *Sclerotinia minor* and *Uromyces loti*. *Rhizoctonia solani* attacks *L. tenuis* in warmer areas of the United States. Several insects attack foliage and seeds, including: meadow spittlebug (*Philaenus spumarius*), alfalfa plant bug (*Adelphocoris lineolatus*), potato leafhopper (*Empoasca fabae*), and clover seed chalcid (*Bruchophagus platypterus*).

CONTRIBUTORS: J. A. Duke, M. B. Forde, J. D. Miller, C. F. Reed.

Lotus uliginosus Schkuhr

FAMILY: Fabaceae

COMMON NAME: Big trefoil

SYNONYMS: *Lotus pedunculatus* auct.
See Lotus Newsletter, Vol. 5

USES: Big trefoil is prominent pasture, hay, and nonbloating forage legume valued in the Pacific Northwest coastal regions of the United States and in New Zealand.

FOLK MEDICINE: No data available.

CHEMISTRY: Superior to *L. corniculatus* in having a higher protein content and no prussic acid. Contains flavonol polymers in roots and leaves (0–4% dry weight).

DESCRIPTION: Perennial legume with creeping shallow rhizome; stems 20–100 cm tall, hollow, erect or ascending; leaves compound, the leaflets pale or glaucous beneath, the lower pair sessile, broadly ovate, acuminate, 8–25 mm long, 6–15 mm broad, the upper ones obovate, acuminate; veins conspicuous; peduncles 7–10 cm long; flowers 5–15, in axil of upper trifoliolate often deciduous leaf, on pedicels 1–2 mm long; calyx 5–6 mm long, with linear-lanceolate acute teeth as long as tube, sometimes hirsute, stellate before anthesis; standard 11–12

Lotus uliginosus Schkuhr

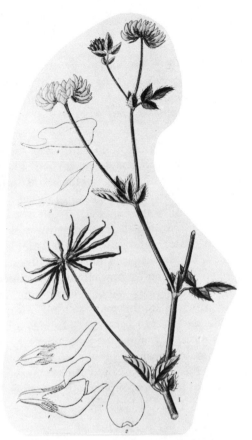

Figure 65. *Lotus uliginosus* Schkuhr

DISTRIBUTION: Native to the Balkans, Caucasus, Transcaucasus west to Central and Atlantic Europe. Introduced in Australia, New Zealand, South America, and United States.

ECOLOGY: Big trefoil is suited for marshes, wet grasslands, and boggy meadows. It lacks drought resistance, and grows best on continuously moist soil, and in regions where summer rains are frequent. Once recommended for the acid flatwood soils of southeastern United States, because its shallow root systems form solid stands (*Rhizoctonia* infestations have dampened this recommendation). Does well on acid coastal soils of Pacific Northwest, especially in areas that are frequently flooded during winter and on relatively moist uplands. Thrives on low-fertility hill pastures and peat areas in New Zealand, where grazing is light. Grows well on acid soils without liming; nodulation is best at pH 6.0–6.5. Moderately salt tolerant; yield is not significantly reduced until salinity reaches 3–6 mmhos. Ranging from Cool Temperate Moist to Wet through Tropical Thorn to Wet Forest Life Zones, big trefoil is reported to tolerate annual precipitation of 3.5–13.6 dm (mean of 12 cases = 8.5 dm), annual mean temperature of 5.9°–21.3°C (mean of 12 cases = 12.0°C), and pH of 5.5–8.2 (mean of 10 cases = 6.6).

mm long, the ovate limb as long as the claw; wings 10–11 mm long; keel 9 mm long; corolla 10–18 mm long, brilliant yellow; pod slender, cylindrical, dark brown, 10–35 mm long, 2–2.5 mm broad; seeds spherical, yellowish to olive-green, unspeckled. Seeds about 2,205,000/kg; wt. kg/hl = 77. Fl. July; Fr. summer.

GERMPLASM: Several cvs have been developed. In Oregon two are well suited to wet winters and cool summers of the Northwest coastal areas; 'Beaver' has hairy stems and leaves, giving the plant a grayish appearance; 'Columbia' is smooth-leaved, with relatively few hairs, dark green foliage, resembling that of birdsfoot trefoil. 'Marshfield' is based on a New Zealand introduction. 'Greensland Maku' is a tetraploid cv incorporating winter-active Portugese germplasm. Assigned to the Eurosiberian Center of Diversity, big trefoil or cultivars thereof is reported to exhibit tolerance to disease, fungus, high pH, and waterlogging. ($2n$ = 12, 24.)

CULTIVATION: Crop is mainly propagated by seed, and scarification may be necessary before planting. A fine firm seedbed is required to produce vigorous trefoil seedlings, as seedling plants lack vigor and do not compete with nurse crops, weeds, or grasses. Plowing should be well in advance of seeding to permit rains or irrigation to settle seedbed. Seedbed may then be disked, harrowed and rolled before seeding. Amount of preparation needed depends on kind of soil, moisture content, and season. Seeding rate 3–12 kg/ha broadcast, band-seeded or drilled. To band seed, drill fertilizer in. Phosphate and potassium may be needed in some areas. Row spacing 25–45 cm. Inoculation is necessary for successful stands in areas where trefoils have not been grown before. Bacteria used to inoculate alfalfa, red clover and other common legumes are not effective on trefoils. Strains of bacteria that inoculate birdsfoot and narrowleaf trefoil may not be effective on big trefoil. Herbicides are often useful to control weeds to permit seeding establishment with reduced competition. If seeded alone, EPTC, applied preplant, controls weeds. For postemergence control, Dalapon may be used for weedy grasses, and 2,4DB for broad leaf weeds. Weed specialists should be consulted.

HARVESTING: Big trefoil is used for hay and pasture in the coastal regions of the Pacific Northwest and Southeastern areas. It may be grazed from May or June onward, or if cut for hay or silage in early June, give an aftergrowth that can be grazed to early September. If cut for hay, the stubble must contain sufficient green leaves for regrowth. All cvs set seed over an extended period of time, with both ripe pods and new flowers on plant at same time. Some pods dehisce before flowering and pod set are completed. Seed should be harvested when 70–80% of pods are mature. Seed shattering is less serious than with *L. tenuis* or *L. corniculatus*. Most common method is to mow, windrow and allow to dry before combining. Crop may also be harvested and stored like hay, and then threshed, but loss of seed is excessive.

YIELDS AND ECONOMICS: Year to year differences in seed set and difficulty in harvest often cause wide variation in annual seed yields, which average 55–160 kg/ha. An important and valuable forage crop used throughout Central Europe and Western Asia, as well as in southeastern and northwestern United States, New Zealand, and South America. No longer recommended in southeastern United States because of devastating infections with *Rhizoctonia solani*.

BIOTIC FACTORS: Big trefoil is self-incompatible, and requires insects, usually Hymenoptera, to fertilize the flowers. This trefoil is not so susceptible to crown and root diseases which cause serious losses of birdsfoot trefoil in southeastern United States. Following fungi cause damaging diseases to big trefoil: *Cercospora loti, Fusarium roseum, Leptodiscus terrestris, Rhizoctonia solani, Sclerotinia trifoliorum, Stemphylium loti,* and *Verticillium* spp. Insects attacking the plants include: meadow spittlebug *(Philaenus spumarius)*, alfalfa plantbug *(Adelphocoris lineolatus)*, potato leafhopper *(Empoasca fabae)*, and trefoil seed chalcid *(Bruchophagus gibbus)*. Shows resistance to the grass grub *(Costelytra novae-zelandiae)* in New Zealand.

CONTRIBUTORS: J. A. Duke, M. B. Forde, J. D. Miller, C. F. Reed.

Lupinus albus L.

FAMILY: Fabaceae
COMMON NAME: White lupine
SYNONYMS: *Lupinus graecus* Boiss. & Sprun.
Lupinus jugoslavicus Kazim. & Now.
Lupinus termis Forsk.

USES: White lupine has been cultivated for more then 3,000 years and has been carried wherever southern Europeans have migrated. Plants used for late winter and early spring grazing, forage, green manure, and soil improvement. Fresh seeds are poisonous, but prolonged treatment by boiling or steeping in water removes bitter alkaloids and makes them edible. Lupines are easily digested by most animals. In Peru, lupine flour is added to biscuits, bread, noodles, sauces, and soups. Some forms are grown in southern and central Europe for their edible seeds and for forage. Lupines are also good as sources of honey. Some years back, a German botanist, pushing lupines as an economic crop, is said to have given a dinner served on lupine-fiber tablecloth of lupine soup, lupine steak

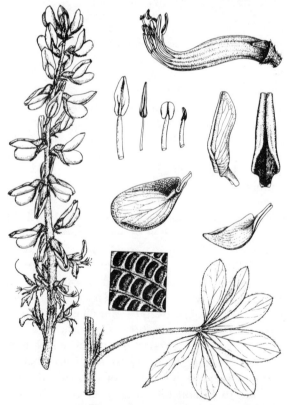

Figure 66. *Lupinus albus* L.

roasted in lupine oil seasoned with lupine extract, bread containing 20% lupine, lupine margarine, and cheese of lupine albumin, topped off with lupine liqueur and coffee. Hands were washed and dried with lupine soap and towel. Lupine paper and envelopes with lupine adhesive were provided the guests so they could write home about lupine.

FOLK MEDICINE: A decoction of seeds increased the sugar tolerance in diabetic patients. Medicinally, it is used as alterative, anthelmintic, carminative, deobstruent, depurative, discutient, diuretic, emmenagogue, pectoral, and tonic. Sometimes bruised seeds are soaked in water and applied to sores. In old England lupine meal was mixed with goat gall and lemon juice to form an ointment. Burning lupine seeds is supposed to drive gnats away.

CHEMISTRY: Seeds are reported to contain 11.0% moisture, 32.8% protein, 8.9% fat, 44.7% N-free extract (including fiber), 6.3% sugar, and 2.6% ash. On a dry basis they contain 0.75% K, 0.33% Ca, 0.28% Mg, 0.01% Na, 0.25% P, and 0.01% Fe. Another analysis attributes the whole seed with 38% crude protein, 9% oil, and 12% crude fiber; the meal with 44% crude protein, 11% oil, and 4% crude fiber. Freshly cut forage of the so-called "sweet" cv contains 2.8% protein, 2.7% fiber. Seeds contain β-sitosterol, and Egyptian seed are said to be good sources of vit. A, B, and C. Amino acids (percentage of crude protein) are: 5.2% lysine, 1.0% methionine, 1.8% cystine, and 1.2% trytophane. The component fatty acids are: saturated 10%, oleic 61%, linoleic 20%, linolenic 2%, and erucic 7%. Seedlings contain allantoin and asparagine. Lupine poisoning of cattle is attributed to quinolizidine alkaloids or their N-oxides. Five important toxins in lupines are, from most to least toxic: d-lupanine, sparteine, lupinine, spathulatine, and hydroxylupanine. Effects of these alkaloids do not appear to be cumulative. Seeds of both "sweet" and "bitter" strains are reported to contain trypsin inhibitors. Lupinosis is not clearly understood, and may be due to metabolites (but not aflatoxins) from the fungus Phomopsis. The liver damage with lupinosis appears to be associated with abnormal iron uptake. Acute cases exhibit dullness, poor appetite, jaundice, photosensitization occasionally, and unmanageability. If removed from the lupines early enough, most animals recover. In reviewing these writeups on Lupinus, the Poisonous Plant Research Laboratory had the following reservations: "Insuffient coverage is given to the toxicosis and teratogenic effects of the quinolizidine alkaloids from lupine. Both are of major importance when certain lupines are grazed by livestock and consequently signal potential problems from human use of lupines." The genus *Lupinus* is reported to contain dopa, hydrocyanic acid, hypoxanthine, lysine, malonic acid, pachycarpine, pectin, piperidine, saponin, sparteine, trigonelline, tryptophane, and xanthine.

DESCRIPTION: Short-hairy annual up to 120 cm tall; leaflets of lower leaves 25–35 mm long, 14–18 mm wide, of upper leaves, 40–50 mm long and 10–15 mm wide, obovate-cuneate, all mucronate, nearly glabrous above, sparsely villous beneath; stipules setaceous; racemes 5–10 cm long, sessile, flowers alternate; calyx 8–9 mm long, both lips shallowly dentate; corolla white to blue, 15–16 mm long; fruit flattened, 60–100 mm long, 11–20 mm broad, becoming longitudinally rugulose, short-villous, glabrescent, yellow; seeds 3–6 per legume, 8–14 mm in diameter, orbicular-quandrangular, compressed or depressed, smooth, and dull, light yellow, sometimes with dark variegation; cotyledon yellowish; germination phanerocotylar; stipules absent. Fl. May–June. Seeds 3,308/kg; weight 77 kg/hl.

GERMPLASM: White lupine was probably derived from wild forms native to the Balkan Peninsula, now incorporated in ssp. *graecus*. Major botanical cvs are: *L. albus* ssp. *albus* (*L. termis*) with white corolla, the keel pale blue at apex, legume 8–100 mm long by 17–20 mm broad, and seeds 12–14 mm and unspotted; ssp. *graecus* (Boiss. & Spruner) Franco & P. Silva, with deep blue corolla, legume 6–70 mm long by 11–13 mm broad, and seeds 8–9 mm and dark variegated, found mainly in Balkan Peninsula and Aegean region. Alkaloid-free or "sweet" strains of *L. albus* were selected by van Sengbusch and others during the 1930s. The most important center of alkaloid-free lupine production outside Europe is southwest Cape in South Africa, where lupines are grown for grazing as mature standing crops in summer, or for harvesting and stubble grazing; the seed is fed to livestock as protein concentrate. Cv 'Hope,' especially developed for use as a winter cover and green manure crop, has high alkaloid content that acts as

Lupinus albus L.

a natural herbicide when plant material is decomposing in the soil. The so-called 'Egyptian' cv is suited to tropical and saline conditions. Assigned to the Mediterranean Center of Diversity, white lupine or cultivars thereof is reported to exhibit tolerance to frost, fungus, high pH, low pH, salt, photoperiod, virus, weeds, and waterlogging. ($2n = 26, 22$.)

DISTRIBUTION: Presumably from the Balkan Peninsula, white lupine is widely cultivated around the Mediterranean, the Canary Islands, Madeira, and the Upper Nile. Occasional in central and southeastern Europe, USSR, Ethiopia, S. Africa, Australia, southeastern United States, and S. America.

ECOLOGY: Adapted to lower southern United States as winter annual on well drained, fertile neutral soils, such as alluvial soils of lower Mississippi Delta. Native in acid soils, also grows on mildly acid and slightly calcareous loamy sandy loams. Requires more fertility and nutrients than either *L. angustifolius* or *L. luteus*. Phosphorus deficiency may reduce yields. It is the most winter-hardy of the three lupines; frost resistance is good, but varies with the biotype; some show excellent frost and cold resistance and others thrive only in mild climates. Temperatures of $-6°$ to $-8°C$ are harmful at germination, $-3°$ to $-5°C$ at flowering and fruiting. There also seems to be a physiological difference in cold tolerance among plants. Responds to vernalization, but extent of response to photoperiod is uncertain. Requires at least a 5-month period with mean monthly temperatures $15°-25°C$, optimum $18°-24°C$ and annual rainfall, 4–8 dm. A long-day plant, intolerant to waterlogging, but more tolerant than other lupines. In the Sudan and Egypt, it grows on flooded lands that are too hard or saline for other crops. Some cvs are intolerant of salt. Ranging from Cool Temperate Steppe to Wet through Subtropical Moist Forest Life Zone, white lupine is reported to tolerate annual precipitation of 3.6–17.8 dm (mean of 38 cases = 8.4 dm), annual mean temperature of $5.7°-26.2°C$ (mean of 38 cases = $12.7°C$), and pH of 4.8–8.2 (mean of 35 cases = 6.4).

CULTIVATION: Most successful lupine-growing areas have growing season of at least 5 months free from serious moisture stress, during which mean monthly maximum temperatures are between 15° and 25°C. In winter-growing areas a period of lower temperature is tolerated in midwinter if growing periods before and after are long enough to compensate. Propagation is by seed, planted on a well-prepared seedbed. Time of sowing depends on the locality. In Kenya seed sown after rains begin. In Australia, sown early before mid-May; seeding with the first rains or dry seeding gives the best results as lupines grow rapidly during the warm early part of season and compete with weeds. In regions with mild winters, seeding should be done from mid-September to late October; however, plantings as late as early December may give good results. In northern United States and Europe should be seeded in early April to mid-May. Lupine is often grown in mixture with oats, other cereals, or with forage legumes such as serradella (*Ornithopus sativus*). Seeding rate is 56–180 kg/ha. Low-moisture seed retains good germination for 2 yr or more under good storage conditions. Seeds should be inoculated before sowing with *Rhizobium lupini* bacteria of the "slow-growing" type. Some recommend inoculation every year even on the same land. In Australia seed is not always inoculated if it is planted where lupine has been grown before. Rhizobial growth is optimal at pH 6.5–7.5. They can be grown on new or old land, but do best on old land following a cereal crop. Higher seeding rates reduce branching and give more even ripening. A grain drill can be used for planting, or seed can be broadcast and then covered by disking. Seed should be planted to depth of 2.5–5 cm in firm soil with a cultipacker or by other means to insure contact of seed with moist soil. For seed production, seed sown in rows 30–90 cm apart depending on the region, to allow weed control and to make roguing easier. In southeastern United States fall seedings mature in May and early June; in northern States early spring seedings mature in August and early September. Preemergence herbicides are used in some regions to control weeds. Lupines usually need extra phosphorus, usually at rates of 300–500 kg/ha of a 0–10–20 fertilizer, unless the previous crop in rotation has been heavily fertilized. In the United States and South Africa 300–400 kg/ha of superphosphate is applied before sowing or later. In South Africa, on poor sandy soils, 28–56 kg/ha potash is also added. In Kenya 100 kg/ha of double superphosphate is used. Nitrogen is normally not applied, but hastens ripening of seed in USSR; in Europe boron is sometimes used to hasten seed ripening. In Australia superphosphate is applied at rate of 373–1,119 kg/ha, depending on soil, plus trace elements and potassium and sulfur if needed. Contact of seed with fertilizer reduces inoculation and germination. Lupines are generally planted in rotation following a cereal. In Europe most often rotated with rye, oats, barley, or potatoes; in the United States with corn, and not with peanuts because of diseases common to both. If previous lupine crop is healthy, lupines may follow themselves, and need not be inoculated. They should not follow themselves more than 4 yr.

HARVESTING: Grain lupines should not be grazed at any stage during growth. Seed losses may be serious unless nonshattering cvs are used. Freshly combined lupine seed should be cleaned immediately to remove large green seeds and trash. Small lots of seed may be spread out 15 cm deep on a floor to dry. Seed should be stirred twice daily until dry. In wet, humid weather, artificial heat may be necessary. Large lots of seed should be taken to a commercial drying plant and dried to 12% moisture at temperatures below 46°C. Dried seed may be stored under good conditions at least 2 yr without serious loss in germinability. Seeds mature in 100–160 days, depending on growing conditions and cv. All lupine cvs are best harvested in moderately cool weather to prevent shattering pods and damaging seeds. In Australia a suitably adjusted normal header type harvester is used. For seed production in the United States and Europe, chemical defoliants are applied before harvest without detriment to seed germination. However, germination is lower if seed are not ripe when harvested. Speed of threshing drum should be low (about 400–450 rpm) to avoid seed injury. Coarse material is removed from seed with fanning mill. Grazing of stubble is important in the economics of grain lupine production. Residues of lupine crops are normally 4.5–7.5 MT/ha. Heavy lupine crops leave up to 10 MT/ha, some of which is coarse, inedible stems. Stock-carrying capacity is better for lupine stubble than for most cereal stubbles and probably comparable with stubble of other seed legume pastures. Grazing the stubble removes most of the residue and may remove diseased material.

YIELDS AND ECONOMICS: White lupine yields 5–7.5 MT of dry herbage/ha, containing 250–450 kg of nitrogen. Seed yields average about 800–1,000 kg/ha. Major producers of lupines are Australia, Europe, South Africa, USSR, and the United States. In the United States yields fluctuate from 1,000 to 2,200 kg/ha. Western Europe, the leading importer, buys some seed from South Africa. World markets grow faster for high-protein foodstuffs than for other agricultural products including cereal feed grains. Alkaloid-free lupine production may increase, perhaps for processing into meat substitutes. Yields of seed might be increased by breeding.

BIOTIC FACTORS: Though self-pollinated, lupines are bee pollinated. Placing bee hives in lupine fields greatly increases seed yield. *Rhizobium lupini*, slow-growing group, is used to inoculate seed. Rabbits may eat seedlings of sweet cvs and may pose a serious problem. Following fungi have been reported on white lupine: *Erysiphe polygoni* (powdery mildew), *Fusarium avenaceum, F. moniliforme, F. oxysporum, F. scirpi, F. solani, Glomerella cingulata* (anthracnose), *Laveillula taurica, Phytophthora parasitica, Pleiochaeta setosa* (brown leaf spot), *Pythium vexans, Rhizoctonia solani, Sclerotium rolfsii, Stachybotrys lobulata, Uromyces anthyllidis, U. lupini, U. lupinicola, U. renovatus*. The brown leaf spot, growing at temperatures as low as 5°C, can limit the commercial cultivation of white lupine. Viruses known to attack this lupine include: Bean yellow mosaic virus (BYMV), *Pisum* virus 2, mosaic and pea mosaic. Nematodes include: *Heterodera glycines, H. goettingiana, H. marioni, H. schachtii trifolii, Meloidogyne arenaria thamesi, M. hapla, M. incognita acrita, M. javanica*. Since lupines are highly susceptible to 2,4-D, preemergence herbicides such as diallate (Avadex), simazine, and trifluralin have been suggested. Aphids often infest the sweet cvs, and transmit virus diseases. Budworms or earworms (*Heliothis* spp.) are serious pests on winter grown lupines and cause losses as drastic as 90% in Australia. Thrips often reduce yields in the United States. Lucerne fleas, *Sminthurus viridis* are bad on sweet cvs in Australia, the lupine maggot, *Hylemya lupini*, in the United States. Grasshoppers and white-fringed beetles (*Graphognathus*) are troublesome in southeastern United States.

REFERENCES:

Dukhanin, A. A., 1975, Root system of lupin and its importance as a green manure. Selektsiyz, Semenovodstvo P Priemy

Lupinus albus L.

Vozdelyvaniya Lyupina: Orel, *USSR Ref. Zh.* **9. 55. 652**:79–86 [Russian] (orig. not seen).

Gladstones, J. S., 1967, 'Uniwhite'—A new lupin variety, *W. Aust. Dep. Agric., Bull.* **3502**:2–6, illus.

Gladstones, J. S., 1970, Lupins as crop plants, *Field Crop Abstr.* **23**(2):123–148.

Greenwood, E. A. N., Farrington, P., and Beresford, J. D., 1975, Characteristics of the canopy, root system, and grain yield of a crop of *Lupinus angustifolius* cv Unicrop, *Aust. J. Agric. Res.* **26**:497–510.

Gross, U., Reyes, J., Gross, R., and Baer, E. von, 1976, Die Lupine, ein Beitrag zur Nahrungsversorgung in den Anden. III. Die Akzeptabilitaet des Mehles von *Lupinus albus*, *Z. fur Ernaehrungswiss.* **15**(4):396–402 (orig. not seen).

Henson, P. R., and Hollowell, E. A., 1960, Winter annual legumes for the South, *U.S. Dep. Agric., Farmers Bull.* **2146**:3–12, 17.

Henson, P. R., and Stephens, J. L., 1958, Lupines: Culture and use, *U.S. Dep. Agric., Farmers Bull.* **2114**:1–12.

Naimark, L. B., and Brantsevich, S. F., 1974, Effect of phosphorus–potassium fertilizers on yield of yellow lupin. Sbornik Nauchnykh Trudov, *Belorusskaya Sel'-Skokhozyaistvennaya Akademiya: Ref. Zhur.* **6. 55. 622.**:1974, **133**:58–71 [Russian] (orig. not seen).

Rahman, M. S., and Gladstones, J. S., 1974, Effects of temperature and photoperiod on flowering and yield components of Lupin genotypes in the field, *Aust. J. Exp. Agric. Anim. Husb.* **14**(67):205–213.

Weimer, J. L., 1952, Diseases of cultivated lupines in the Southeast, *U.S. Dep. Agric., Farmers Bull.* **2053**:1–18.

CONTRIBUTORS: J. A. Duke, R. F. Keeler, C. F. Reed, J. K. P. Weder.

Lupinus angustifolius L.

FAMILY: Fabaceae

COMMON NAMES: Blue lupine, Narrow-leaved lupine, New Zealand blue lupine

SYNONYMS: *Lupinus leucospermus* Boiss.
Lupinus linifolius Roth.
Lupinus reticulatus Desv.
Lupinus varius L.

USES: Blue lupine is used as forage and silage and for late winter and early spring grazing. The species is poisonous, especially to sheep. However, alkaloid-free or "sweet" cvs are grown and the seeds used as a protein additive in mixed animal feed. Lupines are easily digested by most animals. Bitter types are used solely for soil improvement, whereas sweet types also are used for grazing and seed.

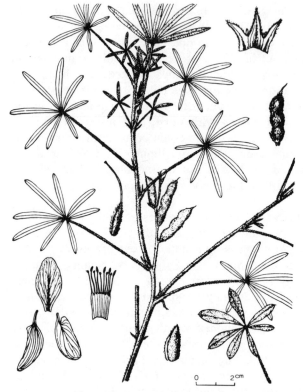

Figure 67. *Lupinus angustifolius* L.

Lupines are also good honey plants. In India they are planted as green manure between successive crops of potatoes, increasing potato yields nearly 4 MT/ha.

FOLK MEDICINE: The germinating seeds, rich in asparagine, have been used in culture media for commercial production of tuberculin.

CHEMISTRY: Seeds are reported to contain 13.8% moisture, 31.0% protein, 6.4% fat, 34.9% N-free extract, 11.5% fiber, and 2.4% ash. The ash contains 31.9% K_2O, 0.81% Na_2O, 9.8% CaO, 10.9% MgO, 0.73% Fe_2O_3, 39.0% P_2O_5, 5.6% SO_3, 0.34% SiO_2, and 0.59% Cl. They also contain lecithin, A-amino adipic acid, and lupeose. Amino acids are distributed in the two globulins of the seed proteins as follows: *globulin-1* (N = 17.1%), 2.3 threonine, 3.5 valine, 6.0 isoleucine, 7.6 leucine, 4.8 phenylalanine, 2.3 histidine, 4.3 lysine, 13.4% arginine: *globulin-2* (N = 18.3%), 3.7 threonine, 4.3 valine, 0.50 methionine, 5.6 isoleucine, 8.4 leucine, 5.3 phenylalanine, 3.0 histidine, 3.9 lysine, 14.3 arginine, and 1.6% tryptophane. With high digestibility values (90%) and a biological value of 53%, they are deficient in methionine. The fatty acids of the seed are 47.5% oleic, 33.7% linoleic, 1.8% linoleic, and 7% erucic acid. β-sitosterol is also reported from the seed. Bitter cvs contain up to 2.25% d-lupanine. A quintal of seeds, germinated 12–20 days, can yield 10–16 kg asparagine. Fresh-cut lupine forage contains 0.98% N, 1.18% P_2O_5, and 0.53% K_2O. Immature hay (88.7% DM) contains on a moisture-free basis: 32.5% crude protein, 5.2% fat, 20.1% crude fiber, 3.7% ash, and 38.5% N-free extract. Mature green forage (8.8% DM) contains on a moisture-free basis: 15.5% crude protein, 2.2% fat, 21.9% crude fiber, 15.3% ash, and 45.1% N-free extract. The immature hay contains: 0.55–0.62% Ca, 0.46–0.47% P, and 0.36–0.37% Mg.

DESCRIPTION: Short-hairy annual, 20–150 cm tall; leaflets 10–50 cm long, 2–5 mm wide, linear to linear-spathulate, glabrous above, sparsely villous beneath; stipules linear-subulate; racemes 10–20 cm long; upper lip of calyx about 4 mm long, 2-dentate, lower lip 6–7 mm long, irregularly 3-dentate to subentire; corolla 11–13 mm long, blue, occasionally pink, purple or white; legume shortly hirsute, yellow to black, to 6 cm long, 1.5 cm broad; seeds 3–7 per pod, ellipsoid to subglobose, smooth and dull, 6–8 mm long, yellow-brown, dark brown, or gray with yellow spots. Germination phanerocotylar. Fl. April–June. Seeds 5513/kg; weight 77 kg/hl.

GERMPLASM: Botanical cvs include the following: *L. angustifolius* ssp. *angustifolius* (incl. *L. leucospermus* Boiss.), with plants 50–80 cm tall, leaflets 30–40 mm long, 4–5 mm wide, flat, linear-spatulate; legume 40–60 mm long, 8–13 mm wide; seeds 6–8 mm long, 4–7 mm wide, on inland loamy soils throughout range of species except Sardinia: *L. angustifolius* ssp. *reticulatus* (Desv.) Coutinho, with plants 20–40 cm tall, leaflets 10–20 mm long, 2 mm wide, conduplicate, linear, legume 35–45 mm long, 6 mm wide, seeds 4.5–5 mm long, 3–3.5 mm wide, on maritime sands, rarely inland, throughout southwestern and western Mediterranean region of Europe. Alkaloid-free or "sweet" cvs include: 'Borre' (Sweden); 'Blanco' with white flowers; 'Frost' (1969) resistant to gray leafspot and anthracnose, blue flowers, and equal to all previous cvs in palatability, digestibility, and protein content; 'Rancher' combines characters of 'Blanco' with resistance to gray leafspot and anthracnose. 'Richy' is a blue bitter cv grown only for soil improvement. 'Uniwhite' and 'Uniharvest,' grown in Australia, are frost-resistant, and are fairly free of shattering seeds. 'Unicrop' with nonshattering pods, white flowers, matures 2–4 weeks earlier than 'Uniharvest,' and is expected to be one of Australia's major cvs shortly. Assigned to the Mediterranean Center of Diversity, blue lupine or cultivars thereof is reported to exhibit tolerance to anthracnose, disease, frost, fungus, high pH, insects, photoperiod, sand, virus, and weeds. Some Yugoslavian germplasm is said to tolerate maritime sands. ($2n = 40, 48$.)

DISTRIBUTION: Native to Mediterranean Basin and southwestern France. Introduced and now widely grown in Australia, Tasmania, New Zealand, South Africa, northern Europe, and southeastern United States. Used in many areas for green manuring orchards and vineyards. Occasionally escaping from cultivation.

ECOLOGY: Grows as winter annual on well-drained soils in the deep South. More winter hardy than yellow lupine but less than white lupine. In vegetative stages it tolerates temperatures down to about $-6°C$, minimum temperature for germination is lower than that for other economic lupines. Flowers drop when temperatures are high. Flowering behavior in commercial cvs depends primarily on a high vernalization requirement and only slightly on photoperiod. Considered to be a long-day plant. Plants intolerant of waterlogging and low levels of 2,4D. Rainfall requirement for cvs grown in Australia is 450–1,000 mm/yr; in Kenya, about 900 mm/yr. Adapted to neutral or moderately acid loamy sands or sandy loams, especially on the coastal plain in southeastern US, blue lupine is less tolerant than others to low fertility; susceptibility to K, P, and Co deficiencies. 'Unicrop' took 10 weeks to reach an LAI (Leaf Area Index) of 1 and 9 more weeks to maximum LAI (3.75), at which time only 33% of daylight reaches the pods on the main axis. Only 23% of the flowers set pods, an important yield limitation. Ranging from Cool Temperate Steppe to Wet through Subtropical Dry to Moist Forest Life Zones, blue lupine is reported to tolerate annual precipitation of 3.5–16.6 dm (mean of 39 cases = 8.4 dm), annual mean temperature of

Lupinus angustifolius L.

5.6°–26.2°C (mean of 39 cases = 12.3°C), and pH of 4.9–8.2 (mean of 12 cases = 6.6).

CULTIVATION: Seeded at rate of 65–90 kg/ha, blue lupine is cultivated and harvested much as the white lupine.

YIELDS AND ECONOMICS: Yields average 5–7.5 MT/ha of dry herbage containing 250–450 kg of N. Green manure yields of 7.5–12.5 MT/ha are reported in India; seed yields are only 500–600 kg/ha. Seed yields average 800–1,000 kg/ha. 'Unicrop' gave 2100 kg/ha dry seed. Of this, 30%, 60%, and 10% came from the main axis, first and second apical axis, respectively.

BIOTIC FACTORS: Some lupines are mainly pollinated by bees and hives in fields increase seed yields. This species, however, is almost entirely self-pollinating. Blue lupine tends to smother weeds. Fungi attacking blue lupine include: *Ascochyta pisi*, *A. gossypii* (stem canker), *Botrytis cinerea*, *Colletotrichum dematium*, *Diplodia theobromae*, *Erysiphe martii*, *E. pisi*, *E. polygoni* (powdery mildew), *Fusarium avenaceum*, *F. equiseti*, *F. moniliforme*, *F. orthoceras*, *F. scirpi*, *F. solani*, *Glomerella cingulata* (anthracnose), *Phytophthora cinnamomi*, *Pleiochaeta setosa* (brown leaf spot), *Pleospora herbarum*, *Pythium debaryanum* (root rot), *P. intermedium*, *P. rostratum*, *P. ultimum*, *Rhizoctonia solani* (root rot), *Sclerotinia sclerotiorum* (stem rot), *Sclerotium rolfsii* (southern blight), *Stemphylium botryosum*, *Stemphylium solani* (gray leaf spot), *Thielaviopsis basicola*, *Uromyces lupinicola*, *U. renovatus*, and *Verticillium albo-atrum*. Anthracnose may be more troublesome on blue lupine than the other lupine cvs. Viruses attacking blue lupine include: bean yellow mosaic, pisum mosaic 2 and pea mosaic virus. Nematodes attacking blue lupine (which seems more susceptible than other species) include: *Aphelenchoides bicaudatus*, *Belonolaimus gracilis*, *Ditylenchus dipsaci*, *Helicotylenchus dihystera*, *Meloidogyne hapla*, *M. incognita*, *M. incognita acrita*, *M. javanica*, *M. thamesi*, *Pratylenchus brachyurus*, *P. coffeae*, *P. penetrans*, *P. zeae*, *Tylenchus costatus*. Aphids and lucerne fleas are troublesome in Australia, the root-weevil, *Sitona explicata*, and the lupine maggot, *Hylemya lupini* in the United States.

CONTRIBUTORS: J. A. Duke, C. F. Reed

Lupinus luteus L.

FAMILY: Fabaceae

COMMON NAMES: European yellow lupin, European yellow lupine

USES: Yellow lupine grown for late winter and early spring grazing, forage, and silage. Important grain and forage crop on the Baltic Coast. In Chile, a bland yellow flour, 52–60% protein, from seed of *Lupinus luteus* and *L. albus*, is added as a protein supplement to the gruel known as "ulpo." In Australia lupine meal is used as a protein supplement in pet foods. Also grown for soil improvement and green manure, especially in fruit orchards. Lupines are used for grazing as mature standing crops in summer, or for harvesting and stubble grazing; seed is fed to livestock as a protein concentrate. Lupines are also good honey plants. Introduced to England as a fragrant ornamental.

FOLK MEDICINE: Said to be used for poison.

Figure 68. *Lupinus luteus* L.

Lupinus luteus L.

CHEMISTRY: Fresh seeds are poisonous and contain an alkaloid lupinine. Alkaloid-free ("sweet") strains have been developed for grazing and forage. Seeds of some sweet types are reported to contain trypsin inhibitors, but the bitter types investigated contain no trypsin inhibitors. Whole seeds contain: 42% crude protein, 5% oil, and 17% crude fiber; germ meal contains: 54% crude protein, 6% oil, and 5% crude fiber. From seed with 5.0% oil and 43.0% protein, the amino acid composition of the proteins is per 16 g N: 4.8 g lysine, 0.5 g methionine, 9.9 g arginine, 4.0 g glycine, 2.5 g histidine, 3.9 g isoleucine, 7.5 g leucine, 3.6 g phenylalanine, 3.1 g threonine, 1.0 g tryptophane, 3.5 g valine, 3.4 g alanine, 10.1 g aspartic acid, 24.4 g glutamic acid, 3.2 g proline, and 4.9 g serine. The Wealth of India (CSIR, 1962) reports that the seeds contain: 88.9% DM, 39.8% protein, 5.0% fat, 25.7% N-free extract, 14.0% fiber, 4.5% mineral matter, 0.23% Ca, 0.39% P, 0.81% K. Fresh plant contains: 10.4% DM, 8.4% organic matter, 2.3% protein (2.0% digestible protein), and 1.7% fiber. Hay (91% DM) from immature yellow lupine, on a moisture-free basis, contained: 35.6% crude protein, 5.2% fat, 19.6% crude fiber, 5.0% ash, 34.6% N-free extract, 0.64% Ca, 0.82% P, and 0.38% Mg.

DESCRIPTION: Hairy annual, 25–80 cm tall, the stems hairy; leaflets 40–60 mm long, 8–12 mm wide, obovate-oblong, mucronate, sparsely villous; stipules dimorphic, those of the lower leaves 8 mm long, subulate, those of upper leaves 22–30 mm long, 2–4 mm wide, linear-obovate; racemes 5–16 cm long; flowers verticillate, scented (violet-scented); petals bright yellow; legume 4–6 cm long, 1–1.5 cm wide, densely villous, black; seeds 4–6 per pod, 6–8 mm long, 4.5–7.0 mm broad, orbicular-quadrangular, compressed, smooth and dull, black marbled white, with a white curved line on each side, or all white. Fl. March–July. Seeds 8,820/kg; wt. 77 kg/hl.

GERMPLASM: Among cvs: 'Florida Speckled,' is superior in seed yield but shattering makes yields unsatisfactory; 'Weiko III' has low alkaloid content, is nonshattering, has permeable white seed coat and is rapid in early growth. It is taller and quicker to mature than 'Weiko II,' another similar cv. Yields of seeds generally are lower than those of other lupines, but seed protein is more than 40%. Its weaknesses, at least in Australia, are its susceptibility to insect pests, especially the redlegged earth mites in the seedling stage, aphids in the bud stage, and budworms or climbing cutworms in the green pod stage. It also suffers from competition with weeds. Assigned to the Mediterranean Center of Diversity, and long cultivated in the Iberian Peninsula and North Africa, *L. luteus* may have been selected from closely related (perhaps conspecific) *L. hispanicus* and *L. rothmaleri*. Yellow lupine or cultivars thereof is reported to exhibit tolerance to aluminum, high pH, photoperiod, poor soil, and virus. ($2n = 52, 104$.)

DISTRIBUTION: Native to the Mediterranean basin from the Iberian Peninsula and Italy, through the Islands of the Mediterranean to the Middle East and Israel. Introduced and widely cultivated, especially on sandy soils, in northern Europe, South Africa, Australia, and southern United States.

ECOLOGY: Adapted to the lower south of United States as a winter annual, especially on drained soils. Is under investigation as summer annual feed grain legume on sandy soils of the Great Lakes region. Requires a mild growing-season temperature. Tolerates light frosts, but may be damaged in frost-prone locations in colder areas. Temperature of $-4°C$ may be harmful at germination, $-20°C$ at flowering. Long-day plant; flowering behavior influenced both by photoperiod and vernalization, but their respective importance is unknown. In Kenya, requires about 9 dm rainfall/yr; in Australia, 'Weiko III' requires 6.5–13.5 dm/yr and grows best in high-rainfall areas. Tolerates temporary waterlogging. Adapted to strongly to mildly acid soils of low fertility. Where native, it grows on acid, sandy loams. Highly susceptible to Fe and Mn deficiencies on neutral or calcareous soils, often suffering from lime chlorosis. Tolerates high Al levels. Grows in Kenya up to 2,000 m altitude. Ranging from Cool Temperate Steppe to Wet through Subtropical Dry to Moist Forest Life Zones, yellow lupine is reported to tolerate annual precipitation of 3.5–20.9 dm (mean of 42 cases = 8.5 dm), annual mean temperature of 6.6°–26.2°C (mean of 42 cases = 13.0°C), and pH of 4.5–8.2 (mean of 41 cases = 6.5).

CULTIVATION: Seeded at rates of 45–80 kg/ha, yellow lupine is cultivated and harvested much like *Lupinus albus*.

Lupinus luteus L.

YIELDS AND ECONOMICS: Yields average 5–7.5 MT/ha of dry herbage containing 250–450 kg of nitrogen. Seed yields average 800–1,000 kg/ha. Roots penetrate to 2 m in dernopodzolic sandy loam soils in Russia. With fresh forage yield of 15–50 MT/ha, root DM yields were 3–8.7 MT/ha. In good years, roots accumulated 147–160 kg N/ha, 40 kg P_2O_5/ha, and 35 kg K_2O/ha. On dernopodzolic clay loam soil in Belorussian SSR, application of 60 kg P_2O_5 and 120 kg K_2O/ha increased fresh forage yields by 17.9 MT/ha, DM yields by 4.83 MT, seed yields by 1.04 MT, and crude protein by 93% compared with unfertilized controls. Application of P alone was less effective than application of K alone.

BIOTIC FACTORS: Lupines are mainly bee pollinated; placing bee hives in lupine fields greatly increases seed yield. *Rhizobium lupini*, slow-growing group, is used to inoculate seed. Following fungi have been reported on yellow lupine: *Cylindrocladium scoparium*, *Erysiphe pisi*, *E. polygoni* (powdery mildew), *Fusarium moniliforme*, *F. oxysporum*, *F. solani*, *Glomerella cingulata* (anthracnose), *Phytophthora parasitica*, *Pleioscaeta setosa* (brown leaf spot), *Pythium debaryanum*, *P. graminicola*, *Rhizoctonia solani*, *Sclerotium rolfsii*, *Thielaviopsis basicola*, *Uromyces anthyllidis*, and *U. renovatus*. Susceptibility to bean yellow mosaic virus severely limits seed yields of yellow lupine and reduced its use in Florida. Other viruses attacking yellow lupine include: common pea mosaic, *Pisum* virus 2 mosaic, and pea mosaic virus Doolittle & Jones. Nematodes reported attacking yellow lupine include: *Belonolaimus gracilis*, *Heterodera glycines*, *H. goettingiana*, *H. marioni*, *H. schachtii*, *Pratylenchus penetrans*, and *Radopholus similis*. Aphids can be troublesome, especially with sweet cvs of yellow lupine. Of the lupines, *L. luteus* is most susceptible to budworms (*Heliothis* spp.), *L. angustifolius* less susceptible, with *L. albus* and *L. cosentinii* intermediate. Redlegged mites (*Halotydeus destructor*) and lucerne fleas (*Sminthurus viridus*) infest sweet cvs in Australia.

CONTRIBUTORS: J. A. Duke, C. F. Reed, J. K. P. Weder.

Lupinus mutabilis Sweet.

FAMILY: Fabaceae

COMMON NAMES: Tarhui, Altramuz, Chochos, Chuchus, Muti, Pearl lupine, Tarwi, Ullus

USES: A protein-rich (up to 49%) pulse, tarhui could provide protein to many high-altitude, middle-latitude developing countries, if the poisonous properties were bred out. For human consumption, the seed need extensive preparation. With its very attractive flowers, it often borders AmerIndian vegetable and cereal plots in the Andes. Natives suggest that it keeps cattle out of the plots. It is grown in Argentina strictly as an ornamental.

FOLK MEDICINE: Said to be used for poison.

CHEMISTRY: Although some estimate the protein content of the seeds at nearly 50%, analyses performed by WARF for the Plant Taxonomy Laboratory showed 7.4% moisture, 39.1% protein, 30.9% carbohydrates, 10.0% fats, 9.3% fiber, and 3.3% ash, with 370 calories per 100 g. Per 100 g

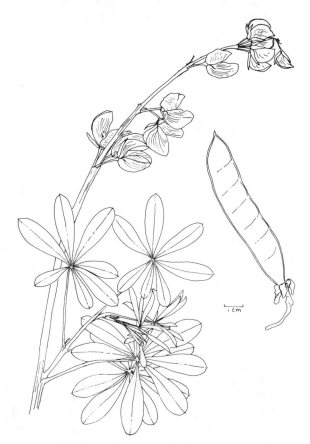

Figure 69. *Lupinus mutabilis* Sweet.

there were 4.5 mg vitamin C, 20 IU vitamin A, 0.60 mg thiamin, 0.18 mg riboflavin, 4.4 mg niacin, 122 mg Ca, 6.7 mg Fe, 0.58 mg Cu, 234 mg Mg, 3.7 mg Zn, 605 mg P, 0.06 ppm I, 2.1 IU vitamin E, 132 μg folic acid, 37.9 μg biotin, 0.22 mg vit. B_6, 0.36 μg vitamin B_{12}, and 0.67 mg pantothenic acid. Information supplied by K. W. Pakendorf indicates that per 16 g N there are: 5.7 g lysine, 2.6 g histidine, 7.7 g arginine, 12.7 g aspartic acid, 4.8 g threonine, 6.6 g serine, 18.1 g glutamic acid, 4.0 g proline, 4.8 g glycine, 4.3 g alanine, 4.6 g valine, 5.6 g isoleucine, 8.5 g leucine, 4.9 g tryosine, 4.8 g phenylalanine, 1.5 g cystine, 0.64 g methionine, and an estimated 1.2 g tryptophane. Analyses of a Peruvian sample with ca. 21.5% oil and a Chilean sample with 11.7% oil had component acids as follows: oleic 55.6–59.5%, and 37.7%; linoleic, 21.5–25.1% and 40.1%; palmitic 9.1–10.1 and 11.6%; stearic 6.2–7.5 and 7.0%; linolenic 1.7–2.2 and 1.9; arachidic 0.5–0.6 and 1.0%; and behenic 0.4–0.6 and 0.7%. *Tabla de composicion de alimentos para uso en America Latina* reports the following values for immature and mature green seeds (per 100 g edible portion): 126 calories, 71.4% moisture, 13.7 g protein, 5.1 g fat, 9.0 g total carbohydrate, 0.6 g fiber, 0.8 g ash, 28 mg Ca, 168 mg P, 1.9 mg Fe, 15 μg vitamin A, 0.09 mg thiamin, 0.15 mg riboflavin, and 0.8 mg niacin; and for semidried seeds, without skin (per 100 g edible portion): 276 calories, 46.3% moisture, 17.3 g protein, 17.5 g fat, 17.3 g total carbohydrate, 3.8 g fiber, 1.6 g ash, 54 mg Ca, 262 mg P, 2.3 mg Fe, 0.16 mg thiamin, 0.29 mg riboflavin, 1.1 mg niacin, and 5 mg ascorbic acid.

DESCRIPTION: Annual herb 0.8–1.8 m tall; leaves digitate, with 5 or more leaflets; flowers blue and/or pink and white, with yellowish eye, 1–1.5 cm long, in terminal racemes; tender fruits very hairy, 5–9 cm long, 2 cm broad, almost indehiscent; seeds 2–6, biconvex, flattened, 6–10 mm in diameter, pure black, brownish-black, or white, or white with a black or gray "halo" around the hilum.

GERMPLASM: Tarhui has been tested less than other lupines, but might contribute resistance for the various fungi that decimate some lupines. Because it is valued for its bitterness, in keeping animals out of other crops, the farmers apparently have not yet selected sweet cvs. On the contrary, they apparently have selected the nonshattering habit. Tarhui is assigned to the South American Center of Diversity. There is some speculation that North American lupines, e.g., *L. douglassi* and *L. ornatus,* were carried to South America by migrating Indians, and may have contributed germplasm to *L. mutabilis* by hybridization. *L. mutabilis* is not known from the wild. If this is a true story of the origin, some explanation must be found for the inability of the low-latitude derivatives to flower at high latitudes.

DISTRIBUTION: Tarhui appears to be an Andean native growing at high altitudes (1,800–4,000 m) and low latitudes (0°–22°S), ranging from Chile to Venezuela, but most frequently encountered in Bolivia and Peru.

ECOLOGY: The short-day tarhui seems to be relatively tolerant to acid and sandy soils, to disease and animal predators, and to drought and frost.

HARVESTING: Usually harvested in May in Bolivia.

YIELDS AND ECONOMICS: In USSR, experimental yields were 50 MT WM/ha with 1,750 kg protein.

BIOTIC FACTORS: Unlike the Mediterranean lupines, the highly aromatic tarhui is not self pollinating. Tarhui is relatively resistant to the lupinosis-inducing fungus *Phomopsis leptostromiphormis* as well as BYMV and mildew and rots, caused by *Erysiphe, Fusarium,* and *Rhizoctonia* on other lupines.

CONTRIBUTOR: J. A. Duke.

Macroptilium atropurpureum (DC.) Urb.

FAMILY: Fabaceae
COMMON NAME: Siratro
SYNONYM: *Phaseolus atropurpureus* DC.

USES: Siratro, primarily used for grazing purposes, is among the most important pasture legumes in Australia, Brazil, Mexico, and some of the Pacific Islands. Because of its twining habit, it is capable of competing with rapid grass growth, thus making it an ideal legume to use in greenchop, or hay programs or for reserve winter pasture (in Florida). It does not make good silage without liberal addi-

Macroptilium atropurpureum (DC.) Urb.

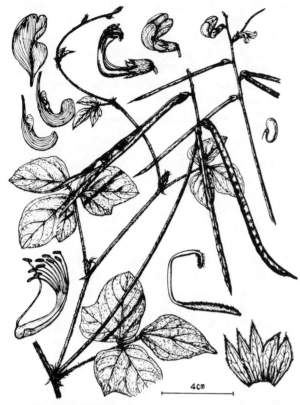

Figure 70. *Macroptilium atropurpureum* (DC.) Urb.

DESCRIPTION: Long-lived perennial with trailing stems of medium thickness, these often conspicuously pubescent and rooting at the nodes; taproots large, penetrating to the watertable on flatwoods type soils (Florida); leaves trifoliolate, the terminal leaflet lanceolate to ovate, rhomboid and entire to variably 3-lobed; lateral leaflets entire or bilobed, sometimes with a small lobe on one or both sides in lower half; upper surface green, undersurface grayish-green; terminal leaflet 2.5–7.5 cm long, 2–5 cm broad; lateral leaflets 2.5–8 cm long, 2.5–5 cm broad; stipules 4–5 mm long, hairy; peduncles elongated, with 6–10 flowers, often paired, with new flowers produced near the elongating apex; racemes 4–8 cm long, with the flowers usually crowded at anthesis, but not in fruit; flowers 15–20 mm long; calyx 5–8 mm long, rather equally 5-lobed, the lobes much shorter than the tube; corolla red-purple or maroon distally, turning black-purple with age; pods linear, nearly straight, 4–8 cm long, 4–6 mm in diameter; seeds 12–13, ca. $4 \times 2.5 \times 2$ mm, ovoid, light brown to black (75,000/kg). Fl. spring to fall, depending on locality in the subtropics, may flower year-round or during short days in the drier tropics.

GERMPLASM: Siratro was derived from a hybrid of two strains of *Macroptilium atropurpureus* from Mexico. It is a result of bulking equal proportions of seeds of three F4 families. A symposium on siratro was dedicated to its breeder, E. M. Hutton, who developed siratro because of deficiencies in the annual *M. lathyroides* (susceptible to root knot, bean fly, and bean virus 2). Assigned to the Middle America and South America Centers of Diversity, siratro or cultivars thereof is reported to exhibit tolerance to aluminum, drought, fire, frost, grazing, heat, low pH, nematode, salt, sand, slope, and short periods of waterlogging. ($2n = 22$.)

DISTRIBUTION: Grows naturally from southern Texas through Central America to Colombia, Argentina, and Peru at altitudes from sea level to 2,000 m. Widely introduced, but most important in Australia, Brazil, and Mexico.

ECOLOGY: Siratro is well adapted to subtropical and drier tropical regions, but grows to altitude of 2,000 m. It is drought resistant due to its large taproot. Grows well on excessively drained soils

tions of molasses. It is also an excellent wildlife feed. Seeds are eaten by quail, dove, and turkeys; plants and buds, by deer. Commercial plantings used as quail and cattle feed. Siratro is an excellent drought-resistant covercrop. Used in Australia for covering road and railway banks, and reclaiming coal spoil. Used to reclaim recent deposits of lava elsewhere.

FOLK MEDICINE: No data found.

CHEMISTRY: Crude protein content is ca. 17–20% with dry matter content 24–32%. Mixed with pangola, the crude protein content varied from 6–9%. Notes on flavonoids, arabinogalactans, and oligosaccharides were published by Ford (1971, 1972a, b). The genus *Phaseolus* is reported to contain acetone, choline, fumaric acid, hydrocyanic acid, inositol, malonic acid, maltose, methyl-cysteine, oxalic acid, pectin, rutin, saponin, and succinic acid. Many of these may be expected also in *Macroptilium*.

(e.g., deep sandy soils), but not on land with long periods of flooded conditions. Grows well on flatland soils where summers are hot, but winters are cold and frosts are common. Cold weather and frost cause leaf drop. In Guadelupe and the Bahamas, it is one of the most promising legumes for alkaline soils. On extremely poor sandy soils in subtropical climates, produces an abundance of high quality forage if soil nutrient deficiencies are corrected. Generally considered more tolerant of low fertility then *Desmodium intortum* and *Glycine wightii*, less tolerant than *Stylosanthes humilis*. Brief periodic flooding in summer does no permanent damage but retards growth rate. In Australia is considered well adapted to areas with 6.5–18 dm annual precipitation, but disease precludes vigorous growth in the humid tropics. Survived well under coconut in Bali grazing experiment (13 dm precipitation). Siratro is a quantitative short-day plant with a critical maximum day length greater than 12 hr. Low temperatures tend to delay flowering but do not alter the basic short-day response. Thus, in the subtropics it flowers mainly in autumn through spring. Tolerates low pH, salinity, low soil calcium, and high aluminum but does not tolerate high manganese as well as some other tropical legumes. Ranging from Warm Temperate Dry to Moist through Tropical Dry to Moist Forest Life Zone, siratro is reported to tolerate annual precipitation of 3.8–28.2 dm (mean of 21 cases = 11.8 dm), annual mean temperature of 12.5°–26.8°C (mean of 21 cases = 22.0°C), and pH of 5.0–8.0 (mean of 18 cases = 6.2).

CULTIVATION: Propagated by seeds that can be planted after March 15 in southern Florida, Dec.–April in Australia and practically anytime in frost-free areas. In the subtropics, should be planted in the warm season at the beginning of the rainy season. Siratro is easy to maintain and persistent. Seedling growth, rainy season nodulation and establishment are rapid. Maintains satisfactory populations vegetatively and by self-sown seeds. These characteristics also contribute to its success when oversown into native pastures. Stand can be established with seeding rate of 2–3 kg/ha at the same time grass is planted. Solitary plants may be erect, but the habit usually ranges from prostrate through climbing to suberect, depending on the associated vegetation. If seeded to clean fallow conditions or at the same time that grass is seeded, growth is prostrate and by winter some internode secondary roots form from stolons. After the first year, these stolons and secondary roots are quite evident. Plants live up to 4 years on flatwoods soils where the watertable fluctuates from 0–30 cm during summers. Best to plant seed in a well-prepared seedbed when grass is planted. Seeds can be broadcast on established pastures (e.g., pangola in Florida), lightly disked or chopped, and then rolled or dragged. Up to 2.5 MT/ha of lime and a normal white clover fertilizer are satisfactory to establish a stand on the virgin flatwoods soils of Florida. After this, yearly application of 350 kg/ha of 0–10–20 or similar fertilizer should be adequate in such soils. In general, fertilizer should be added to correct any local deficiencies. In many areas, only phosphate is necessary. Siratro is not specific in its bacterial requirements and nodulation is effective when seeds are treated with commercial cowpea inoculant. About twice the amount of inoculum recommended by manufacturer is used under most conditions. In Australia, however, commercial seedings are rarely inoculated because the native rhizobium is adequate. Vegetative growth begins slowly in early spring, peaks in May, and continues until autumn.

HARVESTING: Yield of siratro increased linearly from 1.5 to 9.5 MT/ha as the cutting interval increased from 4 to 16 weeks. For a vigorous stand siratro should not be grazed during the summer or fall. This rest period is important since siratro–grass pastures are grazed heavily during the winter or dry season. For Florida suggested management is: graze heavily during Nov.–April, remove cattle until June or July to permit seed production, graze moderately until September, then remove cattle until seed accumulates in November and December. Crop planted prior to July and not severely grazed produces some seed the first winter in southern Florida. Flowering is heavy during the winter and spring unless frost damages plants. Second year plants produce abundant seed. Since mature pods shatter, crop is harvested before full maturity and pods are dried under natural conditions. Seed heads should be mowed and windrowed in April or May, during maximum seed production (Florida). When pods are dry, the windrows can be combined. Siratro–grass mixtures can be handled in the same manner since grass grows little prior to seed harvesting. In Australia the rest period is unnecessary because siratro produces abundant pasture even when grazed. In the midsummer cattle apparently prefer grass to siratro, which then accumulates for fall grazing. In Australia, most siratro pastures are grazed year long, and still produce enough seed for stand maintenance. There foliage is removed and fallen seeds are collected by suction.

YIELDS AND ECONOMICS: Dry matter yields average 5–7 MT/ha with 2 cuts per season and up to 11 MT/ha with 4 cuts per year. Siratro, which has been an important forage and covercrop in tropical America and Australia for more than 10 yr, was introduced into southern Florida, and Africa (Natal and Zululand) and other areas. Grown with pangola or Pensacola bahiagrass in Florida it increases the dry matter and crude protein yields. It is readily eaten by cattle and, under moderate grazing, perennates either vegetatively or by reseeding.

BIOTIC FACTORS: Following fungi have been reported on siratro: *Cercospora* sp., *Erysiphe polygoni, Rhizoctonia microsclerotia* (Africa), *Rh. solani* (Florida, causing leafdrop, most trouble when annual rainfall exceeds 16 dm), *Synchytrium* sp. (said to be serious in Brazil with rainfall 10–20 dm), *Uromyces phaseoli* (rust, bad in Colombia), and *U. appendiculatus*. Also attacked by the bacterium *Pseudomonas phaseolicola* (Bean halo blight), the common bean mosaic virus, and the nematode *Helicotylenchus dihystera*. Sometimes susceptible to rootknot nematodes (*Meloidogyne* sp.). Insect pests include: leaf damage by Bean leafroller (*Urbanus protus*), other caterpillars, and mites (*Tarsonemus confusus*) which cause leafdrop. Susceptibility to pests and diseases is reviewed by Jones and Jones (1977) and Hutton and Beall (1977). It persists and produces well in most areas in spite of some disease and insect problems.

REFERENCES:

Ford, C. W., 1971, Flavonoids from *Phaseolus atropurpureus*, Phytochemistry **10**:2807.
Ford, C. W., 1972a, Arabinogalactan from *Phaseolus atropurpureus* leaves, Phytochemistry **11**:2559.
Ford, C. W., 1972b, Oligosaccharides from *Phaseolus atropurpureus*, Aust. J. Chem. **25**:889.
Hutton, E. M., 1962, Siratro—A tropical legume bred from *Phaseolus atropurpureus*, Aust. J. Exp. Agric. Anim. Husb. **2**:117–125.
Hutton, E. M., and Beall, L. B., 1977, Breeding of *Macroptilium atropurpureum*, Trop. Grassl. **11**:15–31.
Jones, R. J., 1967, Effects of close cutting and nitrogen fertilization on growth of a siratro (*Phaseolus atropurpureus*) pasture at Samford, south-eastern Queensland, Aust. J. Exp. Agric. Anim. Husb. **7**:157–161.
Jones, R. J., and Jones, R. M., 1977, The ecology of siratro-based pastures, in: *Plant Relations in Pastures* (J. R. Wilson, ed.), CSIRO, Melbourne.
Kretschmer, A. E., Jr., 1966, Four years' results with siratro (*Phaseolus atropurpureus* DC.) in south Florida, Soil Crop Sci. Soc. Fla., Proc. **26**:238–245.
Kretschmer, A. E., Jr., 1972, Siratro (*Phaseolus atropurpureus* DC.) a summer-growing perennial pasture legume for Central and South Florida, Fla. Agric. Exp. Stn., Circ. S-214:1–21, illus.
Minson, D. J., and Milford, R., 1966, The energy values and nutritive value indices of *Digitaria decumbens, Sorghum almum*, and *Phaseolus atropurpureus*, Aust. J. Agric. Res. **17**:411–423.
Shaw, N. H., and Whiteman, P. C., 1977, Siratro—A success story in breeding a tropical pasture legume, Trop. Grassl. **11**(1):7–14.
Vallis, I., and Jones, R. J., 1973, New mineralization of nitrogen in leaves and leaf litter of *Desmodium intortum* and *Phaseolus atropurpureus* mixed with soil, Soil Biol. Biochem. **5**(4):391–398.

CONTRIBUTORS: J. A. Duke, E. M. Hutton, A. E. Kretschmer, Jr., H. Pollard, C. F. Reed, N. H. Shaw, J. Smartt.

Macrotyloma geocarpum (Harms) Marechal & Baudet

FAMILY: Fabaceae

COMMON NAMES: Kersting's groundnut, Groundbean, Potato bean

SYNONYMS: *Kerstingiella geocarpa* Harms
Voandezia poissonii Chev.

USES: Indigenous grain legume cultivated in parts of Tropical Africa for food. Plant produces its seed underground much like peanuts. Sometimes use is restricted to chiefs or men (as in Dahomey); restriction may indicate its high quality as food and scarcity. Seeds have agreeable flavor and are highly nutritious with high nitrogen content. Leaves sometimes eaten in soup. Seeds are boiled with shea butter, palmoil, and other local vegetable oils and eaten with certain condiments alone or with starchy foods such as yams and rice.

FOLK MEDICINE: No data available.

Macrotyloma geocarpum (Harms) Marechal & Baudet

DESCRIPTION: Prostrate herbaceous annual, the main stem to 10 cm long, creeping, rooting at the nodes; leaves trifoliolate, rising on long slender petioles, the leaflets broadly ovate or obovate, obtuse, 5–7.5 cm long, 4–5 cm broad; flowers small, in pairs or solitary, subsessile in leaf axils of branches which rest on the ground; corolla papilionaceous, greenish-white with standard pale violet at tip, or tinted bluish-purple; pods maturing underground (after fertilization of chasmogamous flowers, the solid base or stipe of pistil lengthens into a carpopodium and turns toward the ground; then the corolla and style are thrown off, and the ovary is pushed out of the calyx and by the rootlike carpopodium gradually driven into the ground where finally the ovary grows and matures into the seed-bearing pod), indehiscent, 1.2 cm long, ca. 1 cm broad, usually divided by 1 or 2 constrictions into 2 or 3 joints; seeds 1–3, usually 2, oblong or oblong-ovoid, about 0.6–1.3 cm long, kidney-shaped with a white hilum, white, red, brown or black, or mottled, resembling seeds of *Phaseolus vulgaris* but smaller.

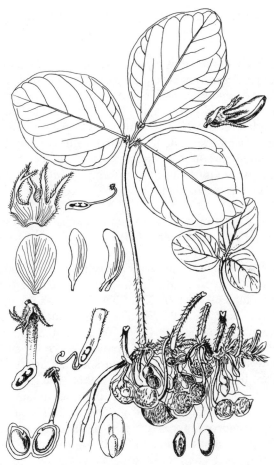

Figure 71. *Macrotyloma geocarpum* (Harms) Marechal & Baudet

CHEMISTRY: Dried seeds contain, per 100 g: 348 calories, 9.7% moisture, 19.4 g protein, 1.1 g fat, 66.6 g total carbohydrate, 5.5 g fiber, 3.2 g ash, 103 mg Ca, 392 mg P, 15 mg Fe, 0.76 mg thiamine, 0.19 mg riboflavin, 2.3 mg niacin, and 0 mg ascorbic acid. Fresh seeds contain, per 100 g: 9.3–12.5 g moisture, 19.0–20.4 g protein, 0.9–2.3 g fat; 66.6 g total carbohydrate, 4.6–5.6 g fiber, 2.8–3.3 g ash, 85-227 mg Ca; 285–425 mg P, and 0.67–0.77 mg thiamine. The fatty acids, as a percentage of the total fatty acids are: 31.5 palmitic, 5.0 stearic, 7.0 oleic, 31.5 linoleic, 21.3 linolenic, 3.1 arachidic, and 0.6 behenic. The amino acid content is (mg/g N): arginine 404, cystine 63, histidine 173, isoleucine 281, leucine 479, lysine 413, methionine 86, phenylalanine 363, threonine 238, tyrosine 219, valine 390, aspartic acid 717, glutamic acid 1088, alanine 279, glycine 279, proline 340, serine 367, and tryptophane 50.

GERMPLASM: Native plants are classified as var *tisserantii* (Pellegrin) Hepper (Syn.: *K. tisserantii* Pellegrin); cultivated plants are var *geocarpum*. White-flowered cvs produce white seeds. Assigned to the Africa Center of Diversity, Kersting's groundnut or cvs thereof is reported to exhibit tolerance to drought, poor soil, sand, and savanna. ($2n = 20, 22$.)

DISTRIBUTION: Native to tropical West Africa (Gold Coast, Togo, Dahomey, from Senegal to Nigeria): known mainly from cultivated plants. Wild ancestors found in Ubangi, Shari, and Cameroons.

ECOLOGY: Flourishes in savanna and rainforest/savanna transition (derived savanna) areas of West Tropical Africa. Requires an annual rainfall of 5–15 dm. Grows well on sandy loams rich in lime, but tolerates poor, sandy loams and moderately acid to neutral soils. Needs plenty of sunshine and temperatures 18°–34°C, but tolerates drought. Reported from Tropical Dry Forest Life Zone (near homesteads), alone or in mixtures with cowpea, cassava, and other crops on mounds, beds, or ridges, Kersting's groundnut is reported to tolerate annual pre-

Macrotyloma geocarpum (Harms) Marechal & Baudet

cipitation of 5–14 dm, annual mean temperature of 26°–27°C and pH of 4.9–5.2.

CULTIVATION: Plants grown in small plots. Seed sown in late May or June, in the beginning or middle of the rainy period. Seeds spaced 15 cm apart in rows 30–40 cm apart when grown alone. Require the usual cultivation for garden crops.

HARVESTING: Plants mature in 90–150 days as the rains end. Seeds ripen as the leaves turn yellow. Plants are dug and left on the ground to dry. Pods are hand-picked and later beaten with sticks or in a mortar to remove the seeds. Small pods and seeds and low yield explain why this groundnut is not grown extensively although it is of good food value. Seeds must be thoroughly dried before storage. Dusting with an insecticide such as BHC formulation may prevent bruchid attack.

YIELDS AND ECONOMICS: Dry seed yields average about 500 kg/ha. A minor crop in areas of West Tropical Africa. Because it is very nutritious, production should be increased.

BIOTIC FACTORS: Kersting's groundnut is not seriously attacked by pests or diseases, except bruchid beetles and weevils, *Piezotrachelus*. Weevils get into improperly stored seeds. Susceptible to leaf molds or blights, and leaf insects especially in savanna areas with high rainfall.

REFERENCES:
Chevalier, A., 1933, Monographie de l'arachide, *Rev. Bot. Appliq.* **13**:697–711.
Harms, H., 1908, *Kerstingiella geocarpa* Harms, n.sp., *Ber. Deutsch. Bot. Ges.* **26a**:230–231, vol. 3.
Hepper, F. N., 1963, *Kerstingiella* Harms, *Kew Bull.* **16**:404–405.
Marechal, R., and Baudet, J. C., 1977, Transfert du genre africain *Kerstingiella* Harms a Macrotyloma (Wight & Arn.) Verdc. (Papilionaceae), *Bull. Jard. Bot. Nat. Belg.* **47**:49–52.

CONTRIBUTORS: J. A. Duke, B. N. Okigbo, C. F. Reed.

Macrotyloma uniflorum (Lam.) Verdc.

FAMILY: Fabaceae

COMMON NAMES: Horsegram, Madras gram, Wulawula

SYNONYMS: *Dolichos biflorus* auct., non L.
Dolichos uniflorus Lam.

USES: Cultivated as a poor man's pulse in southern India, horsegram seeds are parched, boiled or fried, then eaten either whole or ground as a meal. Seeds have an earthy aroma and are not palatable. Boiled seeds are often fed to cattle and horses. In Burma dry seeds are boiled, pounded with salt, and fermented; product is similar to soya sauce. Stems, leaves and husks used as a forage, and entire plant grown as green manure crop.

FOLK MEDICINE: Said to be used as astringent (sd), diuretic (sd), and tonic (sd).

CHEMISTRY: One analysis of seeds gave the following values: moisture 11.8, crude protein 22.0, fat 0.5, mineral matter 3.1, fiber 5.3, carbohydrates 57.3, Ca 0.28, P 0.39%; Fe 7.6 mg, nicotinic acid 1.5 mg, carotene 119 IU per 100 g. Another analysis of red seeds showed 8.4% moisture, 59.3% carbohydrates, 4.8% fat, 24.7% protein, 2.8% ash, 0.02% iron, 0.34% Ca, 0.27% P, 40–119 IU/100 g vitamin A, 1.5 mg nicotic acid/100 g, 0.40 mg thiamin/100 g, 0.15 mg riboflavin/100 g. The globulins of horsegram account for nearly 80% of the total nitrogen. They contain: arginine (6.0-7.1%), tyrosine (6.68%) and lysine (7.64%), but are deficient in cystine and tryptophane. Red-seed amino acids are (mg/g N): arginine 519, threonine 212, leucine 506, isoleucine 344, valine 356, histidine 156, phenylalanine 419, lysine 537, methionine 106, tryptophane 60, cystine

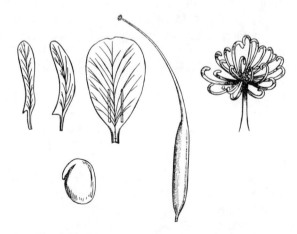

Figure 72. *Macrotyloma uniflorum* (Lam.) Verdc.

62, and tyrosine 331. At 10% level of protein intake the biological value and digestibility coefficient are 66 and 73, respectively. Germinated seeds and seedlings are fair source of 1-asparagine. Horsegram is a rich source of urease. Analysis of hay gave the following values (dry-matter basis): crude protein, 10.6; crude fiber, 16.2; N-free extr., 58.3; fat, 1.8; total ash, 13.1; ash sol. in HCl, 8.0; CaO, 2.5; P_2O_5, 0.42; MgO, 1.0; K_2O, 1.2%. Dried seeds contain: 333 calories, 9.7% moisture, 22.5 g protein, 1.0 to 2.0 g fat, 56 to 60.5 g total carbohydrate, 4.7 g fiber, 6.3 ash. Trypsin inhibitors are reported from the seed.

DESCRIPTION: Low-growing, slender, suberect annual herb, with twining, downy stems and branches, 30–50 cm tall; leaves trifoliolate; leaflets pilose, entire, membranous, broadly ovate, acute, 2.5–5 cm long, lateral leaflets oblique; stipules minute; flowers 1–3 together in leaf-axils, about 1 cm long, each on a short peduncle; calyx downy with lanceolate teeth; corolla pale yellow, standard ovate, emarginate, slightly reflexed, longer than wings, stamens diadelphous, style filiform and persistent; pod 3–5 cm long, linear, recurved, beaked, downy, dehiscent, 5–7-seeded, valves chartaceous; seeds small, rhomboidal, flattened, 3–6 mm long, light red, brown, gray or black, or mottled, with shining testa and small inconspicuous hilum.

GERMPLASM: Several cvs differ in period of maturity and seed color. The cultivated crop usually is a mixture of several cvs. Most cvs require short days, but day-neutral cvs with high yield potential are being sought. Assigned to the Hindustani Center of Diversity, horsegram or cvs thereof is reported to exhibit tolerance to drought, poor soil, and sand. ($2n$ = 20, 22, 24.)

DISTRIBUTION: Native to Old World Tropics, horsegram is extensively cultivated, especially in dry areas, in Australia, Burma, India, Sri Lanka, and the Himalayas.

ECOLOGY: A short-day annual, horsegram thrives in diverse soils (except strongly alkaline), even poor soils with little rain. Does well on light sandy soils, red loams, black cotton loams, and stony and gravelly upland soils but does not tolerate waterlogging. Thrives on newly cleared land. Requires average temperature of 20°–30°C. Grown as a dryland crop in areas with less than 3 dm rainfall; sown in high rainfall areas after the rains cease. Ranging from Subtropical Dry to Moist through Tropical Dry to Moist Forest Life Zones, horsegram is reported to tolerate annual precipitation of 7.0–42.9 dm (mean of 8 cases = 19.8), annual mean temperature of 23.4°–27.5°C (mean of 8 cases = 25.3), and pH of 5.0–7.5 (mean of 6 cases = 6.4). Grown as high as 1800 m in India.

CULTIVATION: Seeds are broadcast or sown 1–2.5 cm deep in rows ca. 0.9 m apart at rate of 25–45 kg/ha; in India sown in Sept.–Oct. Intercropped with *Guizotia*, sown at ca. 45 kg/ha. When mixed with ragi *(Eleusine)* or lentil *(Lens)*, rows should be 0.9–1.8 m apart. Usually grown monoculture, but may be intercropped with castor beans, niger, ragi, or peanuts. Crop requires little attention. In Australia, ca. 375 kg/ha superphosphate is applied. In India 30 kg/ha N and 60 kg/ha phosphoric acid are considered economically productive rates. For seed off-types should be rogued before flowering. In India, horsegram follows *Eleusine, Sesamum*, or *Sorghum* in rotations.

HARVESTING: Crop matures in 4-6 months for seed, in 6 weeks for forage. For seed, crop cut when pods are dry. Plants pulled or cut, stacked and dried in sun. For best quality seeds, only ripe pods are picked from standing crop. In India threshed by flailing plants with sticks, by stone rollers, or trodden by oxen. Seed requires thorough winnowing and sieving as weed content is usually high. Seed stored like cowpeas.

YIELDS AND ECONOMICS: For seed, yields average 130–300 kg/ha but yields to 900 are reported in India; for green forage, 5–12 MT/ha. In Bengal, seed yields of 330 kg/ha, forage yields of 11 MT are reported. In Australia seed yield averaged 1,120–2,240 kg/ha. In Madras rainfed forage yields are 2–6 MT, irrigated forage yields 9–14 MT/ha. Horsegram is grown extensively only in India.

BIOTIC FACTORS: In India, most serious diseases are due to *Rhizoctonia* sp., causing root rot (Kulthis bean blight) and *Glomerella lindemuthianum* (anthracnose). The rust *Uromyces appendiculatus* affects the leaves. Anthracnose causes black sunken

Macrotyloma uniflorum (Lam.) Verdc.

spots on stems, leaves, seeds, and pods. *Vermicularia capsici* induces a dieback, causing the flowers to droop and dry up gradually. Bacterial leaf spot caused by *Xanthomonas phaseoli* var *sojensis* causes speckling. Hairy caterpillars and grasshoppers commonly affect horsegram. The gram-leaf caterpillar *Anticarsia irrorata* is destructive. A green pod-boring caterpillar, *Etiella zinckenella* ("lima-bean pod borer"), causes some damage.

CONTRIBUTORS: J. A. Duke, C. F. Reed.

Medicago arabica (L.) Huds.

FAMILY: Fabaceae
COMMON NAMES: Spotted or Southern burclover
SYNONYM: *Medicago maculata* Sibth.

USES: This burclover is used mostly as pasture for hogs, cattle, sheep, and poultry. Farm animals do not eat it readily at first, but adapt and eat it freely. On Pacific Coast from Oregon south to Arizona and on the Great Plains to eastern Texas and southern Oklahoma, it is valuable agriculturally only when winters are mild. Crop is highly regarded because it readily maintains itself with little or no reseeding, adds humus and nitrogen to soil, maintains high feeding value and still permits the growing of a regular summer crop. In southern United States it is an economical legume that survives as a winter crop, prevents erosion of soil, furnishes pasture and improves soil. Often mixes with bermudagrass, which provides grazing during hot summer months after burclover has disappeared. Yields of cotton were increased each season by use of burclover.

FOLK MEDICINE: No data found.

CHEMISTRY: The genus *Medicago* is reported to contain canavanine, choline, citric acid, hydrocyanic acid, ketoglutaric acid, limonene, malic acid, malonic acid, malvidin-3,5-diglucoside, medicagenic acid, *d*-ononitol, oxalic acid, pantothenic acid, pectin, phylloquinone (vitamin K), quercetagetin-7-methyl ether, quinic acid, saponin, shikimic acid, tannin, trigonelline, and tryptophane. Seeds are reported to contain trypsin inhibitors.

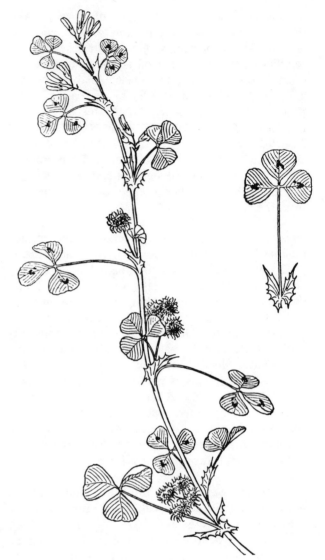

Figure 73. *Medicago arabica* (L.) Huds.

DESCRIPTION: Winter annual herb, usually branched at crown, with 10–20 spreading or decumbent branches 15–75 cm long; stems sparsely pubescent or glabrous; leaves stipulate, trifoliolate, leaflets obcordate, cuneate, dentate near apex, with a dark spot; racemes 1–4 (–10) flowered; flowers small, yellow, 5–7 mm long; legume in lax spiral of 4–7 turns, 4–6 mm in diameter, subglobose to shortly ellipsoid, flattened at both ends, glabrous, usually spiny; transverse veins anatomosing, curved and forming elongated network near margin; margin with 3 grooves, the lateral deeper and wider than the middle one; spines 2–4 mm long; more than 1,000 burs per plant in well-developed plants. Roots fibrous, shallow. Fl. summer and fall.

GERMPLASM: Several cvs of spotted burclover have been developed and differ in general appearance and date of maturity. 'Manganese' and 'Giant burclover' are early-maturing with the reddish-brown spot in center of leaflet enlarged. At least one variety has spineless burs. Assigned to the Eurosiberian Center of Diversity, spotted burclover or cultivars thereof is reported to exhibit tolerance to adobe, alkali, high pH, heavy soil, salt, virus, and weeds. ($2n = 16$.)

DISTRIBUTION: Native to southern Europe, extending from Britain and the Netherlands eastward to Crimea, and a frequent casual further north. Introduced, cultivated and sometimes escaped in southern United States and along West Coast west of Cascade and Sierra Nevada Mountain ranges. Often escaping and becoming naturalized in California, along streams and in shady places. Cultivated in Argentina and Australia.

ECOLOGY: Crop adapted to areas with mild, moist winters. Succeeds in practically any type of soil, but grows best in loams. In South, grows best in soils rich in lime (pH 7.0–8.1), but thrives in lime-poor slightly acid soils. Favored by moist, well-drained soils. In California, it is sometimes grown in poorly drained adobe soils. Plants mature later on moist soils than on well drained soils. Soil adhering to burs often carries enough inoculum to inoculate new stands. Since sun may weaken or destroy the inoculum, burs should be harrowed without delay. Ranging from Cool Temperate Steppe to Wet through Warm Temperate Steppe to Moist Forest Life Zone, spotted burclover is reported to tolerate annual precipitation of 3.1–12.9 dm (mean of 14 cases = 7.7 dm), annual mean temperature of 7.0–26.2°C (mean of 14 cases = 14.5°C), and pH of 5.6–8.2 (mean of 12 cases = 6.9).

CULTIVATION: In native areas, plants are winter annuals. When cultivated, seeds are planted and germinate in fall, grow during the fall, winter and early spring, and mature in early summer. Crop succeeds further north when seed is sown in spring but rarely maintains itself by reseeding from year to year. Further south seed should be sown in late summer or early fall. In Cotton Belt, planting time is August and September when seed is in bur. In South may be planted as late as November and, under favorable conditions, still produce a good crop. Hulled seeds germinate more readily than those in bur, and can be planted 2–4 weeks later. In dry areas, as California, seedlings may have to be irrigated once or twice before fall rains. Hulled seeds sown by grain drill with a press-wheel attachment or by any method of broadcasting at rate of about 15 kg/ha. A firm seed bed is essential. Broadcasting is best in most areas. Seeds should be covered thinly and harrowed lightly. For seed in bur, only broadcasting is practical; for a full stand 5–10 bu/ha are necessary, followed by harrowing. In areas well-adapted for burclover, it is practical to begin with a light seeding and depend on the volunteer crop in subsequent years to fill in crop. Seed in the bur retain viability for several years; hulled seed deteriorate more readily but germination remains high for 3 yr or more. Percentage of hard seed (which do not germinate for 2 yr or more), is high for seed in bur; hulled seed germinate readily. Germinability is improved by treating seed in bur with boiling water. Seed then should be sown immediately or spread to dry as rapidly as possible in airy shady place. Boiling may destroy the inoculum. On Pacific Coast where burclover is established, natural inoculation is adequate. In Cotton Belt, seeds are inoculated with superior legume inoculant before each planting. On pasture lands where established, burclover reseeds indefinitely. Also reseeds on cultivated lands if some pods mature before the land is plowed. For summer crop or with corn and cotton, it is impractical to allow burclover to remain long enough to produce seed each season. Sometimes when these crops are planted in rows 1.3–1.6 m apart, an area of burclover can be left for maturing seed and worked down later. With rows of cotton 2–2.3 m apart, cowpeas or soybeans can be interplanted after matured burclover has been plowed down. Burclover retains viability in the soil for many years, and often volunteer stands develop. Burclover grows poorly on infertile soils. Liberal use of stable manure or commercial fertilizer is recommended. Superphosphate is most essential fertilizer; about 400 kg/ha is needed. Potash and, on very poor

Medicago arabica (L.) Huds.

soils, nitrogen can be beneficial. Boron is needed in some areas (e.g., southern Alabama).

HARVESTING: In United States, North Carolina burclover can be grazed by mid-February. Near the Gulf Coast, it furnished almost continuous winter pasture. Best to pasture for a few hours a day to avoid trampling crop. Few cases reported of bloat from burclover. Sheep like the pods as well as herbage, and much of its value in California depends on a large crop of pods that remain in edible condition for a long time. When burs are abundant, sheep fatten rapidly. In South, permanent pastures are established with burclover and bermudagrass by broadcasting about 2.5 bu of burclover seed (in bur) per hectare before disking the bermudagrass. Burclover alone is commonly used as green manure crop in orchards in California and often provides volunteer crops year after year. Under favorable conditions, burclover makes a dense stand 45–60 cm tall. Unless stand is very dense, plants lie close to ground and are difficult to mow. For hay production, burclover is sown in mixture with erect crops, oats or wheat, at rate of 12.5 bu of burclover seed (in bur)/ha to 5 bu of winter oats or 4.5 bu of wheat/ha. Under such conditions, hairy vetch or crimson clover give higher yields than burclover. Burclover hay is not highly regarded. Seed contaminates wheat or other small grains from which it is separated at mill or warehouse.

YIELDS AND ECONOMICS: Hay yields range from 5–7.5 MT/ha. Seed yields of ca. 400 kg/ha are reported from Chico, California.

BIOTIC FACTORS: Following fungi have been reported on southern burclover: *Cercospora medicaginis, C. zebrina, Colletotrichum trifolii, Oidium balsamii, Phyllosticta medicaginis, Phymatotrichum omnivorum, Pleospora herbarum, Pseudopeziza medicaginis, Pseudoplea trifolii, Sphaerulina trifolii, Sporonema phacidioides, Stagonospora meliloti, Stemphylium botryosum, Uromyces medicaginis, U. medicaginis-orbicularis*, and *U. striatus*. The nematode, *Heterodera marioni*, has been isolated from the crop.

REFERENCE:
McKee, R., 1949, Burclover, cultivation and utilization, *U.S. Dep. Agric., Farmers Bull.* **1741**:1–12, illus.

CONTRIBUTORS: J. A. Duke, C. F. Reed, J. K. P. Weder.

Medicago falcata L.

FAMILY: Fabaceae

COMMON NAMES: Yellow-flowered alfalfa, Sickle medic

SYNONYMS: *Medicago borealis* Grossh.
Medicago romanica Prodan
Medicago sativa ssp. *falcata* (L.) Archangeli

USES: Grown extensively in Europe and Asia, south into India and Himalayas, for forage, but not usually as a regular crop. Plants are useful for covering banks, slopes and borders. Used in breeding to impart frost resistance.

FOLK MEDICINE: No data found.

CHEMISTRY: Neutral and acid saponins have been reported in the plants.

DESCRIPTION: Perennial herb, with widely branching roots and deep-set crown; stems branched, decumbent or prostrate, 30–120 cm long, hairy,

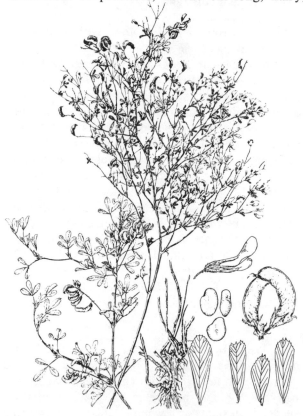

Figure 74. *Medicago falcata* L.

woody at base; leaves pinnately trifoliolate; leaflets narrow, oblong-wedge shaped, up to 1.5 cm long; stipules lanceolate, not toothed; flowers in dense clusters in axillary racemes up to 2.5 cm in diameter; corolla yellow, 5–8 mm long; pods oblong or narrowly oblong, 5–15 mm long, 3–4 mm broad, sickle-shaped or sometimes curved or near straight, glabrous or often somewhat hairy, 5–10 seeds per pod. Fl. May–Sept. (U.S.).

GERMPLASM: This is one of the more cold-hardy species of alfalfa. Variable in many aspects, to the extent that in the "Flora of the USSR," 18 botanical varieties are named and described; 8 more varieties are given for *M. romanica* therein. Most of these are based on whether the stems are erect, decumbent or procumbent, glabrous or pubescent, short or tall, and on size of leaflets, length of racemes, number of flowers per raceme and shape of pod. These probably represent local ecological populations, widely grown in their areas of adaption. However, *M. falcata* is freely interfertile with *M. sativa*, and many cvs of alfalfa in the United States contain some *M. falcata*, although none of them is solely of *M. falcata* origin. *M. X varia* Martyn is a hybrid between the two species and shows much variation, with flowers yellow, purple or yellow becoming dark purple, and pods almost straight to spired with 2–3 turns. Variegated alfalfas are thought to have arisen from natural crosses between *M. sativa* and *M. falcata*. 'Travois' and 'Rambler' are cvs bred for creeping habit, developing adventitious stem shoots that develop at enlarged points of the root, principally along lateral roots. Assigned to the Eurosiberian and Near East Centers of Diversity, yellow-flowered alfalfa or cultivars thereof is reported to exhibit tolerance to drought, frost, high pH, salt, and slope. ($2n = 16, 32$.)

DISTRIBUTION: Native throughout Europe, east to Siberia, Ladakh, Kumaon, Kashmir, Nepal, and Central Asia. Occasionally introduced and escaped as a weed. Extensively used in hybridizing with *M. sativa* because of the xerophytic nature of some of its cvs.

ECOLOGY: Grassy places, waysides, banks, wet or dry meadows, forest borders and cultivated areas are some natural habitats. It is frost-resistant and very cold-hardy, having survived temperatures of −27°C in Alaska. The most cold-resistant cvs grown successfully in northern United States and Canada are variegated, especially 'Vernal.' Ranging from Boreal Moist to Wet through Subtropical Dry Forest Life Zones, yellow-flowered alfalfa is reported to tolerate annual precipitation of 3.1–13.6 dm (mean of 27 cases = 6.9 dm), annual mean temperature of 5.7°–19.9°C (mean of 27 cases = 10.4°C), and pH of 4.9–8.2 (mean of 25 cases = 6.8).

CULTIVATION: Propagation by seeds. Planted, cultivated, and harvested like *Medicago sativa*.

HARVESTING: Large scale harvesting of seeds is difficult because of shattering. Harvesting for forage or pasture is much the same as for common alfalfa.

YIELDS AND ECONOMICS: Yields of forage are usually greater in hybrids and cvs. Extensively cultivated as a forage plant throughout the temperate to cold areas of the world. Because of its wide adaptability to cold and xerophytic conditions, its cvs and hybrids with *M. sativa* are becoming more widely utilized.

BIOTIC FACTORS: Fungi attacking yellow-flowered alfalfa include: *Erisyphe communis* f. *medicaginis*, *E. pisi*, *E. polygoni*, *Leveillula leguminosarum* f. *medicaginis*, *L. taurica*, *Peronospora aestivalis*, *P. romanica*, *Pseudopeziza jonesii*, *Selenophoma murashkinskyi*, *Sporonema phacidioides*, *Uromyces striatus* var *medicaginis*, and *Urophylyctis alfalfae*. It is also attacked by the bacterium *Corynebacterium insidiosum*, and the parasitic flowering plant, *Orobanche purpurea*.

REFERENCE:
Hanson, C. H. and Barnes, D. H., 1973, in: *Forages* (M. E. Heath, D. S. Metcalfe, and R. F. Barnes, eds.) 3rd ed., pp. 136–147, illus., Iowa State University Press, Ames, 755 pp.

CONTRIBUTORS: J. A. Duke, C. F. Reed.

Medicago lupulina L.

FAMILY: Fabaceae

COMMON NAMES: Black medic, Yellow trefoil

USES: Plant used in northern Europe and elsewhere as green manure. Empty seed pods are fed to cattle

Medicago lupulina L.

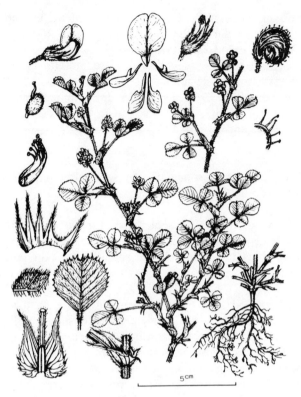

Figure 75. *Medicago lupulina* L.

and poultry. Commonly grown as a forage plant in pastures, mixing well with grasses and other clovers to make good forage. It is said to impart color and good flavor to butter. Seeds are edible, and sometimes used to adulterate alfalfa.

FOLK MEDICINE: Aqueous extracts of plant have antibacterial properties against mycobacteria. The plant is said to be lenitive.

CHEMISTRY: Seeds are reported to contain trypsin inhibitors. One analysis of the plant gave: moisture, 72.76; fiber, 4.29; fat, 0.70; N-free extr., 12.48; nitrogenous matter, 7.00; and ash, 2.77%. Miller's (1958) analyses of the forage (22.7% DM) reveal, on a moisture free basis: 16.7–26.6% crude protein (avg. 23.3), 2.9–3.5% fat (avg. 3.3), 24.7% crude fiber, 10.1–10.5% ash (avg. 10.2), 38.5% N-free extract, 0.6–1.33% Ca, 0.20–0.44 P, 2.28% K, and 0.45% Mg. Analysis of empty seed pods gave: moisture, 10.65; protein, 15.66; fat, 0.8; fiber, 24.63; soluble carbohydrates, 40.21; and ash, 8.05%.

DESCRIPTION: Annual (or perennial) herb; stems pubescent, prostrate or ascending, four-angled, up to 60 cm tall; leaves pinnately trifoliolate; leaflets cuneate-obovate, up to 20 mm long, 12 mm wide, pubescent on both surfaces, denticulate; stipules green, cordate on free side at base, lanceolate, entire or dentate, the teeth about as long as broad; petiole very short in upper leaves, up to 2.5 cm in lower ones; rachis above lateral leaflets 2–3 mm long; penducle up to 3.5 cm long; racemes dense, axillary, 5–50-flowered; bracts white, filiform, about 0.5 mm long; pedicels about 1 mm long; calyx pubescent, 1–1.5 mm long, the teeth subulate at tip, longer than the tube; corolla yellow, 2.5–3 mm long; pod 1-seeded, strongly veined, curved through 180°, ca. 3 mm long, 2 mm wide, sparsely pubescent, black when ripe; seeds small, ovoid-oblong, smooth, pale buff to greenish yellow. Fl. May–Sept. (U.S.).

GERMPLASM: No commercial cvs have been developed, but as the species is quite variable in appearance and adaptation, naturalized strains have been selected in a particular area. These are usually superior for local use and are often given preference whenever they are available. Assigned to the Eurosiberian Center of Diversity, black medic or cultivars thereof is reported to exhibit tolerance to bacteria, high pH, limestone, low pH, mycobacteria, slope, virus, and weeds. ($2n = 16, 32$.)

DISTRIBUTION: Native to Europe and temperate Asia, North Africa, south to Ethiopia, Kenya, and Tanzania; introduced in other temperate regions of world, becoming weedy, as from Nova Scotia to British Columbia, south to Florida, Haiti and Mexico; also in South America.

ECOLOGY: Black medic grows on a wide variety of soils with a good supply of lime. Often found in upland grasslands, up to 2,900 m in East Africa, and 3,600 m in the Himalayas. Requires 500–800 mm rainfall annually. Widely adapted to reasonably fertile, but not distinctly acid, soils. Does not tolerate salinity. Withstands frost when well established. Ranging from Boreal Moist to Wet through Subtropical Moist Forest Life Zones, black medic is reported to tolerate annual precipitation of 3.1–17.1 dm (mean of 52 cases = 8.4 dm), annual mean temperature of 5.7°–22.5°C (mean of 52 cases = 11.7°C), and pH of 4.5–8.2 (mean of 48 cases = 6.5).

CULTIVATION: Crop propagated by seed, sown broadcast at rate of 10–15 kg/ha, usually in spring.

Most hectarage is volunteer in pasture, waste places and in open grasslands. After seed is sown, no particular cultivation is required. Plants reseed themselves and maintain the stand.

HARVESTING: Value is greatest as pasture, used throughout its growing season. Plants may also be used for green manure during the summer and early fall.

YIELDS AND ECONOMICS: As most plants develop from volunteer seed, like a weed, it is extensively available to domestic as well as wild animals for food. Only infrequently cultivated.

BIOTIC FACTORS: Seems to be self-pollinated with some cross pollination. Seed set seems to improve with cross pollination. Seed production is increased by bees. Following fungi have been reported on black medic: *Ascochyta imperfecta, Cercospora medicaginis, C. zebrina, Colletotrichum trifolii, Coniothyrium medicaginis, Erysiphe communis* f. *medicaginis, E. pisi, E., polygoni, Fusarium culmorum, Kabatiella caulivora, Leptosphaerulina briosiana, Parodiella perisporioides, Peronospora aestivalis, P. medicaginis-orbicularis, P. meliloti, P. romanica, P. trifoliorum, Phymatotrichum omnivorum, Pleospora herbarum, P. rehmiana, Pseudoplea trifolii, Pseudopeziza medicaginis, Rhizoctonia solani, Sclerotinia trifoliorum, Septoria medicaginis, Stagonospora meliloti, Uredo lupulina, Uromyces straitus, U. s.* var *medicaginis*, and *Urophylctis alfalfae*. Plants are also attacked by the Common Pea mosaic virus. *Orobanche purpurea* is known to parasitize black medic. Nematodes isolated from this crop include: *Ditylenchus dipsaci, Heterodera schachtii, Meloidogyne artiellia, M. hapla, M. incognita*, and *Pratylenchus penetrans*.

CONTRIBUTORS: J. A. Duke, C. F. Reed.

Medicago orbicularis (L.) Bartal.

FAMILY: Fabaceae

COMMON NAMES: Buttonclover, Large disc-medic

USES: Cultivated and used for forage in Mediterranean region and in southwestern Asia. Forage quality is considered rather low. For sheep, buttonclover is preferred to spotted or toothed burclover,

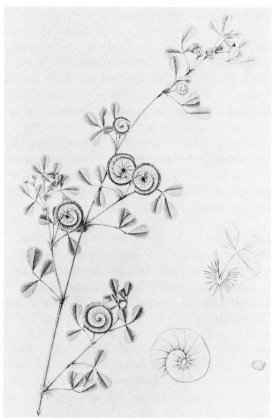

Figure 76. *Medicago orbicularis* (L.) Bartal.

since their fruit becomes tangled in the wool. Of some value as a green manure, about equal to the burclovers.

FOLK MEDICINE: No data found.

CHEMISTRY: No data found.

DESCRIPTION: Annual herb with short, slender roots; stems slender, procumbent, decumbent or ascending, 10–90 cm tall, glabrous or sparsely hairy, branched from base; leaflets glabrous or sparsely glandular-hairy, 5–18 mm long, obovate-cuneate to almost obtriangular, denticulate at apex or in upper two-thirds, often slightly notched at apex with a tooth in notch; stipules 5–6 mm long, laciniate nearly to base, with slender subulate-linear segments; racemes 1–5-flowered, on peduncles shorter than subtending leaf, usually a single pod developing from each inflorescence; corolla 2–5 mm long, orange-yellow; standard large, about twice as long as the equal keel and wings; calyx broadly turbinate-campanulate, obscurely veined, glabrous or sparingly appressed-pubescent; teeth

Medicago orbicularis (L.) Bartal.

subulate-triangular, acute, prominently 1-veined on back, usually as long as tube; pedicels filiform, nodding, as long as or longer than calyx, elongating and becoming stouter in fruit; bracteoles obsolescent; pods lenticular, slightly convex on both faces, 10–18 mm in diameter, in 3–6 coils, each coil containing 4–5 seeds, unarmed, glabrous or glandular-hairy, stramineous when mature, later blackening; transverse veins several from ventral suture, crossing entire surface, prominent, terminating in dorsal suture; seeds brown, ovate-triangular, convex, finely tuberculate-punctate; radicle about as long as cotyledons. Fl. April–July (U.S.).

GERMPLASM: *M. orbicularis* forma *pilosa* Benth. has glandular-hairy pods. Ranging from Warm Temperate Steppe to Moist through Subtropical Dry Forest Life Zones, buttonclover or cvs thereof is reported to exhibit tolerance to high pH, slope, and virus. ($2n = 16, 32$.)

DISTRIBUTION: Native to the Mediterranean region of southern Europe from Portugal and Spain east to Asia Minor, Crimea, Caucasus, Transcaucasus, Turkmania, and Central Asia; Ethiopia. Introduced to Central Europe and infrequently introduced in waste places around seaports in eastern United States. Common in several regions of Australia.

ECOLOGY: Occurs naturally on grassy slopes, open scrub, steppe sites, in olive groves and cultivated places, usually in lower mountain zone up to 800 m in altitude. Occasionally a weed in lawns, growth habit changes in relation to dryness of habitat. Becomes almost erect in grass stands. Requires a warm temperate climate with mild winters, is intolerant of excessive heat and cold. Fares best on fairly well-drained rich loam soils, but grows on almost any soil containing sufficient moisture. It has done well on heavy moist low land and fairly under rather arid conditions. Assigned to the Mediterranean and Central Asia Centers of Diversity, buttonclover is reported to tolerate annual precipitation of 4.4–12.8 dm (mean of 8 cases = 8.0), annual mean temperature of 14.2°–21.8°C (mean of 8 cases = 18.7), and pH of 6.5–8.0 (mean of 8 cases = 7.3).

CULTIVATION: Propagated from seed, sown either broadcast or drilled in rows. Cultivation is about the same as for alfalfa. Broadcasting is usually best, followed by light harrowing. Often grown in mixtures of forage plants. Can be sown in spring, as a summer crop in areas with cold winters, but it is best sown in fall and grown as a winter annual in mild climates. On land new to buttonclover or burclover, inoculation may be essential. Sowing seed in the bur seems to ensure inoculation.

HARVESTING: Used like alfalfa for pasture and forage. Should not be grazed closely in late spring, especially by sheep that eat green burs so that few seed mature. The harvesting of seed from decumbent plants is difficult. In the past, seed were allowed to ripen, then the plants were mowed and piled up in windrows. Burs were then "swept" up and winnowed. An ordinary clover huller separates the seed satisfactorily.

YIELDS AND ECONOMICS: Under favorable conditions it produces much more seed than burclover. At Chico, California, seed yields run ca. 1,000 kg/ha, more than double yields of toothed burclover. Widely used in areas of adaptation for fodder, especially in Mediterranean region, southwestern Asia and warm temperate regions elsewhere.

BIOTIC FACTORS: Seems to be self-pollinated. Following fungi have been reported on buttonclover: *Erysiphe communis*, *E. polygoni*, *Leveillula leguminosarum*, *Peronospora medicaginis-orbicularis*, *Pseudopeziza trifolia*, *Septoria medicaginis*, *Uromyces anthyllidis*, *U. medicaginis-orbicularis*, *U. striatus*, and *Vermicularia dematium* var *macrospora*. It is also attacked by the Wisconsin pea streak virus.

REFERENCE:
McKee, R., 1916, Button clover, *U.S. Dep. Agric., Farmers Bull.* **730**:9 pp.

CONTRIBUTORS: J. A. Duke, C. F. Reed.

Medicago polymorpha L.

FAMILY: Fabaceae
COMMON NAMES: California or Toothed burclover
SYNONYMS: *Medicago denticulata* Willd.
Medicago hispida Gaertn.

Medicago lappacea Desr.
Medicago nigra (L.) Krocker
Medicago polycarpa Willd.

USES: California burclover, used as green forage and pasture, is relished by cattle, sheep, hogs, horses, camels, and poultry. Rarely used as hay. Burs provide a valuable concentrated fodder for animals during dry season, and are most readily eaten when softened by rains. Sheep fatten rapidly when fed on burs, which may get entangled in wool, causing problems. Burs may cause bloating in some animals. Plants are sometimes eaten as leafy vegetable. Useful for soil renovation and green manuring, as winter cover and for preventing erosion.

FOLK MEDICINE: No data found.

CHEMISTRY: Has caused a photosensitized dermatitis in cattle. On the basis of ca. 50 analyses, Miller (1958) reported the hay (89.7% DM) contains, on a moisture-free basis: 13.2–26.5% crude protein (avg. 19.0), 2.0–3.3% fat (avg. 2.8), 22.8–33.2% crude fiber (avg. 25.1), 6.3–13.7% ash (avg. 9.7), 35.0–48.8% N-free extract (avg. 43.4). Green forage (20.8% DM) contained, on a moisture-free basis: 16.7–32.9% crude protein (avg. 24.1) 2.5–8.2% fat (avg. 4.9), 12.8–30.7% crude fiber (avg. 20.4), 6.1–11.1% ash (avg. 9.3), and 37.5–44.5% N-free extract (avg. 41.3).

DESCRIPTION: Winter annual herb, glabrous or pubescent; stems semierect or prostrate, up to 40 cm long; leaves pinnately trifoliolate, leaflets obovate to obcordate, cuneate, dentate near apex; stipules lanceolate to ovate-lanceolate, laciniate; racemes compact, 1–8-flowered, very small; corolla yellow, 3–4.5 mm long; legume (bur) 4–10 mm in diameter, in 1.5–6 lax spirals, usually glabrous and spiny; transverse veins prominent and anastomosing freely, at least near the conspicuous submarginal vein separating it from the marginal vein by a deep groove, the margin with 3 keels separated by 2 grooves; seeds 3–5 per bur, kidney-shaped, brownish yellow. Fl. summer and fall.

GERMPLASM: Varies in ornamentation of the burs and pubescence. Assigned to the Eurosiberian and Hindustani Centers of Diversity, California burclover or cvs thereof is reported to exhibit tolerance to adobe, alkali, high pH, heavy soil, shade, slope, and weeds. ($2n = 14, 16$.)

DISTRIBUTION: Southern Europe, north to Britain, east to Russia and Crimea; a frequent casual northward in Europe; eastward into Punjab, Kumaon, North Bengal, Madras, ascending to 1,500 m in the Himalayas; western and central Asia, China, Japan, North Africa, south to Ethiopia, Eritrea, Kenya, and Tanzania (ascending to 1,900 m); introduced and cultivated in other dry warm temperate regions as Argentina, Australia, and Pacific Coast of the United States.

ECOLOGY: Primarily adapted to regions with mild winters. Thrives in many soil types, does best in moist, well-drained, slightly alkaline, loamy soil. Found naturally along streams, in shady spots, often as weed in cultivation and pasture where dry upland forest has been cleared. Requires rainfall about 8 dm annually. In California grows on adobe soils which are often poorly drained. Plant is about as tolerant of alkali as barley, but does not tolerate soils heavily charged with salts. Not frost tolerant.

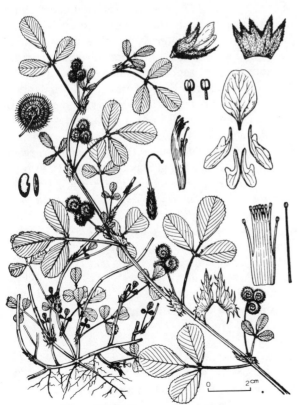

Figure 77. *Medicago polymorpha* L.

Medicago polymorpha L.

Seems to be more sensitive to 2,4-D than is Trifolium. Ranging from Warm Temperate Dry to Moist through Tropical Very Dry Forest Life Zones, California burclover is reported to tolerate annual precipitation of 3.1–19.1 dm (mean of 36 cases = 8.6), annual mean temperature of 10.5°–27.5°C (mean of 36 cases = 16.3°C), and pH of 5.3–8.2 (mean of 30 cases = 6.8).

CULTIVATION: Propagated by seeds, either hulled or in the bur. Hulled seeds germinate readily. Treatment of burs in boiling water hastens germination. Seeds are sown broadcast or drilled. Seed should be sown in late summer or fall. California burclover grows poorly on soils of low fertility; liberal use of stable manure or commercial fertilizer is recommended. Grows luxuriantly in alkali soil containing 12.5 MT/ha carbonate of soda in the top 75 cm of soil. Superphosphate is most often needed, at about 400 kg/ha. Sometimes potash and, on poor soils, nitrogen are beneficial. Boron is needed in some areas. In dry areas of California, seedlings may need irrigation once or twice before fall rains. Hulled seed are sown by grain drill or broadcast, at rate of 15 kg/ha. A firm seedbed is essential. Seed should be covered lightly. Broadcasting, followed by light harrowing, is most successful method in most areas. In areas well adapted to burclover a light seeding is adequate because reseeding in the following years establishes the stand. Seed in the bur retain viability for several years; hulled seed deteriorate more readily, but germination remains high for 3 years or more. Seeds in the bur contain hard seeds that do not germinate for 2 or 3 yr, but hulled seeds germinate readily. On Pacific Coast of North America where burclover is established natural inoculation is adequate. In other areas, superior legume inoculants are used before each planting. Burclover reseeds indefinitely in established pastures and on cultivated lands, if some pods mature before the land is plowed. For summer crops like corn and cotton, it is impractical to allow burclover to remain long enough to produce seed each season. When these crops are planted (rows 1.3–1.6 m apart), an area of burclover is left for maturing seed and worked down later.

HARVESTING: Plants widely used for pasture. Best to pasture for a few hours a day to avoid trampling of crop. Sheep especially like the pods as well as the herbage. When burs are abundant in a pasture, sheep fatten very rapidly. Burclover alone is commonly used as green manure in orchards in California and is often managed to provide a good volunteer crop year after year. Under favorable conditions burclover makes a dense stand 45–60 cm tall. Unless stand is very dense, it lies close to ground and is difficult to mow. For hay production, burclover is sown in mixture with erect crops, like oats or wheat, at rate of 12.5 bu/ha burclover seed (in bur) to 5 bu/ha winter oats or 4.5 bu/ha wheat. Much seed is harvested as impurity in wheat or other small grains, from which it is separated at mills and warehouses. Occasionally some seed is harvested from pasture lands, where burs in pure stands are often produced in abundance. Simplest method of harvesting is by sweeping by hand with large stiff brooms.

YIELDS AND ECONOMICS: Yields average 5–7.5 MT/ha hay. Extensively grown and cultivated in southern Europe, northern India, Argentina, Australia, and the Pacific Coast of North America for pasture and fodder.

BIOTIC FACTORS: Apparently self-pollinated. Most serious insect pest is the clover seed chalcid wasp *Bruchophagus platypterus*, which also attacks red clover and alfalfa. It lays its eggs through the green pod into the soft seeds. In California about 10% of the early-maturing seed is destroyed, and about 75% of the late seed. No practical method for control of this insect is known. Nematodes reported include: *Ditylenchus dipsaci, Heterodera glycines, H. trifolii. Meloidogyne incognita,* and *Pratylenchus brachyurus*. Fungi reported include: *Cercospora medicaginis, Colletotrichum trifolii, Erysiphe graminis, Erysiphe polygoni, Ischnochaeta pisi, Leptosphaerulina briosiana, Peronospora aestivalis, Phoma melaena, Pleosphaerulina sojaecola, Pseudopeziza medicaginis, Pseudopeziza trifolii, Pythium irregulare,* and *Uromyces striatus*. The bean yellow mosaic virus, pea mosaic virus and the parasitic flowering plant *Orobanche ramosa* have been reported as pests of this species.

REFERENCE:
McKee, R., 1949, Bur-clover: Cultivation and utilization, *U.S. Dep. Agric., Farmers Bull.* **1741**:1–12, illus.

CONTRIBUTORS: J. A. Duke, C. F. Reed.

Medicago sativa L.

FAMILY: Fabaceae

COMMON NAMES: Alfalfa (en. sp), Lucerne (en)

USES: Alfalfa or lucerne is a highly valued legume forage, extensively cultivated in warm temperate and cool subtropical regions. It has been heralded as having the highest feeding value of all commonly grown hay crops, producing more protein per hectare than any other crop for livestock. In some areas it is used in combination with corn for silage. In parts of China and Russia tender alfalfa leaves serve as a vegetable. Sprouts are consumed in many countries. An excellent pasture for hogs, cattle, and sheep, often in mixtures with smooth bromegrass, orchardgrass, or timothy. Supplemental feeding of grain to dairy cows, sheep and fattening cattle reduces bloating and balances the high protein level of the alfalfa pastures with energy and extends the usefulness of the pasture. Pelleted alfalfa meal is used in mixed feeds for cattle, poultry, and other animals. Alfalfa may be grown as a cover crop to reduce soil erosion and often increases yield of succeeding crops, as potatoes, rice, cucumber, lettuce, tomatoes (increased by 10 MT/ha), corn, apples, and oranges. It is valued as a honey plant. Extracts produce antibacterial activity against gram-positive bacteria. Powdered alfalfa is used as a diluent to adjust strength of digitalis powder, and the root has been used as an adulterant of belladonna root. Seeds yield 8.5–11% of a drying oil suitable for making paints and varnish. Seed screenings are ground and used to a limited extent in feeds for ruminants. The seeds also contain a yellow dye. Alfalfa fiber has been used in manufacturing paper.

FOLK MEDICINE: Seeds contain the alkaloids stachydrine and 1-homostachydrine, and are considered emmenagogue and lactigenic. They are used as a cooling poultice for boils in India. For weight gain, a cupful of 1 : 16 alfalfa : water infusion has been recommended. Alfalfa contains saponins but is sold in many "health food" stores. It is considered antiscorbutic, aperient, diuretic, ecbolic, estrogenic, stimulant, and tonic, and said to aid peptic ulcers, as well as urinary and bowel problems. In Colombia, the mucilaginous fruits are used for "tos."

CHEMISTRY: Tender shoots of alfalfa are reported to contain per 100 g 52 calories: 82.7% moisture, 6 g protein, 0.4 g fat, 9.5 g total carbohydrate, 3.1 g fiber, 1.4 g ash, 12 mg Ca, 51 mg P, 5.4 mg Fe, 3,410 IU vitamin A, 0.13 mg thiamine, 0.14 mg riboflavin, 0.5 mg niacin, and 162 mg ascorbic acid. Green forage of *Medicago sativa* is reported to contain per 100 g: 80.0% moisture, 5.2 g protein, 0.9 g fat, 3.5 g fiber, and 2.4 g ash. Silages contain per 100 g: 69.5% moisture, 5.7 g protein, 1.0 g fat, 8.8 g fiber, and 2.4 ash. Alfalfa whole meal and leaf meal are reported to contain per 100 g: 66 and 77 calories, 7.5% and 8.0% moisture, 16.0 and 20.4 g protein, 2.5 and 2.6 g fat, 27.3 and 17.1 g fiber, 9.1 and 11.5 g ash, respectively. The genus *Medicago* has been reported to contain the following chemicals, of which relative toxicities are tabulated in Duke's "Phytotoxin Tables": choline, citric acid, hydrocyanic acid, limonene, malic acid, malonic acid, oxalic acid, pantothenic acid, pectin, quinic acid, saponin, shikimic acid, tannin, trigonelline, and tryptophane. The "betaine fraction" of alfalfa contains: 0.785% stachydrine, 0.063% choline, 0.0069 trimethylamine, and 0.0052% betaine. The following purines have been identified: adenine,

Figure 78. *Medicago sativa* L.

guanine, xanthine, and hypoxanthine; the pyrimidine, isocytosine; and the ribosides adenosine, guanosine, inosine, and cytidine. These factors stimulate the growth of *Bacillus subtilis*. The three most abundant compounds in alfalfa juice were adenine (0.17%), adenosine (0.25%), and guanosine (0.36%). Nitrate concentrations greater than 0.2% can harm livestock. Prebud alfalfa in spring growth has 0.18% nitrate gradually decreasing to 0.12% at green pod stage. Prebud alfalfa in summer growth has 0.32% nitrate, gradually decreasing to 0.17% at green pod stage. Heavy N fertilization may increase nitrates to 1.0%. One study (4th Annual Alfalfa Symposium) cites 5.1 g arginine/16 g N; 3.1 g cystine and methionine, 1.5 g histidine, 4.6 g isoleucine, 7.2 g leucine, 5.6 g lysine, 4.6 g phenylalanine, 4.1 g threonine, 1.5 g tryptophane, and 4.6 g valine for alfalfa hay. Another study (WOI) gives 3.5 arginine, 1.5 histidine, 4.2 lysine, 1.5 tryptophane, 4.1 phenylalanine, 1.3 methionine, 5.0 threonine, 7.9 leucine, 4.3 isoleucine, and 4.9 valine. One report gives total crude lipids as 5.2% of total dry weight, with 11% fatty acids, 10% digalactolipids, 16% monogalactolipids, 8% phospholipids, 44% neutral lipids, and 12% others. In the chloroplasts, linoleic, linolenic, and palmitic acids are the predominant acids, whereas stearic and oleic are low. The triglyceride fraction of alfalfa meal contains 16.9% linoleic acid, 32.2% linolenic acid, 31.0% oleic acid, and 19.9% saturated fatty acids. The phospholipid fraction (0.24% dry alfalfa meal) contained 35.2% linolenic acid, 36.8% oleic acid, 14.7% linoleic acid, and 13.3% saturated acids. Good alfalfa hay contains 138 to 198 mg choline/100 g hay. Alfalfa meal contains 21 mg A-spinasterol/100 g. Five xanthophylls comprise 99% of the xanthophyll fraction of fresh alfalfa viz. 40% lutein, 34% violaxanthin, 19% neoxanthin, 4% cryptoxanthin, and 2% zeaxanthin. Alfalfa volatiles include acetone, butanone, propanal, pentanal, 2-methylpropanal, and 3-methylbutanal. Australian herbage samples contain 0.3 and 0.4% glucose, 0.3 and 0.3% fructose, 2.5 and 2.7% sucrose, and 0.6 and 0.6% oligosaccharides at early and later flowering stages. Tannin concentrations between 2.7 and 2.8% have been reported. Several coumestans occur in alfalfa, 4'-0-methylcoumestrol, 3'-methoxycoumestrol, lucernol, medicagol, sativol, trifoliol, and 11, 12-dimethoxy-7-hydroxycoumestan. Four isoflavones are reported in alfalfa (daidzein, formononetin, genistein, and biochanin A) and they, like coumestrol, produce an estrogenlike response, perhaps contributing to reproductive disturbances of cattle on high-estrogen forage. Seeds are reported to contain trypsin inhibitors. Some people are allergic to the dust generated when alfalfa is milled. Mineral constituents include (avg. values, dry basis): calcium (CaO) 2.80, phosphorus (P_2O_5) 0.74, potassium (K_2O) 4.11, sodium (Na_2O) 0.35 and magnesium (MgO) 0.44%: trace elements; barium, chromium, cobalt, copper, iron, lead, manganese, molybdenum, nickel, silver, strontium, tin, titanium, vanadium, and zinc. Alfalfa is a valuable source of vitamins A and E; it contains: β-carotene 6.24, thiamine 0.15, riboflavin 0.46, niacin 1.81, and α-tocopherol, 15.23 mg/100 g; pantothenic acid, biotin, folic acid, choline, inositol, pyridoxine, vitamin B_{12}, and vitamin K are also present. Fresh lucerne is rich in vitamin C (1.78 mg/g) but it loses 80% on drying. A flavone, tricin, that inhibits the movements of smooth muscle has also been isolated. Lucerne is reported to contain citric, malic, oxalic, and malonic acids; succinic, fumaric, shikimic, and quinic acids are present in minor quantities. Enzymes reported in alfalfa are amylase, emulsin, coagulase, peroxidase, erepsin, lipase, invertase, and pectinase. Among the miscellaneous constituents present in lucerne are: saponins (0.5–2% or more), an alkaloid 1-stachydrine (0.14%), which also occurs in seeds, and two ketones, myristone $(C_{27}H_{54}O)$, and alfalfone $(C_{21}H_{42}O)$. Hay contains the following vitamins (avg. values): vitamin A, 3013 IU; thiamine, 0.29 mg; riboflavin, 1.37 mg; niacin, 3.84 mg; pantothenic acid, 1.78 mg; biotin, 0.018 mg; vitamin D, 199.5 IU; and vitamin E (as α-tocopherol) 2.60 mg/100 g. Lucerne and mixtures of lucerne and grass make good silage. When alfalfa alone is used, fermentation is enhanced by the addition of molasses. Wilted cuttings make good silage. Cattle and sheep fed on alfalfa as the sole source of forage occasionally suffer from bloat or tympanitis. The cause of bloat has not been identified but bloat may be prevented by feeding corn or sorghum silage with alfalfa hay or using mixed alfalfa–grass pasture. Forage cut at flowering and dried in the field rarely causes bloating. Several methods have been developed for processing concentrates and beverages for human consumption; they involve the removal of fiber and the odor and taste characteristic of alfalfa. A fiber-free concentrate is prepared by steeping tender leaves in cold water, blending the mixture extracting with water, concentration and drying under vacuum. The concentrate (yield, 13%) contains: protein 44.2, fiber 0.86, ether extr. 3.55, N-free extr. 30.69, ash, 13.2, calcium 1.90, and phosphorus 0.52%, carotene 110.1 mg, ascorbic acid 51.6 mg, and thiamine 1.15 mg/100 g. Alfalfa honey gave the following average values: water 16.56, invert sugar 76.90, sucrose 4.42, dextrin 0.34, protein 0.11, acid (as formic) 0.08, and ash 0.07%. Alfalfa seeds contained: moisture 11.7, protein 33.2, fat 10.6, N-free extr. 32.0, fiber 8.1, and mineral matter 4.4%.

DESCRIPTION: Perennial herb; stems erect or sometimes decumbent, 0.3–1 m long, 5–25 or more per crown, much-branched, 4-angled, glabrous or the upper part hairy; rhizome stout, penetrating the soil as much as 7–9 m; stipules united ⅓ to ½ length, free portion triangular-lanceolate, tapering, basally entire or with 1–2 teeth, glabrous or sparingly appressed-hairy; leaves pinnately trifoliolate; leaflets obovate-oblong, ovate or linear, tapering to base, crenate above middle mostly retuse and mucronate, 10–45 mm long, 3–10 mm broad, glabrous or appressed hairy, paler green beneath; racemes oval or rounded, 1–2.5 cm long, 1–2 cm broad, axillary, 5–40-flowered; peduncle slender, firm, always exceeding the subtending leaf, glabrous or appressed-hairy; calyx tubular, with linear-subulate teeth longer then tube; corolla yellow or blue to purple or violet, 6–15 mm long; bracteoles whitish, linear-subulate, mostly equaling the pedicel; pod slightly pubescent or glabrous, 3–9 mm in diameter, with 2–3 spirals, prominently reticulate-veined; seeds 6 or 8 per pod, yellow, castaneous or brown, ovoid, irregularly cordate or reniform. Fl. May–July (U.S.); Fr. late summer to fall.

GERMPLASM: Demand for better adapted, disease- and insect-resistant cvs has prompted extensive alfalfa breeding programs. *Medicago sativa* easily hybridizes with *Medicago falcata* (which is sometimes treated as a subspecies of *M. sativa*). Most newer cvs are synthetics, i.e., they contain a set of selected clones or seed lines produced and maintained under conditions specified by the originator. Promising plants are vegetatively propagated to obtain clonal lines and then the polycross progenies are tested to determine the combining ability of each clone. Of about 50 named cvs grown in the United States, about 13 make up 74% of alfalfa acreage. 'Vernal' and 'Ranger' account for ca. 52%. Other cvs include: 'Buffalo,' 'Hairy Peruvian' (7,250 ha in 1969), 'Indian' (2,450 ha), 'Cayuga,' 'Cody,' 'Cossack' (100,000 ha), 'Du Puits,' 'Grimm' (180,000 ha), 'Ladak' (425,000 ha), 'Lahontan,' 'Moapa' (resistant to spotted alfalfa aphid), 'African' (5000 ha), 'Narragansett,' and 'Sonora.' 'Buffalo' is a bacterial wilt-resistant selection from 'Kansas Common.' 'Cayuga' and 'Ranger' are also wilt resistant. 'Cody' and 'Zia' are resistant to spotted alfalfa aphid. 'Lahontan' and 'Washoe' are resistant to stem nematodes. There are many new cvs especially selected for specific areas, resistances and other agronomic purposes. Each cv is adapted to a limited area or to a specific set of climatic conditions. Assigned to the Central Asia, Near East, and Mediterranean Centers of Diversity, alfalfa or cvs thereof is reported to exhibit tolerance to aluminum, anthracnose, bacteria, disease, drought, fungus, hydrogen fluoride, high pH, heat, insects, low pH, mine, mycobacteria, nematode, salt, slope, smog, SO_2, virus, weeds, waterlog, and wilt. ($2n = 16, 32, 64$.)

DISTRIBUTION: Native to southwest Asia as indicated by occurrence of wild type in the Caucasus and in mountainous regions of Afghanistan, Iran, and adjacent regions. The cultivated forms probably arose in western Persia and then spread, to become widely cultivated, often a "weed" throughout Asia, Europe, and America.

ECOLOGY: Alfalfa shows considerable variation in form and adapation to environment. Form of plant varies from erect to decumbent from southern subtropical areas to northern temperate regions and from lower to higher elevations. Length of vegetative period decreases from south to north and from low to high elevations. Strains are adapted up to about 2,400 m (to 4,000 in Bolivia); can withstand high temperatures of 39°–41°C and low temperatures. Degree of adaptibility varies among strains. Thrives particularly well in semi-arid regions under irrigation. Some clones that are self-sterile at low temperatures may be partially self-fertile at high temperatures. Annual rainfall of 5–6 dm is optimum (in the temperate zone), but crop survives less. Soil moisture can be reduced to 35% of the water-holding capacity of a soil before affecting photosynthesis. In areas of high rainfall, 10 dm or more, alfalfa does not grow well as a perennial. Grows on a variety of soils, but thrives on rich, friable, well-drained loamy soil with loose topsoil supplied with lime; does not tolerate waterlogging and acid soils. Alfalfa may be a bit more tolerant of frost and salt than wheat. Salinities of 3 mmhos reduce yields by 10%, 5 mmhos by 25%, and 8 mmhos by 50%. Deep penetrating roots make alfalfa drought resistant. Alfalfa is reported to range from Boreal Moist to Boreal Wet through Tropical Desert (irrigated or alluvial situations) to Tropical Dry Forest Life Zones, annual precipitation ranging from 0.9–27.2 (mean of 214 cases = 10.3 dm), annual temperature ranging from 4.3°–28.5°C (mean of 213 cases = 17.7°C), and pH ranging from 4.3–8.7 (mean of 136 cases = 6.5). One source (*Agr. Gaz NSW*, 1972) suggests that a pH of 6–7 is optimum, another (Parodi, 1964) suggests that 7–8 is optimum.

CULTIVATION: For cultivation of alfalfa, land should be well plowed. Farm manure could be applied 6 weeks before planting and as top dressing after every third cutting. Crop is propagated by seed. Seeds have hard coats and should be scarified or soaked in water before sowing. Fresh seeds do not germinate as well as those 2–3 yr old. Seed should be inoculated with the proper strain of bacteria. Crop may be sown pure or in mixture with grasses or other legumes. Seeds broadcast or drilled in rows or on ridges 55–72 cm apart. Sowing on ridges facilitates weed control. When broadcast, seed rate is 12–20 kg/ha, when sown on ridges, 10–12 kg/ha. Crop may be grown with or without irrigation. In some areas, frequent irrigation is necessary until seedlings become established. When crop is well established, a single irrigation is sufficient. Crop requires frequent hoeing or cultivation to control weeds and stir the soil. Especially in the seedling stage, weed competition for light, nutrients, and water reduces alfalfa yields. Weed competition is reduced by using clean seed, selecting planting dates, clipping weeds, and using companion crops, but herbicide use is increasing in the US. Alfalfa is sensitive to some herbicides. Results are good with farm manure and any fertilizer containing phosphates, such as bone meal, fish meal, basic slag, rock phosphate, or superphosphate (ca. 250 kg/ha annually or 500 kg in alternate years). Alfalfa is often used in rotation with other crops, e.g., wheat, oats or flax. Liming to pH 7 is often recommended. Generally alfalfa and alfalfa/grass mixtures with at least one-third alfalfa do not respond markedly to N, but respond well to P and K. Many soils lack adequate K for high yields. Apparently deficiencies of S and Mg are becoming more common. Addition of micronutrients, especially B, may be necessary locally.

HARVESTING: Alfalfa hay is harvested when the first flowers have opened. One cut at the prebloom stage will not seriously damage the stand and subsequent production. Number of cuttings per year varies from one in drier northern areas to nine or more under irrigation in southern areas; 3–4 cuttings in central areas are normal (U.S.). For high-quality hay, leaves must be retained since, at early bloom stage, leaves constitute about half of plant weight. All processes from field to storage (mowing, conditioning, windrowing, baling, bale handling, etc.) are mechanized in the United States. Seed production is most efficient under irrigation, but rainfed seed is produced. Row spacings of 60–110 cm are superior to closer plantings. In dense stands, nectar production, insect visitation and subsequently, seed production often are depressed. Nearly all seed is directly combined. Use of spray defoliants has facilitated harvesting. Proper adjustment of combine is very important to avoid injury to seed. Alfalfa tolerates rotational grazing but weakens rapidly under continuous grazing. 'Ladak' is superior on rangelands, and creeping or decumbent alfalfas are better adapted for dryland conditions.

YIELDS AND ECONOMICS: Forage yields are 5–75 MT/ha per year (with 8–12 cuttings per year). Seed yields are 186–280 kg/ha annually. Alfalfa is estimated to fix 83–594 kg N/ha/yr (Miller, 1976). About 10 years ago, this most valuable of common hay crops had an estimated hectarage of 33 million, the United States, USSR, and Argentina producing 70% of world production, France, Italy, Canada, and Australia producing another 20%. Alfalfa accounts for ca. 60% of total U.S. hay production.

BIOTIC FACTORS: Flowers of nearly all alfalfas must be tripped for pollination to take place; this amounts to release of the sexual column from the keel of the flower. Many types of bees can serve as trippers, e.g., the alfalfa leaf-cutter bee (*Megachile rotundata*), alkali bee (*Nomia melandri*), honeybees (*Apis mellifera*), and bumblebees. In North America, alfalfa-pollinating genera, listed with the number of effective species, are as follows: Apis (1), Bombus (7), Xylocopa (1), Anthophora (1),

Tetralonia (1), Osmia (2), Megachile (10), Hoplitis (1), Nomia (1), Agapostemon (2), Halictus (3), Evylaeus (1), Lasioglossum (1), Andrena (2), and Calliopsis (1). More than 100 species of fungi have been reported to cause diseases on alfalfa. Among the most serious are the following: *Colletotrichum trifolii*, (anthracnose), *Pseudopeziza medicaginis*, *P. jonesii*, *Leptosphaerulina briosiana*, *Stemphylium* spp., *Uromyces striatus* (rust), *Peronospora trifoliorum* (downy mildew), *Phoma medicaginis* (spring black stem and leafspot), *Ascochyta imperfecta* (leafspot), *Sclerotinia trifoliorum* (stem rot), *Fusarium* spp., and *Phytophthora megasperma* (root rot). Bacterial wilt, caused by *Corynebacterium insidiosum*, is one of the most destructive alfalfa diseases. Virus diseases include: Alfalfa mosaic, Lucerne mosaic, Yellow-green stripe mosaic, Rugose leaf curl, and Witches broom virus of lucerne. Parasitic plants attacking alfalfa include: *Orobanche lutea*, *Cuscuta australis*, *C. arvensis*, *C. campestris*, *C. chinensis*, *C. epithymum*, *C. gronovii*, *C. indecora*, *C. pentagona*, *C. planiflora*, *C. racemosa*, *C. suaveolens*, and *C. trifolia*. Nematodes attacking alfalfa are numerous and belong to several genera. Some are found in the roots and others only in the soil near the roots. Stem nematodes *(Ditylenchus dipsaci)* and rootknot nematodes *(Meloidogyne* spp.) cause much damage; the cultivars, 'Lahontan' and 'Washoe' are resistant to stem nematodes. Among important insect pests of alfalfa are the following: alfalfa weevil *(Hypera postica)*, to which some cvs are resistant; spotted alfalfa aphid *(Therioaphis maculata)*, and pea aphid *(Acrythosiphon pisum)*, to both of which some cvs are resistant; Potato leafhopper causing alfalfa yellows *(Empoasca fabae)*; and meadow spittlebug *(Philaenus spumarius)*. Lygus bugs, seed chalcids and alfalfa weevils are most harmful to seed production.

REFERENCES:
Barnes, R. F., and Gordon, C. H., 1972, Feeding value and on-farm feeding, in: *Alfalfa Science and Technology* (C. H. Hanson, ed.), pp. 601–630, Agronomy Series No. 15, American Society of Agronomists, Madison, Wisc.
Hanson, C. H. (ed.), 1972, *Alfalfa Science and Technology*, Agronomy Series No. 15, American Society of Agronomists, Madison, Wisc., 812 pp.
Hanson, C. H., and Barnes, D. K., 1973, in *Forages* (M. E. Heath, D. S. Metcalfe, and R. F. Barnes, eds.), 3rd ed., pp. 136–147, illus., Iowa State University Press, Ames, 755 pp.
Heinrichs, D. H., 1963, Creeping alfalfas, *Adv. Agron.* **15**:317–337.
Miller, P. K., 1976, Breakthrough in alfalfa research, *Minn. Sci.* **32**:4–7.
Ries, S. K., Wert, V., Sweeley, C. C., and Leavitt, R. A., 1977, Triacontal: A new naturally occuring plant growth regulator, *Science* **195**:1339–1341.

CONTRIBUTORS: J. A. Duke, M. B. Forde, C. F. Reed, J. K. P. Weder.

Medicago scutellata (L.) Mill.

FAMILY: Fabaceae
COMMON NAME: Snail medic

USES: Grown to a limited degree in the Mediterranean region east into southwest Asia for forage.

FOLK MEDICINE: No data found.

CHEMISTRY: No data found.

DESCRIPTION: Annual or perennial herb; stems glandular-pubescent, long, slender, weak, 15–60 cm tall; leaves trifoliolate, stipules ovate-lanceolate

Figure 79. *Medicago scutellata* (L.) Mill.

Medicago scutellata (L.) Mill.

to lanceolate, incise-dentate; leaflets obovate to oblong or elliptic, 10–25 mm long, dentate in upper part, cuneate; racemes 1–3-flowered, much shorter than the subtending leaf; pedicels shorter than calyx-tube and lanceolate bracteoles; corolla yellow or rarely orange-yellow, 5–7 mm long, standard about twice length of both wings and keel; pods 7–15 mm in diameter, flat on one side, convex on other, unarmed, glabrous, with 5–8 coils, each coil containing 2 seeds, veins from ventral suture 10–14, oblique, very prominent, on other half of pod surface mostly reticulately anatomosing and finally confluent with dorsal suture. Radicle less than half length of cotyledons. Germination phanerocotylar. Fl. May–June.

GERMPLASM: One of the least variable annual species of *Medicago;* presumably not autotetraploid. Assigned to the Mediterranean Center of Diversity. ($2n = 32$.)

DISTRIBUTION: Native to Mediterranean region, from Portugal to Spain, east to Balkans, Asia Minor, Crimea, and Russia.

ECOLOGY: Natural habitats in meadows and wood margins. Thrives in warm temperate climate with mild winters. Grows on nearly any type of good soil, especially loamy or heavy soils, mostly in cultivated or fallow fields. Ranging from Cool Temperate Moist through Subtropical Dry Forest Life Zone, snail medic is reported to tolerate annual precipitation of 4.4–11.9 dm (mean of 7 cases = 8.0 dm), annual mean temperature of 11.5–26.2°C (mean of 7 cases = 18.9°C), and pH of 5.5–7.5 (mean of 7 cases = 6.8).

CULTIVATION: Propagated by seed. Apparently not widely cultivated.

YIELDS AND ECONOMICS: No data available.

BIOTIC FACTORS: Pods and viable seeds are obtained without pollination agents, not surprising in that most annual species of *Medicago* are devoid of floral nectaries. Following fungi have been reported on snail medic: *Cercospora zebrina, Uromyces genistae-tinctoriae* and *U. striatus.* The nematode, *Meloidogyne javanica,* has been isolated from this species.

CONTRIBUTORS: J. A. Duke, C. F. Reed.

Melilotus alba Medik.

FAMILY: Fabaceae

COMMON NAMES: White sweetclover, Hubam, White melilot

USES: Cultivated for forage, hay and ensilage, white sweetclover is more useful for pasture and soil improvement. Also grown for green manure and cover crop. Sweetclovers are among the most valuable plants for honey production. Fiber from cortex is suitable for paper pulp manufacture. Seed oil used in paint and varnish. Seed meal, after removal of toxic substances, used as protein supplement in cattle feeds.

FOLK MEDICINE: Plant has astringent and narcotic properties. Sometimes used as substitute for *Melilotus officinalis*. Considered aromatic, antiasthmatic, anticoagulant, antispasmodic, bactericide, carminative, digestive, emollient, vulnerary. The plant

Figure 80. *Melilotus alba* Medik.

has been used as a poultice or plaster for abdominal and rheumatic pains.

CHEMISTRY: Nutritive value of white sweetclover is as follows: green feed, (protein 4.1%, moisture 79.2%, crude fiber 4.9%, total digestible nutrients 12.8%; hay (16.5% protein, 8.2% moisture, 24.6% crude fiber, 50.3% total digestible nutrients); silage (4.5% protein, 72.0% moisture, 9.6% crude fiber, 15.7% total digestible nutrients); and seed (37.4% protein, 7.8% moisture, 11.3% crude fiber, 64.9% total digestible nutrients). Analysis of green material showed: N, 0.83; P_2O_5, 0.16; K_2O, 0.83; and CaO, 0.69%. Another analysis of green forage showed: dry matter, 31.5; crude protein, 5.2; ether extr, 1.2; carbohydrates, 14.3; and ash, 3.1%. Mature hubam seeds gave the following values (dry basis): crude protein, 39.4; oil, 7.6; starch, 3.0; pentosans, 8.1; total sugars (as glucose), 2.7; crude fiber, 12.9; and ash, 4.1%. The seeds are reported to contain small amounts of resins, and glucosides, 2 unidentified red and yellow pigments, and trypsin inhibitors and chymotrypsin inhibitors. The genus *Melilotus* is reported to contain coumarin, hydrocyanic acid, malonic acid, and melilotin.

DESCRIPTION: Annual or biennial, erect or decumbent herb; stem 1 m or more tall; stipules entire; leaves trifoliolate, leaflets lanceolate or oblanceolate to narrowly oblong, rarely ovate, 1–2.5 cm long; rachis of larger leaves, excluding terminal petiolule, often prolonged more than 4 mm beyond lateral leaflets; racemes numerous, 5–20 cm long, on peduncles up to 4 cm long; pedicels 1–2 mm long; calyx about 2 mm long; corolla white, 4–6 mm long; style 1.7–2.3 mm long; pod with weak irregular network of veins. Fl. summer and fall.

GERMPLASM: *M. alba* var. *annua* Coe is an annual form, grown on poor soils, yielding good silage rich in protein; often grown in mixture with wheat, or in rotation with wheat or corn. 'Hubam' and 'Emerald' are cvs of this variety. 'Floranna' has greater forage yields, especially during winter months, than 'Hubam' and is adapted to Florida. 'Israel' is a tall, late-maturing cv, developed in Israel and introduced to Texas. 'Denta', a low-coumarin, late-maturing biennial cv, is vigorous, producing flowers 3–4 weeks later than does the common variety. 'Arctic,' an early-maturing, winter-hardy cv and 'Polara,' a low-coumarin cv, are adapted to Canada. An unproductive wild species, *M. dentata* (Waldst. and Kit.) Pers. provided a low-coumarin gene transferred to *M. alba* by crossing the two species and grafting the resulting albino seedlings on normal green plants. The same gene was later transferred to *M. officinalis* from *M. alba* by excising and culturing the immature hybrid embryos. *Melilotus infesta* Guss. appears to be immune to weevil feeding, but attempts to transfer the resistance have been unsuccessful. Assigned to the Eurosiberian and Mediterranean Centers of Diversity, white sweetclover or cultivars thereof is reported to exhibit tolerance to alkali, bacteria, disease, drought, frost, high pH, heat, insects, limestone, mine, mycobacteria, poor soil, slope, smog, SO_2, virus, weed, and waterlog. ($2n$ = 16, 24, 32.)

DISTRIBUTION: Native to Europe and west and central Asia, south to India; introduced elsewhere, especially in temperate areas.

ECOLOGY: Occurs in grasslands, arable soils, wastes, and along roadsides, especially in calcareous soils. Grows well from sea level up to 2000 m altitude. Highly adaptive to wide range of environmental conditions, but sensitive to soil acidity; requires near neutral pH. Heavy clays and light sands will produce a successful crop. Fertile, well-limed soils with adequate drainage and organic material are best. Comes up well under irrigated conditions, but does not give a good ratoon. Said to fare better with annual rainfall from 7–11 dm. Crop is extremely drought-resistant and considered satisfactory crop in areas, such as the Great Plains, where rainfall is ca. 4 dm/yr. Growth is superior in regions with abundant rainfall. Crop withstands both high and low temperatures, and is seldom winter-killed. Sweetclover is unusually susceptible to a number of herbicides, particularly 2,4-D. Ranging from Boreal Moist to Wet through Subtropical Dry to Moist Forest Life Zones, white sweetclover is reported to tolerate annual precipitation of 0.9–16.0 dm (mean of 54 cases = 7.8 dm), annual mean temperature of 5.7°–24.3°C (mean of 54 cases = 12.7°C), and pH of 4.8–8.2 (mean of 50 cases = 6.7).

CULTIVATION: Propagation is by seeds or cuttings. Due to hard testa, seeds must be scarified to hasten germination. Seed rate varies from 11–17 (–25) kg/ha, depending on soil and climatic conditions. When seeded in a companion grain crop in northern latitudes, it seldom produces a crop of hay or

Melilotus alba **Medik.**

pasture in first year. Grazing or mowing in first season prevents adequate root development and food accumulation. In southern areas, fall grazing is less damaging, especially if it is not close or prolonged. Sweetclover grows vigorously in second year and when tops are 20–25 cm tall heavy grazing is essential to keep plants in palatable stage. If not grazed off, plants may become woody and unpalatable to cattle. Late-maturing cvs produce more forage than early cvs, and maintain pasture later in season. In some areas sweetclover is planted with a spring or winter small grain as a companion crop to help keep weeds down during establishment of crop. When first-year growth is for pasture, it is usually planted alone. When companion crop is harvested, the binder or combine should be set high. During first year, top-growth for pasture or hay should be removed before mid-August or after mid-October in the United States. As moisture becomes available after harvesting the companion crop, biennial sweetclovers grow rapidly both in roots and tops. Plants may be plowed under in late fall of first year, but results are best when plants are plowed the following spring when top-growth is about 15 cm tall. Such soil improvement practices have led to as much as 13.8% increases in corn yields. In southern areas, good-quality hay, equal in palatability and feeding value to alfalfa, may be produced from first year's growth. Second year's growth is less satisfactory for hay as leaves become brittle and shatter in handling. Quality of silage is best from crop cut prior to flowering. Mixtures of grasses and small grains also make high-quality silage.

HARVESTING: Grazing should begin when tops are 20–25 cm tall. During period of rapid growth, grazing should be heavy to prevent forage from becoming coarse and unpalatable. Good-quality hay from first year's growth, cut in bud stage and cured properly, has about same food value as alfalfa. Second year's growth is often converted into silage, and for this, plants are cut before flowering begins. Most sweetclover grown for seed is harvested just as pods turn black or dark brown. Crop may be turned under at any time for green manure.

YIELDS AND ECONOMICS: Yields of hay from 2.2–3.5 MT/ha during first year to 2.2–8.1 MT/ha during second year in the United States. For three cvs, the following yields are reported: 'Common White,' first year 1.2–1.5 MT; second year 2.56–3.7 MT; 'Grundy County,' 0.87–1.13 MT, and 2.0–3 MT; and 'Williamette,' 1–1.95 MT and 2.6–4.49 MT. In India, 'Hubam' has yielded in 3 cuttings 9–10.5 MT/ha protein-rich green forage. As a green manure crop, 'Hubam' adds ca. 80 kg/ha N.

BIOTIC FACTORS: Seed production requires pollination by insects, ca. 5 colonies of honeybees/ha, recommended but 25 colonies necessary for maximum pollination. Promising bee pasture is combination of annual sweetclover ('Hubam' or *M. alba*) immediately followed by buckwheat. Annual sweetclover blooms after native biennial sweetclovers, thus extending the clover–honey production. Cvs and strains of *M. alba* and *M. officinalis* may intercross unless sufficiently isolated. Attempts to transfer the resistance to sweetclover weevil *(Sitona cylindricollis)* from *M. infesta* to *M. alba* have not been successful. Fungi attacking white sweetclover include: *Alternaria tenuis, Aphanomyces euteiches, Ascochyta caulicola, A. imperfecta, A. meliloti, A. pisi, Cercospora davisii, C. meliloti, C. zebrina, Colletotrichum dematium, C. destructivum, C. trifolii, Erysiphe communis, E. martii, E. polygoni, Fusarium acuminatum, F. avenaceum, F. equiseti, F. oxysporum, F. poae, F. sambucinum* var *coeruleum, F. solani* and var *aduncisporum, Gloeosporium vexans, Kabatiella caulivora, Leptosphaeria meliloti, L. pratensis, Macropsorium meliloti, Mycosphaerella davisii, M. lethalis, Peronospora meliloti, P. trifoliorum, Phoma galagae* var *melilota, Ph. meliloti, Ph. oleracea* var *meliloti, Phymatotrichum omnivorum, Phytophthora cactorum, Ph. megasperma, Plenodomus meliloti, Pleospora kansensis, Pseudopeziza meliloti, Ps. medicaginis, Rhizoctonia solani, Sclerotinia sclerotiorum,* and var *minor, S. trifoliorum, S. sativa, Stagonospora carpathica, S. meliloti,* and *Uromyces baumlerianus*. The bacterium, *Phytomonas lathyris*, has been isolated from this species. Viruses attacking it include: cucumber mosaic, 'Steinklee virus,' and ringspot. *Orobanche purpurea* often parasitizes the plant. Nematodes isolated from white sweetclover include: *Belonolaimus gracilis, Ditylenchus dipsaci, Heterodera glycines, H. trifolii, Meloidogyne hapla, M. javanica, Paratylenchus projectus, Pratylenchus penetrans, P. pratensis, P. vulnus,* and *Tylenchorhynchus martini*. Sweetclover weevil *(Sitona cylindricollis)* is major insect pest, destroying crop by defoliating newly emerged seedlings. Some attempts are being made to develop resistant cvs. High levels of nitrate deter the weevils from feeding on certain noneconomic species. Sweetclover root borer *(Walshia miscecolorella)*, and sweetclover aphid *(Therioaphis riehmi)* to which some resistance is being included in breeding, have damaged sweetclover in the Great Plains. Blister beetles *(Epicauta* spp.), feed on leaves of low-coumarin cvs; grasshoppers, leafhoppers, clover leaf weevil *(Hypera punctata)*, cutworms, webworms, green cloverworms *(Platypena scabra)*, and other caterpillars are often pests, but rarely cause serious damage. For disease and pest control local agents should be consulted for most effective local methods. Breeding resistant cvs seems to be the best approach to disease control.

REFERENCES:

Gorz, H. J., and Smith, W. K., 1973, in: *Forages: The Science of Grassland Agriculture* (M. E. Heath, D. S. Metcalfe, and R. F. Barnes, eds.), 3rd ed., pp. 159–166, illus., Iowa State University Press, Ames, 755 pp.

Gross, A. T. H., and Stevenson, G. A., 1964, Resistance in *Melilotus* species to the sweetclover weevil *(Sitona cylindricollis)*, Can. J. Plant Sci. **44**:487–488.

Smith, W. K., and Gorz, H. J., 1965, Sweetclover improvement, Adv. Agron. **17**:163–231.

Stevenson, G. A., 1969, An agronomic and taxonomic review of the genus *Melilotus* Mill, Can. J. Plant Sci. **49**:1–20.

CONTRIBUTORS: J. A. Duke, H. J. Gorz, C. F. Reed, J. K. P. Weder.

Melilotus indica (L.) All.

FAMILY: Fabaceae

COMMON NAMES: Indian sweetclover, Senji, Sourclover

SYNONYM: *Melilotus parviflora* Desf.

USES: Long cultivated as forage in India, Indian sweetclover is also grown for hay and silage. Often occurs as a winter weed in many areas. Plant used for green manure, for improving alkaline soils, reclaiming saline areas and as a cover crop. Used as green forage especially for cattle and cows and as a maintenance ration for heifers.

FOLK MEDICINE: Medicinally, plant is discutient and emollient, and used as a fomentation, poultice, or plaster for swellings. Also considered astringent

Figure 81. *Melilotus indica* (L.) All.

and narcotic. Seeds used in bowel complaints and infantile diarrhea.

CHEMISTRY: Chopped forage should be mixed in with other dry forage. Fed alone or in excess, it causes lethargy, tympanitis, and paralysis and is reported to taint milk of dairy cows. Plants have a distinct odor and bitter taste of coumarin. However, nonbitter low-coumarin cvs have been developed through selection. Nutritive value of Indian sweetclover is as follows: green feed (protein 3.3%, moisture 78.4%, fiber 6.4%, total digestible nutrients 14.4%); silage (2.1% protein, 70.2% moisture, 14.1% crude fiber, 15.1% total digestible nutrients); hay (protein 18.6%, moisture 10.0%, crude fiber 23.8%); pods (dry basis) (25.3% protein, 0.0% moisture, 14.8% crude fiber). Analysis of the green manure gave the following percentage values (dry basis): N, 3.36; P_2O_5, 0.49; K_2O, 1.53; and CaO, 2.49%.

DESCRIPTION: Annual herb; stems erect, up to 50 cm tall, branching, sparingly pilose; leaves trifo-

Melilotus indica (L.) All.

liolate; stipules often with a small tooth near base; petiole up to 2 cm long; rachis about 5 mm long; leaflets cuneate-oblong or lanceolate, up to 25 mm long, 9 mm wide, coarsely toothed; rounded at apex; racemes many-flowered, up to 10 cm long, including peduncle up to 3 cm long; bracts subulate, ca. 0.5 mm long; pedicels ca. 1 mm long, reflexed after flowering; calyx ca. 1.5 mm; corolla yellow, ca. 2.5 mm long; style 0.9–1.2 mm long; pod 3–4 mm long, 2–3 mm wide, 1-seeded, with prominent reticulate veins; seeds small, 250,000–275,000/lb.; testa hard.

GERMPLASM: Some cvs have low coumarin content. This species will not intercross under natural conditions with either *M. alba* or *M. officinalis*. Assigned to the Hindustani Center of Diversity, Indian sweetclover or cvs thereof is reported to exhibit tolerance to alkali, drought, frost, high pH, insects, limestone, salt, slope, virus, and weeds. ($2n = 16$.)

DISTRIBUTION: Native to Mediterranean region south to Ethiopia; Central Asia and India. Introduced to other warm temperate regions. Cultivated in southwestern United States for forage, but less productive and less palatable than 'Hubam' (*M. alba* var *annua*). In North America, often a weed, e.g., in the Sonoran Zone from Oregon into Baja California. Occurs along streams even in the desert.

ECOLOGY: Adapted to a wide range of soil conditions. Fares best on well-drained neutral or alkaline soils. Requires soil of medium to high fertility, and is resistant to soil alkalinity. Often occurs as a winter weed of cultivation almost throughout India, ascending to 1650 m. Tolerates 3–6 mmhos salinity, 5–8 in gypsiferous soils. Thrives in regions with 9–13 dm rainfall annually. Ranging from Cool Temperate Steppe to Wet through Subtropical Dry Forest Life Zone, Indian sweetclover is reported to tolerate annual precipitation of 0.9–12.9 dm (mean of 18 cases = 7.6 dm), annual mean temperature of 7.2°–22.5°C (mean of 18 cases = 14.8°C) and pH of 5.0–8.2 (mean of 17 cases = 6.8).

CULTIVATION: In the Punjab, Indian sweetclover is cultivated as an irrigated cold season crop along perennial canals and on irrigated land. It can be cultivated in areas with restricted irrigation. Propagation is by seeds. This land is plowed 3–6 times and soil is brought to a fine tilth, and divided into small plots to facilitate irrigation. Seed is sown in Sept.–Nov. at rate of 40–50 kg/ha. The hard testa should be scarified before sowing to hasten germination. Seeds can be broadcast on moist or dry seedbed and covered with soil. When sown on dry beds, they are immediately irrigated. Seeds can be sown in standing water, allowing 6–8 hr for soaking. Plots are irrigated 7–10 days after sowing and thereafter as needed; 2–3 irrigations are sufficient. Weeding during early stages is beneficial. Manuring is usually not necessary, when the crop follows a manured crop, such as cotton or corn. However ammonium phosphate or a mixed fertilizer containing ca. 22.5 kg/ha N and 67 kg/ha phosphoric acid is considered beneficial. Indian sweetclover is often rotated with wheat, corn or rice among grain crops and with cotton, sugarcane, or tobacco as cash crops. It is also grown in mixtures with barley and oats.

HARVESTING: In India, the crop is harvested for forage during Feb.–April, when plants are in full bloom and seed formation has already started. If harvested earlier and fed to animals, it may cause bloating. Crop may be cut 2–4 times.

YIELDS AND ECONOMICS: In India green forage yields are ca. 4.5–6.0 (–7.5)/ha MT. In the Punjab 'Fo. S. I.' has yielded nearly 10 MT/ha green forage. Yields of ca. 250 kg/ha seed have been reported. Extensively grown in India, Punjab and in some other warm temperate regions, including southwestern United States. Locally important as a forage and pasture and green manure crop, less often used for soil improvement, silage, and hay.

BIOTIC FACTORS: Plants are self-fertile and flowers set seed without insect pollination. In India, the crop is susceptible to powdery mildew (*Erysiphe polygoni*) and downy mildew (*Peronospora meliloti*). Other fungi reported are: *Ascochyta caulicola, A. meliloti, Cercospora davisii, C. zebrina, Colletotrichum trifolii, Entyloma meliloti, Erysiphe communis, Phymatotrichum omnivorum, Ramularia meliloti, Sclerotinia sclerotiorum, S. trifoliorum, Sclerotium rolfsii, Stagonospora meliloti, Synchytrium meliloti,* and *Uromyces bacumelerianus*. Bacteria isolated are *Pseudomonas meliloti*. Pea mosaic virus has also been isolated from plants.

Among the nematodes, the following have been isolated: *Ditylenchus dipsaci*, *Heterodera marioni*, and *Tylenchorhynchus martini*.

CONTRIBUTORS: J. A. Duke, C. F. Reed.

Melilotus officinalis Lam.

FAMILY: Fabaceae

COMMON NAME: Yellow sweetclover

SYNONYMS: *Melilotus arvensis* Walr.
Melilotus bungeana Boiss.
Melilotus diffusa Gaud.
Melilotus expansa Hort.
Melilotus pallida Besser
Melilotus rugosa Gilib.
Trifolium melilotus-officinalis L.
Trifolium petitpierreanum Hayne

USES: Cultivated in Europe and America for forage, hay, pasture, and soil improvement. Plants are excellent for soil improvement and erosion control since roots develop copious nitrogen-fixation nodules. Herb used to flavor cheeses and put in tobacco snuff. Seeds used as substitute for Tonka beans. Plants useful as honeybee pastures (flowers good source of honey). Roots used as food by Kalmuks.

FOLK MEDICINE: Plants considered aromatic, carminative, styptic, and emollient. Also used as moth repellent. The plant is used in asthma cigarettes; as a decoction in lotions and enemas. Seeds have been suggested as a remedy for the common cold. Said to be used for anticoagulant, antispasmodic, arthritis, boils, bronchitis, carminative, colic, digestive, diuretic, expectorant, eyewash, headache, poison, rheumatism, swelling, vulnerary, and wounds.

CHEMISTRY: The seeds are said to poison horses. Cattle are said to be poisoned by eating moldy sweetclover hay. Toxicity is attributed to a fungus and bishydroxycoumarin. Hemorrhages occur both internally and externally, and sudden death may result. (Such hemorrhage may be controlled by quick administration of vitamin K.) The plant contains coumarin (2%), melilotic acid (*o*-hydrocoumaric acid), coumaric acid, hydrocoumarin, and resin.

DESCRIPTION: Biennial or annual herb with strong taproot; stems erect or ascending, 0.3–2.8 m tall, often tinged with red; leaves trifoliolate, the leaflets of lower and middle leaves broadly oval, lanceolate to rhomboid-ovate, 1–5 cm long, rounded at tip, those of upper leaves oblong-lanceolate, rounded or truncate at tip, irregularly dentate or entire; stipules entire, 8–12 mm long, lanceolate or subulate; racemes lax, 30–80-flowered, 4–15 cm long; pedicels 2–2.5 mm long; calyx 2–3 mm long, teeth as long as tube or shorter; corolla yellow, 4.5–7 mm long, standard and wings more or less equal to or up to 1 mm longer than keel; style 1.7–2.3 mm long, often persisting; ovules 4–6, rarely 3 or 8; pods ca. 3.4 mm long, 2–2.5 mm broad, 1.5 mm thick, compressed with strong rugose transverse veins or wrinkled, gray to straw-colored, rarely black, obtuse at tip; seeds 1, rarely 2, smooth, ovoid-elliptical, 1.5–3 mm long, ca. 1.5 mm broad, yellow or greenish-yellow, sometimes with purple spots; radical more than half as long as cotyledons. Fl. summer and fall; Fr. late summer and fall.

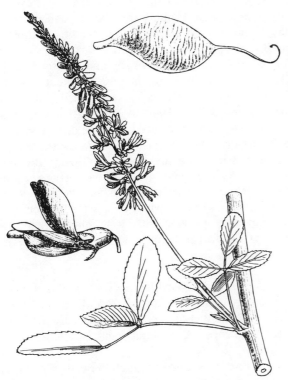

Figure 82. *Melilotus officinalis* Lam.

Melilotus officinalis Lam.

GERMPLASM: A variable species with 3 subspecies and about 16 ecotypes and forms. Annual types are quite rare. Some of the better cvs are: 'Albotrea,' 'Common Yellow,' 'Erector,' 'Madrid,' and 'Switzer.' 'Goldtop,' developed in Wisconsin, has excellent seeding vigor, is 2 weeks later in maturity, gives higher yield of better-quality forage, and has slightly larger seed than 'Madrid.' Leafy cvs are most valuable for hay, pasture, and silage; tall stemmy cvs are better for soil improvement. 'Madrid,' introduced from Spain, is a good seedling cv of good vigor and production, the foliage relatively resistant to fall freezing. *Melilotus infesta* Guss. is a source of resistance to the sweetclover weevil. Assigned to the Eurosiberian and Mediterranean Centers of Diversity, yellow sweetclover or cultivars thereof is reported to exhibit tolerance to alkali, drought, frost, high pH, heavy soil, heat, insects, limestone, mycobacteria, salt, poor soil, slope, smog, SO_2, virus, and weeds. ($2n$ = 16, 32.)

DISTRIBUTION: Native from Europe to central temperate Asia and western China, south to North Africa, Ethiopia, Eritrea, and Kenya. Introduced in North and South America and other temperate regions of world; cultivated in some parts of India.

ECOLOGY: Adapted to a wide range of environmental conditions from sea level to 4,000 m in altitude, but highly sensitive to soil acidity, faring better in a near-neutral pH. Soil types range from heavy clays to light sands, and those that are fertile, well-limed, with adequate drainage and organic matter are best. Plants are extremely drought-resistant, and are considered a satisfactory crop in areas where rainfall reaches only 42 cm/yr. However, production is excellent in areas where rainfall is abundant. Withstands both high and low temperatures, and seldom winter-kills. Often becomes weedy on wastelands and in fallow fields; it is often the first plant to invade newly exposed soil, especially on nonacid soils. Grows well on limestone soils. Tolerates 3–6 mmhos salinity. Ranging from Boreal Moist through Warm Temperate Thorn to Moist Forest Life Zones, yellow sweetclover is reported to tolerate annual precipitation of 3.1–16.0 dm (mean of 47 cases = 7.8 dm), annual mean temperature of 4.9°–21.8°C (mean of 47 cases = 10.9°C), and pH of 4.8–8.2 (mean of 43 cases = 6.6).

CULTIVATION: Crop propagated from seed. Testa are hard, and seed should be scarified before sowing. Requirements for establishment are similar to those for alfalfa. Seed sown 11–17 kg/ha. In Corn Belt of the United States, seeded in spring with a spring or winter small grain as a companion crop to help control weeds during establishment. Planted without a companion crop for use in pasture the first year.

HARVESTING: Yellow sweetclover matures 10–14 days earlier than biennial white sweetclover. Harvest is most efficient if crop is windrowed when damp and threshed after 4–5 days with combine equipped with pickup attachment. The binder or combine should be set high. First-year top growth for pasture or hay should be removed before mid-August or after mid-October in the United States. As moisture becomes available after harvest of the companion crop, biennial sweetclover grows rapidly both in tops and roots. May be plowed in late fall of first year, but results are best when plowed in following spring after top growth is more than 14 cm tall. Such soil improvement practices have led to significant increases in corn yields. In southern areas, substantial amounts of pasture may develop in the first year. First year's growth can produce hay, equal in palatability and feeding value to alfalfa. Second year's growth is less satisfactory for hay. Leaves become brittle and shatter in handling. Best quality silage is from crop cut prior to flowering. Mixtures of grasses and small grains also make high-quality silage. In North America, most sweetclover grown for seed is harvested in an area from the prairies of western Canada south through western Minnesota and Iowa to Texas. Biennial cvs produce large amounts of forage early in second year, becoming less palatable as plants mature. Grazing should begin when tops are 20–25 cm tall. During rapid growth, grazing should be heavy to prevent the forage from becoming coarse and unpalatable.

YIELDS AND ECONOMICS: 'Madrid' yields 4.2–4.5 MT/ha hay in first season, 5.5–8.5 MT/ha in second season. Widely cultivated for forage, hay, and pasture in Europe, Asia, India, and North America.

BIOTIC FACTORS: Seed production needs pollinating insects. Seed set where flowers are not pollinated is about 5.1%; when selfed, 16.4%; and when cross-pollinated, 47.1%. Although 25 colonies/ha give maximum pollination and seed production, 2–5 colonies of honeybees are recommended per hectare. Following fungi cause diseases in yellow sweetclover: *Ascochyta caulicola, A. meliloti, Cercospora davisii, C. zebrina, Entyloma meliloti, Erysiphe communis, E. martii, E. polygoni, Leptosphaeria pratensis, Leveillula leguminosarum, Mycosphaerella davisii, Peronospora trifoliorum, Phoma medicaginis, Phymatotrichum omnivorum, Phytophthora cactorum, Pythium debaryanum, P. ultimum, Pleospora herbarum, Septoria meliloti, Stagonospora meliloti,* and *Uromyces baumlerianus*. Viruses isolated from this legume include: alfalfa mosaic (3 strains), alsike clover mosaic, bean chlorotic ringspot, bean mosaic, bean necrosis, white clover mosaic, wound tumor, yellow bean mosaic, and ringspot virus. Nematodes isolated include: *Heterodera glycines, H. schachtii, H. trifolii,* and *Meloidogyne hapla*. Sweetclover weevil *(Sitona cylindricollis)* is a major pest, destroying stands by defoliating newly emerged seedlings. Others include the sweetclover root borer *(Walshia miscecolorella)*; sweetclover aphid *(Therioaphis riehmi)*, to which resistance is included in breeding; blister beetles *(Epicauta* sp.*)*, which feed on leaves of low-coumarin cultivars; grasshoppers, leafhoppers, cloverleaf weevils *(Hypera punctata)*, cutworms, webworms, and other caterpillars. For control of all diseases and pests local agricultural agents should be consulted.

REFERENCES:

Gross, A. T. H., and Stevenson, G. A., 1964, Resistance in *Melilotus* species to the Sweetclover weevil *(Sitona cylindricollis)*, Can. J. Plant Sci. **44**:487–488.

Smith, W. K., 1954, Viability of interspecific hybrids in *Melilotus*, Genetics **39**:266–279.

CONTRIBUTORS: J. A. Duke, H. J. Gorz, C. F. Reed.

Melilotus suaveolens Ledeb.

FAMILY: Fabaceae
COMMON NAME: Sweetclover
SYNONYM: *Melilotus graveolens* Bunge

USES: Infrequently cultivated for forage in northern Asia and northern Great Plains of the United States.

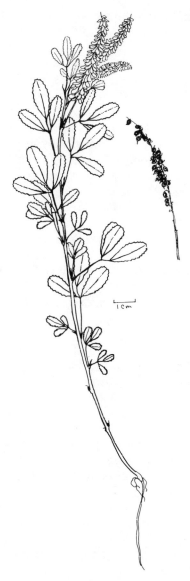

Figure 83. *Melilotus suaveolens* Ledeb.

FOLK MEDICINE: In Vietnam, a decoction is used for eye diseases.

CHEMISTRY: No data found.

DESCRIPTION: Biennial or annual herb, short-pubescent or nearly glabrous; stems erect, 1–1.6 m tall, upright or somewhat spreading, terete, branched above; stipules entire or subulate, dilated at base; leaves unequally denticulate, with 6–11 teeth on each margin; lower leaves obovate-orbicular, the upper ones oblanceolate, obtuse; leaflets

Melilotus suaveolens Ledeb.

narrowly elliptic to broadly lanceolate, 1.5–3.5 cm long, axillary, dense, 25–50-flowered, elongating after flowering; peduncles 2–4 cm long, bracts linear-subulate, ca. 1 mm long, slightly longer than pedicels; pedicels 2–2.5 mm long; flowers yellow or pale yellow, 3.5–4.5 mm long; calyx 2 mm long, teeth narrowly triangular, usually shorter than the tube; standard, wings and keel about equal; pods elliptic, 3–3.5 mm long, reticulately veined or wrinkled, nearly smooth, dark gray or black; seeds broadly oval, greenish-yellow, sometimes flecked with purple. Radicle more than half as long as cotyledons. Fl. June; Fr. August (U.S.).

GERMPLASM: Cv 'Redfield Yellow' was selected and developed at Redfield, South Dakota; 'Golden Annual' is a selection of the annual form. Assigned to the Eurosiberian, China–Japan, and Indochina–Indonesia Centers of Diversity. ($2n = 16$.)

DISTRIBUTION: Native to Eastern Siberia, Mongolia, Manchuria, China, and Japan, south to Korea, Indochina, and India. Introduced in many temperate areas as Tanzania and Great Plains region of North America.

ECOLOGY: Thrives on sandy loams and sands, riverbanks and meadows, and is rather common on waste ground and fields in lowlands; rarely found in cultivated fields. Does best on alkaline alluvial soils near lakes and water courses, but grows up to 1,700 m in altitude. Said to fare well with average annual rainfall of 7–8 dm. Ranging from Cool Temperate Steppe to Moist through Subtropical Dry Forest Life Zone, sweetclover is reported to tolerate annual precipitation of 4.9–11.6 dm (mean of 2 cases = 8.3 dm), annual mean temperature of 8.4°–12.5°C (mean of 2 cases = 10.5°C), and pH of 5.6–7.3 (mean of 2 cases = 6.5).

CULTIVATION: This sweetclover is only occasionally cultivated, mainly in northern Siberia, Great Plains of North America, Tanzania, and other areas where it has been introduced. Cultivation is much like that for other species of *Melilotus*. Seed sown in spring, broadcast and lightly harrowed into soil. Crop requires little care.

HARVESTING: Crop used mainly for forage and may be pastured by midsummer in most areas. When treated as a biennial, it is also pastured in following spring, or may be cut for dry feed during the winter.

YIELDS AND ECONOMICS: Average hay yields ('Redfield') are ca. 1.2–1.4 MT/ha the first year and 1.8–2.9 MT/ha the second year. A sweetclover of minor importance, it is grown to a limited extent in Great Plains, North America, and in Japan, Siberia, and Tanzania.

BIOTIC FACTORS: Following fungi have been reported on this species: *Cylindrocarpon ehrenbergii, Erysiphe pisi, E. polygoni, Fusarium avenaceum, F. culmorum, Phymatotrichum omnivorum, Phytophthora cactorum,* and *Sclerotinia sativa*.

CONTRIBUTORS: J. A. Duke, H. J. Gorz, C. F. Reed.

Mucuna spp.

FAMILY: Fabaceae
COMMON NAME: Velvetbean

USES: Velvetbean is grown primarily as a soil-improving crop, a "smother" crop to control weed pests such as bermudagrass and nutgrass, a cover crop for a summer green manure, and as a forage plant. In the southern United States it is often grown as an ornamental vine. In southern Georgia and northern Florida, it is grown for late fall and winter pasture. In Brazil, the seed starch has been studied as a food thickener and adhesive base. Elsewhere seeds from unripe pods are used as human food, usually after soaking, boiling, roasting, or fermenting to remove toxic principles. Immature pods and leaves are said to be boiled as vegetables.

FOLK MEDICINE: Levodopa (L-Dopa) extracted from velvetbean is used to provide symptomatic relief in Parkinson's disease. L-Dopa also repels insects. Species of Mucuna are said to be aphrodisiac, emetic, poison.

CHEMISTRY: Analysis of green material gave the following values: N, 0.56%; K, 0.37%; and P, 0.06%. Analysis of green forage from velvetbean gave the following values (dry basis): protein, 15.1%; ether extr., 2.1%; N-free extr., 48.5%;

Mucuna spp.

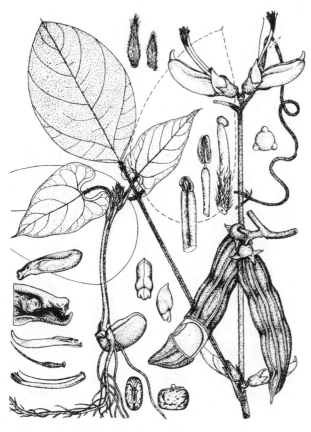

Figure 84. *Mucuna* pruriens (L.) DC.

crude fiber, 19.3%; and ash, 14.9%: digestible protein, 10.7%; digestible carbohydrates, 49.6%; digestible ether extr., 1.4%; and total digestible nutrients, 63.4%; it is rich in Ca, P, Fe, and I. Whole pods contained: moisture, 10.0%; protein, 18.1%; fat, 4.4%; N-free extr., 50.3%; fiber 13.0%; ash, 4.2%; digestible protein, 13.4%; total digestible nutrients, 73.8%. Seeds contained: moisture, 10.0%; protein, 23.4%; fat, 5.7%; total carbohydrate, 59.5%; N-free extr., 51.5%; fiber, 6.4%; and ash 3.0%: Ca, 0.18%; P, 0.99%; K, 1.36%; vitamin A 50 IU/100 g; thiamine 0.50 mg/100 g; riboflavin 0.20 mg/100 g; niacin 1.7 mg/100 g; digestible protein, 19.0%; total digestible nutrients, 81.7%. The amino acids are (mg/g N): isoleucine 300; leucine 475; lysine 388; methionine 75; cystine 56; phenylalanine 300; tyrosine 319; threonine 250; valine 344; arginine 494; histidine 131; alanine 219; aspartic acid 794; glutamic acid 763; glycine 288; proline 369; and serine 306. When fed to pigs in excessive quantities, the seeds cause severe vomiting and diarrhea; the toxic principle is reported to be dihydroxyphenylalanine. Seeds are also reported to contain trypsin and chymotrypsin inhibitors. The genus *Mucuna* is also reported to contain dopamine, nicotine, physostigmine, and serotonin. Screening 11 accessions of Mucuna, Daxenbichles et al (1971) reported L-Dopa content was 3.1–6.7% and was highest from *Mucuna holtonii*.

DESCRIPTION: Strong-growing, short-day annual, with stems slender, as bushes or vines 3–18 m long or more; leaves trifoliolate with large, ovate, membraneous leaflets shorter than the petioles; flowers borne singly or in twos or threes in long pendant clusters, varying from white to dark purple, 2.5–3.2 cm long; pods pubescent with dense black velvety hairs, up to 15 cm long, with 3–6 seeds per pod; seeds subglobose, usually marbled, but sometimes full-colored, white, brown, or black; roots numerous, rather fleshy, 7–10 m long, abundantly nodulated, near surface of ground. Fl. summer; Fr. Nov.–Dec.

GERMPLASM: The taxonomy of the cultivated species is confused, but recent authors (Kay, 1978) have accepted the easy way, and perhaps the proper way out. Five epithets are involved in velvetbean, *Mucuna pruriens* (L.) DC. var *utilis* (Wall. ex Wight) Baker ex Burck. (distinct from the larger horse-bean, *Mucuna sloanei* Fawcett & Rendle alias *M. urens* auctt.): (1) *nivea* the Lyon bean, cultivated for the immature pods as a vegetable in the Philippines and southeastern Asia; (2) *hassjoo*, the Yokohoma velvetbean, an early maturing type thought to have originated in Japan; (3) *aterrima* or *mauritius* velvetbean, grown in Australia, Brazil, Mauritius, and the West Indies as a drought resistant cover crop or rotation crop with sugarcane; (4) *utilis*, the Bengal velvetbean, grown in India; and (5) *deeringiana*, the Florida or Georgia velvetbean. Several cvs are grown in the United States, some bushy, others vining; some early-maturing, others late-maturing; with varying size, color, and number of seeds; and varying color of flowers. A late-maturing cv was first used in Florida and southern Georgia, but through introduction of new species, hybridization and selection, several earlier maturing cvs were developed. Some of the best cvs now grown are: 'Osceola' (a cross between Florida and Lyon cvs), a heavy seed producer, flowers white, rarely purple, in short racemes; pods 10–12 cm long, flat, ridged lengthwise, covered with black velvety hairs, nearly sting free, 4–5 seeds per pod,

Mucuna spp.

seeds larger than most velvetbeans, usually marbled with brown, rarely white, maturing in 150 days. 'Bunch Velvetbean,' a nonvining sport of a single plant of Florida cv, vineless with short branches at base, seed in clusters of pods near base, both seed and forage yields low, seeds often get watersoaked and rot. 'Victor,' seed medium-sized, marbled, nearly round, maturing in 190 days, about 1,400 per kg. '120-Day Florida,' a popular cv, seed medium-sized, marbled, nearly round, usually require more than 120 days to mature. 'Early Jumbo,' seed large, ovate, marbled, with occasional white or brown, matures in 175 days, easy to pick because pods produce in large clusters and the entire cluster is gathered; about 260 seeds/kg. Assigned to the Indochina–Indonesia Center of Diversity, velvetbean or cultivars thereof is reported to exhibit tolerance to drought, laterites, poor soil, sand, and to smother seeds. Some types of Mauritius velvetbean, Florida velvetbean, and Lyon bean, tolerating drought, show promise for dryland farming. ($2n = 22$.)

DISTRIBUTION: Derived from southern Asia or Malaysia, the various velvetbeans probably represent cvs of some species or several species. It was introduced to Florida ca. 1876, and was then reintroduced to tropical and subtropical areas. At present it is grown especially as a cover crop in Hawaii, Australia, the Philippines, and Malaysia.

ECOLOGY: Velvetbean is tropical or subtropical, but matures seeds as far north as Maryland and Kansas in the US. Adapts to well-drained, sandy soils of the eastern and southern Coastal Plains, and to clay soils in the more northern cotton belt. It requires a long growing season to produce pasture and is grown in the south for late pasture and soil improvement. Plants are sensitive to frost, require a frost free period of 180–240 days. Does not succeed on cold, wet soils and should not be planted before soils are thoroughly warm. Exposure to temperatures below 5°C for 24–36 hr may be fatal to Florida cvs. During the growing season, temperatures of 20°–30°C are recommended. Night temperatures of 21°C are said to stimulate flowering. Ranging from Warm Temperate Dry to Moist through Tropical Dry Forest Life Zone, velvetbean is reported to tolerate annual precipitation of 3.8–31.5 dm (mean of 12 cases = 15.3 dm), annual mean temperature of 18.7°–27.1°C (mean of 12 cases = 24.1°C), and pH of 4.5–7.7 (mean of 10 cases = 5.9). Yields are optimum at pH 5–6.5 on light sandy loams.

CULTIVATION: Germination of good hand-picked seeds is highest (ca. 99%) when seeds are taken from green pods. Seeds should be sown 15–90 cm apart in rows ca. 90–180 cm apart in a well prepared seedbed. Broadcast plantings, often not satisfactory, are sown at 45–90 kg/ha. When planted alone, 10–35 kg/ha is suggested, sown after soil has become warm. Can be interplanted with corn at the same time, about 90–120 cm apart in rows, requiring 4–15 kg/ha depending on width of rows between the corn. Late cvs can be grown in the lower South, but earlier cvs should be chosen for the upper South. Planting time varies: Florida, Mar.–Apr.; Hawaii, Feb. When grown alone for green manure or as a shade crop, 2–3 cultivations are necessary to control weeds until plants start vining; eventually they form a dense mat about 60 cm deep and smother out weeds. However, such growth produces very few seed. When grown for seed or feed, velvetbean should be interplanted with corn or some other strong upright crop for vines to climb upon. Flowers require open-air circulation for pollination and good seed set. Cultivation should be the same as for corn grown alone. Millet and sorghum are also used for support in Australia. In the United States, 50–225 kg/ha of superphosphate is recommended, but seed yield seldom increases with fertilizers. However, lime often does increase seed yield. Inoculation is often beneficial in tropical soils, but usually not needed in temperate areas. Strains of organisms used to inoculate lima beans, cowpeas, and lespedeza can be used on velvetbeans. Velvetbeans can be rotated with corn or cotton.

HARVESTING: Seeds mature in 100–130 days in southeastern United States, depending on cv; other cvs in some areas may take up to 150 days; in the tropics 7–9 months are required. As pods shatter easily, they must be harvested as soon as mature. November to early December is best for seed harvest. Velvetbeans are harvested by hand. No machine has been devised to gather them from the matted vines, stalks, and leaves in the field. When grown with maize, cutting just below the maturing cob facilitates hand harvesting. Mature and dry pods thresh easily. After drying, beans may be threshed or hulled with a regular grain thresher, removing several teeth from the cylinder and concaves of the machine to prevent excessive cracking of the beans. In Australia a corn sheller is used; in Africa, they are placed in sacks and beaten with sticks. Seed stored in a cool dry place remains viable for two years or more. Grown for forage, the crop may be harvested in 90–120 days. Velvetbean is seldom used for hay because the tangled vines are difficult to handle and vines and leaves turn black when cured and appear unpalatable. Velvetbean is used as a frosted pasture crop for cattle. Since pods absorb moisture from the winter rains, seeds are soft enough for cattle to chew. Sometimes they are ground into a meal and mixed with ground corn or cottonseed meal and the meal fed dry. Hogs, poultry, and horses do not do well on velvetbeans or velvetbean meal.

YIELDS AND ECONOMICS: On good soils yields were 1,000–2,000 kg/ha in the United States, and 900–1,500 kg/ha in Hawaii. In India, seed yields are ca. 240–1,120 kg/ha of seed and 900–3,600 kg/ha of hay. Green forage yields are 3–6 MT/ha, 90–100 days after sowing and 18 MT/ha over the growing season. At New Delhi, corn–velvetbean mixtures gave ca. 50% more forage than corn–cowpea. The green manure crop is estimated at 18–19 MT/ha. Velvetbean seed crop is estimated at ca. 895,000 MT/yr. From decorticated seed meal yields of 4.8% L-Dopa were reported.

BIOTIC FACTORS: Velvetbeans are largely self-pollinating; natural crossing is rare. Perhaps because of the L-Dopa content, velvetbean is subject to few insect problems. In Florida, larvae of the velvetbean caterpillar *Anticarsia gemmatalis* eat the leaves. L-Dopa in the seeds is a chemical barrier to attack by insects and small mammals (Rehr *et al.*, 1973). Velvetbean is immune to most diseases, including fusarium wilt, but in southern Rhodesia is very susceptible to a vine-rot disease that can wipe out the crop. Fungi known to attack velvetbean include: *Cercospora stizolobii, Mycosphaerella cruenta, Phyllosticta mucunae, Phymatotrichum omnivorum, Phytophthora dreschsleri, Rhizoctonia solani, Sclerotium rolfsii* (southern blight), *Uromyces mucunae*. Among the bacteria are bacterial leaf-spot, *Xanthomonas stizolobiicola, Pseudomonas stizolobii,* and *Ps. syringae. Striga gesnerioides* parasitizes the plant. Yellowing is due to zinc deficiency. A mosaic virus attacks it. Velvetbean is resistant but not immune to rootknot nematodes, and is attacked by *Meloidogyne thamesi, M. hapla., M. incognita acrita,* and *M. javanica.* Other nematodes isolated from this crop include *Belonolaimus gracilis, Pratylenchus brachyurus,* and *Rotylenchulus reniformis.*

REFERENCES:
Bort, K. S., 1909, The Florida velvet bean and its history, *U.S. Dep. Agric., Bur. Pl. Ind., Bull.* **14**(3):25–32, illus.
Daxenbichler, M. E., VanEtten, C. H., Hallinan, E. A., Earle, F. R., and Barclay, A. S., 1971, Seeds as sources of L-Dopa, *J. Med. Chem.* **14**:463–465.
Piper, C. V., and Tracy, S. M., 1910, The Florida velvet bean and related plants, *U.S. Dep. Agric., Bur. Pl. Sci., Bull.* **179**:1–26, illus.
Rehr, S. S., Janzen, D. H., and Feeny, P. P., 1973, L-dopa in legume seeds: A chemical barrier to insect attack, *Science* **181**:81–82.
Stephens, J. L., 1970, The velvetbean, *U.S. Dep. Agric., Agric. Res. Ser., Crops Res. Div.* **CA-34-162**:1–6.
U.S. Department of Agriculture, 1959, The velvetbean caterpillar. How to control it, *U.S. Dep. Agric. Leaflet* **348**:1–4, illus.

CONTRIBUTORS: J. A. Duke, C. F. Reed, J. K. P. Weder.

Myroxylon balsamum (L.) Harms

FAMILY: Fabaceae
COMMON NAMES: Balsam of Tolu, Tolu balsam, Quinoquino
SYNONYMS: *Myroxylon toluiferum* H.B.K.
Toluifera balsamum L.

USES: Tolu balsam, a gum collected from wounded trees, is used mainly as a pleasant flavoring in cough syrups, soft drinks, confectionaries, ice cream, and chewing gums. Oil of Tolu balsam is used in perfumes, cosmetics and soaps. The tree is sometimes grown for timber, as an ornamental, or as a shade tree for cultivated crops. In the United States, oil of Tolu balsam is used mostly as an ingredient of tincture of benzoin.

FOLK MEDICINE: Medicinally, it is used as a feeble expectorant in cough mixtures, and as an inhalant for catarrh and bronchitis. Said to be used for: antiseptic, bactericidal, fumigatory, pectoral, tonic, and vulnerary. An ethanol extract is antibiotic against *Mycobacterium tuberculosis.*

Myroxylon balsamum (L.) Harms

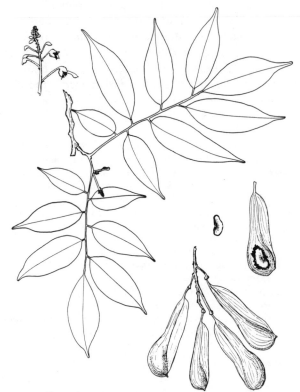

Figure 85. *Myroxylon balsamum* (L.) Harms

CHEMISTRY: Tolu balsam is a shining, translucent, yellow or reddish-brown, viscous mass and consistency varies with temperature; on aging it becomes a plastic solid that softens at 30°C and melts at 60–65°C. With prolonged aging it becomes hard, brittle, and rosinlike. Used in high-grade perfumes because its pleasant hyacinthlike scent blends well with floral and oriental-type compounds. Tolu balsam contains resin esters (75–80%), chiefly toluresinotannol cinnamate; volatile oil (7–8%), mainly benzyl benzoate; free cinnamic acid (12–15%); free benzoic acid (2–8%); and vanillin. A volatile oil distilled from the wood contains l-cadinol and d-cadinene, farnesol, and traces of nerolidol. Tolu balsam has the following constants (on the basis of dry-alcohol soluble matter): acid value, 97–160; ester value, 47–95; sap. value 170–224; balsamic acids, 35–50%; it is soluble in 90% alcohol, ether, and choloroform.

DESCRIPTION: Tall tree to 34 m, with spreading crown, branching ca. 15 m aboveground; glabrous except racemes; bark gray, more or less spotted with yellow rough areas; wood with hardness of mahogany or cedar; leaves alternate, evergreen, oddly pinnate, leaflets without stipules, several to 13, ovate or ovate-oblong, acuminate, often obtuse, subcoriaceous, reticulate veiny, lustrous, 6–9 cm long, 3–4 cm wide; racemes closely tomentulose, pedicels 1.5 cm long, calyx 4–6 mm long, banner 12 mm long, 8 mm wide; pod usually about 12 mm long, 8 mm wide; pod about 8–13 cm long, 2.5 cm broad, style-tip noticeably situated below curved upper edge; seed portion with 2 globose balsam pits between the corky mesocarp and coriaceous endocarp, leaving seeds smooth, not sticky with balsam; testa if present dry and thin.

GERMPLASM: Two varieties grow wild; var *punctatum* (Klotzsch) Baill. ex Harms, has thicker leaves than typical plants or the var *pereirae* (Royle) Baill. of Central America, which is without uniform mixture of pellucid lines with dots. Assigned to the South America Center of Diversity, Tolu balsam is reported to exhibit tolerance to mycobacteria. ($2n = 28$).

DISTRIBUTION: Native to Venezuela, Peru, and Colombia, growing abundantly along the Magdalena and Cauca Valleys, Prov. Tolu, Colombia.

ECOLOGY: Trees grow wild, scattered in forests of northern South America, where annual rainfall is ca. 200 cm, and the mean annual temperature is 21°–28°C. Ranging from Subtropical Dry to Wet through Tropical Dry Forest Life Zone, Tolu balsam is reported to tolerate annual precipitation of 13.5–40.3 dm (mean of 6 cases = 26.4 dm), annual mean temperature of 23.1–27.3°C (mean of 6 cases = 25.4°C), and pH of 5.0–8.0 (mean of 5 cases = 6.1).

CULTIVATION: Trees grow wild in the forests.

HARVESTING: For harvest of gum, a **V**-shaped notch is cut through the bark. Below the point of the notch the bark and wood are hollowed out to hold a cup for the drippings. Incisions are made at many points up and down the trunk, but in such a way as not to girdle the tree. When tree is scarred as far as one can reach from the ground, scaffolds are sometimes rigged around it, so that the same process can be repeated farther up. As many as 20 incisions may be made up the side of a tree. Gum is collected from the cups by gatherers who make their rounds of the trees with a pair of rawhide bags slung over the backs of burros. Contents of the cups are emptied into bags, the cups replaced for more gum. Later the bags are emptied into tin export containers. The collection season continues over an 8-month period from July to March or April.

YIELDS AND ECONOMICS: Steam distillation of gum yields 1.5–7% of a volatile oil which may be lighter or heavier than water. Because the number of incisions per tree and the length of collection of gum are indefinite, data vary. Averages are 8–10 kg balsam/tree per season. Tolu balsam is produced in Colombia, the chief source, Venezuela and the West Indies. About 50 MT are imported by the United States annually from Colombia. Some is imported from Great Britain where some of the volatile oil has been removed.

BIOTIC FACTORS: Following fungi are known to attack Tolu balsam trees: *Meliola xylosmae*, *Phyllosticta myroxyli*, *Phomopsis* sp., and *Tabutia xylosmae*.

REFERENCES:
Guenther, E., 1945, Survey of balsam Tolu and oil of balsam of Tolu, *Am. Perfum. Essent. Oil Rev.*, 2 pp.
Weir, J., 1864, On *Myroxylon toluiferum* and the mode of processing the balsam of Tolu, *Technologist* 5:67–71.
CONTRIBUTORS: J. A. Duke, C. F. Reed.

Myroxylon balsamum var *pereirae* (Royle) Harms

FAMILY: Fabaceae
COMMON NAMES: Peru balsam, Balsam of Peru
SYNONYMS: *Myrospermum pereirae* Royle
Myroxylon pereirae (Royle) Klotzsch
Toluifera pereirae (Royle) Baill.

USES: Peru balsam, produced by injured trees, exudes from trunk and limbs or is extracted from bark. Peru balsam is not produced in Peru, but received its name because it was originally assembled and shipped to Spain from the Port of Callao, Peru. Oil (cinnamein) is used in perfume, cosmetic, and soap industries. Balsam and its essential oil

Figure 86. *Myroxylon balsamum* var *pereirae* (Royle) Harms

have been used as flavorings for baked goods, candy, chewing gum, gelatin, ice cream, pudding, soft drinks, and syrups. Balsam is an excellent fixative, blending very well into perfumes of the oriental type and floral perfumes of the heavier type. Balsam wood is close-grained, handsomely grained, nearly of a mahogany color, but redder, with a very agreeable odor, which it retains for a long time, takes a good polish and is highly esteemed for cabinetwork. Tree is sometimes cultivated as shade for coffee plantations or as an ornamental. At one time balsam was so popular as an incense that papal edicts forbade the destruction of the trees. Seeds are used to flavor aguardiente, a popular Latin American alcoholic beverage.

FOLK MEDICINE: Peru balsam used extensively as a local protectant, rubefacient, parasiticide in certain skin diseases, antiseptic and vulnerary, and applied externally as an ointment, or in alcoholic solutions; internally rarely used as an expectorant. Dried fruits are sold in Guatemala for itch. An alcohol infusion is rubbed on in Cuba to alleviate headache and rheumatism. The resin is used for asthma, catarrh, rheumatism, gonorrhea, and to heal cuts and wounds. The balsam is widely used, by physicians, to treat venereal sores. Blended with castor oil, or prepared as a tincture, it is used for chilblains, pediculosis, ringworm, and scabies. In suppositories, it is applied for hemorrhoids and anal pruritis. Formerly it was used internally for amenorrhea, bronchitis, diarrhea, dysentery, dysmenorrhea, laryngitis, and leucorrhea. It is no longer used internally, at least in the United States. Alcoholic extracts inhibit *Mycobacterium tuberculosis*.

CHEMISTRY: Peruvian balsam occurs as a dark brown, viscid liquid, transparent in thin layers, free of stringiness and stickiness, with an aromatic, empyreumatic, vanillalike odor and an acrid, bitter, persistent taste. It contains about 60% cinnamein, a volatile oil, consiting mainly of benzyl cinnamate with some benzyl benzoate, resin esters (30–38%), vanillin, free cinnamic acid and peruviol. Once the balsam is removed, the spent wood yields an essential oil containing 68–70% nerolidol. The seeds yield a "balsam" composed of 67.7% resin, 14.8% wax, 11.9% acid resin, 0.4% coumarin, 0.4% tannin, and 4.6% water.

DESCRIPTION: Trees to 35 m tall, 1 m in diameter; branches terete, smooth, ash to ash brown in color; leaves evergreen, alternate, petiolate, compound, the leaflets 5–11, alternate, with short petioles, blades 5–8 cm long, 2.5–4.0 cm wide, oblong or oval-oblong, sometimes ovate, rounded or very slightly tapering, but not cordiform, at base; apex abruptly contracted into an emarginate point; sparsely hairy along petioles and midribs with short hairs with a glossy or a resinous appearance; leaflets marked by rounded and linear pellucid spots; flowers in axillary racemes 10–20 cm long, white; fruit a one-celled, one-seeded, winged, indehiscent pod (samara), the fruit-stalk naked at base, but amply winged above; fruit 5–10 cm long, with a fibrous mesocarp, and a yellow oleoresinous or balsamic juice, which on aging hardens and resinifies, in receptacles (vittae) immediately exterior to the endocarp; seed loose and dry in the cell of the pericarp, covered with a thin, white, mebranous coat; cotyledons yellowish and oily, with the odor of melilot or tonka-bean, and a bitter taste similar to bitter almond; fruits said to be like almonds with a golden-colored juice.

GERMPLASM: Assigned to the Middle America Center of Diversity, Peru balsam or cultivars thereof is reported to exhibit tolerance to high pH, mycobacteria, poor soil, and slope. ($2n = 28$.)

DISTRIBUTION: Native and localized along the Pacific Coast jungles of Central America, El Salvador, Nicaragua, Guatemala. Introduced to southern Florida, Ceylon, India, and west Africa.

ECOLOGY: Grows well on poor but well-drained soils, at altitudes up to 600 m. In the balsam forests region of El Salvador, mean annual temperature is about 21°C and annual rainfall of 20–25 dm. Ranging from Subtropical Dry to Wet through Tropical Dry to Wet Forest Life Zones, Peru balsam is reported to tolerate annual precipitation of 13.5–40.3 dm (mean of 4 cases = 24.1 dm), annual mean temperature of 23.3–26.5°C (mean of 4 cases = 24.6°C), and pH of 5.0–8.0 (mean of 3 cases = 6.9).

CULTIVATION: Peru balsam is obtained entirely from wild trees, which grow singly or in groves in forests of Central America.

HARVESTING: Trees can be tapped when 5–6 yr old. They have been reported to withstand severe treatment for many years, and can be tapped for 40–60 yr or more. Gum is collected by two methods: Method 1: *Cascara* or Bark Process: A fire is made on the root of the tree and tree is scorched 10 min. After about 8 days the bark becomes soft and pieces 30–60 cm long and 30 cm wide are cut off with machete. Additional incisions are made on the opposite side of the tree or above. Bark is crushed and hot water poured over it to soften the balsam and facilitate its flow. The cooled balsam is heavy and sinks to bottom and can be separated. Method 2: *Panal* (Cloth) or *Trapo* (Rag) Process: Trees are beaten on four sides, then scorched with a torch to cause bark to separate from trunk; 4 intermediate strips are left uninjured so the tree survives. Within 1 week the bark drops from the trunk and the balsam begins to exude freely from the exposed wood. Areas are then wrapped with rags that are removed from time to time when they become saturated with balsam. They are boiled with water, and as the water cools, the balsam settles out. It is recovered, strained and packed, usually in tin cans, averaging 31.5 kg net and 42–45 kg gross. These are called 'Bark balsam' and 'Rag balsam.' In Nicaragua, a 15-cm square is cut in the bark and rags applied and secured. In 2 weeks the rags are removed, and the oil pressed and boiled out. Each year the tree is tapped about 1 m above the scar of the previous year. In other areas gatherers climb the trees to various heights, and cause the gum to exude by pounding patches of bark with wooden mallets. The outer bark of the pounded areas is then removed. After 4–5 days gum begins to flow and is soaked up with cotton or linen rags applied to the wound. Flow is stimulated further by firing the wood with torches. When the rags are well soaked, they are put through crude presses and then boiled in water to remove the remainder of the gum. When gum ceases to flow, fresh cuts are made in the bark and the process repeated. A lower grade balsam is obtained by cutting up the bark that is removed in the wounding process and boiling it in water to extract the gum.

YIELDS AND ECONOMICS: Properly worked and tapped twice a month throughout the year, a healthy, fully grown tree yields 112–170 g of balsam each extraction. Trees over 60 yr old produce 226 g. Trees less than 25 yr old give low yields of balsam of inferior quality. Average yields are ca. 1–4 kg. With the *Cascara* method an incision is tapped only twice and then a new one is made above or on the opposite side of the tree. In the *Trapo* method, the same incisions may be treated every 2 weeks for 40–60 years. Commercial supplies of Peru balsam come mainly from El Salvador, Nicaragua, and Honduras. Annual world production is about 65 MT, ca. half of this exported to the United States. Other importing countries are Germany, France, and England. El Salvador exports about 48 MT/yr. Price of Peruvian balsam is $1.30–$2.40/lb.

BIOTIC FACTORS: Following fungi are known to attack Peru balsam trees: *Meliola xylosmae, Myiocopron pereirae, Peckia pereirae,* and *Tabutia xylosmae.*

REFERENCES:
Attfield, Dr., 1963, Note on the Gum-resin of the Balsam of Peru Tree, *Pharm. J.* **5.2.5**:248.
Bennett, C. T., 1928, Balsam of Peru, *Perfum. Essent. Oil Rec.* **19**:423–424.
Guenther, E., 1940, Balsam Peru, a survey, *Drugs Cosmetic Ind.* **July**:6 pp., illus.
Hanbury, D., 1863–1864, On the manufacture of Balsam of Peru, *Pharm. J.* **5.2.5**:241–248; Additional note, **5.2.5**:315–317, illus.
Martinez, Alfredo, *et al.*, 1940, El balsamo negro de El Salvador, *Cafe El Salvador* **10**:5–72, illus.

CONTRIBUTORS: J. A. Duke, C. F. Reed.

Onobrychis viciifolia Scop.

FAMILY: Fabaceae
COMMON NAMES: Common or Giant sainfoin, Esparcet, Holy clover
SYNONYMS: *Onobrychis sativa* Lam.
Hedysarum onobrychis Neck.

USES: Grown as forage for livestock, especially in Europe, western Asia, western Canada, and northwestern United States, sainfoin is recommended for dryland pasture or hay in areas with 33 cm or more of precipitation, or areas where a shortage of water limits irrigation to one application early in season. Yield of hay is high. Hay and green material readily eaten by livestock; does not cause bloat when fed green, as alfalfa does. Excellent supplement for swine. Regarded highly in Europe and Canada for honey production.

FOLK MEDICINE: No data available.

CHEMISTRY: Cut at 10% flowering, sainfoin hay contains 5.9% water, 14.9% protein, 1.1% ether extract, 6.7% ash, 22% crude fiber, 0.7% Ca, and 0.3% P; at 50% flowering, contains 5.9% water, 12.1% protein, 1.2% ether extract, 5.7% ash, 27.6% fiber, 0.8% Ca, and 0.1% P; at 100% flowering; 4.8% moisture, 10.6% protein, 1.1% ether extract, 5.5% ash, 32.0% fiber, 0.8% Ca, and 0.2% P (Ditterline and Cooper, 1975). Green sainfoin forage contains per 100 g: 81.0% moisture, 4.9 g protein, 0.6 g fat, 3.1 g fiber, and 2.4 g ash. At 16% moisture sainfoin hay contains 10.5% protein, 2.5% fat, 19.7% fiber, and 7.1% ash. Leaves contain 5–7% dry weight proanthocyanandins with molecular weights 17,000–28,000. Sainfoin seeds comprise ca. 70% of the weight of unmilled pods, which are composed primarily of crude fiber. Sainfoin seed contain: 9.2% water, 37.7% protein, 0.9% ether extract,

Onobrychis viciifolia Scop.

Figure 87. *Onobrychis viciifolia* Scop.

3.9% ash, 0.2% Ca, and 0.4% P, after oil extraction. Before extraction it contained 6.8% water, 35.6% protein, 4.3% ether extract, 4.4% ash, 0.2% Ca, and 0.4% P. Amino acid contents of the unextracted sainfoin seed were (% of recovered protein): 6.4 lysine, 4.2 histidine, 11.8 arginine, 3.6 threonine, 3.9 valine, 1.0 methionine, 3.4 isoleucine, 6.6 leucine, 4.0 phenylalanine, 11.1 aspartic acid, 5.1 serine, 20.8 glutamic acid, 4.3 proline, 4.3 glycine, 3.6 alanine, and 2.9 tyrosine (Ditterline *et al.*, 1977). Seed are reported to contain trypsin inhibitors and chymotrypsin inhibitors.

DESCRIPTION: Subglabrous to pubescent, erect perennial herb, 10–80 cm tall; leaves odd-pinnately with compound, 5–14 pairs of leaflets, 10–35 mm long, 3–7 mm broad, linear to obovate, with appressed hairs beneath; racemes up to 80-flowered, dense, elongate, up to 9 cm long in flower; calyx 5–9 mm long, pubescent, teeth 2–3 times as long as tube; corolla 10–15 mm long, bright pink with purple or reddish veins; keel and standard about equal, wings very short; legume 5–8 mm long, pubescent, 1-seeded, sides toothed, margin usually with 6–8 teeth up to 1 mm long, with network of pits on both surfaces, often appearing reticulate, almost semicircular; seed kidney-shaped, 2.5 × 2.0–3.5 × 1.5–2.0 mm, olive, brown or black, 1.32–1.68 g/100 seed. Taproot deep, penetrating 1–2 m, occasionally to 10 m. Fl. June–Aug. (U.S.).

GERMPLASM: Several botanical forms are recognized, each with its particular advantages and disadvantages. *O. viciifolia* forma *persica* Shiryaev (many-cut sainfoin) gives twice as much forage, or hay as the more widely cultivated forma *europaea*. Its lime requirements are more moderate, but its nectar-yielding capacity is about the same (moderate). 'Eski,' developed in Montana, is derived from three Turkish selections. 'Eski' and 'Remont' are recommended for Montana, 'Melrose' for Canada. 'Single Cut' is a slow-growing long-lived strain that rarely flowers; 'Giant' and 'Double Cut' grow quickly, flowering and seeding even when two cuts of forage are taken annually. Other species of Onobrychis are cultivated for hay in Europe and western Asia. These include: *O. transcaucasia* Grossh (most productive for hay), *O. arenaria* (Kit.) DC., *O. vaginalis* C. A. Mey. (highest in protein content, 20.2–20.4%), *O. tanaitica* Sprengel, *O. inermis* Stev., and *O. altissima* Grossh. Assigned to the Near East and Eurosiberian Centers of Diversity, sainfoin or cvs thereof is said to exhibit tolerance to alkali, drought, frost, grazing, high pH, limestone, salt, poor soil, and slope. ($2n = 28, 22, 29$.)

DISTRIBUTION: Native to Europe and western Asia; now cultivated there and in North America (Montana, northern Idaho, Nevada, and Canada). Sainfoin was grown for forage in Russia more than 1,000 years ago, in France during the 14th century, in Germany during the 17th century, and in Italy during the 18th century. Naturalized in many areas.

ECOLOGY: Grows on banks, waste places, grassy places, often escaped from cultivation. Adapted to wide variety of soils, doing well on gravelly, high lime soils, even poor, shallow, rocky soils; less suitable to wet soils with high level of ground water. Requires weed-free fields, especially first year. Competes well in swards for 3–5 yr, then begins to thin out. An important crop because it survives in dry alkaline soils. Also it has frost tolerance to −4.4°C but does not tolerate much frost if snowcover is poor. Germination is optimum at 15°–23°C. Grows as well at 18°C as at 27°C. The well-developed root system makes it quite drought resistant. Its salt tolerance is about like that of oats, rye, and wheat. Sainfoin seed can germinate at higher salinity levels than alfalfa seed. If a more effective inoculum could be found, sainfoin could replace alfalfa where manganese is a problem. Ranging from Boreal Wet through Warm Temperate Thorn to Moist Forest Life Zones, sainfoin is reported to tolerate annual precipitation of 3.5–10.7 dm (mean of 24 cases = 6.6), annual mean temperature of 5.9°–18.6°C (mean of 24 cases = 10.2), and pH of 4.9–8.2 (mean of 20 cases = 7.1).

CULTIVATION: Uniform, firm and well-prepared seed beds are essential for good stands. Seeds may be sown in early spring in one operation under winter or spring crop, or along with a companion spring crop. Seeding rates vary. In Montana, the recommended seed rate is 38 kg/ha for irrigated stands, 25 kg/ha for rainfed cropping spaced 15–20 cm apart in rows with no companion crops. Seeds should be planted 1–1.5 cm deep. In dry steppe regions, 75–90 kg/ha is recommended; in park and other regions and sometimes under irrigation, 80–100 kg/ha. For seed production, 90–110 kg/ha is sown with a small-grain companion crop to control weeds. General practice is to remove grain crop at maturity and irrigate sainfoin. Sometimes in combination with another crop, seeding is reduced by 20–30%. New stands require careful weeding. Advisable to leave a high stubble companion crop to conserve moisture. On poor soils, organic and mineral fertilizers may be desirable. Under some circumstances sainfoin responds favorably to N and P. There has been no response to P in Canada, Montana, or North Dakota. Potassium is needed in some areas. On established sainfoin, broadcast applications of phosphate–potash are more satisfactory than drill methods. Crop has low boron requirement. Nodules form on roots 20–30 days after inoculation. Nodules of sainfoin are 5–8 times as large as those of alfalfa. Optimum temperatures, for germination and for seedling growth are 15°–20°C and 20°–30°C, respectively. At all temperatures, presence reduces speed of germination and speed of seedling elongation. Seeds from brown pods germinate better than those from green pods. Pods have a water-soluble inhibitor easily removed by washing. Field emergence of seedlings does not differ between shelled and unshelled seed.

HARVESTING: Sainfoin is cut at the 50–100% flower stage. Delayed harvest results in a loss of forage value. Stems are coarse but readily consumed by livestock. Around 1950, cut and dried hay was raked into small cocks and then stacked. Harvested like alfalfa, sainfoin dries faster and more intact. For seed, stands in their second or third year of growth are used. Sainfoin is swathed at 40% moisture and dried in the windrow before threshing (5–10 days) with a combine. Swathing prevents seed shatter. Airdried seed store best at 12% moisture.

YIELDS AND ECONOMICS: In Montana, yield was significantly higher from mixtures of sainfoin-grass than from grass alone but not always higher than from sainfoin alone. Sainfoin yielded ca. 11 MT/ha hay (12% moisture) the first year, about 8.5 the second, and about 7 the third. Seed yields range from 6–12 centners/ha. Seed yield is maximum when seed are harvested at about 40% moisture; quality is poor from earlier harvests at higher moisture and losses from seed shatter are greater at later harvests. Second year seed crops average 500 kg/ha; third year crop, 220 kg/ha. Important in Europe and western Asia, sainfoin is being developed as a fodder and forage crop in North America, but has never become a major forage crop in the United States largely due to disease problems.

BIOTIC FACTORS: Pollination may be effected by honeybees that prefer sainfoin to white clover. Provide minimum of 2–3 colonies/ha. Fungi reported include: *Ascochyta onobrychidis* (anthracnose or stemspot), *Botrytis cinerea* (chocolate spot on stems and flower bud killer), *Cylindrosporium onobrychidis, Erysiphe communis, E. martii, E. polygoni* (powdery mildew), *Fusarium oxysporum, F. solani* (root rot, the most important pathogen in Montana), *Peronospora ruegeriae, Placosphaeria onobrychidis, Pleospora herbarium* (ringspot), *Ramularia onybrichidis* (leaf spot), *Sclerotinia trifoliorum* (root, crown, and stem rot), *Septoria orolina*

Onobrychis viciifolia Scop.

Sacc. (leaf spot), *Stemphylium botryosum*, and *Verticillium albo-atrum* (verticillium wilt). The crown- and root-rot pathogen complex *(Fusarium solani)* is the most important limiting factor in sainfoin production. Within 3 months of planting, more than 50% of the seedlings are infected, but symptoms do not usually develop until the second year. In years 3–5 plants begin to die and stands become so depleted that forage yields are reduced (Auld *et al.,* 1977). Sainfoin inhibits *Aspergillus* spp. Nematodes reported on sainfoin include: *Ditylenchus dipsaci,* (seedlings especially susceptible) *Heterodera glycines, H. radicicola, Meloidogyne* sp., and *Tylenchulus dipsaci*. Although sainfoin is resistant to alfalfa weevils, insects may be problems in various areas, and should be controlled. Insect pests of sainfoin may be divided into three groups: root feeders, as weevils (*Sitonia scissifrons*, larvae feed on nitrogen nodules); stem and leaf feeders, as garden webworm, sugarbeet webworm, alfalfa butterfly (*Colias eurytheme*), beetles (*Phytonomus farinosus*, damages leaves, buds, and flowers, winters in soil, lays eggs in May, emergence of beetle in mid-June); and flower and seed feeders, as sesiid moth of sainfoin (*Dipsesphecia ichneumonifer*), sainfoin seed chalcid (*Eurytoma onobrychidis*, very serious, attacks some areas, attacks seeds), weevils (*Bruchidius unicolor*, very serious, attacks seeds), *Perrisia onobrychidis, Odontothrips intermedius, Meligethes erythropus, Contarunia onobrychis* (sainfoin midge), *Oriorrhynchus lingustici*, and potato leafhopper (*Empoasca fabae*, most damaging in the United States.

REFERENCES:

Auld, D. L., Ditterline, R. L., and Mathre, D. E., 1977, Screening sainfoin for resistance to root and crown rot caused by *Fusarium solani* (Mart.) Appel and Wr, *Crop Sci.* **17**:69–73.

Carleton, A. E., and Cooper, C. S., 1968, A compilation of abstracts on sainfoin literature, 93 pp.

Cooper, C. S., and Carleton, A. E. (eds.), 1969, Sainfoin Symposium at Montana State University, Dec. 12–13, 1968, *Mont. Agric. Exp. Stn., Bull.* **627**:1–109, illus.

Ditterline, R. L., and Cooper, C. S., 1975, Fifteen years with sainfoin, *Mont. Agric. Exp. Stn., Bull.* **681**:23 pp.

Ditterline, R. L., Newman, C. W., and Carleton, A. E., 1977, Evaluation of sainfoin seed as a possible protein supplement for monogastric animals, *Nutr. Rep. Int.* **15**(4):397–405.

CONTRIBUTORS: C. S. Cooper, J. A. Duke, M. B. Forde, C. F. Reed, J. K. P. Weder.

Ornithopus sativus Brot.

FAMILY: Fabaceae
COMMON NAME: Serradella
SYNONYM: *Ornithopus roseus* Dufour

USES: Cultivated for forage, as a cover crop, and for green manure. Grown for seed in Kenya. In southeastern United States grown as a winter annual; in northern United States may be of value interplanted with oats and barley to provide cover or forage after grain is harvested.

FOLK MEDICINE: No data available.

CHEMISTRY: Seeds are reported to contain trypsin inhibitors and chymotrypsin inhibitors. With 89% DM, the hay contains, on a moisture-free basis 18.4% crude protein, 3.6% fat, 33.5% crude fiber, 8.5% ash, and 36.0% N-free extract. Based on 8 analyses the green forage (21.3% DM) contains (moisture-free): 11.0–16.4% crude protein (avg. 14.8%), 3.0–4.1% fat (avg. 3.7), 20.7–25.3% crude fiber (avg. 23.0), 11.1–17.8% ash (avg. 15.3), and 40.3–48.1 N-free extract (avg. 43.2).

DESCRIPTION: Pubescent annual herb, procumbent in thin stands, more ascending in thick stands; root system extensive, estimated to be twice that of tops; stems 20–70 cm tall; leaves compound with 9–18 pairs of leaflets, these lanceolate or elliptic to ovate; stipules small; heads 2- to 5-flowered, subtended by bracts with 5–9 leaflets, shorter than the flowers; calyx-teeth slightly shorter than equal to

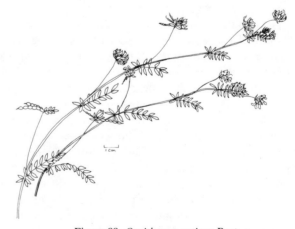

Figure 88. *Ornithopus sativus* Brot.

the tube; 2–2.5 mm broad, compressed, contracted between the segments; segments 3–7, elliptic-oblong, strongly reticulate. Fl. April to June (U.S.). Seeds 352–800/kg; wt. kg/hl = 46, unhulled.

GERMPLASM: Two subspecies are recognized: ssp. *sativus*, with legumes 12–25 mm long, straight, the beak usually not more than 5 mm long, straight, sometimes hooked at tip. Native to Azores, sw France and northern Iberian Peninsula, and cultivated elsewhere; subsp. *isthmocarpus* (Cosson) Dostal, with legumes 20–40 mm long, curved, usually with a long, cylindrical constriction between the segments, the beak 10 mm long or more, curved. Southwestern part of Iberian Peninsula. The two subspecies are linked by intermediate forms (var *macrorrhynchus* Willk.) in Portugal and western central Spain. Assigned to the Mediterranean Center of Diversity. ($2n = 14, 16$.)

DISTRIBUTION: Native from Azores to sw Europe, Morocco, and Algeria; also known in Germany, Poland, and Central and Western Russia. Cultivated in Kenya and introduced to the United States several times, but never established commercially.

ECOLOGY: A temperate plant, growing best on sandy loams of good to high fertility. In Kenya grown for seed with rainfall of 7–8 dm annually at altitude of 2,350 m or more. In United States made good growth in only a few places, perhaps due to unfavorable soil, climate or other factors. Ranging from Cool Temperate Moist to Wet through Warm Temperate Dry to Moist Forest Life Zones, serradella is reported to tolerate annual precipitation of 5.7–11.9 dm (mean of 11 cases = 7.0), annual mean temperature of 7.0°–18.3°C (mean of 11 cases = 11.1), and pH of 5.5–8.2 (mean of 11 cases = 6.6).

CULTIVATION: Propagated by seed. Under good storage conditions, seed remain viable for at least 5 years; few or no hard seed. Seeding rates broadcast are 40–60 kg/ha, in Europe and 17–22 kg/ha, in the United States. Seed requires inoculation of specific strain of rhizobium. For good seed yields, seed must be preinoculated when sown on new land after 4–5 years' growth of other species. Fall seeding matures in June; spring seeding, in August or September (U.S.). Phosphorus–potassium fertilizer is recommended; a little lime is beneficial in acid soils, but heavy liming is detrimental.

HARVESTING: For forage, serradella can be mowed and handled like any common legume hay. For seed, it is mowed when pods are brown, but before they are fully mature, and windrowed. If not handled with care, the jointed seed pods break and drop from the plant upon ripening. For threshing, an ordinary combine is used but does not hull the seed. For hulling, a clover huller can be used.

YIELDS AND ECONOMICS: In the United States good forage yields average 2–2.5 MT/ha (Massachusetts, southern Georgia, and northern Florida). In sandy soils of New Zealand, diploid and tetraploid serradella yielded 10.9 and 11.7 MT DM/ha, compared with 8.9 for *O. compressus* cv Pitman and 8.4 for subterranean clover. Seed yields average 350–1,200 kg/ha in Europe, but only about 100 kg/ha in Kenya. A forage crop of minor importance. Introduced but not commercially established in southern United States.

BIOTIC FACTORS: Serradella is largely self-pollinated. Fungi reported on serradella include the following: *Ascochyta ornithopi, Colletotrichum trifolii, Corticium solani, Fusarium avenaceum, F. equiseti, Helicobasidium purpureum, Olpidium trifolii, O. viciae, Peronospora ornithopi, P. viciae, Rhizoctonia solani, Rh. violacea, Thielavia basicola,* and *Uromyces ornithipodioides*. Viruses known to attack this crop include rosette disease and roseeten krankeit (Phaseolus virus 2). It is parasitized by *Orobanche minor*. Nematodes isolated from serradella include the following: *Ditylenchus dipsaci, Heterodera gottingiana, H. marioni, H. radicicola, Meloidogyne hapla, M. arenaria, M. incognita acrita, M. javanica, Pratylenchus penetrans, Trichodorus minor,* and *Tylenchus devastatrix*.

REFERENCES:
USDA, ARS, Forage and Range Section, Field Crops Research Branch, Serradella, 1954, 2 pp. mimeo.
Williams, W. M., DeLautour, G., and Stiefel, W., 1975, Potential of serradella as a winter annual forage legume on sandy coastal soil, *N. Z. J. Exp. Agric.* 3(4):339–342.

CONTRIBUTORS: J. A. Duke, C. F. Reed, J. K. P. Weder.

Pachyrhizus erosus (L.) Urban

FAMILY: Fabaceae

COMMON NAMES: Yam bean, Manioc Pea, Jicama de agua (Mex.), many local names in China, India, and Central America

SYNONYMS: *Cacara erosa* (L.) Kuntze
Dolichos articulatus Lam.
Dolichos bulbosus L.
Dolichos erosus L.
Pachyrrhizus angulatus Pich. ex DC.
Pachyrrhizus articulatus (Lam.) Duchass. ex Walp.
Pachyrrhizus bulbosus (L.) Kurz
Pachyrrhizus jicamas Blanco
Robynsia macrophylla Mart. et Gal.
Stizolobium bulbosum (L.) Sprengel
Stizolobium domingense Sprengel
Taeniocarpum articulatum (Lam.) Desv.

USES: The large starchy tubers of the yam bean are widely cultivated in Central America, China, and India, where they are eaten raw or cooked in soup. They are also sliced and made into chips. Tubers are highly nutritious and may be fed to cattle. Yam beans contain about 0.1% rotenone and other toxins and are toxic to insects and fish. Rotenone on dry weight basis: stems 0.03%; leaf 0.11%; pod 0.02%; seed 0.66%; tuber none. Young pods are eaten whole, but mature pods may be poisonous. The stems yield a fiber used for making fishing nets. Animals rarely graze the plant so it makes good green manure.

FOLK MEDICINE: Powdered seeds are used in Java for skin afflictions like prickly heat. One-half seed works as a laxative, but greater doses can be toxic. Seeds are also said to serve as a vermifuge.

CHEMISTRY: An unnamed alkaloid has been reported, and the coumarinlike compound pachyrhizin. Cyanogenic glycosides have been reported in the leaves, bark, and root. Saponins (pachysapogenin A & B), and isoflavanones (dehydroneotenone) have also been reported from the genus *Pachyrhizus*. Raw tubers per 100 g contain: 47 calories, 86.9% moisture, 1.4 g protein, 0.1 g fat, 11.0 g carbohydrate, 0.6 g fiber, 0.6 g ash. Peeled tubers from Bangladesh contained per 100 g: 82.4% moisture, 1.47 g protein, 0.09 g fat, 9.72 g starch, 2.17 g reducing sugars, 3.03 g nonreducing sugars, 0.64 g fiber, 0.50 g ash, plus 0.43 mg Cu, 1.13 mg iron, 16 mg calcium, There are ca. 0.05 mg thiamine, 0.02 mg riboflavin, 0.2 mg niacin, and 14 mg ascorbic acid per 100 g. Adenine, arginine, choline, and phytin also occur in the tubers. From tuber meal extracted with 70% ethanol, Evans *et al.* (1977) reported per 100 g recovered amino acid: 12.8–55.1 g asparagine, 2.7–5.0 threonine, 3.3–6.3 serine, 5.4–11.7 glutamic acid, 2.5–4.4 proline, 1.9–5.0 glycine, 2.5–5.6 alanine, 3.2–8.2 valine, 0.8–1.5 g methionine, 2.1–4.8 isoleucine, 2.9–7.9 leucine, 1.7–4.7 tyrosine, 2.3–5.1 phenylalanine, 2.5–3.2 histidine, 3.3–7.8 lysine, 5.0–7.4 arginine, and 0.4–1.0 g cystine. Young pods contain 86.4% moisture, 2.6 g protein, 0.3 g fat, 10.0 g carbohydrates, 2.9 g fiber, and 0.7 g ash, with 121 mg Ca, 39 mg P, 1.3 mg Fe, and 575 IU vit. A, 0.11 mg thiamine, 0.09 mg riboflavin, and 0.8 mg niacin. Dried seeds contain 6.7% moisture, 26.2 g protein, 27.3 g fatty oil, 20.0 g carbohydrates, 7.0 g fiber, 3.6 g ash, with a high content of phosphorus and calcium. Of

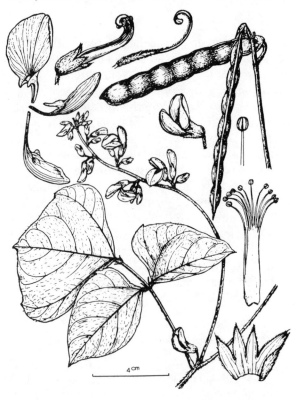

Figure 89. *Pachyrhizus erosus* (L.) Urban

the fatty acids 27.6% are saturated, 62.4% unsaturated.

DESCRIPTION: Herbaceous perennial vine, trailing or climbing, with tuberous root, the tubers solitary or several, simple or lobed, variously shaped, light-brown outside, white inside, up to 30 cm in diameter; stems usually herbaceous, sometimes woody below, spirally striated, somewhat angulate, finely or coarsely strigose, 2–6 m long; leaves alternate, trifoliolate, with linear-lanceolate stipules, 5–11 mm long and strigose; petioles 2.5–19 cm long; terminal leaflet rhomboidal or ovate-reniform, cuneate basally, acuminate apically, obscurely dentate or lobed, 3.8–18.5 cm long, 4–20 cm wide, its stalk 1.3–6 cm long; lateral leaflets ovate or rhomboidal, entire, dentate or palmately lobed, 2.7–15.5 cm long, 2.3–16 cm wide; flowers in racemes on erect axillary peduncles, the racemes 4–71 cm long, the peduncles 3.8–43.5 cm long; 1–5 flowers in a cluster, deep violet or pale violet to white, except the medial basal spot which is green, with blade 12–13 mm long and claw 5–8 mm long; fruits (pods) 7.5–14 cm long, 1.1–1.8 cm wide, finely strigose, nearly smooth at maturity, constricted with the segments 0.9–1.2 cm long; seeds square or rounded, mostly flattened, 5–11 mm long, 5–11 mm wide, yellow, red or brown. Seeds ripen in fall and winter (in warmer climates).

GERMPLASM: Several "varieties" are based on leaf-cut (unlobed to prominently lobed, entire to coarsely dentate), violet or white flowers, seeds light tan or red, seeds square or rounded, size, luster, and plumpness of seed, and large or small seeds; size and shape of tuber, varying from turnip-shaped to long slender; tubers with a sweet, watery taste to a milky taste. Sinha *et al.* (1977) report on the phenotypic variability in 34 cvs. Assigned to the Middle American Center of Diversity, yam bean or cvs thereof has been reported to tolerate grazing, insects, laterite, and virus. ($2n = 22$.)

DISTRIBUTION: Native to Mexico and northern Central America south to Guatemala and western Nicaragua. Introduced and widely cultivated in the tropics and subtropics of the world; recorded from Costa Rica, West Indies, Hawaiian Islands, Guam, Philippine Islands, Java, Sumatra, s China, Indochina, India, upper Guinea, Nile River Valley, Cameroons, and Pretoria; naturalized in China, Thailand, Indochina, Reunion, and southern Florida.

ECOLOGY: Yam bean grows best in rich light sandy loam, with good drainage. Soil should be liberally manured about 1 month before planting the seeds. For flowering and fruiting, the plants require a short day. They do not flower and fruit outdoors in the long days of the North Temperate area. Best tubers for eating are from plants from which the flowers are removed as they form, a process that hastens the development of the tuber, making it larger and sweeter (Mexico). Farther south in Guatemala and El Salvador, where the days are even shorter, the plants produce sizable tubers before they begin to flower, hence flower buds are not removed. In Mexico yam bean grows up to 1,400-m elevations, but wild specimens have been collected as low as 10-m elevation. This yam bean ranges ecologically from Subtropical Dry to Moist through Tropical Moist to Wet Forest Life Zones; reported to tolerate annual precipitation of 6.4–40.3 dm (mean of 7 cases = 21), annual temperature of 21.3°–27.4°C (mean of 7 cases = 24.2), and pH of 5.0–6.8 (mean of 5 cases = 6.0).

CULTIVATION: Yam bean is easily propagated from seed, sown at rate of 45–50 kg/ha, spaced at intervals of 30–40 cm in rows 60–75 cm apart, sufficient for cultivation in the early stages. For good development of the tubers several prunings are needed and the flowering buds removed as they appear to check the vegetative growth. Except for a good manuring of the soil before planting the seed, the crop requires little attention, except pruning. In Mexico jicama beans are intercropped with cabbage and onions and produce a root crop and a seed crop in a single season.

HARVESTING: The crop matures 6–8 months after planting. Tubers are white, and vary in size and shape from rounded or lobed to elongated. They are dug when growth has ceased and plant has withered. In dry areas they are left in the soil and dug when needed (not to exceed 3 months). They are marketed like potatoes, or used locally. Pods, when used as a vegetable, are picked through the season. Mature pods and seeds for extraction of rotenone are separated from the tubers when the whole plant is harvested. The seeds are threshed

Pachyrhizus erosus (L.) Urban

out of the pods and sent to a center for extraction of rotenone.

YIELDS AND ECONOMICS: Market-size tubers weigh from 0.2–1 kg each, but fully developed tubers measure ca. 40 cm in diameter and may weigh ca. 5–15 kg. Yields average 7,000–10,000 kg/ha, with up to 78 MT reported in some parts of Indonesia and Philippines. In the United States root yields of 5,600 kg/ha (valued at $100) and ca. 600 kg/ha of seed were produced experimentally.

BIOTIC FACTORS: The following fungi attack yam bean: *Cercospora pachyrrhizi, Colletotrichum pachyrrhizi, C. pachyrhizicola, Corticium solani, Phakopsora pachyrrhizi,* and *Pythium aphanidermatum.* Also, the nematode *Heterodera marioni* has been isolated from this crop. In the Philippines a systemic virus is reported and may be transmitted by the mealy bug, *Ferrisia virgata,* a nuisance to yam beans grown in greenhouses.

REFERENCES:
Clausen, R. T., 1944, Yam bean, warm-climate plant is a possible new insecticide, *N.Y. State Agric. Exp. Stn., Farm Res.* **July**:14.
Clausen, R. T., 1944, A botanical study of the yam bean (*Pachyrrhizus*), *Cornell Univ. Agric. Exp. Stn., Mem.* **264**:3–38, illus.
Huart, A., 1902, La jicama, su classficacion, su cultivo, sus usos, *Soc. Agric. Mex. Bot.* **26**:555–558.
Norton, L. B., 1943, Rotenone in the yam bean (*Pachyrrhizus erosus*), *J. Am. Chem. Soc.* **65**(11):2259–2260.
Sinha, R. P., Prakash, R., and Hauque, M. F., 1977, Genetic variability in yam bean, *Trop. Grain Leg. Bull.* **7**:21–23.

CONTRIBUTORS: J. A. Duke, C. F. Reed, and R. P. Sinha.

Pachyrhizus tuberosus (Lam.) Spreng.

FAMILY: Fabaceae
COMMON NAMES: Yam bean, Jicama
SYNONYMS: *Cacara tuberosa* (Lam.) Britt
Dolichos tuberosus Lam.
Stizolobium tuberosum (Lam.) Spreng.

USES: Large tuberous root provides a pure white starch used in custards and puddings. Young pods have irritating hairs and usually are not eaten. Plants contain rotenone.

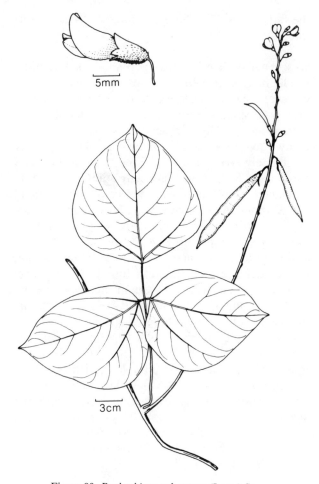

Figure 90. *Pachyrhizus tuberosus* (Lam.) Spreng.

FOLK MEDICINE: Said to be insecticidal and piscicidal.

CHEMISTRY: The genus *Pachyrhizus* has been reported to contain adenine, choline, rotenone, and saponin (for relative toxicities see Duke, 1977b).

DESCRIPTION: Herbaceous perennial vine, 2–6 m long, from a large tuberous root several dm long and up to 30 cm in diameter; leaves compound, leaflets rhomboid, entire but slightly angled, the lateral ones strongly oblique; flowers in clusters of 2 or 3, usually crowded along the upper half of the very long racemelike inflorescence; calyx 1 cm long, appressed rusty pilose; petals white or yellow, 1.5–2 cm long, the basal auricles of standard directed upwards; pods 13–30 cm long, ca. 2 cm broad, with reddish irritating hairs, with triangular beak to 3 mm long; seeds large, reniform, 11–14 mm long, usually red. Fl. Nov.–May; Fr. Apr.–Sept.

GERMPLASM: Two main variations: flowers violet or white, rarely yellow; seeds, red to black, or black and white. Assigned to the South America Center of Diversity, tuberous yam bean or cvs thereof has been reported to tolerate insects and low pH. ($2n = 22$.)

DISTRIBUTION: Native and often cultivated in the West Indies and western South America. Introduced and cultivated in many parts of South America, other West Indies, and China, this is the yam bean commonly cultivated many places in South America.

ECOLOGY: Yam bean grows best in rich, light, sandy soils with good drainage; often grown in clearings in forests in tropical regions. Soil should be liberally manured about a month before planting time. For flowering and fruiting, plants require short days. Best tubers for eating are obtained from nonflowering plants. Debudding hastens the development of the tuber, making it larger and sweeter. Tuberous yam bean is reported from Subtropical Dry to Moist through Tropical Very Dry to Wet Forest Life Zones, where annual precipitation ranges from 6.4–41.0 dm (mean of 5 cases = 18.9), annual temperatures from 21.3°–27.4°C (mean of 5 cases = 25.4), and pH from 4.3–6.8 (mean of 4 cases = 5.2).

CULTIVATION: Yam bean is easily propagated by seed. Seed sown at rate of 45–50 kg/ha, spaced 30–40 cm apart, in rows 60–75 cm apart. For good development of tubers, plants are pruned several times and flower buds are removed as they appear. Except for the manuring before planting time, the crop requires little attention.

HARVESTING: Tubers mature in 6–8 months; dug when growth ceases and plants wither. In dry areas they may be left in the soil and dug when needed. They are marketed like potatoes, or used locally. Pods seldom used as a vegetable because of irritating hairs. Rotenone is extracted from mature pods and seeds. They are separated from the roots when the whole plant is harvested. Seeds are threshed from the pods and sent to a center for extraction of rotenone.

YIELDS AND ECONOMICS: Market-size tubers weigh ca. 1 kg and may measure 30 cm in diameter. Yields average 7,000–10,000 kg/ha. An estimated yield of 600 kg/ha of seed was produced experimentally. Yam bean is grown and consumed locally, mainly in the South American and West Indian tropics. No data are available concerning its economic importance.

BIOTIC FACTORS: This yam bean is attacked by the following fungi: *Colletotrichum pachyrrhizi, Phakopsora pachyrrhizi,* and *Pythium aphanidermatum.* It is also infested by the rootknot nematode *Meloidogyne arenaria.*

REFERENCE:
Schroeder, C. A., 1967, The jicama, a rootcrop from Mexico, *Am. Soc. Hortic. Sci.* **11**:65–71.

CONTRIBUTORS: J. A. Duke, C. F. Reed.

Pentaclethra macrophylla Benth.

FAMILY: Mimosaceae
COMMON NAMES: Owala oil tree, Oil bean tree, Atta bean

USES: Owala oil tree is the source of Owala butter or oil. Beans yield 30–36% of a semisolid pale yellow oil; kernels yield 40–45%. The oil has an unpleasant aroma and burning taste, but is used for cooking, lubrication and manufacture of soaps and candles. Kernels, which are rich in proteins but poor in starch, are edible after roasting or boiling for 12 hours; then they are used as a condiment for seasoning sauces rather than as a food. Seeds can be ground and used to make bread. In Nigeria, seeds are boiled, husks removed, cotyledons shelled, salted, and allowed to ferment for 3 days and then eaten. Seed cake is used as a manure. Seeds are strung as toys. Pod-ashes are used as cooking salt and in the dye industry; insoluble part is used in snuff. Trees are often planted along roadsides in Nigeria. The hard wood is used for turnery and general carpentry.

Pentaclethra macrophylla Benth.

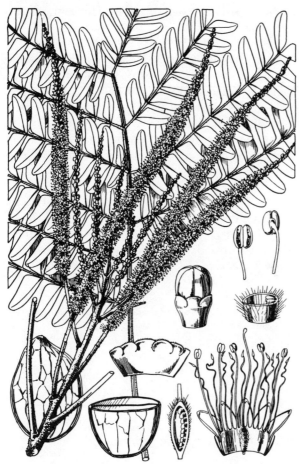

Figure 91. *Pentaclethra macrophylla* Benth.

FOLK MEDICINE: Bark is used as an enema for dysentery and a liniment for itch; it is anthelmintic and is used by women as a lactogenic food and for sores. Powdered bark is applied in leprosy. Said to be used for: astringent, ecbolic, and piscicide.

CHEMISTRY: Seeds contain a poisonous yellow flaky alkaloid, pancine. Used in Ivory Coast to poison or stupefy fish and in arrow poisons. B. N. Okigbo (personal communication, 1976) sent data indicating that the seeds contain 20.8% protein (5 g lysine/100 g; 1.2 g tryptophane/100 g), 45.9% oil, and 19.0% starch. Dried, shelled seeds contain (per 100 g edible portion): 558 calories, 6.2% moisture, 22.6 g protein, 46.3 g fat, 22.6 g total carbohydrate, 2.5 g fiber, 2.3 g ash, 190 mg Ca, 172 mg P, 16.0 mg Fe, 0.07 mg thiamin, 0.32 mg riboflavin, 0.9 mg niacin; fermented seeds contain: 362 calories, 46.8% moisture, 11.7 g protein, 36.1 g fat, 4.8 g total carbohydrate, 2.0 g fiber, 0.6 g ash, 110 mg Ca, 3.3 mg Fe, 0.07 mg thiamin, 0.30 mg riboflavin, and 0.3 mg niacin. Kar and Okechukwu (1978) reported high ash content (13.2% in seed with 9.3% moisture) indicating high mineral content. They reported only 38% of a pale yellow fixed oil with ca. 75% saturated fatty acids, 25% unsaturated. Defatted seeds contained 46.8% protein.

DESCRIPTION: Tree to 40 m tall and 6.6 m in girth, tending to branch low down, buttresses low and blunt or absent, the crown very dense, bole smooth, finely cross-furrowed; twigs rusty-tomentose; leaves dark green, reddish when young, bipinnate with 10–12 pinnae in opposite pairs; pinnae with small leaflets in 12 opposite pairs, unequal, obliquely oblong, base unequal, truncate, 2.5 cm long, 1.2 cm wide; flowers numerous, yellow, fragrant, in long paniculate spikes, sepals and petals short, stamens 5, alternating with staminodes; pods woody, 45 cm long, 5–7.5 cm broad, ridged around edge, surface slightly veined, strongly elastic, bursting violently, and curling up after dehiscence; seeds red, flattened, up to 7.5 cm long, ovoid, striate, up to 8 per pod. Fl. Feb.–Apr. in Ghana.

GERMPLASM: Assigned to the Africa Center of Diversity, Owala oil or cvs thereof is reported to tolerate low pH. ($2n = 26$.)

DISTRIBUTION: Native to tropical Africa from Senegal, Liberia, Sierra Leone, and Togo to Cameroons, Congo, Ghana, and Angola.

ECOLOGY: A tropical tree of closed forests and river banks. Commonly distributed and usually left as a standard tree when forests are cleared for farming. Often cultivated in Ghana. Occurs in rainforests with 17.7–21.5 dm annual rainfall and on soils with pH 4–5. Ranging from Subtropical Moist through Tropical Dry to Wet Forest Life Zone, Owala oil is reported to tolerate annual precipitation of 13.6–25.8 dm (mean of 4 cases = 20.2 dm), annual mean temperature of 21.3°–26.6°C (mean of 4 cases = 24.4°C), and pH of 5.0–6.5 (mean of 3 cases = 5.6).

CULTIVATION: Easily propagated from seed, sown wild or in containers for transplant. Young trees planted along roadsides, about dwellings or in forest. Trees kept branching from near base or as evergreen shade trees.

HARVESTING: As an oilseed-bearing tree, it produces from the 10th year and bears regularly thereafter. Wood is exported from Congo under name of Congo acacia or Gabon yellow-wood; it is durable and used for heavy construction, railroad construction, flooring and house-building, as well as for fuel and charcoal.

YIELDS AND ECONOMICS: The oil is an important product in West Tropical Africa, where it is mainly consumed. Wood is exported from Congo. Other parts of plants are used extensively where tree grows. Not an item of international markets.

BIOTIC FACTORS: No data available.

REFERENCES:
Ahn, P., 1961, Soil vegetation relationships in the western forest areas of Ghana, *Trop. Soils Veg. Proc. Abidjan Symp.*, p. 78, Oct. 20–24, 1959.
Kar, A., and Okechukwu, A. D., 1978, Chemical investigations of the edible seeds of *Pentaclethra macrophylla* (Benth.), *Qual. Plant.—Plant Foods Hum. Nutr.* **28**(1):29–36.

CONTRIBUTORS: J. A. Duke, B. N. Okigbo, C. F. Reed.

Figure 92. *Phaseolus acutifolius* A. Gray

Phaseolus acutifolius A. Gray

FAMILY: Fabaceae
COMMON NAME: Tepary bean

USES: Tepary beans are used as dry shelled beans. Mexicans soak the beans overnight to produce a gelatinous soup base. In Uganda, the beans are boiled and ground, then added to soups. The plant has been considered as hay and cover crop in the United States. Has possibilities as a catch-crop where a rapid food supply is needed in areas of poor rainfall.

FOLK MEDICINE: No data available.

CHEMISTRY: Dried beans contain 9.5% moisture, 24.5% protein, 1.5% fat, 65.5% carbohydrate, 3.7% crude fiber, 4.6% ash, 112 mg Ca, 310 mg P, 0.33 mg thiamin, 0.12 mg riboflavin, and 2.8 mg niacin. Purseglove reports 9.5% water, 22.2% protein, 1.4% fat, 59.3% carbohydrate, 3.4% fiber, and 4.2% ash. Nabham and Felger (1978) report 280–310 mg isoleucine/g N, 480–530 mg leucine, 410–420 mg lysine, 80–170 mg methionine plus cystine, 520–530 mg phenylalanine plus tyrosine, 250 mg threonine, and 360–380 mg valine. The hay is said to contain: 6.6% moisture, 9.9% protein, 1.9% fat, 43.1% N-free extract, 29.3% fiber, 9.2% ash. After seed harvest, haulms and pods are used for forage. The pods contain: 8% moisture, 4.1% protein, 0.5% fat, 43.6% N-free extracts, 37.0% fiber, and 6.8% ash. Miller (1958) reports that immature hay (89.2% DM) contains (moisture-free): 17.0–21.0% crude protein (avg. 18.5), 2.8–3.5% fat (avg. 3.2), 25.7–29.9% crude fiber (avg. 28.1), 9.8–12.3% ash (avg. 11.7), and 37.6–41.2% N-free extract. The genus *Phaseolus* is reported to contain acetone, choline, fumaric acid, hydrocyanic acid, inositol, malonic acid, maltose, methyl-cysteine, oxalic acid, pectin, rutin, saponin, and succinic acid. Seeds of the genus *Phaseolus* are reported to contain trypsin inhibitors and chymotrypsin inhibitors. Alkaloids and cyanogenic compounds are reported as absent in the tepary bean.

Phaseolus acutifolius A. Gray

DESCRIPTION: Ephemeral annual herb, suberect to 30 cm high and bushy on poor land, otherwise recumbent, spreading or twining, to 2 m long; first leaves simple, about 5.5 cm long, 3.5 cm wide, truncate base, on short petioles 4.5 cm long, with lanceolate stipules, appressed to the stem; leaflets ovate, later leaves trifoliolate; 4–8 cm long, 2–5 cm wide, entire, acute; inflorescences axillary with 2–5 flowers; flowers pea-like, white, pink, or pale lilac, calyx 4-lobed, 3–4 mm long, standard partly reflexed, broad, emarginate, 8–10 mm long, wings 10–15 mm long, keel narrow with 2–3 turns to the spiral; pods compressed, 5–9 cm long, 0.8–1.3 cm wide, 2–7-seeded, rimmed on margins, sharp beaked, hairy when young; seeds roundish to oblong, about 8 mm long, 6 mm wide, not glossy, seed coat white, yellow, brown, or deep violet, full-colored or variously flecked. Germination phanerocotylar.

GERMPLASM: *Phaseolus acutifolius* var *lactifolius* Freem. is the cultivated variety. Many strains are known in southwestern United States and Mexico. Freeman isolated 47 cvs in Arizona in 1918. Some varietal names and colloquial names are treated by Nabham and Felger (1978). Assigned to the Middle America Center of Diversity, tepary bean is reported to exhibit tolerance to disease, drought, heat, sand, slope, and virus. ($2n = 22$.)

DISTRIBUTION: Native to southwestern United States and Mexico, where wild form is still extant, it was domesticated by Indian races over 5,000 years ago. Introduced and now cultivated in West Africa, East Africa, and Madagascar. Thrives as an upland crop in sub-Sahara areas of the Sudan savanna; also introduced to Asia and Australia.

ECOLOGY: Tepary beans are day-neutral or short-day plants, suited to arid regions; they withstand heat and a dry atmosphere, are quite drought-resistant, and some cvs flower readily at high latitudes (e.g., U.K.). They are very susceptible to waterlogging and frost and are not suited to the wet tropics. Cannot be grown where night temperatures fall below 8°C. In humid regions where dormancy is absent, seeds may germinate immediately. Diseases cause more serious problems under humid conditions. This bean thrives on very light sandy soils, on flat or shallow ridges, and can be grown on residual soil moisture. Produces crop under conditions where other beans fail completely. Grows where annual rainfall is only 5–6 dm, makes excessive vegetative growth with rainfall at 10 dm. Ranging from Warm Temperate Dry to Moist through Tropical Very Dry Forest Life Zone, tepary bean is reported to me from areas with annual precipitation of 6.4–17.3 dm (mean of 9 cases = 10.6 dm), annual mean temperature of 17.0°–26.2°C (mean of 9 cases = 22.2°C), and pH of 5.0–7.1 (mean of 7 cases = 6.3).

CULTIVATION: Seeds absorb water rapidly; in moist soil testa wrinkle within 5 min; in water, in 3 min. Seeds may be broadcast at rate of 28–33 kg/ha; or drilled 2.5–10 cm deep, in rows 90 cm apart, at rate of 11–17 kg/ha; or grown in hills, 45 cm apart with 2–4 seeds per hill. When grown for hay, usually seeded at more than 60 kg/ha. In Kenya, crop is sown after rains start, 15–20 kg/ha in rows 60 cm apart, 30 cm apart in rows; in Arizona, seed planted in May or June, 2–5 seeds per hill, in rows 45 cm apart. Crop may be dry-farmed or irrigated, but water should be applied to more productive pulses. Often grown as a catch-crop and requires little weeding if grown as end-of-season crop. Tepary bean responds favorably to applications of N and 75 kg/ha of potash as K_2O. Algerian studies suggest that fertilization at planting time may be deleterious. In Arizona and Mexico "mixed farms," tepary has been intercropped with *Allium, Brassica, Capsicum, Cicer, Cucurbita, Phaseolus, Pisum, Proboscidea, Sorghum,* and *Zea.*

HARVESTING: Tepary bean is grown mainly as a dry-shell bean. Whole plants are pulled up as soon as the first pods start to ripen. Harvesting should not be delayed where wet periods are likely to follow. Usually harvested when mature and drying, and are promptly threshed. Tepary beans produce a crop in only 2 months (60–120 days).

YIELDS AND ECONOMICS: Under dry farming conditions yields average 500–800 kg/ha dried beans; under irrigation, 900–1,700 kg/ha. Domesticated white teparies can produce 2,000 kg/ha with irrigation in the Sonoran Desert, 4,500 in more suitable areas. Yields of 4,000–5,000 kg/ha are reported in Algeria. Oven-dry hay yields are ca. 5.5–10.5 MT/ha. Tepary beans are grown year-round in east and west Tropical Africa as a food crop. Spraying with nicotine controlled *Aphis fabae* in Algeria.

BIOTIC FACTORS: Tepary bean is presumably self-pollinated. Following fungi cause diseases in tepary beans: *Erysiphe polygoni, Fusarium solani, Macrophomina phaseoli, Nematospora coryli, Rhizoctonia* sp., *Sclerotium rolfsii, Uromyces appendiculatus, U. phaseoli,* and *Vermicularia capsici.* Several viruses cause diseases also: alfalfa mosaic (3 strains), alsike clover mosaic, bean chlorotic ringspot (4 strains), bean local chlorosis (6 strains), bean mosaic, bean necrosis, cucumber mosaic, white clover mosaic, yellow mosaic of mung, and curly top. Tests show that tepary (Nebr. Acc. No. 10) is tolerant to three of four strains of *Xanthomonas phaseoli,* whereas P.I. 207262 from Colombian *Phaseolus vulgaris* was tolerant to all four strains. Plants are practically disease-free under savanna conditions. Nematodes isolated from tepary beans include: *Heterodera glycines, H. marioni,* and *Tylenchorhynchus dubius.*

REFERENCES:
Freeman, G. F., 1918 (Revised), Southwestern beans and teparies, *Univ. Ariz., Agric. Exp. Stn., Bull.* **68**:1–55.
Nabham, G. P., and Felger, R. S., 1978, Teparies in southwestern North America, *Econ. Bot.* **32**(1):2–19.
Schuster, M. L., Coyne, D. P., and Hoff, B., 1973, Comparative virulence of *Xanthomonas phaseoli* strains from Uganda, Colombia, and Nebraska, *Plant Dis. Rep.* **57**(1):74–75.

CONTRIBUTORS: J. A. Duke, C. F. Reed, J. Smartt, J. K. P. Weder.

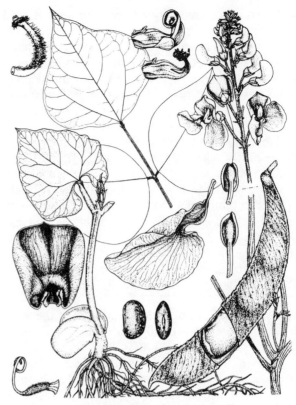

Figure 93. *Phaseolus coccineus* L.

Phaseolus coccineus L.

FAMILY: Fabaceae
COMMON NAMES: Scarlet runner bean, Multiflora bean
SYNONYM: *Phaseolus multiflorus* Willd.

USES: In Europe and other temperate areas only the tender green pods are eaten—sliced and cooked as a vegetable. In Central America, green and dry seeds are used and the fleshy tubers are boiled and eaten. In Ethiopia, the seeds are eaten. In North America plant is grown mainly as an ornamental climber, with scarlet flowers and dark green foliage. In South Africa seed of dwarf white-seeded forms are produced as a pulse under the name "butter bean" (cf. *Ph. lunatus*). The large white "elephant" or "soissons," largely from Argentina, attract large prices in certain European countries, especially Switzerland.

FOLK MEDICINE: No data available.

CHEMISTRY: Seeds contain trypsin inhibitors and chymotrypsin inhibitors. Dried seeds contain (per 100 g edible portion): 338 calories, 12.5% moisture, 20.3 g protein, 1.8 g fat, 62.0 g total carbohydrate, 4.8 g fiber, 3.4 g ash, 114 mg Ca, 354 g P, 9.0 mg Fe, 0.50 mg thiamine, 0.19 mg riboflavin, 2.3 mg niacin, and 2 mg ascorbic acid. Fruits contain (per 100 g edible portion): 250 calories, 34.2% moisture, 16.4 g protein, 0.3 g fat, 46.3 g total carbohydrate, 12.2 g fiber, 2.8 g ash, 61 mg Ca, 277 mg P, 4.1 mg Fe, 15 µg vitamin A, 0.54 mg thiamine, 0.14 mg riboflavin, and 2.3 mg niacin. At 58.3% moisture, 7.4 g protein, 1.0 g fat, 29.8 g carbohydrate, 1.9 g fiber, 1.6 g ash, 50 mg Ca, 160 mg P, 2.6 mg Fe, 0.34 mg thiamine, 0.19 mg riboflavin, 27 mg vitamin C, and 57 IU vitamin A per 100 g are reported.

DESCRIPTION: Perennial in tropics with thickened tuberous roots, annual in temperate regions; per-

Phaseolus coccineus L.

ennation is common in western Europe in mild winters, and tubers are produced. Difference in life form is related to growth habit; dwarf (determinant) forms are invariably annuals while climbing (indeterminant) forms may be perennials. Stems twining, 4 m or more long; leaves trifoliolate, leaflets ovate, 7.5–12.5 cm long; stipules triangular, stipels to 5 mm long; flowers several to many on long axillary peduncles; petals bright scarlet, occasionally white, 4 free petals (a large upper standard, 2 lateral wings and a boat-shaped keel); pods varying from 10–60 cm long, variable in color, from shades of pink, spotted or blotched with black, or sometimes nearly entirely black; seeds 6–10, broad-oblong, convex, flattened, 1.8–2.5 cm long, 1.2–1.6 cm wide, dark purple or pink with red markings, sometimes white, cream, brown, or black; germination cryptocotylar. Roots are thickened dahlialike, but smaller.

GERMPLASM: Several varieties and cvs of the scarlet runner are known, some are climbing, others dwarf. Bailey's Manual lists four forms of *Ph. coccineus*: white dwarf (*albonanus*), scarlet dwarf (*rubronanus*), white climber (*albus*) in addition to the typical scarlet runner type. Some cvs have white flowers with white seeds ('Czar'), others have red and white flowers—red standard, white wings ('Painted Lady'); some are dwarf plants ('Hammond's Dwarf Scarlet'); 'Princeps' is medium height, early and heavy fruiting. Other cvs are: 'Goliath,' 'Kelvedon marvel,' 'Streamline,' and 'Prizewinner.' Of interest to breeders because of its resistance to many of the root-rot organisms affecting other *Phaseolus* spp. Assigned to the Middle America Center of Diversity, scarlet runner bean or cultivars thereof is reported to exhibit tolerance to anthracnose, sand, slope, and virus. ($2n = 22$.)

DISTRIBUTION: Probably native to Mexico, or Central America, or both, as beans found in Mexico are dated 7000–5000 B.C. Perennial forms have been domesticated and are still cultivated in Central America. Now cultivated in many tropical areas, as Africa, Mauritius, and South America, and, as an annual crop, in many temperate areas, especially the United Kingdom.

ECOLOGY: In the tropics scarlet runner is grown in humid uplands (ca. 1,500–2,000 m), but does not set fruit in lowlands. Killed by frost, it is grown as an annual in temperate areas and requires a frost-free period of 120–130 days. Grows well in Britain because it is less sensitive than most species of *Phaseolus* to cool summers. Most cvs are day neutral, but some types are short-day. Plant tends to drop flowers under dry conditions and is sensitive to wind. Requires deep rich soil. Ranging from Cool Temperate Steppe to Wet through Tropical Dry Forest Life Zone, scarlet runner bean is reported to tolerate annual precipitation of 3.1–17.3 dm (mean of 31 cases = 9.1 dm), annual mean temperature of 5.6°–26.6°C (mean of 31 cases = 13.2°C), and pH of 4.9–8.2 (mean of 27 cases = 6.5), although faring best at pH 6–7.

CULTIVATION: Scarlet runner beans require a well-prepared clean seedbed. Seed sown 2.5–4 cm deep on heavy soils, deeper on light soils. When grown for seed, it should be sown pure; in tropical areas, best sown late in season to avoid rain during harvest. Successive sowings are made from mid-April to early June to produce continuous crops from July to early October. In Great Britain, seed is sown in April and May, in rows 90 cm apart, at rate of 60–125 kg/ha; in Mauritius, in October, varying with the variety, in rows 100 cm apart and 30 cm apart in the row. Occasionally interplanted with supporting crop, as corn, at 0.9×0.9 m spacing. A vigorous climber, grows 3–4 m; usually grown on poles 2–3 m long. Tips of leading shoots should be pinched out when they reach the top of the support to encourage lateral growth. For good nodulation seeds should be inoculated when introduced. Cultivation should be shallow as roots are near surface. Cultivating then wetting spreads diseases. Tuberous roots often sprout for 1–2 yr after fields have gone fallow. In tropics, planted in garden plots near houses and treated as perennial. Scarlet runner requires rich soil and some manure or compost. Excessive nitrogen should be avoided, and if necessary be balanced by addition of proper amounts of potash or phosphates. In Mauritius, a fertilizer of 400 kg/ha phosphate guano is applied before planting, and 1 kg applied per 15-m drill.

HARVESTING: Scarlet runner seeds mature in 115–120 days. Pods may be picked whenever mature enough to eat (80–90 days). Crops for seed are harvested when pods are mature and yellow, but not open. Since the crop is usually grown on trellises or on a supporting crop, pods are hand-picked.

YIELDS AND ECONOMICS: Yields of beans average 1,148–1,687 kg/ha and of green pods 9–37 MT/ha. Outside Central America, scarlet runner is seldom found in the tropics, but is more widely grown as an annual in temperate regions. In all areas it is grown for local consumption; rarely enters international or even local markets. Argentina led in production in 1973 and harvested more than 40,000 MT, which was comparable to British production in the 1960s.

BIOTIC FACTORS: Flowers are self-pollinating to some degree, but yields of seed improve when bee colonies are provided, with about 29% cross-pollination (flowers require insect visitation, even for self-pollination). Following fungi cause diseases in scarlet runner bean: *Aphanomyces euteiches, Ascochyta boltshauseri, A. phaseolorum, Botrytis cinerea, Cercospora cruenta, Colletotrichum lindemuthianum, Corticium solani, Erysiphe polygoni, Fusarium acuminatum, F. culmorum, F. solani, F. oxysporum, Macrosporium phaseoli, Melanospora papillata, Pleospora herbarum, Sclerotinia sclerotiorum, Sclerotium rolfsii, Phyllochora phaseoli, Uromyces appendiculatus,* and *Uromyces phaseoli.* Bacteria known to cause diseases in this plant are: *Pseudomonas phaseolicola* (sprayed with copper oxychloride), *Xanthomonas phaseolus* and its var *indicus*. Viruses isolated from scarlet runner include: *Phaseolus* Virus I, bean common mosaic and bean yellow mosaic. Nematodes isolated from scarlet runner include: *Heterodera glycines, H. schachtii, Meloidogyne arenaria, M. hapla, M. incognita acrita, M. javanica,* and *Pratylenchus penetrans.* For weed control in the United Kingdom, diphenamid, trifluralin, and bentazon are recommended, also preemergence paraquat-diquat, or oil immulsions of dinosebamine or dinoseb 4–7 days prior to emergence.

REFERENCES:
Kaplan, L., 1965, Archeology and domestication in American *Phaseolus* (beans), *Econ. Bot.* **19**:353–368.
Masefield, G. B., et al., 1969 (repr. 1971), *The Oxford Book of Food Plants*, p. 36, Oxford University Press, London.

CONTRIBUTORS: J. A. Duke, C. F. Reed, J. Smartt, J. K. P. Weder.

Phaseolus lunatus L.

FAMILY: Fabaceae

COMMON NAMES: Lima bean, Sieva bean, Butter bean, Madagascar bean, Sugar bean, Towe bean

SYNONYM: *Phaseolus limensis* Macf.

USES: Lima bean is grown for green or dried shelled beans that are eaten cooked and seasoned or mixed with other vegetables or foods. Lima beans are marketed green or dry, canned or frozen. Green immature pods of some cvs may be eaten cooked as a green vegetable. Sprouts are also eaten. Deeply colored red or black testas have in the past been associated with high levels of cyanogenic glucosides in the seed, but cyanogenic glucosides have been reduced to safe levels by selection in the United States and elsewhere. The supposed absence of HCN in white cvs may represent wishful thinking. There seems to be no reliable correlation between seed color and cyanide content.

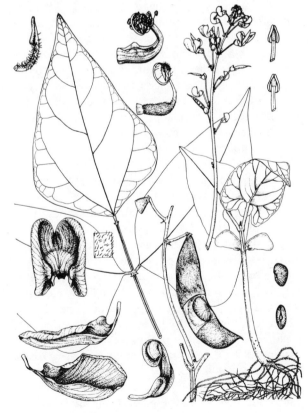

Figure 94. *Phaseolus lunatus* L.

Phaseolus lunatus L.

FOLK MEDICINE: Seeds are astringent and are used as diet food in fevers. A decoction of green pods, seeds, and stems reportedly has been used for Bright's disease, diabetes, dropsy, and eclampsia. In Java, seeds are poulticed into the abdomen for stomachache.

CHEMISTRY: The root, considered poisonous by some, has been reported to cause colic, giddiness, nausea, prostration, purgation, and rapid pulse. Addition of raw limas to the diets of rats is said to diminish the rate of growth and the apparent digestibility of protein and fat. The edible portion of green lima beans analyzed at: moisture, 69.2; N, 1.3; ether extr., 0.3; crude fiber, 0.5; and ash, 1.5%; Ca, 9; P, 97; Fe, 1.3; carotene 0.06; thiamine, 0.03; riboflavin, 0.09; nicotinic acid, 1.6; and ascorbic acid, 30.8 mg/100 g. Raw leaves contain (per 100 g edible portion): 8 calories, 97.2% moisture, 0.6 g protein, 1.7 g total carbohydrate, 0.5 g ash, 8 mg Ca, 36 mg P, 2.3 mg Fe. Cooked leaves contain per 100 g: 112 calories, 70.5% moisture, 7.0 g protein, 0.2 g fat, 21.2 g total carbohydrate, 1.1 g fiber, 1.1 g ash, 36 mg Ca, 98 mg P, 0.7 mg Fe, 0.14 mg thiamine, 0.04 mg riboflavin, 0.9 mg niacin, and 18 mg ascorbic acid. Dried seeds contain: 13.2% water, 20.0% protein, 1.5% fat, 58.0% carbohydrate, 3.7% crude fiber, and 3.4% ash. Seeds are reported to contain trypsin inhibitors and chymotrypsin inhibitors. The seeds may contain: 0.62% lecithin, 0.09% cephalin, a papainlike protease, carotene oxidase, gun, and tannin. Maturing seeds are reported to synthesize *S*-methylseleno cysteine. Raw immature seeds are reported to contain per 100 g: 123 calories, 67.5% moisture, 8.4 g protein, 0.5 g fat, 22.1 g total carbohydrate, 1.8 g fiber, 1.5 g ash. Raw dried mature seeds contain per 100 g, 345 calories, 10.3% moisture, 20.4 g protein, 1.6 g fat, 64.0 g total carbohydrate, 4.3 g fiber, 3.7 g ash. Lima bean flour is reported to contain per 100 g: 343 calories, 10.5% moisture, 21.5 g protein, 1.4 g fat, 63.0 g total carbohydrate, 2.0 g fiber, 3.6 g ash. The average amino acid composition is per 16 g N: 11.9 g asparagine, 5.1 g threonine, 8.1 g serine, 14.9 g glutamic acid, 4.6 g proline, 4.4 g glycine, 4.7 g alanine, 5.8 g valine, 1.2 g methionine, 5.3 g isoleucine, 8.9 g leucine, 3.4 g tyrosine, 6.4 g phenylalanine, 3.2 g histidine, 7.5 g lysine, 6.3 g arginine, and 1.1 g cystine. Silage made from the leaves and stems is reported to contain 27.3% DM, 3.3% protein, 2.1% digestible protein, 14.2% total digestible nutrients (nutritive ratio 5 : 8).

DESCRIPTION: Annual or perennial herb, bush forms to 0.6 m tall, climbing forms to 4 m; leaves variable, trifoliolate, usually hairy; petiole 8–17 cm long; stipules minute, broadly triangular; leaflets ovate to lanceolate, acuminate, usually short hairy beneath, 5–13 cm long, 3–9 cm broad, lateral leaflets oblique, thin or thick; inflorescence an axillary raceme to 15 cm long, many-flowered, with 2–4 flowers at each node; calyx-bracts oval, nearly as long as the calyx, with strong central and usually 2 lateral veins; calyx campanulate with very short teeth; corolla 0.7–1.0 cm broad, the standard pale green, occasionally violet, yellowish-white or pinkish-purple, wings white, when standard is pigmented wings are usually also violet or pinkish-purple, keel prolonged into a complete spiral; stamens 10, style coiled; pods flat, 5–12 cm long, up to 2.5 cm wide, oblong, usually curved, 2–4-seeded, eventually dehiscing; seeds variable in size, color, and shape, 1–3 cm long, ranging from flat to rounded in potato types, white, cream, brown, red, purple, or black, full-colored or speckled, thin or plump; hilum white with translucent lines radiating from it to outer edge of testa. Weight of 100 seeds, 45–200 g, depending on cv. Roots usually fibrous. Fl. summer.

GERMPLASM: Cvs are selected for early maturity, vigor, heat-resistance, and resistance to diseases. 'Fordhook 242' and 'Burpee Bush' are bush beans recommended for the Tropics (yet 'Fordhook 242' is not recommended where monthly temperatures average more than 24°C). Rachie's efforts with tropical limas suggest that the best bush cvs rarely yield more than 1,500 kg/ha, whereas some climbers yield 3,000–4,000 kg/ha, coupled with better tolerance to disease, allic soils, and drought. 'Fordhook 242' and 'Henderson' are the principal cvs grown in the United States for processing and marketing of fresh limas. 'King of the Garden' is the pole bean recommended for the tropics and is also popular in the United States, along with 'Carolina,' 'Florida Speckled Butter,' 'Ventura,' 'Wilbur,' and 'Western.' *P. lunatus* L. var *silvester* Baudet is the wild type of the lima bean, native to central America (s Mexico and Guatemala). From early domestication, three directions of dispersal have been traced (Mackie, 1943) and constitute the original varietal stocks that are clustered into three cv divisions named "cultigroups" (cvgr):

1. Cvgr "Sieva" corresponding to *P. lunatus* L. *sensu stricto,* and to the northern branch or HOPI dispersal line.
2. Cvgr "Potato" corresponding to *P. bipunctatus* Jacq., and to the West Indies branch or CARIB dispersal line; this is characterized by more-or-less spherical seeds and a relatively high HCN concentration.
3. Cvgr "Big Lima" corresponding to *P. inamoenus* L. and to the southern branch or INCA line of dispersal (large flat seeds).

Java beans, red Rangoon beans, and small white beans are subdivisions issued from the three original stocks. *Phaseolus ritensis* M. E. Jones, which can be crossed with *P. lunatus,* combines some useful agronomic characteristics with disease resistance. Some germplasm from Colombia and Bolivia may contain 14–18 times as much carotene as the commercial cvs grown in the United States. This classification corresponds to that of Van Eseltine, who distinguished three commercial types: (1) the large 'lima' with 4–8 large flat seeds in a thick pod = cvgr "Big Lima," and (2) the 'Potato' lima with few spheroid seeds in a thick pod = cvgr "Potato," and (3) the 'Sieva' with 3–4 small, flat to convex seeds in thin pods = cvgr "Sieva." Westphal suggests cvgr 'Lunatus' for small seeds (= cvgrs "Sieva" and "Potato"), cvgr 'Inamoenus' for the large seeds (= cvgr "Big Lima"). The 'Sieva' or 'Lunatus' group contains better germplasm for hot arid situations. The many cvs may be divided into small bush limas and tall pole-climbing types. Usually four groups of beans may be recognized:

1. Java beans—Medium-size, somewhat purplish-red to black, containing considerable amounts of HCN.
2. Red Rangoon beans—Small, reddish beans which are usually plump, occasionally with purplish spots, containing traces of HCN.
3. White Rangoon or white Burma or small white beans, sometimes called sugar beans—Plump, resembling small haricots, usually containing mere traces of hydrocyanic acid, but sometimes more.
4. Lima beans—Resembling large haricots, quite white, said to contain no hydrocyanic acid.

Assigned to the Middle American and South American Centers of Diversity, (Central America, Caribbean, Peru, and northern Brazil) lima bean or cvs thereof is reported to exhibit tolerance to anthracnose, disease, drought, high pH, heat, laterite, low pH, photoperiod, smog, and virus. 'Cowey Red' and 'Winfield' are reported to be tolerant to cold weather, 'Bixby,' 'Butterbean 1962,' 'Fordhook 242,' 'Peerless,' 'Plump Champion,' and 'Triumph' to heat; 'Piloy' to short day and 'Early Thorogreen' to weather shock at pod set. ($2n = 22$.)

DISTRIBUTION: Native to tropical America, from Mexico and Guatemala, south to Brazil, Peru, and Argentina. Lima beans found in excavations in Peru were dated 6000–5000 B.C., and a small lima bean found in Mexico was dated 500–300 B.C. Carried by Spaniards across Pacific to the Philippines and hence to Asia, and from Africa to Brazil. Lima beans have escaped from cultivation and maintain themselves in the wild state in many tropical areas.

ECOLOGY: Lima bean occurs in most tropical countries from sea level to 2,400 m in altitude. It is the main pulse in wet forest regions of tropical Africa, and is widely grown in India and Burma. Baby limas grow in all sections of the United States; large-seeded cvs are grown along the West Coast. Lima beans require growing season free from frost; some cvs fail to set seed above 27°C. Low temperatures (below 13°C) retard growth and high night temperatures (above 21°C) hasten maturation and limit number and size of seed. Some cvs do not germinate satisfactorily at soil temperatures below 20°C; germination is best at 21°–27°C. According to Knott (1957), optimum monthly temperatures are: average, ca. 15.5°–21°C; maximum, ca. 27°C; minimum, ca. 10°C. However, K. O. Rachie (personal communication, 1977) reports some cvs produced extraordinary yields with considerably higher temperatures. Above 32°C flower shedding and poddrop may occur. In general, lima beans are drought-resistant. They require a well-aerated, well-drained soil, some cvs are sensitive to high acidity, others quite tolerant. They tolerate more rain during the growing season than French beans; however, waterlogging is detrimental to both. Some cvs are short-day, whereas others are day-neutral. Purseglove suggests a pH of 6–7, Knott of 5.5–6.8, Ware and McCollum 5.5–7.5; pH is reported as 4.3–8.3 in the Crop Diversification Matrix (Duke et al., 1975). Ranging from Cool Temperate Steppe to Wet through Tropical Very Dry to Wet Forest Life Zones, lima bean is reported to tolerate annual precipitation of 3.1–42.9 dm (mean of 43 cases = 15.5 dm), annual mean temperature of 8.7°–27.5°C (mean of 43 cases = 20.4°C) and pH of 4.5–8.4 (mean of 36 cases = 6.2).

CULTIVATION: Lima beans are propagated from seed, and germinate (epigeally) only if soil temperature is above 15.5°C. Bush beans are sown by hand or machine, spaced 5–20 cm apart in rows 80–90 cm apart; pole cvs are planted, 3–4 seeds per hill, in hills spaced 90–120 cm apart. At IITA and CIAT, horizontal wires 2 m high are strung on poles to support this and other pulses. For row planting, seeds are planted 2.5–5 cm deep, at rate of 135–170 kg/ha for large-seeded cvs and 56–80 kg/ha for small-seeded cvs. In tropics, seed remain viable, under dry conditions, for 3 years or more. Lima beans require rich, well cultivated soil; flat cultivation is preferred to ridges. Cultivations should be frequent enough to control weeds until plants begin to flower. In temperate areas seed is sown in spring; in tropical areas at beginning of or during first part of the rainy season. Lima beans respond favorably to nitrogenous fertilizer. Nodulation occurs, but inoculation with a specific *Rhizobium* culture is not usually necessary as the group bacteria that infect cowpea also infect lima bean. Plants may be fertilized with manure, 250–15,000 kg/ha. Lima bean seed is very sensitive to excessive soluble salts from either too much fertilizer or application too close to the seed; fertilizer should be at least 7.5 cm from seed to minimize injury. Depending on fertility, 28–280 kg/ha phosphoric acid and 28–280 kg/ha potash are recommended. Where pH exceeds 6.5, 5–10 kg Mn may be needed.

HARVESTING: Early maturing cvs are harvested in 100 days onward. Large lima beans in Peru and Malagasy take up to 9 months to mature. Green lima beans for market are harvested by hand when the seeds are nearly full grown, but before the pods start turning yellow. Usually several pickings are made as the pods do not ripen evenly. The high cost of hand-picking limits acreage of green lima beans. When grown for processing, bean plants are cut by machinery, or the beans are shelled by self-propelled viners in the field, and only the beans are collected for processing. For maximum yield of green lima beans, plants are harvested when only 3–5% of beans are white. Green lima beans usually processed immediately, but may be kept for up to 10 days if kept at ca. 0°C. White seeds are not satisfactory for canning or freezing; content of starch increases and of sugar decreases as the color changes from green to white. For dry lima beans 4 rows at a time are cut mechanically 5–7 cm under the soil, either at night or in early morning when the pods are moistened by dew. Vines allowed to dry for about 10 days in windrows, and then threshed with self-propelled threshers. Beans transported in large boxes.

YIELDS AND ECONOMICS: Yields of 1,000–3,000 (–4,500) kg dry bean/ha are not uncommon. In India, 12–14 pickings may yield 5–8 MT green pods/ha, but only 200–300 kg dry beans. Yields higher than 2,500 kg/ha have been reported from Nigeria. Green lima beans (yields ca. 2,500 kg/ha) are an important crop in the United States, southern Canada, and Latin America for canning, freezing, and the fresh market; dry lima beans are important in many areas as a reserve food. White butter beans are exported from Malagasy.

BIOTIC FACTORS: Lima beans are usually self-fertilized, but up to 18% natural cross-pollination has been reported. Honeybees increase seed yield. Lima beans are attacked by many fungi, including: *Agyriella nigra, Alternaria tenuis, Ascochyta boltshauseri, A. phaseolorum, A. pisi, Botrytis cinerea, Cercospora canescens, Cladosporium herbarum, Colletotrichum lindemuthianum* (anthracnose), *C. truncatum, Corticium microsclerotia, C. somani, Corynespora cassiicola, Dactuliophora tarrii, Diaporthe phaseolorum* (pod blight), *D. sojae, Diplodia phaseolina, Elsinoe phaseoli, Erysiphe polygoni, Fusarium solani, Glomerella cingulata, Isariopsis griseola, Leveillula taurica, Macrophomina phaseoli* (downy mildew), *Macrosporium leguminis, Microsphaera diffusa, M. euphorbiae, Myrothecium roridum, Nematospora phaseoli, N. coryli, Pellicularia filamentosa, Phakopsora vignae, Phoma subcircinata, Ph. terrestris, Phomop-*

sis phaseoli, Phyllosticta phaseolina, Ph. phaseolilunati, Phymatotrichum omnivorum, Phytophthora phaseoli (downy mildew), *Ph. capsici, Ph. cinnamomi, Ph. parasitica, Pullularia pullulans, Pythium aphanidermatum, P. ultimum, Rhizoctonia solani, Rhizopus stolonifer, Sclerotina sclerotiorum, Sclerotium rolfsii, Thielaviopsis basicola,* and *Uromyces phaseoli*. Dressing seed with Asaran, Phygon, or Spergon is supposed to curb root-rot problems. Bacteria causing diseases in lima beans include: *Achromobacter lipopticum, Bacillus lathyri, Corynebacterium flaccumfaciens, Erwinia carotovora, Pseudomonas coadunata, Ps. phaseolicola* (halo blight), *Ps. syringae, Ps. tabaci, Ps. medicaginis, Xanthomonas phaseoli* and var *indicus*. Viruses isolated from lima beans are: alfalfa mosaic, alsike clover mosaic, asparagus-bean mosaic, bean chlorotic ringspot (4 strains), bean local chlorosis (6 strains), bean mosaic, bean necrosis, bright yellow mosaic, cucumber mosaic, white clover mosaic and zonate leafspot. Root-knot nematodes (Meloidogyne) are serious pests; soil fumigation with ethylene dibromide or pene-dichloropropane may give temporary (2 or 3 crops) relief. Other nematodes isolated from lima beans include: *Belonolaimus gracilis, Heliocotylenchus digonicus, H. dihystera, Heterodera glycines, H. schachtii, Meloidogyne arenaria, M. incognita, M. i.* var *acrita, M. javanica, Paratylenchus* sp., *Pratylenchus brachyurus, P. coffeae, P. penetrans, P. scribneri, P. vulnus, Rotylenchus buxophilus, Rotylenchulus reniformis, Trichodorus christie, T.* sp., and *Tylenchorhynchus clarus*. Pod borers and leafhoppers limit production in parts of the tropics. Most insects that attack beans attack lima beans. Larvae of the pantropical lima bean pod borer *Etiella zinckenella* may bore into the pods, destroying seeds. Carbaryl, DDT, and sevin have been recommended for its control. Aphids, thrips, and lygus bug may be sprayed with DDT or organophosphorus compounds. In the United States the corn-seed maggot *Hylemya platura*, which may infest slow-germinating seed, is controlled by seed treatment with lindane and/or thiram. Rotenone or organophosphorus insecticides have been recommended for the Mexican bean beetle, *Epilachna varivestis*. Bruchid beetles, e.g., *Callosobruchus chinensis* and *Acanthoscelides obtectus*, can damage the seeds, especially during storage. DDT has been recommended for cutworms, e.g., the variegated cutworm, *Peridroma saucia*, and the gray cutworm, *Agrotis ipsilon*. Wireworms (*Limonius, Melanotus*) and springtails (*Onychiurus*) may also be troublesome.

REFERENCES:

Baudet, J. C., 1977, Origine et classification des espèces cultivées du genre *Phaseolus, Bull. Soc. R. Bot. Belg.* **110**:65–76.

van Eseltine, G. P., 1931, Variation in the lima bean, Phaseolus lunatus L., as illustrated by its synonymy, *N.Y. Agric. Exp. Stn., Geneva, Tech. Bull.* **182**.

Evans, I. M., and Boulter, D., 1974, Amino acid composition of seed meals of yam bean (*Sphenostylis stenocarpa*) and lima bean (*Phaseolus lunatus*), *J. Sci. Fd. Agric.* **25**:919–922.

Le Marchand, G., Marechal, R., and Baudet, J. C., 1976, Observations sur quelques hybrides dans le genre *Phaseolus* III, *P. lunatus*; nouveaux hybrides et considerations sur les affinités interspecifiques, *Bull. Rech. Agron. Gembloux* **11**(1–2):183–200.

Mackie, W. W., 1943, Origin, dispersal and variability of the Lima bean, *Phaseolus lunatus, Hilgardia* **15**:1–29.

Seelig, R. A., and Roberts, E., 1955, *Lima Beans. Fruit and Vegetable Facts and Pointers*, 4 pp., United Fruit and Vegetable Association, Washington, D.C.

CONTRIBUTORS: J. C. Baudet, J. A. Duke, M. Golden, C. F. Reed, R. Marechal, K. O. Rachie, J. Smartt, J. K. P. Weder.

Phaseolus vulgaris L.

FAMILY: Fabaceae

COMMON NAMES: Bean, Common bean, Caraota, Feijao, French bean, Kidney bean, Haricot bean, Field bean, Poroto, Snap or String bean, Frijol, Wax bean

USES: Common bean is most widely cultivated of all beans in temperate regions, and widely cultivated in semitropical regions. In temperate regions the green immature pods are cooked and eaten as a vegetable. Immature pods are marketed fresh, frozen or canned, whole, cut or french-cut. Mature ripe beans, called navy beans, white beans, northern beans, or pea beans, are widely consumed. In lower latitudes, dry beans furnish a large portion of the protein needs of low- and middle-class families. In some parts of the tropics leaves are used as a pot-herb, and to a lesser extent the green-shelled beans are eaten. In Java, young leaves are eaten as a salad. After beans are harvested, straw is used for fodder.

Phaseolus vulgaris L.

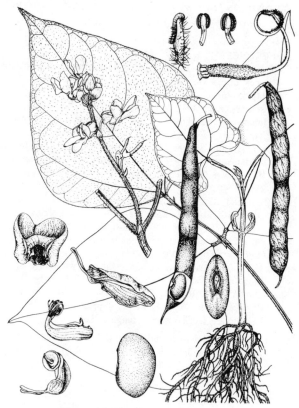

Figure 95. *Phaseolus vulgaris* L.

FOLK MEDICINE: Beans are said to be used for acne, antidiabetic, bladder, burns, cardiac, carminative, depurative, diarrhea, diuretic, dropsy, dysentery, eczema, emollient, hiccups, itch, kidney, resolvent, rheumatism, sciatica, and tenesmus.

CHEMISTRY: Beans are a highly nutritive, relatively low-cost protein food. Green snap beans contain 6.2% protein, 0.2% fat, and 63% carbohydrate. Analysis of a sample of dried beans marketed under the name 'Rajmah' gave the following values per 100 g: moisture, 12.0; protein, 22.9; fat, 1.3; carbohydrates, 60.6; and minerals, 3.2%; Ca, 260; P, 410; and iron, 5.8 mg.; 346 calories/100 g. The vitamin contents of the dried beans are: thiamine, 0.6; riboflavin, 0.2; nicotinic acid, 2.5; and ascorbic acid, 2.0 mg/100. Analysis of dried beans from another source yielded (mg/100 g): Na, 43.2; K, 1160; Ca, 180; Mg, 183; Fe, 6.6; Cu, 0.61; P, 309; S, 166; and Cl, 1.8 mg/100 g. Beans also contain I (1.4 μg/100 g), Mn (1.8 mg/100 g), and arsenic (0.03 mg/100 g). Raw immature pods of green, and yellow or wax snap beans are reported to contain per 100 g, 32 and 27 calories, 90.1 and 91.4 moisture, 1.9 and 1.7 g protein, 0.2 g fat, 7.1 and 6.0 g total carbohydrate, 1.0 g fiber, and 0.7 g ash, respectively. Raw pods of kidney beans contain (per 100 g edible portion): 150 calories, 60.4% moisture, 9.8 g protein, 0.3 g fat, 27.8 g total carbohydrate, 2.3 g fiber, 1.7 g ash, 59 mg Ca, 213 mg P, 3.6 mg Fe, 10 μg vitamin A, 0.38 mg thiamine, 0.12 mg riboflavin, 1.5 mg niacin, 7 mg ascorbic acid. Raw dried mature seeds of white, red, and pinto beans are reported to contain per 100 g: 340, 343, and 349 calories, 10.9, 10.4, and 8.3% moisture, 22.3, 22.5, and 22.9 g protein, 1.6, 1.5, and 1.2 g fat, 61.3, 61.9, and 63.7 g total carbohydrate, 4.3, 4.2, and 4.3 g fiber, 3.9, 3.7, and 3.9 g ash, respectively. Whole seeds of kidney beans contain (per 100 g): 86 mg Ca, 247 mg P, 716 mg Fe, 5 μg vitamin A, 0.54 mg thiamine, 0.19 mg riboflavin, 2.1 mg niacin, 3 mg ascorbic acid. Whole seeds cooked contain: 141 calories, 68.0% moisture, 5.9 g protein, 5.7 g fat, 17.9 g total carbohydrate, 1.1 g fiber, 2.5 g ash, 46 mg Ca, 120 mg P, and 1.9 mg Fe. Raw leaves contain (per 100 g): 36 calories, 86.8% moisture, 3.6 g protein, 0.4 g fat, 6.6 g total carbohydrate, 2.8 g fiber, 2.6 g ash, 274 mg Ca, 75 mg P, 9.2 mg Fe, 3,230 μg β-carotene equivalent, 0.18 mg thiamine, 0.06 mg riboflavin, 1.3 mg niacin, 110 mg ascorbic acid. After harvest, plants can be fed to cattle, sheep, and horses. It is satisfactory as a part of the roughage when fed with good hay and is comparable to corn or sorghum fodder in nutritive value. Analysis of a sample gave the following values: moisture, 10.9; protein, 6.1; fat, 1.4; N-free extr., 34.1; fiber, 40.1; ash, 7.4; Ca, 1.7; P, 0.1; K, 1.0; digestible protein, 3.0; and total digestible nutrients, 45.2%; nutritive ratio, 14.1. After pod removal, silage may be prepared from green vines. Dehydrated bean vine meal prepared from green plants after pod removal is comparable to alfalfa meal as a vitamin supplement for chicks. It contains protein, 18.3; digestible protein, 12.3; and total digestible nutrients, 46.3%; nutritive ratio, 2.8. Meal made from vines with mature leaves is inferior in quality. Leaves contain carotene (178.8 mg/100 g), thiamine, riboflavin, nicotinic acid, folic acid, and pantothenic acid. They also contain a quercetin glycoside. The roots are reported to cause giddiness in human beings and animals. Seeds are reported to contain trypsin inhibitors and chymotrypsin inhibitors. The hull is said to yield 0.13% rubber. The leaves are said to contain allantoin.

Phaseolus vulgaris L.

DESCRIPTION: Highly polymorphic species; annual herb, erect and bushy, 20–60 cm tall, or twining with stems 2–3 m long; with a taproot and nitrogenous nodules (although the germinating bean has a tap root, adventitious roots usually emerge 1–2 days after germination, to dominate the tap root, which remains 10–15 cm long); leaves alternate, green or purple, trifoliolate, stipulate, petiolate, a marked pulvinus at base; leaflets ovate, entire; acuminate, 6–15 cm long, 3–11 cm wide; flowers in lax, axillary few-flowered (–12) racemes, zygomorphic, variegated, white, pink, or purplish, ca. 1 cm long; pods slender, green, yellow, black, or purple, cylindrical or flat, 8–20 cm long, 1–1.5 cm wide; seeds 4–6 (–12), usually glabrous, sometimes puberulent, beak prominent; seeds white, red, tan, purple, grey or black, often variegated, reniform, oblong or globose, up to 1.5 cm long, endosperm absent; 100 seeds weigh 10–67 g, depending on cv; germination phanerocotylar.

GERMPLASM: More than 14,000 cvs are recorded; the major repository and distributor is CIAT in Cali, Colombia. In the United States, they are grouped mainly into early-maturing bush types and later-maturing pole types. Dry-shelled beans are grouped into the following four types: (1) red kidney beans, 1.5 cm or more long, important in Latin America; (2) medium field beans, 1–1.2 cm long, pinkish-buff with brown spots, grown extensively in the United States as 'Pinto'; (3) marrow beans, 1–1.5 cm long, as 'Yellow Eye'; and (4) pea or navy beans, 8 mm or less long, grown extensively in California. Black beans are gaining in importance in some areas. In Latin America and Africa, regional preferences are strong for seed-coat color and brilliance. Venezuela and Guatemala favor black-seeded beans; Colombia and Honduras, red; Peru, cream or tan; Brazil, black or tan. Growth form in *Ph. vulgaris* and other *Phaseolus* spp. depends on two factors—number of nodes produced (oligonodal vs. polynodal) and internode length (long vs. short). Three definite forms are produced for cultivation: (1) polynodal with long internodes—climber; (2) polynodal with short internodes—indeterminant dwarf; and (3) oligonodal with short internodes—determinant dwarf. Two distinct leaf-size grades can also be distinguished in combination with these. With the range of seed sizes and shapes plus testa colors the possible range of distinct types is enormous. Snapbeans cv 'Mild White Giant' is reportedly tolerant to "adverse conditions," 'OSU 949,' 'OSU 2065,' 'Purley King' (British peabean requires temperature of 12°–13°C at soil depth of 10 cm), 'SRS 1884' to cool weather, 'Royalty' to cold soil, 'Longval' to drought, 'Alabama A1,' 'Ashley Wax,' 'Choctaw,' 'Cooper Wax,' 'Logan,' and 'Longval' to heat, 'Pacer' to short season and 'Royalty' to wet soil. Among dry beans cv 'Criolla' is reportedly tolerant to heat and 'Bonita,' 'Borinquen,' and 'Criolla' to tropical conditions. Assigned to the Middle and South American Centers of Diversity, common bean or cvs thereof is reported to exhibit tolerance to aluminum, bacteria, disease, drought, herbicide, hydrogen fluoride, high pH, laterite, low pH, manganese, peat, photoperiod, smog, SO_2, virus, and water excess. Tolerances to pathogens are surveyed in Horsfall *et al.* (1972). ($x = 11$, $2n = 22$.)

DISTRIBUTION: Common beans are native to the New World, probably Central Mexico and Guatemala. Spaniards and Portuguese took them to Europe, Africa and other parts of the Old World. Now they are widely cultivated in the tropics, subtropics, and temperate regions. Roughly 30% of world production is in Latin America. They are less known in India, where other pulses are preferred.

ECOLOGY: Beans tolerate most environmental conditions in tropical and temperate zones, but do poorly in very wet tropics where rain causes disease and flower drop. Rain is undesirable during harvest of dry seeds. Frost kills plant. There are both short-day and day-neutral cvs. Excessive water injures most cvs in a few hours, but some black-seeded cvs tolerate standing water. Beans grow best in well-drained, sandy loam, silt loam or clay loam soils, rich in organic content, but are sensitive to concentrations of Al, B, Mn, and Na. Below pH 5.2 Mn toxicity may be a problem. In calcareous soils, zinc deficiencies can be serious in sandy acid soils; deficiencies of Mg and Mo also may arise. Garden bean yields are 10% lower at EC 1,500, 25% at 2,000 and 50% at 3,500. French or snap beans seem more sensitive to Na than many other cvs. Temperatures of −5° to −6°C are harmful at germination, −2° to −3°C at flowering and −3° to −4°C at fruiting. Some cvs withstand short frosts as low as −3°C. The optimum monthly temperature for growth is 15.6°–21.1°C, the maximum ca. 27°C, the minimum ca. 10°C. Blossom-drop is serious above 30°C, and can completely prevent seed set above 35°C. Beans are traditionally a subtropical or temperate crop. In the tropics they are normally found in montane valleys (800–2,000 m). Very few beans are grown in hot humid tropics where cowpeas fare better. Five different writers give five different pH ranges. Our computer program reported 4.2–8.7; average of 144 cases was 6.4. Other values were 5.5–6.8, 5.5–7.5, 6.0–7.0 and 6.0–7.5. Ranging from Boreal Moist to Wet through Tropical Very Dry to Wet Forest Life Zones, common bean is reported to tolerate annual precipitation of 0.9–42.9 dm (mean of 217 cases = 12.8), annual mean temperature of 5.7°–28.5°C (mean of 216 cases = 19.3), and pH of 4.2–8.7 (mean of 144 cases = 6.4).

CULTIVATION: In temperate areas, seed should be planted about the same time as corn, when soil has become warm. Germination is rapid at soil temperatures above 18°C. In pure stands, bush cvs give good yields at 30 by 30 cm spacings, but wider spacing facilitates weeding. Pole beans are usually planted 4–6 seeds in hills spaced about 1 m apart at a seeding rate of nearly 80 kg/ha. Seed rates are 20–115 kg/ha depending on the cv, seed size, and width of row; 'Red Kidney,' 'Marrow,' and 'Yellow Eye' at 75–100 kg/ha; 'Pea Beans,' 'Black Turtle Soup,' at 30–40 kg/ha; row widths 70–75 or 80 cm. Some pole beans are sown at rates as low as 25 kg/ha. Seed of good quality is essential for production of dry beans. Susceptibility to diseases, mechanical injury, frost damage, and wet weather damage at harvest time, and cracked seedcoats should be considered. With a corn, bean or beet drill with removable plates, beans are usually planted 5–8 cm deep, deep enough to give good coverage and sufficient moisture to promote fast germination and growth. Plants should be cultivated to control weeds; care should be taken late in the season to avoid injuring roots extending out between the rows just beneath soil surface. Inoculation of seed with nitrogen-fixing bacteria is unnecessary for dry beans. Beans should be rotated with other crops to maintain high yields and quality and to reduce the hazard of diseases that may survive in the soil or on plant refuse in the soil. In the tropics beans are often interplanted with such crops as coffee, corn, cotton, and sweet potatoes, and little or no fertilization is used, although plants respond to nitrogen. Still, as much as 25 MT/ha barnyard manure is recommended. In the United States nitrogen and phosphate are applied. Irrigation is beneficial in semiarid regions, with overhead preferred to flood irrigation. Mixtures of cvs are often sown. In Latin America, ca. 70% of the beans are interplanted with corn. Grown alone, they are planted at 200,000–250,000 plants per hectare in 50-cm rows with 5–10 cm between seed. Bush beans are planted 30 × 30 or 50–60 × 5–10 cm, the latter facilitates cultivation. In Latin America beans are produced mainly on marginal soil, nearly always with P deficiency, commonly with inadequate N problems; credit and fertilizer are seldom available.

HARVESTING: Beans mature very quickly and green beans may be harvested 4–6 weeks after sowing. In early snap bean cvs, harvest begins in 7–8 weeks, 1 or 2 weeks after flowering. Beans should be picked every 3–4 days. Bush beans mature over a short time; pole beans continue to bear for a long time. Dry beans should be harvested when most pods are fully mature and have turned color. To minimize shatter, harvesters should not shake the vines. The cutter consists of 2 broad blades set to cut 2 adjacent rows about 5 cm below the ground. Then prongs pull plants from both rows into one windrow in wet weather; plants are forked into field stacks ca. 1.3 m in diameter and 2–3 m high that are supported by a center stake. In the third world, beans are usually hand harvested, or manually gathered and windrowed. Plants are pulled, dried, and threshed; sometimes beans are hand-shelled.

YIELDS AND ECONOMICS: Yields vary widely with cv, culture and region. In the United States yields for dried beans average 1,000–1,500 kg/ha; for Mexican and Colombian hybrids, up to 2,500 kg/ha; in Mauritius, 250–1,500 kg/ha, depending on cv; in Egypt, 500 kg/ha; in Kenya and Malawi, 300–1,000 kg/ha; in Great Britain, 1,200 kg/ha. Yields of green immature snap beans average about 4.5 MT/ha, but higher yields have been reported. Maximum reported experimental yields of dry beans exceed 5.5 MT/ha (bush beans). At CIAT, bush bean yields of 4.5 MT/ha and trellis-bean yields of 5.8 MT/ha were reported for a 100–120-day-growth cycle. Association with corn reduced such yields to ca. 2 MT/ha. In 1975, the world harvest was 13,227,000 MT of dry beans from 24,715,000 ha for an average yield of 535 kg/ha.

Phaseolus vulgaris L.

Yields were highest (2,374 kg/ha) in the Netherlands. Asia had the largest hectarage (11,697,000 ha) followed by South America (4,636,000 ha) and Africa (3,066,000 ha). India with yields of only 313 kg/ha, is estimated to have produced 2,500,000 MT; China, with 937 kg/ha, 2,399,000 MT; Brazil, 563 kg/ha, 2,280,000 MT; and Mexico, with 801 kg/ha, 1,202,000 MT. By contrast, the United States with yields of 1,332 kg/ha produced 780,000 MT (FAO, 1976). Common beans of various types are nearly cosmopolitan, except in tropical Asia, where native pulses are preferred. England imports haricot beans from Japan, Chile, United States, Ethiopia, Mozambique, East Africa, and Malawi. In 1969, Japan imported about 18,000 MT from the United States, and Israel, and the Philippines imported about 1,450 MT (dry beans). Latin America, producing twice as much as the United States still imports beans. The U.S. production in 1969 was about 800,000 MT (dry beans) from about 600,000 ha, bringing about $0.17/kg. Nearly 150,000 MT fresh beans were produced from 36,000 ha at $0.28/kg for immediate consumption. For the processing industry, 500,000 MT at $0.11/kg were produced from 95,000 ha. The major bean-growing areas in the United States are: Michigan (ca. 40%), California, Colorado, Idaho, New York, and Wyoming.

BIOTIC FACTORS: Flowers are self-fertilized. Many diseases are caused by fungi, bacteria, and viruses in beans throughout the world. In Latin America the principal fungal diseases are rust, anthracnose, angular leaf spot, web blight, and various root rots. The chief bacterial diseases are bacterial blight and halo blight; chief viral diseases are bean common mosaic and bean golden mosaic. Following organisms are recorded from beans: *Alternaria atrans, A. brassicae, A. phaseoli-vulgaris, A. tenuis, A. tenuissima, Aphanomyces euteiches, Aristastoma oeconomicum, Ascochyta boltshauseri, A. phaseolorum, A. pisi, A. sojaecola, Ashbya gossypii, Aspergillus ochraceus, Botrydiplodia theobromae, Botrytis cinerea, Brachysporium pisi, Cercospora canescens, C. columnaris, C. cruenta, C. phaseoli, C. phaseolina, C. phaseolorum, C. vanderysti, Chaetomium cochliodes, Choanephora cucurbitarum, Cladosporium fulvum, C. herbarum, C. phaseoli, Colletotrichum lindemuthianum, C. truncatum, Corticium microsclerotia, C. solani, Corynespora cassiicola, Cylindrocarpon radicicola, Dactuliophora tarrii, Diaporthe arctii, D. sojae, Dothiorella phaseoli, Elsinoe phaseoli, Epicoccum neglectum, E. purpurascens, Erysiphe polygoni, E. pisi, Fusarium acuminatum, F. culmorum, F. equiseti, F. martii, F. martii-phaseoli, F. oxysporum, F. poae, F. sambucinum, F. solani, Gloeosporium lindemuthianum, Glomerella lindemuthianum, Helicobasidium purpuream, Helminthosporium carbonum, Hyalodendron album, Isariopsis griseola, I. laxa, Leptosphaeria phaseolorum, Leveillula taurica, Macrophomina phaseoli, Macrosporium commune, M. phaseoli, Microsphaeria diffusa, Mycosphaerella cruenta, M. pinodes, M. sojae, Myrothecium roridum, Nematospora coryli, N. phaseoli, Oidium balsamii, O. erysiphoides, Ophiobolus graminis, Ovularia phaseoli, Parodiella perisporioides, Pellicularia filamentosa, Penicillium cyclopium, Periconia byssoides, Phaeoisariopsis griseola, Phakopsora vignae, Phoma subcircinata, Phyllachora phaseoli, Phyllosticta phaseolina, Phytomonas medicaginis* var *phaseolicola, Phymatotrichum omnivorum, Phytophthora parasitica, Ph. cactorum, Ph. magosperma, Pleosphaerulina phaseolina, P. sojaecola, Pleospora herbarum, Pseudoplea trifolii, Pullularia pullulans, Pythium anandrum, P. acanthium, P. aphanidermatum, P. artotrogus, P. debaryanum, P. helicoides, P. intermedium, P. irregulare, P. myriotylum, P. oligandrum, P. pulchrum, P. rostratum, P. salpingosporum, P. spinosum, P. splendens, P. ultimum, P. vexans, Ramularia deusta, R. phaseolina, Rhizobium phaseoli, Rhizoctonia microsclerotia, Rh. solani, Rhizopus stolonifer, Sclerotinia fructicola, S. fuckelinan, S. homeocarpon, S. libertiana, S. minor, S. sclerotiorum, Sclerotium bataticola, S. rolfsii, Sphaerella phaseolicola, Stagonspora (Stagonosporopsis) hortensis, Thielaviopsis basicola, Uromyces aloes, U. appendiculatus, U. fabae, U. phaseoli, U. phaseolorum, U. viciae-fabae, Vermicularia truncata,* and *Verticillium alboatrum*. Bacteria causing diseases in beans include: *Agrobacterium tumefaciens, Bacillus lathyri, Bacterium carotovora, B. fascians, B. medicaginis* var *phaseolicola, B. phaseoli, B. rubefaciens, Corynebacterium fascians, Erwinia carotovora, Pseudomonas cannabina, Ps. cyamopsicola, Ps. flectens, Ps. marginalis, Ps. medicaginis* var *phaseolicola, Ps. syringae, Xanthomonas phaseoli* var *fuscans*, and *X. vignicola*. Common beans are parasitized by *Striga hermonthica*. Viruses isolated from common beans include: abutilon mosaic, alfalfa mosaic, alsike clover mosaic, Argentina sunflower, asparagus-bean mosaic,

Phaseolus vulgaris L.

bean chlorotic ringspot (4 strains), bean local chlorosis (7 strains), bean mosaic, bean necrosis, beet ringspot, black legginess, Brazilian tobacco streak, broad-bean mottle, carnation mosaic, carnation ringspot, clover yellow mosaic, cucumber mosaic, cucumber necrosis, curly top, euphorbia mosaic, leaf crinkle, lucerne mosaic, oily-pod, pea enation mosaic, peach ringspot, peach yellow butt mosaic, pelargonium leaf-curl, pod mottle, potato bouquet, raspberry leaf-curl, raspberry ringspot, raspberry yellow dwarf, rosette, Rothamsted tobacco necrosis, southern bean mosaic, stipple streak, subterranean clover stunt, subterranean clover, summer death, sweetpea streak, tobacco mosaic, tobacco necrosis, tomato aspermy, tomato black ring, tomato spotted wilt, tomato streak, top necrosis, white clover mosaic, yellow mosaic, yellow bean mosaic, yellow-green mosaic, and yellow spot of nasturtium. Several nematodes infest common beans and may cause problems in certain areas: *Aphelenchoides ritzemabosi, Belonolaimus gracilis, B. longicaudatus, Criconemoides curvatum, Ditylenchus dipsaci, Dolichodorus heterocephalus, Helicotylenchus microlobus, Hemicycliophora parvana, Heterodera glycines, H. schachtii, H. trifolii, Longidorus maximus, Meloidogyne arenaria, M. a. thamesi, M. hapla, M. incognita* and var *acrita, M. javanica, Pratylenchus vulnus, Radopholus similis, Rotylenchus reniformis, Trichodorus christiei, Tylenchorhynchus claytoni*. In India, major insect pests include *Epilachna vigintioctopunctata*, one of the worst enemies, and the flea beetle, *Longitarsus belgaumensis,* an aphid, *Smynthurodes betae,* and the bean fly, *Melanagromyza phaseoli*. The most serious insect attacking bean in the eastern United States is the Mexican bean beetle *(Epilachna varivestis)* (treated with carbaryl, dimethoate, diazinon, malathion, methoxyclor, and parathion). Other pests include: bean leaf beetle *(Cerotoma trifurcata)* (treated with carbaryl, DDT, or rotenone), seedcorn maggot *(Hylemya platura)*, bean weevil *(Acanthoscelides obtectus)*, bean thrips *(Caliothrips fasciatus)*, and bean aphid *(Aphis rumicis)*. Major Latin American pests are *Empoasca* (green leafhopper) and *Diabrotica*. In Africa, the bean fly is very important.

REFERENCES:
CIAT, 1974–1978, Bean production program, *Annual Reports,* 1974–1977.

Graham, P. H., 1978, Some problems and potentials of *Phaseolus vulgaris* L. in Latin America, *Field Crops Res.*
Kaplan, L., 1965, Archeology and domestication in American Phaseolus (beans), *Econ. Bot.* **19**:358–368.
Seelig, R. A., and Roberts, E., 1960, *Green and Wax Snap Beans. Fruit and Vegetable Facts and Pointers,* 18 pp., illus., United Fruit and Vegetable Association, Washington, D.C.
Zaumeyer, W. J., and Thomas, H. R., 1962, Bean diseases—and how to control them, *U.S. Dep. Agric., Agric. Res. Ser., Agric. Handb.* **225**.

CONTRIBUTORS: J. A. Duke, P. H. Graham, C. F. Reed, J. Smartt, J. K. P. Weder.

Pisum sativum L.

FAMILY: Fabaceae

COMMON NAMES: Common or Garden pea

USES: Peas are cultivated for the fresh green seeds, tender green pods, dried seeds and foliage. Green peas are eaten cooked as a vegetable, and are

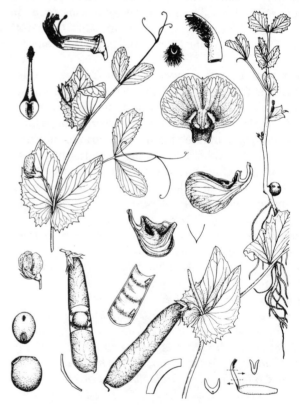

Figure 96. *Pisum sativum* L.

marketed fresh, canned, or frozen. Ripe, dried peas are used whole, split, or made into flour, and eaten by humans and livestock. Leaves are used as a potherb in Burma and parts of Africa. Some cvs are grown for their tender green pods which are eaten cooked or raw. Oil from ripened seed has antisex hormonic effects; produces sterility and antagonizes effect of male hormone.

FOLK MEDICINE: Seeds are thought to cause dysentery when eaten raw. In Spain, flour is considered emollient and resolvent, applied as a cataplasm. Said to be used for contraceptive, ecbolic, fungistatic (sd), and spermicide (sd).

CHEMISTRY: Fresh green peas contain per 100 g: 44 calories, 75.6% water, 6.2 g protein, 0.4 g fat, 16.9 g carbohydrate, 2.4 g crude fiber, 0.9 g ash, 32 mg Ca, 102 mg P, 1.2 mg Fe, 6 mg Na, 350 mg K, 405 μg β-carotene equivalent, 0.28 mg thiamine, 0.11 mg riboflavin, 2.8 mg niacin, and 27 mg ascorbic acid. Dried peas contain: 10.9% water, 22.9% protein, 1.4% fat, 60.7% carbohydrate, 1.4% crude fiber, and 2.7% ash. Raw edible-podded peas contain per 100 g: 53 calories, 83.3% moisture, 3.4 g protein, 0.2 g fat, 12.0 g total carbohydrate, 1.2 g fiber, and 1.1 g ash. Raw immature green peas contain per 100 g: 84 calories, 78.0% moisture, 6.3 g protein, 0.4 g fat, 14.4 g total carbohydrate, 2.0 g fiber, 0.9 g ash. Raw dried mature seeds contain per 100 g: 340 calories, 11.7% moisture, 24.1 g protein, 1.3 g fat, 60.3 g total carbohydrate, 4.9 g fiber, 2.6 g ash. Flour contains: 343 calories, 10.9% moisture, 22.8 g protein, 1.2 g fat, 62.3 g total carbohydrate, 4.2 g fiber, 2.8 g ash, 72 mg Ca, 338 mg P, 11.3 mg Fe, 475 (?) μg β-carotene equivalent, 0.86 mg thiamine, 0.18 mg riboflavin, and 2.8 mg niacin. An average amino acid composition reported is (mg/g N): 267 isoleucine, 425 leucine, 470 lysine, 57 methionine, 70 cystine, 287 phenylalanine, 171 tyrosine, 254 threonine, 294 valine, 595 arginine, 143 histidine, 253 alanine, 685 aspartic acid, 1,009 glutamic acid, 253 glycine, 244 proline, and 271 serine. Methionine and cystine are the main limiting amino acids. Fertilizing peas with S has increased their methionine content from 1.3 to 2.2 g per 100 g protein. Pea hay (at 88.6% DM) contains (zero moisture basis): 10.7–21.6% crude protein (avg. 15.4), 1.5–3.7% fat (avg. 3.0), 16.8–36.1% crude fiber (avg. 28.4), 6.0–9.3% ash (avg. 7.9), and 41.9–50.6% N-free extract (avg. 45.3). Immature field pea (89.3% DM) contains: 16.7–21.6% crude protein (avg. 16.9), 3.0–3.7% fat (avg. 3.7), 16.8–32.2% crude fiber (avg. 27.1), 7.5–9.3% ash (avg. 8.6), 39.9–45.8% N-free extract (avg. 43.7), 1.37% Ca, 0.28% P, 1.4% K, and 0.37% Mg. Seeds are reported to contain trypsin inhibitors and chymotrypsin inhibitors. Pakistani cvs are said to be perorally contraceptive. The genus *Pisum* is reported to contain oxalic acid and saponin.

DESCRIPTION: Annual herb, bushy or climbing, glabrous, usually glaucous; stems weak, round, and slender, 30–150 cm long; leaves alternate, pinnate with 1–3 pairs of leaflets and a terminal branched tendril leaflets ovate or elliptic, 1.5–6 cm long, ca. 1 cm broad, glaucous; stipules large, leaflike to 10 cm long; flowers axillary, solitary or in 2–3-flowered racemes; corolla white, or pink, or purple; pods swollen or compressed, short-stalked, straight or curved, 4–15 cm long, 1.5–2.5 cm wide, 2–10-seeded, 2-valved, dehiscent on both sutures; seeds globose or angled, smooth or wrinkled, exalbuminous, whitish, gray, green, or brownish; 100 seeds/15–25 g; germination cryptocotylar. Taproot well developed with numerous slender laterals. Fl. late spring, Fr. early summer; variable in Tropics.

GERMPLASM: Some botanists treat garden peas as *P. sativum* ssp. *hortense* Asch. & Graebn., field peas as *P. sativum* ssp. *arvense* (L.) Poir., and edible podded peas as *P. sativum* ssp. *macrocarpon*; early dwarf pea as *P. sativum* var *humile*. All peas are completely cross-fertile including two important wild types, *P. sativum* ssp. *elatius* (*P. elatius* Beib.), climbing over shrubs in humid Mediterranean formations, and *P. sativum* ssp. *syriacum* (*P. numile* Boiss. & Noe), in Near Eastern herbaceous formations. More than 1,000 cvs are listed, varying in height, maturity period, pod size and type, and seed characteristics. Gentry (1971) recognized six subspecies: (1) abyssinicum, (2) jomardi, (3) syriacum, (4) elatius, (5) arvense, and (6) hortense, lamenting that few germplasm collections had been assembled in centers of origin or diversity. Two types of garden peas are smooth-seeded and wrinkle-seeded. 'Alaska' has long been the standard type of canning pea; it is early and hardy, but is now considered poor. Other more desirable cvs of 'Alaska' type are: 'Super Alaska,' 'Supergreen,'

Pisum sativum L.

and 'Alaska Wilt Resistant.' Wrinkled-seeded garden peas are sweeter than smooth types and are sometimes called marrow fat. Canning peas are divided into two classes: Alaska and sweet. Canning companies like small peas because the consumer does. Of the sweet type is 'Surprise' (most important) and 'Resistant Surprise' (gaining in popularity because of its disease resistance). Some midsummer varieties are: 'Pride,' 'Early Perfection,' 'Ace,' and 'Wasatch.' Late cvs, which ripen about 14 days after 'Alaska,' are: 'Perfection' and its various strains, as 'Dark-green Perfection,' 'Bridger,' 'Superior,' 'New Era,' and 'Shoshone.' Canning cvs usually have a tough skin that holds its shape during canning. Cvs grown for freezing must have a dark green color and tender skin, size is not important. 'Thomas Laxton' is most widely grown cv and is earliest, but is not always high yielding. Other cvs grown are: 'Laxton 7,' 'Freezonian,' 'Shasta'; later maturing cvs are: 'Victory Freezer,' 'Wyola,' 'Perfection Freezer,' and Dark-green Perfection.' For home garden use, 'Little Marvel,' 'Laxton Progress,' 'World's Record,' 'Wando' holds quality well in hot weather. For the subtropics, the best cvs are: 'Asgrow 40,' 'Canner 75,' 'Greenfeast,' 'Melting Sugar,' 'Ronda,' 'World Record,' and 'Freezer 37.' 'Burpeeana' is best in El Salvador. 'Manoa Sugar,' a dwarf gray sugar type, immune to powdery mildew and somewhat resistant to warm temperatures is preferred in Hawaii. 'Lincoln' is suitable for Senegal. Assigned to the Near Eastern, Mediterranean, and African Centers of Diversity, peas or cvs thereof are reported to exhibit tolerance to aluminum, disease, frost, fungus, hydrogen fluoride, high pH, heat, laterite, low pH, mildew, slope, smog, virus, and wilt. ($2n = 14$, [28, 30])

DISTRIBUTION: Wild prototype of garden pea has never been found, but some writers believe that it was an ancient Egyptian plant. Probably had its origin in Europe and western Asia. At present, it is grown throughout the world.

ECOLOGY: Peas require a cool, relatively humid climate and are grown at higher altitudes in tropics with temperatures from 7°–24°C, with optimum yields between 13° and 21°C. Some peas tolerate frost to −6°C (to −9°C). Temperatures of −7° to −8°C can be harmful at germination, −2° to −3°C at flowering, and −3° to −4°C at fruiting. Minimum temperature for germination is 4°C, maximum 24°C. With little salt tolerance, peas experience yield declines with EC = 3–5 mmhos. Seldom yield well in tropics below 1,200 m or when grown in cool weather as a winter crop. Hot dry weather interferes with proper seeding. Temperatures above 27°C shorten the growing period and adversely affect pollination. A hot spell is more damaging to peas than a light frost. In cool, temperate climes, grown in spring; in the South and warmer parts of California, grown during fall, winter, and early spring. Peas can be grown successfully during midsummer and early fall in those areas having relatively low temperatures and a good rainfall, or where irrigation is practiced. Annual rainfall of 8–10 dm, evenly distributed, is recommended. Peas require a moderate soil fertility, but can be grown on a great variety of soils from light, sandy loams to heavy clays, and chalky and limestone soils (U.K.). For very early crops, a sandy loam is preferred; for large yields where earliness is not a factor, a well-drained clay loam or silt loam is preferred. Peas will not thrive on soggy or water-soaked land. Soil pH should be 5.5–6.5, 5.5–6.8 or 6.0–7.5 according to three authors. In Maine, peas are grown on potato soils with a pH of 5.0–5.3; however, soils for peas should not be limed which would cause scab on potatoes. Ranging from Boreal Moist to Wet through Tropical Very Dry Forest Life Zone, peas are reported to tolerate annual precipitation of 0.9–27.8 dm (mean of 137 cases = 9.2), annual mean temperature of 5.0°–27.5°C (mean of 137 cases = 12.9°C), and pH of 4.2–8.3 (mean of 122 cases = 6.3).

CULTIVATION: Crops propagated only from seed. Seed germinate slowly at temperatures as low as 4.5°C. Time required for emergence decreases rapidly as temperature increases to an optimum of 24°C. At higher temperatures germination is rapid, but seedlings may die from various pathogens in the soil. As temperature rises during growing season, yield drops off rapidly. In New York, yields are highest when seeds are planted during first 2 weeks of April; for each 2-week delay in planting, yield of shelled peas decreased about 400 kg/ha. Thorough preparation of soil is very important, especially when seed is broadcast or planted with a grain drill, as no subsequent cultivation is given. Fall plowing is desirable. Surface of field should be smooth, so that drill plants seeds at same depth.

When time of emergence varies, peas mature at different times, harvest is difficult to time and yield, grade and price could be depressed. When crop is grown for cannery, field surface is rolled smooth either before or after planting, or both, so that harvesting machinery is more effective. For the fresh pea market, seed is sown 5–7.5 cm apart, at depth of 2.5–7.5 cm, in rows spaced 0.6–0.9 m apart. Seed rate is 67–113 kg/ha. For dried peas, seed is sown 18 cm apart at 200–338 kg/ha. Field peas are often intercropped with cereals. In western Washington, growers get earlier peas for market by sowing seed thickly in coldframes or sash houses, and then transplanting them to field when seedlings are 5–7.5 cm high, being careful not to break the long taproot. Peas are relatively unresponsive to fertilizers, especially N. Amount and kind of fertilizer depend on the soil. In some areas drilling 300–400 kg/ha of dolomite with the seed increases yield ca. 600 kg/ha. On soils low in available P, a 1–2–1 ratio is used. Sometimes a 5–10–5 or 4–12–4 ratio may be used, applied 600–800 kg/ha, as for coastal plain soils of South Carolina. N fertilizer can either increase size and yield of peas or decrease yield, depending on the circumstance. Peas also need an abundant supply of K. Method of application of fertilizer is important, as seed can be severely injured by contact with fertilizer. Fertilizer should be placed in bands 6.5 cm to the side of and deeper than the seed. When machinery is not available for side placement, fertilizer should be drilled as a separate operation before planting. N may also be provided by inoculating the seed before planting with nitrogen-fixing bacteria *(Pseudomonas radicicola)*. In other areas green manure and animal manure supply N.

HARVESTING: Peas mature in 52–75 days from planting or 20–30 days from bloom. Peas grown for home use or for fresh market are picked by hand. Some growers make two or three pickings, others only one, in which vines are pulled and all pods picked. Cost of picking is less by this method, but quality and yield of peas are higher from two pickings. Peas for processors are harvested with machines of various types. Sometimes vines are cut with mowing machine, windrowed and loaded on trucks with hay loader. Pea harvesters that mow the peas and load directly into a truck are common in main growing areas. Pea vines are hauled to a vining station, where pods are separated from vines, after which seeds are separated from pods. Peas are then shipped quickly in boxes, hampers or baskets to processing plants; quality, which determines price paid to grower, is determined by tenderness, high sugar content and size. All peas, for the fresh market or processing, benefit by precooling, which retards rate of conversion of sugar to starch. Shelled peas deteriorate more rapidly than do peas in the pod. For shipment to distant markets peas must be cooled quickly (usually in cold water), kept cool with crushed ice during transit, then kept cool until sold. Peas keep 2 weeks or more at 0°C and 85–90% relative humidity; average freezing point is −0.5°C. At 21°C, sugar quickly converts to starch, and pathogens grow rapidly and spoil the peas. For green peas, pods are picked when well filled, but seeds are still immature. Pod-peas are picked when pods are filled but still flat. For dried peas, pods are harvested when yellow, but not open.

YIELDS AND ECONOMICS: Average seed yields vary from 400 kg/ha in Kenya to 1,800 kg/ha in New Zealand. Yields of pods may attain 225–375 bu/ha and of shelled peas, about 2.25 MT/ha. Throughout temperate regions both green and dried peas are an important garden and field crop. In the United States ca. 550,000 MT are produced commercially for food annually, and ca. 200,000 MT of field peas for feed. In 1975, Asia, excluding the USSR, produced 5,575,000 MT dry peas (avg. 997 kg/ha); the USSR, an estimated 3,700,000 MT (avg. 949 kg/ha); Europe 632,000 MT (avg. 1,802 kg/ha); North and Central America, 218,000 MT (avg. 1,655 kg/ha); Africa, 294,000 MT (avg. 620 kg/ha); Oceania 110,000 MT (avg. 1,933 kg/ha); and South America, 93,000 MT (avg. 737 kg/ha). Yield was highest for Israel at 5,200 kg/ha, followed by East Germany at 3,793 kg/ha, the Netherlands at 3,646, and Denmark at 3,500 kg/ha. The United Kingdom averaged 2,923 kg/ha, the United States 1,725. (FAO, 1976).

BIOTIC FACTORS: The most serious diseases of peas are the following: *Ascochyta pisi* (carried by seed; said to be controlled by benomyl), *Cladosporium pisicola* (leaf spot or scab), *Erysiphe polygoni* (powdery mildew; controlled by sulfur), *Fusarium oxysporum* (wilt), *Peronospora pisi* (downy mildew, favored by high humidity and temperatures 10°–21°C), and *Phytomonas pisi* (bacterial blight). Other fungi that attack peas are: *Acremoniella atra*,

Pisum sativum L.

A. verrucosa, Alternaria brassicae, A. tenuis, Aphanomyces euteiches (root rot, bad in wet summers with soil temperatures 22°–28°C), *Aristastoma oeconomicum, Ascochyta pinodella, Aspergillus flavus, Botrytis cinerea, Cercospora lathyrina, C. pisisativae, C. canescens, C. szechuanensis, Chaetomium bostrychodes, Ch. cochliodes, Ch. dolichotrichum, Ch. elatum, Ch. funicola, Ch. globosum, Ch. indicum, Ch. murorum, Ch. reflexum, Ch. succineum, Choanephora cucurbitarum, Ch. conjuncta, Cladosporium herbarium, Collectotrichum pisi, C. lindemuthianum, Coniothyrium diplodiella, Corticium solani, Cunninghamella echinulata, C. elegans, Curvularia geniculata, C. inaequalis, C. lunata, Cylindrocladium scoparium, Cylindrosporium pisi, Didymella pinodes, Didymosphaeria pinodes, Epicoccum neglectum, E. purpurascens, Erysiphe communis, E. pisi, Fusarium acuminatum, F. anguioides, F. avenaceum, F. culmorum, F. equiseti, F. falcatum, F. graminearum, F. javanicum, F.lactis, F. lateritium, F. martii, F. merismoides, F. moniliforme, F. orthoceros, F. poae, F. redolens, F. roseum, F. semitectum, F. scirpi, F. solani, F. sporotrichoides, F. vasinfectum, Fusicladium pisicola, Helminthosporium sativum, Heterosporium maculatum, Hypomyces solani, Ischnochaete pisi, Leveillula taurica, Lichtheimia corymbifera, L. ucrainica, Macrophomina phaseoli, Macrosporium commune, M. nodipes, Melanospora papillata, M. zamiae, Microascus variabilis, Momnoniella echinata, Monilinia fructicola, Mucor plumbens, M. racemosus, M. sphaerosporus, Mycosphaerella phaseolicola, M. pinodes, M. tulasnei, Myrothecium verrucaria, Nigrospora sphaerica, Oidium balsamii, O. erysiphoides, Oospora lactis, Ophiobolus graminis, Papillaria arundinis, P. sphaerosperma, Pellicularia filamentosa, Penicillium cyclopium, P. kapuscinskii, P. terrestris, P. verrucosum, Periconia circinata, P. pycnospora, Pestelotia funerea, Petriella asymmetrica, Phoma terrestris, Phyllosticta pisi, Phymatotrichum omnivorum, Phytophthora parasitica, Ph. cactorum, Pleospora armeriae, P. herbarum, P. hyalospora, Pullularia pullulans, Pythium acanthicum, P. adhaerens, P. aphanidermatum, P. debaryanum, P. dissotocum, P. graminicola, P. helicoides, P. intermedium, P. irregulare, P. myriotylum, P. oligandrum, P. polymorphum, P. pulchrum, P. salpingophorum, P. splendens, P. ultimum, P. vexans, Rhizoctonia solani, Sclerotinia fuckeliana, S. homeocarpa, S. sclerotiorum, S.trifolium, Septoria apii, S. leguminum, Sclerotium rolfsii, Sordaria fimicola, S. inaequalis, Sporocybe byssoides, Stachybotrys atra, Stemphylium botryosum, S. consortiale, S. polymorphum, Syncephalastrum fulginosum, Thielavia basicola, Th. sepedonicum, Tieghemella italica, Trichocladium asperum, Trichoderma viride, Uromyces pisi, U. fabae, Verticillium albo-atrum*. The following bacteria also cause diseases in peas: *Bacillus megatherium, B. lathyri, Bacterium carotovora, B. rubefaciens, Corynebacterium fascians, Erwinia* sp., *Pseudomonas cannabina, Ps. pisi, Xanthomonas alfalfae, X. badrii, X. cassiae, X. pisi. Cuscuta pentagona, Orobanche speciosa* and *Striga hermonthica* are higher plants known to parasitize peas. Many viruses attack peas: Abutilon mosaic, alfalfa mosaic, alsike clover mosaic 1, Argentine sunflower, bean chlorotic ringspot (4 strains), bean local chlorosis (7 strains), bean mosaic, bean necrosis, bean yellow mosaic, broadbean mottle, clover yellow mosaic, common pea mosaic (*Marmor pisi, M. trifolii, M. efficiens, M. leguminosarum*), cucumber mosaic, dwarf mosaic, enation mosaic, pea enation mosaic, lettuce mosaic, lucerne mosaic, pimple pod, spotted wilt (*Lethum*), soybean stunt, streak, subterranean clover stunt, sweet pea streak, tomato spotted-wilt, watermelon mosaic, white clover mosaic, Wisconsin pea streak, pea wilt (*Marmor repens*), and yellow bean mosaic. The following nematodes are known to infest pea plants: *Belonolaimus gracilis, B. longicaudatus, Ditylenchus allii, D. dipsaci, Helicotylenchus dihystera, H. erythrinae, Heterodera cruciferae, H. galeopsis, H. glycines, H. goettingiana, H. schachtii, H. trifolii, Hoplolaimus uniformis, Longidorus maximus, L. pisi, Meloidogyne arenaris, M. artiella, M. hapla, M. incognita, M. i. acrita, M. javanica, M. marioni, Nacobbus batatiformis, Pratylenchus brachyurus, P. coffae, P. crenatus, P. globulicola, P. penetrans, P. thornei, P. vulnus, Rotylenchus reniformis, R. robustus, Trichodorus christiei, T. primitivus, Tylenchoryhnchus brevidens, T. claytoni, T. dubius, T. martini*, and *Xiphinema americanum*. Among the insects and pests the following may be serious: wireworms and garden symphylans may damage peas at germinating period; alfalfa looper (*Autographa californica*); pea weevils (*Bruchus pisorum*); cutworm (*Agrotis segetum*); Lygus bugs (*Lygus* spp.) damage developing pods; pea aphid (*Acyrthosiphon pisum*), most damaging insect; pea leafminers (*Liriomyza* spp.) breed in sugar beet fields;

western yellow-striped armyworm *(Spodoptera praefica)*; pea leaf weevil *(Sitona lineatus)*, in United Kingdom, the pea moth *(Laspeyresia nigricana)* is one of the most serious pests, (azinphosmethyl, carbaryl, fenitrothion, and tetrachlorvinphos have been recommended as controls in United Kingdom).

REFERENCES:

Gentry, H. S., 1971, Pisum resources, a preliminary survey, *Plant Genet. Res. Newsl.* **25**:3–13.

Zohary, D., and Hopf, M., 1973, Domestication of pulses in the Old World, *Science* **182**:887–894.

CONTRIBUTORS: J. A. Duke, C. F. Reed, J. K. P. Weder, D. Zohary.

Psophocarpus tetragonolobus (L.) DC.

FAMILY: Fabaceae

COMMON NAMES: Winged bean, Asparagus pea, Four-angled bean, Goa bean, Manila bean, Princess pea

USES: Winged bean is primarily cultivated for its immature edible pods, and cooked as a vegetable; young pods are even eaten raw. Leaves, young sprouts, flowers, and fruits are also used as vegetables and in soups. Dry seeds are often made into a fermented food product "tampeh" commonly used in Indonesia. The seed oil, similar to that of soybean, can be used for cooking, illumination, and soap. Oilcake can be used as food or forage. In Java, ripe seeds are roasted and eaten with rice. The flowers are added to various dishes to color them blue. Fried flowers are said to taste like mushrooms. In Java, a fungus, *Synchytrium psophocarpi,* makes young shoots swell; swollen parts are steamed as a delicacy. In Burma and Papua New Guinea tuberous roots are eaten raw or cooked. Chop is also used as green manure, cover, forage, and because of exceptional nodulation as a restorative crop. When sugarcane followed winged bean, yields increased by 50%. Recently suggested as a cover and cash crop in rubber plantations.

FOLK MEDICINE: Seeds are said to be aphrodisiac. In the Shan States (Burma) the root is used as a poultice to treat vertigo. Leaves are utilized with other items in a Malayan treatment for smallpox.

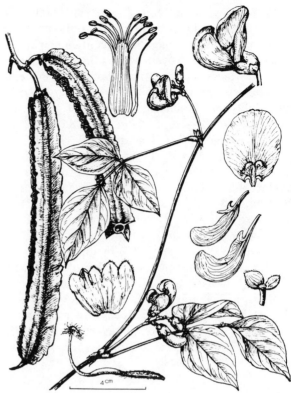

Figure 97. *Psophocarpus tetragonolobus* (L.) DC.

CHEMISTRY: Green pods, eaten like stringbeans, contain per 100 g edible portion: 34 calories, 89.5% moisture, 1.9 g protein, 0.1 g fat, 7.9 g total carbohydrate, 1.6 g fiber, 0.6 g ash, 53 mg Ca, 48 mg P, 0.2 mg Fe, 3 mg Na, 205 mg K, 340 μg β-carotene equiv., 0.19 mg thiamine, 0.08 mg riboflavin, 1.0 mg niacin, and 21 mg ascorbic acid. Raw leaves contain per 100 g: 47 calories, 85.0% water, 5.0 g protein, 0.5 g fat, 8.5 g total carbohydrate, 1.0 g ash, 134 mg Ca, 81 mg P, 6.2 g Fe, 3.1 μg β-carotene equiv., 0.28 mg thiamine, and 29 mg ascorbic acid. Dried seeds contain per 100 g: 405 calories, 9.7% moisture, 32.8 g protein, 17.0 g fat, 36.5 g total carbohydrate, 4.1 g ash, 80 mg Ca, 200 mg P, and 2.0 mg Fe. Seeds are reported to contain trypsin inhibitors and chymotrypsin inhibitors. Dried roots contain per 100 g: 9% moisture, 24.6 g protein, 1 g fat, 56.1 g carbohydrate, 5.4 g fiber, and 3.9 g ash. Stems and leaves, used as forage, contain per 100 g: 78.9% moisture, 6.3 g protein, 4.8 g digestible protein, 1 g fat, 7.9 g carbohydrate, 4.1 g fiber, 1.8 g ash, 0.37 g CaO, and 0.12 g P_2O_5. In oven-dried material, leaf protein of 10 introductions ranged from 24.5 to 31.5%, compared with

Psophocarpus tetragonolobus (L.) DC.

38.5% for Ceylonese material. The trypsin inhibitor is not destroyed by boiling, and only partially destroyed by autoclaving. Seeds contain hemagglutinins that are destroyed with sufficient cooking. HCN has been reported from the stems. Proteins of seeds are comparable to those of soybean in digestibility and composition. Per 100 g protein, seeds contain 1.6–2.6 g cystine, 7.4–8.0 g lysine, 2.7 g histidine, 6.5–6.6 g arginine, 11.5–12.5 g aspartic acid, 4.3–4.5 g threonine, 4.9–5.2 g serine, 15.3–15.8 g glutamic acid, 6.9–7.6 g proline, 4.3 g glycine, 4.3 g alanine, 4.9–5.7 g valine, 1.2 g methionine, 4.9–5.1 g isoleucine, 8.6–9.2 g leucine, 3.2 g tyrosine, 4.8–5.8 g phenylalanine, and 1.0 g tryptophan. Of the fatty acids in the oil, 71% are unsaturated. By weight 0.06% of the fatty acid is myristic, 8.9–9.7 palmitic, 0.83 palmitoleic, 5.7–5.9 stearic, 32.3–39.0 oleic, 27.2–27.8 linoleic, 1.1–2.0 linolenic, 2.5 parinaric, 2.0 arachidic and 13.4–15.5% behenic acid. Seeds are rich in the antioxidant tocopherol (125.9 mg/100 g oil), which improves human utilization of vitamin A, often deficient in the Tropics.

DESCRIPTION: Perennial vine, twining, glabrous, usually grown as an annual; roots numerous with the main long lateral roots running horizontally at a shallow depth, later becoming thick, tuberous, and nodulous; stems produced annually, 2–3 m long, ridged; leaves trifoliolate, the leaflets generally broadly ovate, acute, the margin entire, glaucous beneath, 8–15 cm long, 4–12 cm wide, with small, 2-parted stipules; flowers 2–10, in axillary racemes, up to 15 cm long; corolla large, much exserted, with the broad standard much flexed and deeply emarginate, auricled at base, pale green on back, white or pale blue within, 2.5–4.0 cm in diameter; pods more or less square, 6–30 cm or more long, 2.5–3.5 cm wide, with 4 longitudinal jagged ridges, 2-valved; seeds 4–17 per pod, nearly globular, up to 1 cm long, white, black, brown, or yellow, often with varying markings, smooth; 100 seed weight ca. 25 g.

GERMPLASM: Much variation in pigmentation, pod length, shape of pods, seed and tuber yield, and a host of other characteristics. Papua New Guinea and Indonesian Islands appear to be the center of greatest genetic diversity. Assigned to the Indochina-Indonesian Center of Diversity, winged bean or cvs thereof is reported to exhibit tolerance to heat, laterite, and low pH. ($2n = 18$.)

DISTRIBUTION: The center of origin is not established, but Mauritius or Malagasy was suggested. However, it is widely cultivated in Burma, India, Indonesia, and Papua New Guinea, but not in Africa until recently. In the rest of Southeast Asia it is grown mainly as a backyard vegetable. It has been distributed in West Africa and the West Indies.

ECOLOGY: Winged bean thrives in moist tropics at altitudes from sea level to 2,000 m. Grown in a range of soil types, but good drainage is essential. Said to do better with annual precipitation more than 15 dm and to thrive with 25 dm or more. It is sensitive to drought, frost, salinity, and waterlogging. Malayan plants may develop more than 400 nodules, with no apparent need for inoculation (probably due to a widespread *Rhizobium* of the cowpea type). A single nodule may be 1.2 cm in diameter and weigh 0.6 g. Ranging from Subtropical Dry to Wet through Tropical Very Dry to Wet Forest Life Zones, winged bean is reported to tolerate annual precipitation of 7.0–41.0 dm (mean of 15 cases = 19.3 dm), and annual mean temperature of 15.4°–27.5°C (mean of 15 cases = 25.2°C), and pH of 4.3–7.5 (mean of 14 cases = 5.8).

CULTIVATION: In Burma seed is sown 2.5–7.5 cm deep at the beginning of rains in drills where plants remain, spaced 60 × 120 cm, requiring stakes or trellis for support. Spaced much closer (ca. 10 cm apart) in Papua New Guinea. It is intercropped with bananas, sugarcane, sweet potatoes, taro, and other vegetables or legumes. Plants for tuber production are allowed to trail in Burma and are given short stakes for support in Papua New Guinea. Flowers are plucked to increase root yields. For cover crop, no stakes are required. Trellised plants may yield more than twice as much seed as unstaked plants. The crop grows easily but during first months it requires some attention, such as manuring, weeding, and hoeing. After fruiting, the top dies, but plant is perennial; stored food allows resumption of growth the following year when rains begin. Papuan cvs tend to be annual in growth habit. Many cvs exhibit senescence after pod maturity.

HARVEST: First green pods are ready for consumption 6–10 weeks after sowing (2 weeks after pollination). After 3 weeks, pods become tough; after 3 more weeks seeds are mature. The plant may bear pods indefinitely but production declines, so that the plant, although perennial, is often treated as an annual. Each vine is said to yield a small meal of about 25 pods every 5 or 6 days, but manuring is necessary every 2 or 3 weeks to maintain productivity. Fully mature seed may require 180–270 days from planting. One Ghanan cv is said to mature seed in 114 days.

YIELDS AND ECONOMICS: Dry seed yields generally are 500–1,000 kg/ha but have exceeded 2,000 kg/ha have in Nigeria, Malaysia, Papua, New Guinea, and Western Australia. Tuber yields are 2,000–10,000 kg/ha. Experimental green-fruit yields exceed 35,000 kg/ha in Malaysia (Kay, 1978). Nodules may total 750 kg/ha. Cultivated as a backyard or garden crop in most of Southeast Asia, and consumed locally; grown as field crop in Papua New Guinea. No statistics available on exportation or import. Not now profitable commercially, but useful because it bears pods when winter vegetables are out of season. With improved cvs and agronomic methods, might be grown profitably with intermediate technology.

BIOTIC FACTORS: Winged beans are attacked by the following fungi: *Cercospora arantae, C. canescens, C. cruenta, C. psophocarpi, Corticium solani, Corynespora cassicola, Erysiphe cichoracearum, Meliola erythrinae, Myrothecium roridum, Oidium* sp., *Periconia byssoides, Pythium debaryanum, Sporidesmium bakeri,* and *Synchytrium psophocarpi.* Mosaic of *Crotolaria hirsuta* and yellow mosaic of *Cajanus cajan* are viruses that attack winged bean. Winged-bean nematodes include *Meloidogyne javanica* and *M. incognita.* Attacks by eelworms *(H. radicicola)* do not affect the growth or development of the plant, but *M. incognita* infestation is known to cause severe loss in yield and quality. Occasionally, damage is caused by caterpillars, grasshoppers, leaf miners, and spider mites. Although bumble bees have been regarded as essential to pollination in some areas (Ghana), it has more recently been concluded that bees are helpful, but certainly not essential.

REFERENCES:
Hymowitz, T., and Boyd, J., 1977, Origin, ethnobotany and agricultural potential of the winged bean—*Psophocarpus tetragonolobus, Econ. Bot.* **31**:180–188.
Khan, T. N., 1976, Papua New Guinea: A centre of genetic diversity in winged bean [*Psophocarpus tetragonolobus* (L.) DC], *Euphytica* **25**:693–706.
Khan, T. N., Bohn, J. C., and Stephenson, R. A., 1977, Winged Bean: Cultivation in Papua New Guinea, *World Crops* **29**:208–216.
Levy, J. (ed.), 1977, *The Winged Bean Flyer,* 2 Nos., Department of Agronomy, University of Illinois, Urbana.
Masefield, G. B., 1961, Root nodulation and agricultural potential of the leguminous genus *Psophocarpus, Trop. Agric. (Trinidad)* **38**:225–229.
Masefield, G. B., 1973, *Psophocarpus tetragonolobus*—A crop with a future, *Field Crop Abstr.* **26**:157–160.
National Academy of Sciences, 1975, *The Winged Bean—A High-Protein Crop for the Tropics,* National Academy of Sciences, Washington, D.C., 42 pp.
PCARR, 1978, Proceedings of the Workshop/Seminar on the Development of the Potential of the Winged Bean, Philippine Council of Agriculture and Resources Research, Los Bunos (in press).
Senanayake, Y. D. A., and Sumanasinghe, V. A. D., 1976, Leaf protein content of *Psophocarpus tetragonolobus* (L.) D.C., *J. Natl. Agric. Soc. Ceylon* **13**:119–121.

CONTRIBUTORS: J. A. Duke, T. N. Khan, C. F. Reed, J. K. P. Weder.

Pterocarpus erinaceus Poir.

FAMILY: Fabaceae

COMMON NAMES: African or Senegal rosewood, Apepe, Madobia

USES: Dye from the blood-red resin, mixed with palm oil, is used as a cosmetic by some tribes in West Africa and for dyeing cotton cloth in India. It is dark purple when mixed with shea butter. Tree is not large enough for good timber but the wood is hard, fine-grained, a beautiful rose-red, elastic and is used for making rice mortars and pestles, stools, spindles, keys for xylophone, and another musical instrument called a 'Balangi' in Sierra Leone. The wood is durable enough for fenceposts and for construction of better-class native dwellings, canoes, and some carpentry. Bows are made from the roots. Bark is used in tanning. Foliage makes a fairly good forage for domestic animals, and trees are planted for this purpose on stock farms. Raw seeds are intoxicating. The resin, 'Kino,' is similar to the product from *P. marsupium,* the official Indian 'Kino.' Kino is crushed with a mallet and applied as a cloth glaze in West Africa.

FOLK MEDICINE: Medicinally, it is used as an astringent for severe diarrhea, urethral discharges

Pterocarpus erinaceus Poir.

Figure 98. *Pterocarpus erinaceus* Poir.

and in abortifacients. Dried pulverized bark and kolanut used as restoratives, and applied locally as a dressing for chronic ulcers, and ringworm of the scalp. Infusion of leaves used for fevers. The bark is said to be used for astringent, diarrhea, dysentery, and ecbolic, and the resin for wounds.

CHEMISTRY: Leaves contain 19% crude protein, 35.5% crude fiber, 2.44% K, and small amounts of Ca, Na, Fe, P, Cl, and S. Seeds of the genus *Pterocarpus* are reported to contain trypsin inhibitors.

DESCRIPTION: An erect, small tree, up to 17 m tall and 2.3 m in girth, buttressed when old, crown high, giving little shade; bark almost black, rough and scaly; slash yellow, streaked with deep red and sticky gum or resin; branchlets densely pubescent; leaves pinnate; leaflets up to 12, alternating on rachis, 10 cm long, 5 cm broad, mostly oblong-elliptic, with numerous parallel lateral veins, shortly pubescent below; flowers bright yellow, conspicuous, in masses when tree is almost leafless; calyx softly tomentose; pods papery, ca. 7.5 cm in diameter, persistent, 1-seeded, bristly, flat; seeds covered with prickles and surrounded by a membranous wing. Fl. March (Nigeria); Aug.–Jan. (Ghana); Fr. Dec.–Jan.

GERMPLASM: Assigned to the Africa Center of Diversity, African rosewood is reported to exhibit tolerance to low pH and savanna. ($2n = 22$.)

DISTRIBUTION: Native to West Tropical Africa (Nigeria, Togo, Ghana, Sierra Leone, Gambia, Senegal to Chad, and Gaboon).

ECOLOGY: Trees confined principally to the drier parts, growing in open dry savanna forests and along fringe of mixed forests, along river banks, but more frequently in open forests. Thrives on shallow soils of West Africa. Occurs in small groups scattered throughout the savanna forests. Ranging from Subtropical Moist through Tropical Moist Forest Life Zone, African rosewood is reported to tolerate annual precipitation of 17.3–40.3 dm (mean of 3 cases = 28.5), annual mean temperature of 23.5–26.6°C (mean of 3 cases = 25.2), and pH of 4.5–5.0 (mean of 2 cases = 4.8).

CULTIVATION: Propagated by seeds planted in places where the foliage may be needed as forage, mainly on stock farms. Trees require little attention once established.

HARVESTING: Trees are cut for domestic purposes whenever needed, especially for the wood, bark, and resin. The foliage is grazed or branches are cut for forage when needed; trees regenerate. Timber wood is produced mainly in Togo, Nigeria, and Sierra Leone and is substituted for Camwood (*Baphia nitida*) where that wood is not found. It is often sold as African 'Rosewood' or as teak.

YIELDS AND ECONOMICS: A small tree valued locally in tropical West Africa for its timber, which is mainly produced in Togo and Nigeria and, to a lesser degree, elsewhere in that region of Africa. Sierra Leone exports to France. Woods used extensively locally. Foliage used for forage on stock farms.

BIOTIC FACTORS: The following fungi have been reported on this tree: *Phomopsis pterocarpi, Phyllachora pterocarpi (Catacauma pterocarpi),* and *Pseudothis congensis (Homostegia pterocarpi).*

REFERENCES:

Githins, T. S., 1948, *Drug Plants of Africa,* p. 100, *African Handbooks*, No. 8, University of Pennsylvania Press, Philadelphia, Pa.

Unwin, A. H., 1920, West African forests and forestry, in: *The Nigerian Timber Trees,* Chapter IX, pp. 46, 123, 133, 206, 273, New York.

CONTRIBUTORS: J. A. Duke, C. F. Reed, J. K. P. Weder.

Pterocarpus santalinus L. f.

FAMILY: Fabaceae

COMMON NAMES: Red sandalwood, Lal chandan, Red sanders

USES: Wood valuable as dyewood, contains the red, resinous santalin, (soluble in alcohol, essential oils, and ether, but not in water). As a dye for cloth, produces a fast salmon-pink. Red sandalwood is very valuable for dyeing leather red, staining wood and in calico printing. In India and Pakistan it is used for dyeing silk and cotton. Colors as carmine blue, scarlet red, deep violet, or brown are produced on wool, cotton, and linen according to the mordant used. Wood, which takes a high and lasting polish, is dark, claret-red, almost black, extremely hard, resistant to termites; timber is highly valued and used for cart shafts, yokes, spokes of wheels, plough shafts, carvings, house posts, and agricultural implements. Heartwood used for carving dolls and images.

FOLK MEDICINE: Medicinally, the wood is astringent and is used as cooling agent for external applications for inflammation, headaches, fevers, boils, as diaphoretic, and for scorpion stings. Said to be used for depurative and emetic.

CHEMISTRY: Wood contains the resinous santalin and fresh shoots yield a glucoside coloring material.

DESCRIPTION: Small to medium-size tree; stems erect with rounded crown, bole 5–6.6 m high; branches obscurely gray-downy; leaves compound with 3 (rarely 5) broad-elliptic leaflets, 3.5–10 cm long, rounded at both ends, slightly emarginate; veins fine, pale pubescent beneath; flowers few, in short racemes, on rather short pedicels; calyx 0.5–0.6 cm long, teeth deltoid, minute; limb of standard not longer than the calyx; stamens 2–3-adelphous; pod 3.5 cm long, oblong, gradually narrowed into a short stalk, silky at first, center turgid, winged near the base; stalk of pod much exceeding the calyx.

GERMPLASM: Assigned to the Hindustani Center of Diversity, red sandalwood or cultivars thereof is reported to exhibit tolerance to drought, fire, high pH, heat, laterite, low pH, poor soil, slope, and wind. ($2n = 22$.)

DISTRIBUTION: Native to South India, in e Deccan from Godaveri to Palar River, and in districts of Madras, Bombay, and Bengal.

ECOLOGY: Fares well in dry, rather rocky soils (gneiss, gravel, laterite, quartzite, shale), in a hot fairly dry climate. Also grows well on aluvial soils but does not tolerate waterlogging. Demands strong light, does not tolerate shade. Ascends to 500 m in hills of Deccan. Ranging from Tropical Dry to Tropical Moist Forest Life Zone, red sandalwood is reported to tolerate annual precipitation of 13.6–27.8 dm (mean of 3 cases = 18.9), annual mean temperature of 24.2°–26.6°C (mean of 3 cases = 25.7), and pH of 4.5–5.0 (mean of 2 cases = 4.8).

CULTIVATION: Tree reproduces vigorously by seed or suckers. Seeds are gathered in May or June and

Figure 99. *Pterocarpus santalinus* L. f.

Pterocarpus santalinus L. f.

sown, within easy reach of water, in July in small beds ca. 2.6 m square, that accommodate 700–800 seeds. They are thrust into a light soil perpendicularly or at an angle, and ca. 2.5 cm deep. Seeds should be watered by hand every other day in the evening. After soaking overnight in water, seeds germinate in 20–25 days, otherwise, in 30–35 days. After germination, the beds should be watered moderately for the first 6 months. Too much water can be as destructive as too little or none. Nursery should be kept free of weeds. At 6 months shoots require the support of a forked stick and are transferred to wicker or bamboo baskets, with care not to injure the long taproots. Baskets must be placed in a shady place and watered every 2 or 3 days. When plants are rooted firmly, baskets should be buried in pits and watered until the rains set in. Then plants may be put down in the plantation. In some areas, as in Bombay, seeds are sown in September, plants allowed to grow for only 3 yr, then pulled; small roots cut off, dried in the sun and the dye is extracted. In other regions, as in Ratnagiri District, the tree is not cultivated, but grows in the woods, where the dyestuff is collected for export to Bombay.

HARVESTING: Trees are harvested from mature stands; some hill areas in eastern Cuddapah contain more than 10% of red sandalwood. However, the felling of trees for dye is under strict government control and harvest and prices depend on the quantity in the market. The total area covered by this species in eastern and western Cuddapah is more than 1,000 sq. mi. The wood contains the red coloring matter called santalin that is easily extracted with any alkaline solution. Cloth is dyed by boiling not merely soaking, with the wood; color is said to be fast.

YIELDS AND ECONOMICS: In eastern Cuddapah, the largest producing area in India, yields of the dyestuff average ca. 1,000 MT/yr, and of fuel, about 25,000 MT/yr. South India is the major producer of the dyestuff and timber. Trees are well protected and cultivated. Timber always finds a ready market.

BIOTIC FACTORS: Trees are sometimes parasitized by the flowering plant *Dendrophthoe falcata*. No other serious pests or diseases have been reported for the plant, and the wood is very resistant to termites.

CONTRIBUTORS: J.A. Duke, C. F. Reed.

Pterocarpus soyauxii Taub.

FAMILY: Fabaceae

COMMON NAMES: Barwood, Redwood, Arakpa, Large fruited camwood, African coralwood, Gabon padouk

USES: Tree provides barwood or redwood used by natives of West Africa as a dye. Wood exported from Cameroons as 'muenge' and commercial name 'African Padouk.' Wood is blood-red, medium heavy, difficult to plane, cut for local use for walking sticks, drums, buildings, wooden shovels, yam-pestles, and for making heavy furniture. In some areas considered the best canoe tree. Sapwood is thick and white turning brownish-yellow; heartwood red darkens on drying. Dye, from roots and stems, is used for fabrics and fibers such as raffia. Also mixed with palm oil and used as a cosmetic.

FOLK MEDICINE: Dry dead wood pulverized and used as a fetish medicine in medicomagical rites

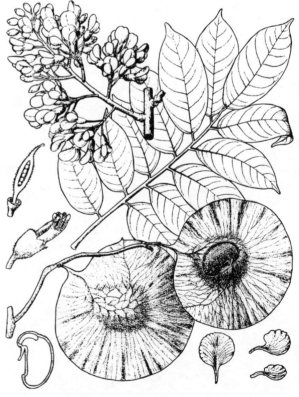

Figure 100. *Pterocarpus soyauxii* Taub.

and in fertility cults. Tannins from wood used as cure for animal skin parasites.

CHEMISTRY: No data available.

DESCRIPTION: Tree 27–34 m tall; bole length of 17 m, girth up to 3.3 m with undivided stem; bark reddish-gray, detaching in flakes and exuding on injury a red gum; leaves compound, unarmed; leaflets 11–13, alternate, linear-oblong to subobovate-oblong, apex obtusely acuminate, slightly emarginate, glabrous, lateral veins of leaflets close together, disappearing before the margin; calyx turbinate, the 2 upper teeth more or less connate, vexillum orbicular or broad-ovate; stamens all connate in a sheath; style curved round to the side of the base; fruits 6–9 cm in diameter, with numerous prickly thorns in center of each surface, indehiscent, obliquely orbicular, compressed. Flowers in pyramidal panicles.

GERMPLASM: Assigned to the Africa Center of Diversity.

DISTRIBUTION: Native to West Tropical Africa (Nigeria, Cameroons, Lower Ubangi, Gabon, particularly abundant in the Cross River area of Congo).

ECOLOGY: Common tree in mixed deciduous forests in tropical West Africa; in Cameroons in dense evergreen forests. Requires much light and deep rich moist tropical soil. Ranging from Subtropical Moist through Tropical Dry Forest Life Zone, barwood is reported to tolerate annual precipitation of 15–17 dm, and annual mean temperature of 23°C.

CULTIVATION: Propagated readily from fallen seeds not eaten by "ground pigs." Seedlings grow rather quickly. Sprouts grow from stump but do not attain any size. Most trees are tended in the forest until they are ready for cutting.

HARVESTING: Trees are cut from the forest and the heartwood is lumbered out. Dye is extracted from the roots and stems. The slash is white, but soon exudes the thin red latex. Dye is prepared by pounding thoroughly dried wood chips in a mortar or in natural holes in rocks. Pulverized material is moistened with water and molded for sale as cakes. Pomade is made by mixing with palm oil, for use as a cosmetic.

YIELDS AND ECONOMICS: Redwood or barwood is an important lumber exported from Cameroons. The wood is used locally in West Tropical Africa for many purposes.

BIOTIC FACTORS: Wood is very hard and not attacked by termites. The following fungi have been reported on this tree: *Coniophora cerebella, Merulius lacrymans, Polystictus versicolor,* and *Poria vaporaria.*

REFERENCE:
Baker, E. G., 1929, *The Leguminosae of Tropical Africa,* Part II, *Suborder Papilionaceae,* p. 542.

CONTRIBUTORS: J. A. Duke, C. F. Reed.

Pueraria lobata (Willd.) Ohwi

FAMILY: Fabaceae

COMMON NAMES: Kudzu, Ko, Ko-hemp, Japanese arrowroot

SYNONYMS: *Dolichos lobatus* Willd.
Pachyrhizus thunbergianus Sieb. & Zucc.
Pueraria thunbergiana (Sieb. & Zucc.) Benth.

USES: Kudzu is primarily grown for pasture, hay, and silage. It is palatable to all types of livestock. Kudzu is nearly equal to alfalfa in nutritive value. Leaves, shoots, and roots are eaten by some humans. Useful fiber is obtained from stems, and starch is obtained from the tuberous root (roots up to 35 kg each). In China and Japan, Ko-fen flour, made from the roots, is used in soups. Said to be cultivated for its tuber in the uplands of New Guinea and New Caledonia. Used for erosion control and soil improvement on banks, slopes, and gullies, where a permanent planting is desired. It is used as shade, planted around buildings. Once established, it may be difficult to eradicate.

FOLK MEDICINE: Chinese reportedly use the plants as a diaphoretic and febrifuge, the root decoction for colds, dysentery, and fever. The root starch is official in the Japanese pharmacopoeia. Shoots are used as a lactagogue.

CHEMISTRY: Raw roots contain per 100 g edible portion: 113 calories, 68.6% moisture, 2.1 g protein,

Pueraria lobata (Willd.) Ohwi

Figure 101. *Pueraria lobata* (Willd.) Ohwi

DESCRIPTION: Perennial coarse herbs with woody base; stems elongated, up to 20 m long, twining or prostrate, whitish-puberulent with coarse spreading or reflexed brown hispid hairs; stipules lanceolate, subacute, medifixed, 15–20 mm long, green; leaves trifoliolate, the leaflets green, loosely appressed, hirsute on upper surface, densely whitish-puberulent beneath, terminal lobe rhombicorbicular, 10–15 cm long and as wide, abruptly acuminate, sometime 3-lobed, lateral lobes often bifid; racemes densely many-flowered, nearly sessile or short-peduncled, 10–20 cm long; flowers reddish-purple, rarely almost white, 18–20 mm long, fragrant; bracts linear, 8–10 mm long, 0.2–0.3 mm wide, long-pilose, caducous; bracteoles caducous, narrowly ovate or broadly lanceolate, acute; lowest calyx-lobe 1.5–2 times as long as the tube; pods flat, densely dark brown, spreading-hispid, linear, 6–8 cm long, 8–10 mm wide. Seed maturing before frost, rarely setting seed northward. Seed small, about 100,000/kg. Fl. July–Sept. (U.S.).

GERMPLASM: Few cvs have been developed, and 'Kudze 23' is the only cv developed that produces more crowns than common; its finer leaves and stems make it especially valuable as forage. Most strains grow 35 m or more in a single season. Assigned to the Indochina–Indonesia and China–Japan Centers of Diversity, kudzu or cvs thereof is reported to exhibit tolerance to drought, frost, grazing, heavy soil, slope, virus, and weeds. ($2n = 24$.)

DISTRIBUTION: Native to Japan and the Orient, areas of Eastern Asia; rarely cultivated in Java. More widely cultivated in southeastern United States, being best adapted south of Virginia and Kentucky, west to Oklahoma and Texas, although it will grow as far north as New York and Lincoln, Nebraska.

ECOLOGY: A warm weather plant, growing from early spring until late fall. Aboveground parts killed by frost. Deep-freezing kills the entire plant. Grows on a wide range of soil types, but does not make good growth on very light poor sand or on poorly drained heavy clay. Cannot stand waterlogging on any soil. Grows best on well-drained loam soil of good fertility. On soils of low fertility, liberal use of manure and light application of superphosphate when plant is first cut is beneficial. Sometimes soil needs a complete fertilizer, other times lack of boron is a limiting factor. (Use borax at rate of 30

0.1 g fat, 27.8 g total carbohydrate, 0.7 g fiber, 1.4 g ash, 15 mg Ca, 18 mg P, and 0.6 mg Fe. Starch of roots contains per 100 g: 340 calories, 16.5% moisture, 0.2 g protein, 0.1 g fat, 83.1 g total carbohydrate, 0.1 g ash, 35 mg Ca, 18 mg P, 2.0 mg Fe, and 2 mg Na. Cooked leaves contain per 100 g: 36 calories, 89.0% moisture, 0.4 g protein, 0.1 g fat, 9.7 g total carbohydrate, 7.7 g fiber, 0.8 fat, 34 mg Ca, 20 mg P, 4.9 mg Fe, 0.03 mg thiamin, 0.91 mg riboflavin, and 0.8 mg niacin. Feeding trials on goats indicated that kudzu hay (protein, 10.3; total dig. nutrients, 28.7; and starch equivalent, 16.1 g/100 g) compared well with cowpea hay, berseem hay, and wheat bran in digestible protein value, but was inferior to legume hays in starch equivalent. Kudzu hay (92.3% DM) contained (moisture-free basis): 9.7–17.9% crude protein (avg. 13.8), 1.8–3.8% fat (avg. 2.5), 30.5–38.5% crude fiber (avg. 33.6), 3.6–9.6% ash (avg. 6.6), 38.2–53.6% N-free extract (avg. 43.5), 1.61% Ca, 0.47% P, and 0.81% Mg. The green forage (22.2% DM) contained (moisture-free basis): 14.4–20.0% crude protein (avg. 17.2%), 1.8–3.3% fat (avg. 2.1), 27.1–35.6% crude fiber (avg. 33.9), 7.2–8.3% ash (avg. 8.0), and 37.4–44.4% N-free extract (avg. 38.8).

kg/ha.) Kudzu is deep-rooted, therefore drought-resistant. However, it does not do well in the Tropics and dies out very quickly. Ranging from Warm Temperate Moist through Subtropical Moist Forest Life Zone, kudzu is reported to tolerate annual precipitation of 9.7 to 21.4 dm (mean of 15 cases = 13.5), annual mean temperature of 12.2°–26.7°C (mean of 15 cases = 18.1°C), and pH of 5.0–7.1 (mean of 13 cases = 6.1).

CULTIVATION: Kudzu is propagated from seed, cuttings, or crowns. Plants are usually started by rooting runners at the nodes and by transplanting 2-year-old plants. Plants can be started from both softwood and hardwood cuttings, but need special conditions and care (greenhouse misting; this method rarely used). The very hard seed coats should be scarified with acid or by mechanical means before planting to insure higher germination; even then, 70% germination is considered excellent. Seed sown very thick when planting. Usually not much seed is set, and then only on parts of the plant that have climbed up on a support. If allowed, kudzu will climb up and over trees eventually smothering them out. Seed planted in a nursery in well-drained soil of good structure, in rows 1 m apart, planting 15–25 seed/20 cm of row, 0.6–1.3 cm deep when soil is warm, depending on the locality, in early or late spring. Seedlings require about 4 months to develop 4–6 true leaves and one or more roots 1.3 cm in diameter and 15 cm long. At this stage they are ready to transplant to field about the time of first fall frost. Where seed is plentiful, it may be directly seeded in the field, allowing 1 kg/ha, with 10–12 seeds per 30 cm in rows 2 m apart. Fertilizer is applied at planting time. New stands must be cultivated and kept free of weeds the first year. Kudzu in the cotyledonous stage will withstand temperatures down to −7°C. It loses this tolerance to cold as the third and fourth leaves develop. Plants should be inoculated with the right strain of bacteria to insure maximum production. Important to protect plants from drying during planting and to tamp moist, well-prepared soil about them to prevent drying after planting. Holes should be deep enough for roots to be spread out to full length. Crown buds should be level with ground surface and very lightly covered with soil. Short-cuts in planting usually result in poor stands. On ordinary good land, kudzu will grow enough in one year to extend 14 m. When field is to be used for hay or grazing, spacing is not as important, as when the field is to be regularly planted or rotated to another crop. Spacing varies from 3 to 10 m between rows and 1.3 to 3.3 m apart in rows, usually requiring about 1,250 plants/ha. Established stands used for grazing or hay should receive 400–600 kg/ha superphosphate every second or third year, or 10 tons of good stable manure or mixture of smaller amounts of manure and a mineral fertilizer.

HARVESTING: Harvested in several ways, depending on usage. Kudzu gives in 2–3 yr a good ground cover which is long-lived if not overgrazed or mowed too often. *Hay:* Makes a good coarse hay, retaining its leaves after cutting, does not shed leaves appreciably during growing season, is palatable to all kinds of livestock, and can be fed with very little waste. Kudzu with its heavy viny growth is difficult to cut, particularly the first time, because the vines catch on the divider board of an ordinary mower; modified mowers have been developed. Hay should be harvested when vines and ground are dry. Leave hay in swath for several hours before windrowing. Following morning when dew is off, cut plants should be put in small stacks or the windrow turned, and in the afternoon it should be put in the barn or baled. Stacks are capped with a waterproof cover, e.g., canvas. *Pasture:* Kudzu makes good pasture, steers gaining more than 0.5 kg/day, averaging 107.5 kg season. Kudzu can be pastured from late spring until frost or even later. It is especially valuable as a reserve feed for periods of drought, but should not be grazed until second or third year. If growth is vigorous, it may be grazed lightly the second year. For maximum production, pasture should be divided into 2 or more plots and grazed alternately or in rotation. In fall, rye, oats, or a winter legume (crimson clover, burclover, or vetch) should be seeded in the kudzu pasture to prevent loss of plant food by leaching and to supply pasture before kudzu growth starts in spring. Livestock should be taken from the pasture before growth starts in spring. If pastured continuously, plants should not be grazed closer than 30–45 cm. If alternate or rotation grazing is practiced, plants can be grazed to 15–25 cm. *Silage:* Good silage can be made of kudzu by mixing it with grass, the mixture containing about 60% moisture. Total moisture content of kudzu at time of cutting is about 75%, so kudzu must be handled rapidly to prevent drying out too much. Cattle readily eat good silage.

Pueraria lobata (Willd.) Ohwi

YIELDS AND ECONOMICS: Forage yield of 5 MT/ha are expected from good stands on fertile soils. About 25,000 plants or crowns can be harvested from an ordinary well-established plantation of 1 ha. Under ideal conditions, twice that number. Planting stock should be left in the field until needed. However, if large numbers of plants are being handled for commercial sale, and it is necessary to dig them before the planting season, they should be stored in a cool, well-ventilated place, and should be heeled-in in moist sphagnum moss or in soil with ample but not too much moisture.

BIOTIC FACTORS: Kudzu is said to be cross pollinated and bees are the reported pollinators. Velvetbean caterpillars *(Anticarsia gemmatilis)* eat the leaves. Nematodes attacking the roots include: *Meloidogyne hapla, M. incognita acrita, M. javanica, M. thamesi,* and *Rotylenchulus reniformis.* Kudzu is attacked by several fungi: *Alternaria* sp. (leafspot), *Colletotrichum lindemuthianum* (anthracnose), *Fusarium* sp. (stem rot), *Macrophomina phaseoli* (charcoal rot), *Mycosphaerella puericola* (angular leaf-spot), *Pellicularia solani* (damping-off). It is also attacked by the bacteria *Pseudomonas phaseolicola* and *Ps. syringae* (bacterial blight and halo blight).

CONTRIBUTORS: J. A. Duke, C. F. Reed.

Figure 102. *Pueraria phaseoloides* (Roxb.) Benth.

Pueraria phaseoloides (Roxb.) Benth.

FAMILY: Fabaceae
COMMON NAMES: Tropical kudzu, Puero
SYNONYM: *Pueraria javanica* Benth.

USES: Tropical kudzu makes good pasture and hay for cattle, and is used for erosion control in tropical areas where common kudzu *(P. lobata)* does not thrive. Tropical kudzu has many of the good qualities of common kudzu, seeds more heavily, establishes readily from seed, and grows year round. With regular rains it thoroughly covers the ground in about 6 months. In Java it is cultivated as green manure; in the Far East, used as ground cover to prevent soil erosion and fertility loss after clearing for cinchona, citrus, coconut, oil-palm, and rubber plantations. Tropical kudzu does not dry out as much as other cover crops, thus reducing fire danger during drought. It persists indefinitely in open areas, but as trees grow and shade gets denser, it thins out and may disappear. When used in Tanzania, it is used as a cover crop to control weeds, permitting increased yields of sisal. The tuberous roots are said to be eaten in parts of Asia. Strong fibers from the stem are used for rope making can be extracted.

FOLK MEDICINE: Used for boils and ulcers in Malaya.

CHEMISTRY: Nodules contain 2½ times, leaves 2 times and roots (minus nodules) 1½ times as much N as the stems. Fresh leaves and stems of tropical kudzu have the following chemical composition and nutritive value: moisture, 80.9; protein 3.8; fat, 0.4; soluble carbohydrates, 7.9; fiber, 5.5; ash, 1.5; calcium, 0.14; and phosphorus, 0.03%; digestible nutrients: protein, 2.9; fat, 0.2; and soluble carbohydrates, 6.5%; nutritive ratio, 3.5; and starch equivalent, 12.9 kg/100 kg. The green feed is rich in vitamins C and A and also contains B vitamins. The manurial constituents of tropical kudzu are

Pueraria phaseoloides (Roxb.) Benth.

(DM basis): N, 1.8; Ca, 0.58; P, 0.24; and K, 1.13%.

GERMPLASM: Many introductions have been made in the United States, mainly from India, Philippines, Java, Liberia, Nigeria, Congo, and several West Indies. Assigned to the Indochina-Indonesian Center of Diversity, tropical kudzu or cvs thereof is reported to exhibit tolerance to disease, drought, fungus, heavy soil, insects, laterite, low pH, poor soil, slope, and weeds. ($2n = 22$.)

DISTRIBUTION: Native to lowland Malaysia, has been cultivated in Far East (Sumatra, Java, Sri Lanka, south China) for many years. More recently it has been introduced in Africa (Zanzibar, where it is the most valuable ground cover; Tanzania), West Indies, and southern United States.

ECOLOGY: Tropical kudzu is suitable for frost-free tropical climates with fair to high rainfall, and moderate to high temperatures. Growing well in full sun or beneath moderate shade, it is difficult to establish in shade and intolerant of dense shade. Relatively drought-resistant. Grows from sea level to 1,000 m in areas where winter night temperatures drop to 10°C. Grows luxuriantly under average rainfall of 20 dm, with a winter dry season of 3–4 months, but not as well under 12 dm. In Liberia, it thrives in rubber plantations with annual average rainfall of 32 dm, with a Nov.–Feb. dry season. Stands considerable waterlogging and grows on low, recently drained swamps where the water table is 15–20 cm below the surface. More difficult to establish on fertile lowland soils, as weeds compete with seedlings; once established, it competes successfully. Grows on red gravelly lateritic soils that are definitely acid, but also on fertile lowland and rocky less fertile upland soils; grows satisfactorily on heavy clays, sending roots down 1.5 m, probably explaining why it continues to grow slowly during dry periods. It has been grown on sandy loams and on Nipe Clay (the most unproductive soil in Puerto Rico) with pH 4.5–5.1; a 10–10–5 fertilizer was used to obtain a good stand. Ranging from Subtropical Moist to Wet through Tropical Dry to Wet Forest Life Zones, tropical kudzu is reported to tolerate annual precipitation of 9.1–42.9 dm (mean of 18 cases = 24.2), annual mean temperature of 22.1°–27.4°C (mean of 18 cases = 25.6), and pH of 4.3–8.0 (mean of 15 cases = 5.8).

CULTIVATION: If allowed to seed, tropical kudzu tends to spread from an original planting, but little evidence indicates that it becomes a serious pest. It does extend 7 m or more and grows up, over and can smother vegetation in its way. Runners are cut every 3 months or oftener to save trees. In Puerto Rico, seeds retain viability above 90% for at least 1 year at room temperature and high humidity; some seeds germinated 68% after 3 yr. Seed 6 months or older germinate more readily than fresher seed. Seed coats are hard and should be scarified mechanically, by acid, or by soaking in water for 24 hr. Seeds are planted at the beginning of the rainy season, depending on the region: Apr.–May in Puerto Rico, Oct. in the Philippines. Seed should be inoculated with the proper *Rhizobium* before planting. Can be propagated by stem cuttings but usually by seed. Tropical kudzu should be planted in strips or hills to help prevent erosion during establishment. Strips should be contour-cultivated, spaced 5–8 m apart, center to center, depending on rapidity with which a complete ground cover is desired, steepness of slope and method of planting. Uncultivated areas between strips help control erosion until establishment. Seed requirements are: for strips 2 m apart, 2.5 kg/ha; strips 6 m apart, 1 kg/ha; hills 1 m apart each way with 10–12 seeds per hill, ca. 5 kg/ha. In strips kudzu covers ground completely in about 1 year; 2 or 3 weedings are necessary to control tall-growing weeds and bushes. As runners reach the edge of the cultivated strips, 2 or 3 additional furrows are plowed toward the vines until the entire hillside has been plowed and covered with kudzu. Plowing reduces weed competition and helps runners take root. The hill system requires more hand labor, but is more convenient on small areas of ground. Within fields, plots 1 m in diameter and spaced 1–6 m on the square are spaded and prepared. Except for initial weedings, no additional attention is needed. Under that system runner growth is considerable in 12 months, and field is covered in 18–24 months. If soil is of low fertility 10–10–5 fertilizer may be used at planting at 400–600 kg/ha, and manure (2.5–7.5 MT/ha mixed with soil), may be applied at planting sites (ca. 56 g per hill). Application to the entire field wastes fertilizer and encourages weeds. During the first 3–4 months, seedlings grow slowly and require special weeding and cultivation to promote growth; once well-established (ca. 5 months), they grow rapidly. Sometimes *Calopogonium mucu-*

Pueraria phaseoloides (Roxb.) Benth.

noides is planted with kudzu; it grows rapidly at first, giving early cover until the more vigorous kudzu takes over. On level ground 2 rows of kudzu may alternate with 4 rows of Para grass or molasses grass as pasture or forage. For competition in mixed plantings with the vigorous elephant, Guinea, and Guatemala grasses, about 4 rows of kudzu are needed. Where grown with field corn, 1 row of tropical kudzu is sown for every 4–6 rows of corn. Kudzu usually takes over about corn harvest and continues to grow, producing excellent pasture. The kudzu can be left or plowed under the following spring for another crop of corn. Stem and root cuttings left from the first planting of kudzu take over gradually after the second corn crop matures.

HARVESTING: For grazing, tropical kudzu pastures should be alternated, allowing at least 6 weeks of regrowth between grazings. During dry season, pastures should be grazed only once. During drought kudzu pasture should be saved until others show burning because it lasts longer than most other forage crops. Dairy cows, work oxen, goats, and poultry all like kudzu, and do well on it. In Zanzibar, zebu stock eat it with relish. Seed ripens over 3–4 months and usually is harvested by hand. The black ripe pods should be picked systematically once or twice a week. One man can pick 1 kg of seed in about 2.5 hr. Harvested pods are laid in the sun and soon dry and open. Pods and seed are then placed in a sack or basket and beaten gently so seed fall to the bottom.

YIELDS AND ECONOMICS: Seed yields average about 150 kg/ha; 0.4 ha of kudzu pasture will feed 1 cow grazing continuously during dry season or 2 cows during the rainy season; 30–50 MT of green forage/ha are possible per year. It is calculated that the plants fix ca. 225 kg N/ha in 5 months in Malaysia.

BIOTIC FACTORS: Few serious insect pests or diseases affect tropical kudzu, either the plant or the stored seed. In Puerto Rico, blackbirds and pigeons eat recently planted seeds.

REFERENCE:
Rajaratnam, J. A., and Ang, P. G., 1972, Nitrogen fixation by *Pueraria phaseoloides* in Malaysia, *Malau. Agric. Res.* **1**(2):92–97.

CONTRIBUTORS: J. A. Duke, C. F. Reed.

Sesbania bispinosa (Jacq.) W. F. Wight

FAMILY: Fabaceae
COMMON NAMES: Canicha, Danchi, Dunchi fiber
SYNONYMS: *Coronilla cannabina* Willd.
Closely related to, if not synonymous with, *S. aculeata* and *S. cannabina*

USES: Stems of danchi are used for pipe-stems, provide strong durable fiber, substituted for hemp in rope, twine, cordage for fish net, and made into a cloth used for sails. The crop is grown as green manure (adding 150 kg N/ha), leaves for forage, and in South Africa, for poultry feed. Plant is eaten during famine.

FOLK MEDICINE: Medicinally, seeds are mixed with flour and applied to ringworm, other skin diseases, and wounds.

CHEMISTRY: Seeds of the genus *Sesbania* are reported to contain trypsin inhibitors and chymotrypsin inhibitors. The genus *Sesbania* is reported to

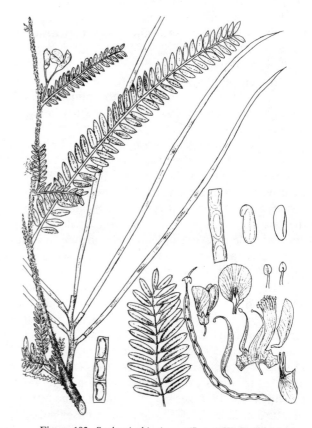

Figure 103. *Sesbania bispinosa* (Jacq.) W. F. Wight

contain saponin. Closely related *S. bispinosa* seed are reported to contain 6.2% of a fixed oil and 32.9% crude protein. Oven-dry fiber is reported as 0.71% ash, 0.94% fat and wax, 2.3% nitrogenous matter, 9.76% pentosan, 16.3% lignin, 85.2% holocellulose (63.6% alpha cellulose), etc. [These figures from Mazumdar *et al.* (1973) add up to more than 100%, and must be evaluated carefully.]

DESCRIPTION: Erect suffruticose low annual subshrub, up to 7 m tall; stems fairly thick, glabrous, branched from the base but soft and pithy; leaves up to 38 cm long, pinnate, leaflets 18–55 pairs, 1.2–2.5 cm long, 0.3 cm wide, glabrous, glaucous; inflorescence 2–8-flowered, 2.5–7.5 cm long; flowers yellow and purple-spotted; pods up to 25 cm long, 0.3 cm thick, curved, many-seeded. Fl. Sept.–Nov. (India).

GERMPLASM: Assigned to the Hindustani Center of Diversity, danchi or cvs thereof is reported to exhibit tolerance to heavy soil, low pH, salt, sandy soil, weeds, and waterlogging. ($2n = 12, 24$.)

DISTRIBUTION: Native to northern India, Pakistan, China, Sri Lanka, and tropical Africa, this crop is widespread in the Old World Tropics, and has been introduced in southern United States and the Philippines; it is a common weed in tropical Africa from Senegal to the Cameroons.

ECOLOGY: Crop adapted to wet areas and heavy soils, which do not require much preparation. Under waterlogged conditions stem produces a spongy mass of aerenchyma. It thrives in low to medium regions, along streams, in open wetlands or often as a weed in rice paddies. Ranging from Warm Temperate Moist through Tropical Dry to Moist Forest Life Zones, danchi is reported to tolerate annual precipitation of 5.7–22.1 dm (mean of 4 cases = 13.4), annual mean temperature of 19.9°–27.3°C (mean of 4 cases = 23.8), and pH of 5.8–7.5 (mean of 3 cases = 6.9).

CULTIVATION: In India seed sown in June–July at onset of southwest monsoon; sowings after September produce poor seed yields. In southern United States seed broadcast after soil has been moistened by rains in April or May and harrowed. In India seed is usually broadcast, but is sometimes drilled in rows 30 cm apart. Seed may be drilled or broadcast at rate of 20–60 kg/ha. Thicker planting facilitates harvest of small plants. The crop is fast-growing, needs little weeding. Usually no fertilizers are applied. In India, grown either as a main crop in rice rotation or as a border crop on the edge of rice fields. On alkali soils (pH 9.2) with added N, P, K, and zinc sulfate rice ('IR8-68') yielded 6.74 MT/ha where danchi was plowed in, only 4.52 MT/ha after fallow. The effect of green manuring was equivalent to the addition of 80 kg N/ha.

HARVESTING: In India ready to cut in September or October, but the fiber does not suffer if left standing until seed ripen in November. In India seed matures in about 5–5½ months; in the United States in about 2 months. Ripe pods normally do not shatter. In India pods are usually hand-picked and threshed with sticks; however, if hand-picking is delayed beyond March, some pods shatter. In the United States crop is harvested by machine, windrowed, and then threshed with an ordinary grain thresher. Seeds are liable to damage by insects and must be treated with insecticides before storage. Processes for steeping and cleaning the fiber are similar to those for sunn hemp (*Crotalaria juncea*). About 2 kg of fiber can be dressed per day.

YIELDS AND ECONOMICS: In India yields of seed are about 600 kg/ha; in Peru, 900 kg/ha; in California, 1,000 kg/ha. Fiber yields are 100–1,000 kg/ha.

BIOTIC FACTORS: This crop is self-pollinating and requires no isolation for pure seed production. Several nematodes attack this Sesbania: *Meloidogyne incognita*, *M. javanica*, and *Trichodorus minor*. In southern United States this crop usually precedes autumn planted vegetables. However, because of nematodes, it should not be grown in sandy soils with other susceptible crops, as cucurbits. Weevils and caterpillars attack seed pods, and the seeds in storage. These may be controlled with insecticides. Plants are attacked by the parasitic flowering plant, *Dendrophthoe falcata*.

REFERENCES:
Chela, K. S., and Brar, Z. S., 1973, Green-manuring popular again, *Prog. Farming (India)* **16**(3):11.
Mazumdar, A. K., Day, A., and Gupta, P. D., 1973, Composition of dhanchia fiber (*Sesbania aculeata* Pers.), *Science Cult.* **39**(10):473–474.

CONTRIBUTORS: J. A. Duke, C. F. Reed, J. K. P. Weder.

Sesbania exaltata (Raf.) Rydb.

FAMILY: Fabaceae

COMMON NAMES: Colorado River hemp, Hemp, Peatree

SYNONYMS: *Sesban exaltata* (Raf.) Rydb.
Sesbania macrocarpa Muhl.
Sesban sonorae Rydb.
Darwinia exaltata Raf.

USES: Colorado River Hemp is used to improve soil on irrigated farms and as cover crop in citrus orchards from Florida to southern California. Sometimes rotated with truck crops on irrigated land. It produces a lustrous, smooth, very strong fiber, used by Yuma Indians for nets and fish lines.

FOLK MEDICINE: No data available.

CHEMISTRY: No data available.

DESCRIPTION: Perennial semiwoody herb (annual at high latitudes), stems striate, glabrous, 0.4–4 m tall, sparingly to widely branching; leaves pinnate, 10 cm long or less, bright green, elongate; leaflets 20–80, linear-oblong, 1–3 cm long, narrow, oblong or elliptic, pale green, villous beneath; flowers yellowish with purple spots, solitary or in axillary few-flowered racemes; corolla pale yellow, streaked or spotted with brown-purple; calyx 4–5 mm long; corolla about 15 mm long; pods 10–20 cm long, 3–4 mm broad, slender, dehiscent with cross-partitions, many (30–40)-seeded; seeds oblong, brown, about 4 mm long. Fl. April–Oct., Fr. persistent on woody stems throughout winter. Seeds 88,200/kg.; wt. 77 kg/hl.

GERMPLASM: Assigned to the North America Center of Diversity. ($2n = 12$.)

DISTRIBUTION: Native in the Coastal Region of southern United States, west to Arizona and southern California, and lower California, Mexico. Escaped from cultivation in irrigated areas of southern Arizona and New Mexico.

ECOLOGY: Thrives on moist soils throughout the warm temperate regions of southern United States. Frequent in overflow lands, and along ditches. Easily killed by frost. Requires a humid area with sufficient moisture, or irrigation. Ranging from Warm Temperate Moist through Tropical Dry Forest Life Zone, Colorado River hemp is reported to tolerate annual precipitation of 6.2–23.2 dm (mean of 12 cases = 13.1), annual mean temperature of 12.5°–27.8°C (mean of 12 cases = 21.6), and pH of 4.5–7.2 (mean of 10 cases = 6.2).

CULTIVATION: Propagated by seeds, usually broadcast at 22–28 kg/ha to provide food for wildlife or as cover crop. Plants grow well when broadcast on rich moist soil; can be maintained by disking and harrowing. Plants demand a long growing season. In some areas seeds are planted in rows in spring with ample fertilizer together with equal parts of browntop millet, proso or hog-millet and beggarweed. Such plantings are given at least one cultivation to reduce weed competition.

HARVESTING: During summer the proso and browntop millet furnish preferred quail foods; after they ripen and die, the Florida beggarweed and sesbania grow and produce high-grade foods for winter and spring use. Seed is available to wildlife throughout the winter and well into spring.

YIELDS AND ECONOMICS: Mainly used in southeastern and southwestern United States.

Figure 104. *Sesbania exaltata* (Raf.) Rydb.

BIOTIC FACTORS: The following fungi have been reported on this crop: *Botryosphaeria minor, Dendrodochium macrosporium, Diaporthe orthaceras, Phymatotrichum omnivorum, Physalospora sesbaniae, Sphaerella sesbaniae*. Nematodes isolated from this plant include the following species: *Belonolaimus longicaudatus, Dolichodorus heterocephalus, Heterodera glycines, H. schachtii, H. trifolii, Meloidogyne arenaria, M. incognita acrita, M. javanica,* and *Pratylenchus pratensis*.

CONTRIBUTORS: J. A. Duke, C. F. Reed.

Spartium junceum L.

FAMILY: Fabaceae
COMMON NAMES: Spanish broom, Weaver's broom
SYNONYM: *Genista juncea* Lam.

USES: Fibers are obtained from stems of Spanish broom, used like flax for making rope, canvas, coarse cloth, mats, for filling mattresses, pillows, and for making paper. Sometimes fiber is spun into yarn, then rubberized and used in manufacture of rubberized belts for conveyors. Broom fiber stated to be superior to that of flax or cotton in that it neither rots nor loses its strength under humid conditions, and has a longer life. Fibers are often blended with rayon or wool. Stems used in basketry. Flowers provide an essential oil with odor of orange flowers, recommended for perfumes; blends well with ylang-ylang. Plants frequently used for hedges.

FOLK MEDICINE: No data available.

CHEMISTRY: Seeds contain cytisine and are poisonous. The genus *Spartium* is reported to contain caprylic acid, saponin, and sparteine.

DESCRIPTION: Spineless shrub, 1–3 m tall, sometimes taller in cultivation; branches sparingly leafy or almost leafless, cylindrical, striate, flexible, glabrous, glaucous-green; leaves alternate, few, simple, small, oblong-linear to lanceolate, 10–30 mm long, 2–5 mm broad, entire, glabrous above, appressed-pubescent beneath, subsessile, glaucous;

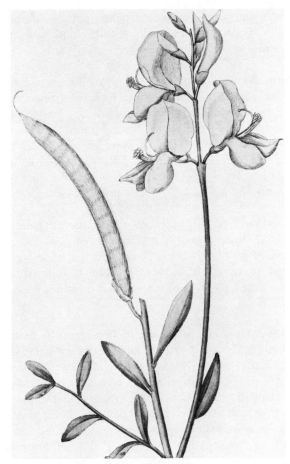

Figure 105. *Spartium junceum* L.

flowers in lax, terminal, leafless, many-flowered racemes, sweet-scented, yellow, 2.5 cm long, with a showy standard, the lower edge of keel hairy, on pedicels with a small bract at base and 2 bracteoles at apex; pod flat, pubescent, becoming glabrous, black, 5–7.5 cm long; seeds 10–18, with a callus appendage at base. Fl. May–Aug. (N.H.) usually, summer and autumn in California, and year round in some tropical areas.

GERMPLASM: Assigned to the Mediterranean Center of Diversity, Spanish broom or cvs thereof is reported to exhibit tolerance to calcareous soils, drought, dunes, high pH, low pH, poor soil, sand, shade, slope, weeds, and wind. ($2n = 48, 52, 54$.)

DISTRIBUTION: Native to southwestern Europe, the Canary Islands, Azores, and the Mediterranean region from Portugal to Turkey and Russia. Intro-

Spartium junceum L.

duced and hardy in southern and central United States. Grown commercially for fiber and perfume on small scale in France since World War II.

ECOLOGY: Grows naturally on dry hills, especially in the Mediterranean climate. Ranging from Cool Temperate Steppe to Wet through Warm Temperate Thorn to Dry Forest Life Zone, Spanish broom is reported to tolerate annual precipitation of 2.9 to 13.7 dm (mean of 11 cases = 7.5 dm), annual mean temperature 8.1°–18.7°C (mean of 11 cases = 13.3°C) and pH of 4.5–8.2 (mean of 8 cases = 6.6).

CULTIVATION: For cultivation, land is completely cleared. Seed is scarified before sowing, and then sown in poor soil in drill 90 cm apart, spaced 5 cm apart in rows. Root development is extensive. Sometimes planted between *Pinus maritima* as a wind break.

HARVESTING: First harvest of fiber may be made when plants are 3 yr old, and thereafter harvests are made every 18 months. After first cutting, plants develop a bush form, which increases the yield of fiber. Branches develop around the main stem, which has been cut. Stalks are cut by hand. Harvesting usually begins in November. Stalks are sent to factory after harvesting, where fiber is extracted by cooking with caustic soda for 2½ hr at 100°C. Stems are then rinsed to remove caustic soda, and passed through machines which beat out the fiber. Fibers are then washed and dried. For perfume production, the large yellow flowers are treated with solvents. Concrete so produced has odor of syringa, tuberose and orange blossom. By extractions of flowers with petroleum ether, 1.29% of concrete yields a yellowish green wax melting at 48°C. The distillate is a golden yellow liquid which solidifies on chilling. Extraction of concrete with alcohol yields 50% of absolute essence, a semisolid maroon mass melting at 28°C.

YIELDS AND ECONOMICS: About 1,200 kg of flowers are source of 1 kg concrete, giving 300 to 350 g of absolute. Yield of fiber is about 1.5 tons/ha (fiber = ca. 10%).

BIOTIC FACTORS: Following fungi are known to attack Spanish broom: *Colletotrichum spartii, Diplodia sarathami, Erysiphe communis, E. martii, Leptosphaeria lusitanica, Mycosphaerella scopulorum, Oidiopsis taurica, Pestalotia poluchaetia, Phoma lupulina, Ph. sarathami, Ph. spartiicola. Phragmothyrium spartii, Phytophthora cinnamoni, Pleospora spartii,* and *Uromyces spartii-juncei.* The nematodes *Heterodera glycines* and *Meloidogyne javanica* have also been isolated from this plant.

REFERENCE:
Kirby, R. H., 1963, *Vegetable Fibers, Botany, Cultivation and Utilization,* pp. 191–192, Leonard Hill, Ltd., London.

CONTRIBUTORS: J. A. Duke, C. F. Reed.

Sphenostylis stenocarpa (Hochst. ex A. Rich.) Harms

FAMILY: Fabaceae

COMMON NAMES: Yam-bean, Akitereku (Twi), African Yam bean, Yam-pea Girigiri (Hausa), Haricot igname (Fr.), Kotonosu (Tschaudjo), Kulege (Ewe), Okpo dudu (Ibo), Sese (Yoruba), Sfenostilo (Esp.), Yam-pea

SYNONYMS: *Dolichos stenocarpus* Hochst. ex A. Rich.
Vigna ornata Welw. ex Bak.
Sphenostylis ornata A. Chev.

USES: Plants cultivated for both seeds and tuber in West Tropical Africa. The tuber is the most valuable part. It resembles a sweetpotato, tastes more like an Irish potato and is rich in starch; dry tubers yield 10.6% protein and 68.4% starch. Seeds are soaked for several hours then cooked. Meal is made from the cooked beans, or the beans (seeds) are eaten whole. Seeds may be eaten with palm oil and spices or mixed with other foods, e.g., corn, yams. Plants have beautiful flowers and are grown as ornamentals in European and other countries. *Sphenostylis erecta* is said to be used as a fish poison.

FOLK MEDICINE: If eaten in excess, seeds are said to cause giddiness and thirst. Seeds mixed with water are claimed to cure drunkenness.

Sphenostylis stenocarpa (Hochst. ex A. Rich.) Harms

1976) analysed several cvs and reported that protein content was 11.47–17.82%.

DESCRIPTION: Perennial herb, cultivated as an annual, with prostrate or climbing viny stems 1–3 m long, produced from a tuberous root, 5–7.5 cm long and rather spindle-shaped; stems often reddish, glabrous or sparsely puberulous; leaves trifoliolate, the leaflets thin, ovate, elliptic, ovate-lanceolate or lanceolate, 2.7–13 cm long, 0.2–5.5 cm broad, shining, acute to acuminate at apex, rounded at base, glabrous or sparsely pubescent on midrib beneath, sometimes with microscopic black speckling; petiole 2–8 cm long; rhachis 0.5–2.2 cm long; petiolules 1.4 mm long; inflorescences 4–9-flowered; rachis 2–6.3 cm long; peduncles 6–27 cm long; pedicels 2–6 mm long; bracts and bracteoles deciduous, small, ovate; calyx-tube 3–7 mm long, the lobes short, broad and rounded, 0.5–2 mm long, minutely ciliolate; standard slightly twisted back on itself, creamy to pink or mauve, sometimes with a darker maroon area at base outside and a greenish-yellow or white basal blotch inside, edges with pink or purple, round, 1.6–3.5 cm long, 1.5–4 cm broad, deeply emarginate; wings cream to pink or mauve; keel slightly twisted, white or greenish, sometimes flushed pink or mauve at apex; pods linear, 10–20 cm long, 5–8 mm broad, margined, glabrous; seeds mostly orange-brown, almost entirely covered with a black suffusion or speckling, subcylindrical or oblong-ovoid, compressed, 4–7 mm by 3.5–5 × 2.5–3 mm finely reticulate or shining, and sometimes with a few scattered scaly hairs. Fl. and Fr. year around; in some areas Fl. in Jan., Fr. Oct.–Mar.

Figure 106. *Sphenostylis stenocarpa* (Hochst. ex A. Rich.) Harms

CHEMISTRY: On a zero-moisture basis, seeds contain per 100 g: 391 calories, 21.1% protein, 1.2 g fat, 74.1 g total carbohydrate, 5.7 g fiber, 3.2 g ash, 61 mg Ca, and 437 mg P. The roots contain 366 calories, 10.8% protein, 0.6 g fat, 86.3% total carbohydrate, 1.1% crude fiber, 2.3% ash, 28 mg Ca, and 227 mg P. Of the amino acids determined in 4 cvs at Ibadan (Evans and Boulter, 1974), ca. 11% was aspartic acid, 4% threonine, 6% serine, 15% glutamic acid, 5% proline, 4% glycine, 4% alanine, 5% valine, 1% methionine, 4% isoleucine, 7% leucine, 4% tyrosine, 6% phenylalanine, 5% histidine, 9% lysine, 6% arginine, and 2% cysteine. In another report (Evans *et al.*, 1977), meal from the tubers is reported to contain, per 100 g of recovered amino acids, 11.3–13.8 g aspartic acid 4.1–4.5 g threonine, 5.1–6.4 g serine, 11.6–12.3 g glutamic acid, 7.5–7.8 g proline, 4.6–5.2 g glycine, 5.3–5.6 g alanine, 5.2–5.9 g valine, 1.5–1.9 g methionine, 4.2–4.9 g isoleucine, 7.0–8.4 g leucine, 3.0–3.4 g tyramine, 4.2–4.9 g phenylalanine, 4.4–5.0 g histidine, 7.1–8.3 g lysine, 5.2–6.5 g arginine and 1.6–2.0 g cysteine. B. N. Okigbo (personal communication,

GERMPLASM: Narrow-leaved variant occurs in Zanzibar and southern Tanzania (Synonym: *S. congensis* A. Chev.). Many variations in seed size and color, from whitish and unmarked to various shades of brown or gray with speckling or marbling. Some cvs have a brown or black ring around the hilum. The yam bean is usually cultivated for its edible seeds and tuberous roots. Its closest relatives are the following wild species which have been reported from various locations on the African continent:

1. *Spenostylis briarti* De Wild—a tuberous climbing herb found in Congo (Zaire).
2. *Sphenostylis marginata* E. Mey. complex, which consists of closely related taxa that exhibit

Sphenostylis stenocarpa (Hochst. ex A. Rich.) Harms

little difference except in habit and frequency of occurrence of acute leaflets with most of the South African material prostrate and most of the East and Central African material either erect or prostrate. This complex consists of (a) *S. marginata* ssp. *marginata*—a climbing or erect plant widespread in Natal and Swaziland; (b) *S. marginata* ssp. *obtusifolia*—a climbing or erect plant reported from the Congo (Zaire), Zambia, Rhodesia, and Angola; and (c) *S. marginata* ssp. *erecta*—an erect plant that appears to be synonymous with *S. erecta* (Bak.f) Hutch. and reported from Congo (Zaire), Tanzania, Mozambique, Malawi, Zambia, Rhodesia, and Angola.

3. *Sphenostylis schweinfurthii* Harms. An erect undershrub found in the open savanna in Ghana, Central Africa, Dahomey, and Nigeria. Seeds reported to be used as food in emergencies and the flowers are also used as vegetable.

4. *Nesphostylis holosericea* (Bark) Verdc. A twining herb with tuberous root stock reported from Senegal, Ivory Coast, Ghana, Nigeria, Tanzania, and Mozambique.

Assigned to the Africa Center of Diversity, yam bean or cultivars thereof reportedly exhibits tolerance to savanna and waterlogging. ($2n = 18$.)

DISTRIBUTION: Both wild and cultivated in West Tropical Africa from Guinea to southern Nigeria, east to northern Ethiopia, Eritrea, and Mozambique, Tanzania, and Zanzibar.

ECOLOGY: Mainly found in grassland, wooded grassland and in woodland, but sometimes also in marshy and disturbed placed, at altitudes of 200–1,950 m. In Nigeria, yam bean production ranges from 5–10°N, from "tropical rain forest" through "savanna" areas, with rainfall ranging from less than 12 dm to more than 20 dm, elevation from 10 to more than 1,000 m, and pH 4.5–6.5. Grows well on a range of soils, not necessarily producing best on rich garden soils. Ranging from Subtropical Dry through Tropical Dry Forest Life Zone, yam bean is reported to tolerate annual precipitation of 8.7–13.6 dm (mean of 2 cases = 11.2), annual mean temperature of 18.7°–26.2°C (mean of 2 cases = 22.4), and pH of 5.0–6.5 (mean of 2 cases = 5.8).

CULTIVATION: Propagated by seeds or tubers. In southeastern Nigeria, the yam bean is usually grown in yam and bean fields on the same supports. Plants perform better interplanted than when planted alone. Requires the same cultural practices for other climbing vegetables, producing well on rich garden soils. Plants sometimes allowed to ramble over the ground, when no supports are available.

HARVESTING: Seeds mature in 150–300 days in lowland Africa. Beans are picked by hand for the pods and seeds when needed, depending on their use. In Nigeria, dry woody pods are harvested and stored in bundles over the fireplace until needed. Tubers harvested from mature plants either for eating or for propagation.

YIELDS AND ECONOMICS: Yields are highest when planted with yam, corn, or okra. Dry seed yields average 300–500 kg/ha ranging up to 3,000 kg/ha. Tubers, produced at about 1,800 kg/ha, are an important source of starch and protein for the natives of tropical Africa.

BIOTIC FACTORS: Reported to show low susceptibility to pests and diseases, yam beans are affected by powdery mildew, leaf spot, stem rust, and viruses which cause misshapen leaves. Gryllids and acridids damage the seedlings. Later lepidopterous larvae, thrips, leaf-rolling caterpillars, and leaf miners damage the plants, which are also very susceptible to nematodes.

REFERENCES:

Evans, I. M., and Boulter, D., 1974, Amino acid composition of seed meals of yam-bean (*Sphenostylis stenocarpa*) and lima bean (*Phaseolus lunatus*), *J. Sci. Food Agric.* 25(8):919–922.

Okigbo, B. N., 1973, Introducing the yambean *Sphenostylis stenocarpa* (Hochst ex A. Rich) Harms, pp. 224–237, Proceeding of the First IITA Grain Legumes Improvement Workshop, Oct. 29–Nov. 2, 1973, London.

Oliver, D., 1871, *Flora of Tropical Africa, Leguminosae*, Vol. 2, (*Vigna ornata*), 613 pp., L. Reeve & Co., London.

CONTRIBUTORS: J. A. Duke, B. N. Okigbo, C. F. Reed.

Stylosanthes erecta Beauv.

FAMILY: Fabaceae

COMMON NAME: Nigerian stylo

SYNONYMS: *Stylosanthes erecta* var *acuminata* Welw. ex Bak.
Stylosanthes erecta var *guineensis*

Stylosanthes erecta Beauv.

(Schumach. & Thonn.) Vogel
Stylosanthes guineensis Schumach. & Thonn.

USES: Use of this plant for forage is increasing in tropical Africa.

FOLK MEDICINE: Medicinally, in Gambia, infusion of plant taken internally for colds. In northern Nigeria, the odorous smoke is blown on arrow wounds as an antidote to their poison. It is also an ingredient in aphrodisiac prescriptions.

CHEMISTRY: No data available.

DESCRIPTION: Woody herb or subshrub, completely prostrate to erect, 0.1–1.5 m tall, usually much-branched; stems glabrous or with sparse pubescence at internodes; leaflets elliptic or lanceolate, 0.5–3 cm long, 1.5–7 mm broad, rounded or acute at both ends, mucronulate, usually rather thick, pubescent or glabrous, or with few bristles along veins beneath; veins sometimes thickened and prominent beneath towards margins; petiole and rachis 2–10 mm long; petiolules 0.5–1 mm long; stipules 0.9–1.6 cm long; inflorescences dense but sometimes elongated; peduncle 0–3.5 cm long, primary bracts glabrous to densely pubescent and bristly, 1.2–1.5 cm long; secondary bracts 5–6 mm long, 1.5 mm broad; bracteoles 1–2, 3–6 mm long; receptacle 6.5–9 mm long; calyx-lobes oblong, 3.5–5 mm long, 1–1.5 mm broad; standard dull yellow or orange, with bronze or orange-red blotch near base, 6–8 mm long, 5–7 mm broad; wings yellow, sometimes darker at base; keel white or yellow; pods 4–7 mm long, 1–2-jointed, the articles 3–5 mm long, 1.5–2.5 mm broad, mostly glabrous; beak 1–2 mm long; seeds pale brown, oblong or compressed-pyriform, pointed at corner nearest hilum, shiny, ca. $2 \times 1.5 \times 1$ mm. Fl. and Fr. Nov. to June.

GERMPLASM: Variants include plants with glabrous or hairy pods, and hairier stems. Some populations in East Africa are possibly hybrids with *Stylosanthes fruticosa* (Retz.) Alston. *S. erecta* was once widespread in East Africa but now grows only in small pockets that are being reduced by the hybridization with *S. fruticosa*. Assigned to the Africa Center of Diversity, nigerian stylo or cvs thereof has been reported to tolerate low pH, poor soil, sand, slope, and savanna. ($2n = 60$.)

DISTRIBUTION: Mainly native in West Africa from Sierra Leone to Nigeria, Senegal, and Gambia, and then scattered across tropical Africa to Congo, Angola, Kenya, Zanzibar, Sudan, Malagasy.

ECOLOGY: Thrives in sandy places near coast, usually just above loose sand dunes, and nearly at sea level. Also grows along shores and on savannas, on dry gravelly hills, in sandy pastures and sandy thickets along rivers. Ranging from Subtropical Moist through Tropical Dry to Moist Forest Life Zone, Nigerian stylo is reported to tolerate annual precipitation of 11.2–27.8 dm (mean of 4 cases = 18.4), annual temperature of 18.3°–26.6°C (mean of 4 cases = 23.9), and pH of 4.5–7.1 (mean of 4 cases = 5.5).

CULTIVATION: Propagated by seed, mainly in the wild. Sometimes seeds are broadcast over areas where plant is to be used for fodder. CPI 2414 (from the Ivory Coast) nodulates effectively with a

Figure 107. *Stylosanthes erecta* Beauv.

Stylosanthes erecta Beauv.

wide range of Rhizobium strains. It required more than 11.5 hr day length for flowering.

HARVESTING: Plant grazed or cut for fodder as needed.

YIELDS AND ECONOMICS: No data available.

BIOTIC FACTORS: No serious pests or diseases have been reported for this plant.

REFERENCE:
't Mannetje, L., 1969, Rhizobium affinities and phenetic relationships within the genus *Stylosanthes, Aust. J. Bot.* **17**:553–564.

CONTRIBUTORS: J. A. Duke, C. F. Reed, L. 't Mannetje.

Stylosanthes guianensis (Aubl.) Sw.

FAMILY: Fabaceae

COMMON NAMES: Stylo, Brazilian Stylo, Brazilian Lucerne, Lengua de Rana

SYNONYMS: *Stylosanthes gracilis* H.B.K.
Stylosanthes surinamensis Miq.
Trifolium guianense Aubl.

USES: Introduced in southeastern Asia and grown in India. Planted in Australia in an effort to impede soil erosion. Introduced in East and West Africa for pasture. Now a widespread native legume in tropical Central and South America. Its main use in cultivation is as a pioneer legume in grass/legume pastures in infertile regions of the humid tropics.

FOLK MEDICINE: No data available.

CHEMISTRY: Like many other tropical legumes, some stylos can accumulate nutrients, especially Ca and P, even when levels of nutrients in soil are low. Crude protein yields range from 13% to 18%.

DESCRIPTION: Plants herbaceous or suffrutescent perennials, erect or procumbent, rarely prostrate, often much-branched, sometimes pendent from banks; stems to 1 m long, pubescent or pilose, with usually long spreading hairs, the pubescence often in 2 lines running down from the stipules, and with some yellow rigid long hairs, often glabrate; stipules

Figure 108. *Stylosanthes guianensis* (Aubl.) Sw.

up to 1.5 cm long; leaflets lanceolate to oblong or linear-lanceolate, densely hirsute or pilose, with some viscid yellow hairs, acute at apex and base; spikes ovoid, up to 1.5 cm long, capitate with 1–40 flowers; primary bracts densely hirsute with rigid yellow hairs; secondary bracts and bractlets linear; calyx about 2.5 mm long, with stalklike receptacle 3 mm long; petals ca. 6 mm long, yellow or red-yellow; pod 1-jointed, the lower joint usually aborted, the upper one glabrous, reticulated, ca. 3 mm long, 2 mm broad, with short beak about 1 mm long; seeds pale brown or purple. Mostly short-day ('Oxley' is long-day).

GERMPLASM: *S. g.* var *gracilis* (H.B.K.) Vog. has leaves few, and leaflets linear-lanceolate; the spikes densely capitate and densely hispid. *S. g.* var *subviscosa* Benth. has a slightly viscid long setulose pubescence. Australian 'Schofield,' 'Endeavour,' and 'Cook,' and Brazilian IRI 1022 are short-day-length sensitive; Australian 'Oxley' is long-day-length sensitive (from plants in Argentina and Paraguay); *Colletotrichum* leaf spot has prevented success in parts of tropical America and is also found in Australia and Africa. 'Oxley' has specific bacterial requirements and is insect susceptible.

'Schofield' is well adapted to wetter, tropical, frost free areas with long growing seasons. It tolerates heat, drought and waterlogging, but is intolerant of fire and heavy frosts. Daylight sensitive, it flowers at daylengths less than 12 hr. Grows on acid low-phosphate soils. 'Schofield' apparently can mine the soil more efficiently than many tropical forage species. 'Schofield' does not withstand close defoliation (grazing) as well as 'Cook' and 'Endeavour.' 'Cook' is superior to 'Schofield' for intensive grazing in cold weather and weed competition. It tolerates poor soils and waterlogging and nodulates readily with commercial 'cowpea' inoculant. 'La Libertad' from Colombia tolerates the llanos, but may be badly attacked by anthracnose. Cv 'Oxley' withstands fires, frosts, and droughts, but not waterlogging. It grows at lower temperatures than 'Townsville' or 'Schofield.' Daylight sensitive, it flowers when daylengths exceed 12 hr. 't Mannetje (1977) recently revised the varietal classification. Assigned to the South American Center of Diversity, stylo or cvs thereof is reported to exhibit tolerance to aluminum, drought, fire, grazing, laterite, light frost, low pH, low phosphate, poor soil, slope, and waterlogging. ($2n = 20$.)

DISTRIBUTION: Widespread in Central and South America, from Mexico to northern Argentina; Trinidad. Known to be introduced into west Africa to east tropical Africa (Rhodesia, Uganda, and Tanzania); also in India.

ECOLOGY: Mainly found in grassland, in open, often rocky areas, on slopes and plains, in moist or dry fields, frequently in oak or pine forests, occasionally along sandy stream beds. Usually at altitudes of 2,000 m or lower. Fares well in well-drained soil but grows in wet places in open fields. Common cvs do not tolerate prolonged flooding or frost. Has been reported from alkaline soils to pH 8.3. Not tolerating shade, yields under oil palm are reduced by as much as 75%. Ranging from Subtropical Dry to Wet through Tropical Dry to Moist Forest Life Zone, stylo is reported to tolerate annual precipitation of 5.3–41.0 dm (mean of 13 cases = 15.6), annual mean temperature of 18.3°–27.4°C (mean of 13 cases = 23.5°C), and pH of 4.3–7.7 (mean of 11 cases = 5.7).

CULTIVATION: Propagated by seed (ca. 3 kg/ha), usually broadcast over ground for pasture or soil erosion. Can be established by drilling in rows 45–60 cm apart, allowing for 1 or 2 weedings. Seed germinate readily, require little attention. Var *intermedia* and *robusta* are less responsive to added P than var *guianensis*.

HARVESTING: Pastures grazed after stand is well established, which requires from 3–5 months, depending on area and soil moisture. If allowed to grow too long, commercial types can be killed by cutting.

YIELDS AND ECONOMICS: Mixed with grass pasture in South Florida, 'Cook' or 'Endeavour' stylos average ca. 15,000 MT/ha with ca. 14% crude protein. Dry weight yields are 3.5–10 MT/ha. Irrigated Queensland pastures carried 3.75 head/ha with a live-weight gain of nearly 0.75 kg/day for short periods. Yields for stylograss pastures were 37 kg N/ha in Zambia, cf. 94 in Australia. In Uganda, yield responses in mixed pastures were equivalent to dressing 165 kg/ha N on the grass alone.

BIOTIC FACTORS: *Colletotrichum* leaf spot and stem canker is the most serious disease of stylo. 'Endeavour,' 'La Libertad,' and 'Schofield' are moderately to severely affected in South America. 'Endeavour' is severely affected by a new strain of *Colletotrichum gloeosporioides* in Australia. Seedlings of 'La Libertad,' 'Schofield,' and accessions from Central and South America are severely affected by Australian and Florida *C. gloeosporioides*. *Botrytis* head blight and *Rhizoctonia* blight also affect stylo in Australia. Other diseases include a *Diplodia* collar rot, a black smut, and *Corticium* and *Rhizoctonia* infections. It is susceptible to little-leaf virus (transmitted by leafhoppers), but resistant cvs occur. Stylo is attacked by psyllids, termites, and larvae of *Hedylepta diemenalis*. Although a reported host of *Meloidogyne hapla*, it has been locally favored as a fallow because it was not susceptible to the local nematodes. *Helicotylenchus pseudorobustus* and *Scutellonema clathricaudatum* do infect this host (occasionally).

REFERENCES:
Anonymous, 1972, Annual Report, Centro Internacional de Agricultura Tropical, CIAT, Cali, Colombia.
Irwin, J. A. G., and Camerou, D. F., 1977, Two diseases of *Stylosanthes* caused by *Colletotrichum gloeosporioides* in Australia, and pathogenic specialization with one of the causal organisms, *Aust. J. Agric. Res.* **28** (in press).
Lenne, J. M., and Sonoda, R. M., 1978, *Colletotrichum* spp. or

Stylosanthes spp., *Trop. Grassl.* **12** (submitted for publication).

't Mannetje, L., 1977, A revision of *Stylosanthes guianensis* (Aubl.) Sw., *Aust. J. Bot.* **25**:347–362.

O'Brien, R. G., and Pout, W., 1977, Diseases of *Stylosanthes* in Queensland, *Queensl. Agric. J.* **103**:126–128.

Tuley, P., 1968, *Stylosanthes gracilis*, *Hetbage Abstr.* **38**(2):87–94.

CONTRIBUTORS: J. A. Duke, J. M. Hopkinson, A. E. Kretschmer, Jr., J. M. Lenne, L. 't Mannetje, C. F. Reed.

Stylosanthes humilis H.B.K.

FAMILY: Fabaceae
COMMON NAME: Townsville stylo
SYNONYM: *Astyposanthes humilis* H.B.K.

USES: Townsville stylo is valuable for grazing cattle when grown in mixtures with suitable permanent grasses with or without white clover. Such mixtures can also be used for green-chop or silage. If properly managed, excellent Townsville stylo-pangolagrass hay can be made after seed production in early winter. Townsville stylo may be used as a cover crop in citrus groves. In Australia, it has increased the carrying capacity of pastures of coastal speargrass (*Heteropogon cortortus*). Its reputation is based on its ability to extract nutrients under low fertility conditions.

FOLK MEDICINE: No data available.

CHEMISTRY: N content of Townsville stylo was always above 1.2% (seed can have nearly 8%). Sulfur content was always above 0.1, except on unfertilized pasture. Potassium/sodium ratios were 68/1 and calcium/phosphorus ratios, 7/1–11/1. Levels of P and Mg (0.3–0.4 and 0.1%) were adequate for lactating cows (Ritson *et al.*, 1971).

DESCRIPTION: Annual legume with ascending or prostrate stems, up to 90 cm or more in height; leaves trifoliolate, the leaflets lanceolate to somewhat elliptical, about 1.3 cm long, 3–5 mm broad; flowers yellowish, terminal in clusters of 6 or more; loment (pod) containing one seed, the biarticulate loment terminated by a persistent style giving a beaked appearance. Flowers open in morning and close about midday. Stems develop adventitious roots when crop is heavily grazed or when they rest on moist soil. Seeds 396,000–484,000/kg.

GERMPLASM: A most variable species; yields vary by tenfold and flowering time up to 3 months. Habit, seed yield, branching, hard seededness and other morphological characteristics also vary. Late flowering 'Katherine' produces greater cattle growth in dry-season than earlier flowering cvs. According to J. M. Hopkinson (personal communication, 1978) there are only three official cvs in Australia, 'Paterson,' 'Lawson,' and 'Gordon,' but genetic instability renders these cvs rather trivial. Assigned to the Middle American and African Centers of Diversity, Townsville stylo or cvs thereof is reported to exhibit tolerance to aluminum, drought, high pH, low pH, poor soil, and slope. ($2n = 20$.)

DISTRIBUTION: Native from central Mexico south into Central and northern South America, in most tropical and subtropical areas. Introduced and naturalized to Australia about 1900, and now growing in Australia over the latitudinal range from 12° to

Figure 109. *Stylosanthes humilis* H.B.K.

28°S. Suitable for South Florida and other Gulf Coast States.

ECOLOGY: This summer-growing, short-day, subtropical legume grows in areas that normally do not have frost before late fall. It flowers better under short-day conditions. Late-flowering cvs do not flower under 13-hr conditions. Long days increase main axis length, main axis node number, erectness and time until flowering. Temperature is less influential on flowering, but higher temperatures may hasten flowering slightly. It is a sun species, uncommon in shaded places, and may dominate in heavily grazed situations. At 74% full daylight, shoot yields are reduced to 47%; at 38% full daylight, 33% mortality occurs. Plants are very responsive to temperature rises; over the range 15/10°C to 27/22°C, an increase of 2.7°C doubled the growth rate. In Australia this legume grows in areas with 5.0–37.5 dm annual rainfall; in Central and South America it grows in areas with 12.5–15.0 dm rainfall, but will tolerate as low as 4.5 dm. Thrives on poor soils, especially in well-drained sandy places. Tolerates acidity (pH 4.5–5.3) and grows better in forest soils than scrub soils. Used on a wide range of soils, it does not establish readily on deep cracking clays or tolerate waterlogging. Ranging from Subtropical Dry to Wet through Tropical Dry Forest Life Zone, Townsville stylo is reported to tolerate annual precipitation of 5.3–23.2 dm (-38) (mean of 11 cases = 12.9), annual mean temperature of 18.3°–26.8°C (mean of 11 cases = 23.4), and pH of 4.5–7.1 (mean of 10 cases = 5.8).

CULTIVATION: More than 90% of the seed may be hard at harvest. Scarification and dehulling improve emergence. Strip-planting in lightly timbered areas has been successful. Native grasses should be weakened by firing or by overgrazing. Shallow well-prepared seedbeds are satisfactory and seed should be broadcast following cultivation. Post cultivation by harrowing reduces establishment. Seed can be broadcast on established pangolagrass, lightly disked or chopped, and then rolled, or dragged and rolled. Sowing 0.6 cm deep gives optimum emergence; rain provides coverage on surface-sown seed. From seeds planted in spring, at rate of 2–5 kg/ha, plants begin to grow rapidly by early summer. Flowering usually begins by early fall and reaches maximum shortly. Flowering may be late if seeds are sown late, or if stands are grazed heavily. After flowering begins, vegetative growth declines. Seed can be planted after frost danger in spring if soil moisture is adequate. During the first year, more dry matter is produced by early than by late plantings. Seed probably can be planted until early summer without much reduction in seed yield if no frosts occur until midwinter. Seeds should be inoculated with commercial cowpea inoculant. Plants require fertilizer for most efficient production. Protein synthesis is enhanced by adequate P, S, K, and Mo. Plants nodulate at pH 4.5, or 4.0 if calcium supply is adequate. Aerial fertilization with superphosphate is increasing in Australia. Mixing seed with superphosphate may lower germination; separation by pelleting may be desirable. Lime cannot be specifically recommended, as Townsville stylo has an alkali-producing rhizobium. Fertilized 400–900 kg/ha with 0–10–20 applied each year; no nitrogen needed.

HARVESTING: Pastures are grazed from early summer to early fall, and then again after the first frost. Mixing with pangolagrass increases dry weight, protein content, and yields of forage. In all stages of growth stylo is readily eaten by cattle and regenerates each year even under heavy grazing. It can also be green-chopped and fed direct or stored as an early winter hay crop.

YIELDS AND ECONOMICS: When harvested by mechanical suction, stylo may yield 1,250 kg/ha of clean seed. When Townsville stylo is used in mixtures with pangolagrass and other tropical grasses and with white clover, forage yields were increased 2–2.5 times as compared with grasses grown alone. Very valuable annual legume in Australia and other tropical and subtropical areas, including the Gulf State area of southern United States.

BIOTIC FACTORS: Self-pollinated. Colletotrichum leaf spot and stem canker have damaged this crop. The nematodes *Meloidogyne javanica* and *Radopholus similis* have been reported.

REFERENCES:
Cameron, D. F., 1965, Variation in flowering time and in some growth characteristics of Townsville stylo (*Stylosanthes humilis*), *Aust. J. Exp. Agric. Anim. Husb.* **5**:49–51.
Humphreys, L. R., 1967, Townsville stylo: History and prospect, *J. Aust. Inst. Agric. Sci.* **March**:3–13.
Kretschmer, A. E., Jr., 1968, *Stylosanthes humilis*, a summer-growing, self-regenerating, annual legume for use in Florida pasture, *Fla. Agric. Exp. Stn., Circ.* **S-184**:1–21, illus.

Stylosanthes humilis H.B.K.

't Mannetje, L., and van Bennekom, K. H. L., 1974, Effect of time of sowing on flowering and growth of Townsville stylo, *Aust. J. Exp. Agric. Anim. Husb.* **14**:182–185.

Norman, M. J. T., 1959, Influence of fertilizers on the yield and nodulation of Townsville stylo (*Stylosanthes sundaica* Taub.), *CSIRO, Aust. Div. Land Res. Reg. Surv., Tech. Paper* **5**.

Ritson, J. B., Edye, L. A., and Robinson, P. J., 1971, Botanical and chemical composition of a Townsville stylo-speargrass pasture in relation to conception rate of cows, *Aust. J. Agric. Res.* **22**:993–1007.

Sillar, D. I., 1967, Effect of shade on growth of Townsville stylo (*Stylosanthes humilis* H.B.K.), *Queensl. J. Agric. Anim. Sci.* **24**:237.

Skerman, R. H., and Humprheys, L. R., 1975, Flowering and seed formation of *Stylosanthes humilis* as influenced by time of sowing, *Aust. J. Exp. Agric. Anim. Husb.* **15**:74–79.

CONTRIBUTORS: J. A. Duke, J. M. Hopkinson, A. E. Kretschmer, Jr., C. F. Reed.

Tamarindus indica L.

FAMILY: Caesalpiniaceae

COMMON NAMES: Tamarind, Tamarindo

USES: Tamarind is cultivated mainly for the pulp in the fruit, used to prepare a beverage and to flavor confections, curries and sauces, and made into preserves and syrups. "Jugo" or "Fresco de Tamarindo" is a favorite beverage in many Latin American countries, and is bottled commercially in some. Some Latins claim that this is the most important "secret" ingredient in sweet and sour sauces and in chutneys. In places the flowers, seedlings, even the leaves are eaten as vegetables in curries. The bark is chewed as a masticatory. Seeds, once boiled and peeled, are used as a starchy foodstuff. In Sri Lanka, the pulp is made into a brine for pickling fish. In Africa, the pods are added to pots to detoxify poisonous *Dioscorea* dishes. Elsewhere it is used to deodorize fish dishes. Seeds are used for food in India, and provide a source of carbohydrate for sizing cloth, paper, and jute products, and a vegetable gum used in food processing. Wood of tamarind is very hard and durable, used locally for tool handles, rice pounders, oil and sugar mills, furniture and turnery, and is said to produce the finest grade of gunpowder charcoal. As firewood, tamarind gives off considerable heat. Frequently grown as a dooryard, roadside or windbreak tree. The seeds are reported to consist of ca. 70% kernels and ca. 30% testa. Seed coats are separated by roasting or soaking and the remaining kernels are boiled or fried before they are eaten.

FOLK MEDICINE: For years, the pulp of the fruit has been used, with good reason, as an antiscorbutic. In Eritrea, the pulp is sold for dysentery and malaria; in Indonesia, for hair ailments; in Madagascar, for worms and stomach disorder; in Mauritius, as a liniment for rheumatism; in Tanganyika, for snakebite; in Sri Lanka, for jaundice, eye diseases, and ulcers; in Cambodia, for conjunctivitis; in Brazil as diaphoretic, emollient, and purgative; in Brazil, for hemorrhoids. The astringent seed is used as a dysentery remedy, and cooked bark is used as an antiasthmatic in Mauritius. Said to be used for: alexeritic, antiseptic, aperient, apoplexy, febrifuge, hangover, lenitive, and leprosy.

CHEMISTRY: Per 100 g the pulp contains: 214 calories, 38.7% water, 2.3 g protein, 0.2 g fat, 56.7 g carbohydrate, 1.9 g fiber, 2.1 g ash, 81 mg Ca, 86 mg P, 1.3 mg Fe, 3 mg Na, 570 mg K, 10 μg β-carotene equivalent, 0.22 mg thiamine, 0.08 mg riboflavin, 1.10 mg niacin, and 3 mg ascorbic acid. The leaves contain, per 100 g: ca. 75 calories, 78% water, 4 g protein, 12 g fat, 16 g carbohydrate, 3 g fiber, 1 g ash, 250 g Ca, 40 mg P, 2 mg Fe, 8 mg

Figure 110. *Tamarindus indica* L.

Na, 270 mg K, 2500 μg β-carotene, 0.1 mg thiamine, 0.1 mg riboflavin, 1.5 mg niacin, and 6 mg ascorbic acid. The inflorescence, eaten in some parts of the world, is rather similar to the leaves in nutritional value. The fruits are reported to contain 55% pulp, 33.9% seeds, and 11.1% shells and fiber. The edible portion of the ripe pod consists of 63.3–68.6% water, 1.6–3.1% protein, 0.27–0.69% fat, 22.0–30.4% total sugars, 0.1–0.8% sucrose, 2.0–3.4% cellulose, 1.2–1.6% ash, pH 3.15. The dried fruit contain per 100 g edible portion: 270 calories, 21.3% moisture, 5.0 g protein, 0.6 g fat, 70.7 g total carbohydrate, 18.3 g fiber, 2.4 g ash, 166 mg Ca, 190 mg P, 2.2 mg Fe, 60 μg β-carotene equivalent, 0.18 mg thiamin, 0.09 mg riboflavin, 0.6 mg niacin, and 9 mg ascorbic acid. The kernels are reported to contain 15.4–22.7% protein, 3.9–7.4% fat, 0.7–8.2% crude fiber, 65.1–72.2% nonfiber carbohydrate, and 2.4–3.3% ash on dry basis. "The carbohydrate moiety consists of D-glucose, D-xylose, and D-galactose as free sugars and a polysaccharide fraction containing the same sugars in a molar ratio of 3 : 2 : 1. The polysaccharide shows a linear chain of poly(1→4)β-D-glucopyranose with (1→6) linked α-D-xylopyranose units and (1→2)β-D-galactopyranosyl-α-D-xylopyranose units attached alternately to each glucose unit. Findings on L-arabinofuranosyl residues are controversial." Seeds are reported to contain trypsin inhibitors and chymotrypsin inhibitors. The genus *Tamarindus* is reported to contain citric acid, hydrocyanic acid, malic acid, pectin, and tartaric acid.

DESCRIPTION: Long-lived (more than 120 yr), large evergreen tree, spreading and open, to 33 m tall, to 8 m in circumference; bark brownish-gray, somewhat shaggy; leaves 5–12 cm long, abruptly pinnate, the leaflets 20–40 pairs, narrow, oblong, rounded on both ends, 1.3–2 cm long; flowers in few-flowered racemes at the ends of the branchlets; calyx 1.3 cm long, with 4 sepals, ovate-lanceolate, tip acute; petals only 3 developed, ovate-lanceolate; somewhat wrinkled along the margin, yellow, pink-lined; stamens only 3 developed; fruit a straight or curved pod, brown, 5–18 cm long to 2.5 cm wide, somewhat flattened, scurfy, constricted at intervals, with a thin brittle shell containing a soft brownish pulp traversed by a few strong branched fibers; seeds 1–12 per pod, ovate-oblong, truncate at both ends, glossy and smooth, flattened. Fl. April–May in northern hemisphere, Fr. ripening in late autumn and winter.

GERMPLASM: There is much variation in the African savannas, Center of Diversity for tamarind. Cvs from India produce the largest fruits with the best flavor. Some cvs are sweet-fruited. Selection could improve the quality and earliness of fruiting. Tamarind or cultivars thereof is reported to exhibit tolerance to drought, grazing, high pH, low pH, poor soil, salt spray, slope, savanna, waterlogging, and wind. ($2n = 24$.)

DISTRIBUTION: Tamarind, thought to have spread from Tropical Africa (Ethiopia and Central Africa), has been grown in India since prehistoric times. Now spread throughout the tropics and subtropics of the world, it is self-sown in all regions it grows in; also cultivated for shade and as an ornamental, as well as for its fruit. It is one of the most common of village trees, often planted along roads and around villages and dwellings, and frequently growing wild in many countries.

ECOLOGY: Tamarind grows on limestone, alluvial, shale, and poor soils, is able to withstand drought and frequent flooding, but thrives and produces best on deep friable, fairly fertile loam. Growth may be impeded by shallow soils. It grows well where annual precipitation is 10–13 dm and mean monthly temperatures do not drop below 21°C. It can withstand some frost and survive brief exposures to −2°C. It tolerates but is not recommended for humid areas. Like several other woody members of the Caesalpiniaceae, the tamarind shows no nitrogen-fixing nodules. Widely planted in the dry zones of the Tropics, and thriving in the warmer parts of India, it sets little fruit in Punjab. Similarly, the tree grows well in parts of California but does not set fruit. Ecologically, the tamarind ranges from Subtropical Dry to Wet through Tropical Very Dry to Wet Forest Life Zones. It is reported to tolerate annual precipitation of 0.9–42.9 dm (mean of 110 cases = 14.5), annual mean temperature of 17.3°–28.5°C (mean of 109 cases = 25.3), and pH of 4.5–8.7 (mean of 42 cases = 6.8).

CULTIVATION: Seed for planting should be taken from trees yielding heavy crop of well formed rounded pods. Fresh seeds are not suitable for planting. Seeds gathered from April crop may be planted in September. Seeds are soaked 4–5 days in water, then planted about 4 cm deep in baskets, bamboo pots, or a nursery bed. When the seedlings are 60–70 cm tall, in about 9 or 10 months, they are transplanted during the rainy season to groves, where they are spaced 3–7 m apart each way. The richer the soil the further distance they should be planted. Often the seedlings are planted in pits previously prepared with well decomposed manure and allowed to "weather." Once established, the plants require little attention, except watering during prolonged droughts, loosening the soil and an occasional weeding. Budding on young seedlings is also possible.

HARVESTING: Under favorable conditions tamarind usually bears at 5 years and may bear for 60–120 years. Pods are gathered when ripe, and the hard pod shell is removed. The pulp is preserved by placing it in casks and covering with boiling syrup or packing it carefully in stone jars with alternate layers of sugar. When used as a food in India, seeds are roasted and soaked in water until the hard seed coat splits and comes off. The cotyledons are then boiled and eaten.

YIELDS AND ECONOMICS: Average yield of tamarind is 80–90 kg of prepared pulp per mature tree. Up to 170 kg of prepared pulp per tree has been reported in India and Sri Lanka. From 100 trees/ha, each yielding 100 kg, 10 MT pulp/ha is possible (12–16 MT have been reported). Practically all products of tamarind are consumed locally; some pulp is imported by the United States (mostly from the West Indies) and European countries.

BIOTIC FACTORS: The following fungi attack tamarind: *Ascochyta* sp., *Dendrophthoe falcata*, *Exoporium tamarindi*, *Hypoxylon nectrioides*, *Meliola tamarindi*, *Oidium oblongisporum*, *Pestalotia poonensis*, *Pholiota gollani*, *Phyllosticta tamarindina*, *Polyporus calcutensis*, *Schizophyllum commune* and *Xanthomonas tamarindi*. The following nematodes also have been isolated from tamarind: *Heterodera radicicola*, *Longidorus elongatus*, and *Meloidogyne* sp.

REFERENCES:
Fisch, B. E., 1974, The tamarind, *California Rare Fruit Growers Yearbook* **1974**:221–250.
Mell, C. D., 1921, The tamarind as a fruit tree, *Bull. Pan Am. Union* **Feb.**:168–170, illus.

CONTRIBUTORS: J. A. Duke, C. F. Reed, J. K. P. Weder.

Tephrosia candida (Roxb.) DC.

FAMILY: Fabaceae
COMMON NAME: White tephrosia
SYNONYM: *Cracca candida* Kuntze

USES: White tephrosia is used to increase soil nitrogen in tropical agriculture. Often used as cover crop for coconut, coffee, rubber, and tea plantations. Said to improve tobacco and hasten ripening of coconut when intercropped. One of the most satisfactory green manures in Java, Sumatra, and Hawaii. Often grown as a contour hedge in citrus area, around rubber plantings and cinnamon orchards. Cuttings used for mulching. Occasionally grown as an ornamental plant.

FOLK MEDICINE: Said to be insecticidal and piscicidal.

CHEMISTRY: The genus *Tephrosia* is reported to contain HCN, rotenone, and rutin. Leaves have a slight rotenone value and are used as fish poison in India. Cattle avoid the plant, perhaps because it is poisonous. Seeds grown in Sri Lanka contained on an air-dried basis, 21.9% moisture, 5.29% ash, 81.8% organic matter. Seeds contain 0.9–1.15% rotenoids on a dry weight basis, root bark 0.35%.

Figure 111. *Tephrosia candida* (Roxb.) DC.

One MT of seed contains ca. 56 kg of N, 11 kg of phosphoric acid, 12 kg of K, and 5 kg of lime. Grown for manure, the plant contains (on a dry basis): CaO, 0.78%, MgO, 0.25%, P_2O_5, 0.58%, K_2O, 1.51%, and N, 2.36%, Mn 38 ppm, Cu 11.2 ppm, Bo 32.0 ppm, Zn 25.8 ppm, Fe 78.0 ppm, and Mo 1.25 ppm.

DESCRIPTION: Shrub 1–3 m tall; often much-branched, the branches tomentose, angular; stipules setaceous, up to 1 cm long; leaves 1–2 cm long, compound; leaflets 13–27 in 6–13 pairs, lanceolate, acute, and mucronate at apex, acute at base, glabrous above, appressed-pubescent beneath, 2.5–7.5 cm long, up to 1 cm broad, middle ones larger; racemes terminal and in the upper axils, many-flowered; flowers solitary or 3–6 together; calyx pubescent, the tube 4 mm long, the teeth rounded, shorter than the tube; petals white to pale rose, 2.5–3 cm long, limb of standard 1.75–3 cm long, sericeous outside, pod thin, linear, straight, densely clothed with appressed rufous, brown or grayish pubescence, 6–10 cm long, about 8 mm broad, 10–15-seeded, apiculated by long style. Fl. and Fr. year round.

GERMPLASM: Assigned to the Indochina–Indonesian Center of Diversity, white tephrosia or cvs thereof is reported to exhibit tolerance to drought, grazing, low pH, mine spoil, poor soil, shade, slope, waterlogging, and wind. ($2n = 22$.)

DISTRIBUTION: Native to southeast Asia, tropical Himalayas from Garhwal to Khasia and Assam, Bengal; Indochina, Sumatra, Java, rarely adventive in Java. Introduced to West Indies, Surinam, and Hawaii. Cultivated in Sri Lanka, Malaysia, Java, and India for many years.

ECOLOGY: Grows on the plains up to 1,650 m in altitudes. Thrives in poor soils, even mine spoils, in tropical climates. Tolerates no frost. Ranging from Subtropical Moist to Wet through Tropical Very Dry to Moist Forest Life Zones, white tephrosia is reported to tolerate annual precipitation of 7.0–26.7 dm (mean of 9 cases = 16.1), annual mean temperature of 18.0°–27.5°C (mean of 9 cases = 24.2), and pH of 5.0–8.0 (mean of 7 cases = 6.4).

CULTIVATION: Propagated by seeds planted *in situ*. Germination is improved if seed is soaked in concentrated sulfuric acid for 10–12 min. Plants are slow to establish, but thereafter grow steadily. Regular leaf-fall and periodic prunings add appreciable vegetative matter to the soil. For green manure under young tea shrubs, seed sown at rate of 3–3.5 kg/ha. It establishes well on slopes of terraces planted in tea. Nodules form on the roots of both young and full-grown plants. On account of its thick woody main stem, the plant may prove costly to uproot when an area is cleared for other crops.

HARVESTING: Plant must be cut back before it flowers and sets seed if it is to persist for any length of time. Treated this way, plants are said to last 4–7 yr in Ceylon, and for 3 years in Assam. Two cuttings are made the first year; 3–4, the second; 4, the third year.

YIELDS AND ECONOMICS: Average yields of green manure are about 15 MT/ha, containing about 6% N. A valuable cover crop and plant for green manure in tropical Asia, especially in tea, coffee, coconut, and rubber plantations, and to some extent in citrus orchards.

BIOTIC FACTORS: Fungi reported as causing diseases in this plant are the following: *Armillaria mellea, Auricularia polytricha, Clitocybe tabescens, Corticium invisium, C. salmonicolor, Cylindrosporium tephrosiae, Fomes lamaoensis, Ganoderma pseudoferrum, Irpex subvinosus, Lasionectria tephrosiae, Nectria haematococca, Phytophthora parasitica, Poria hypobrunnea, Ravenelia tephrosiicola, Rosellinia arcuata, R. brumodes, Sclerotium rolfsii, Stachylidium bicolor, Stigmina tephrosiae, Rhizoctonia solani, Rh. bataticola,* and *Uredo tephrosiicola*. It is also attacked by the virus, Brazilian tobacco streak. Nematodes isolated from this plant include: *Meloidogyne incognita, M. i. acrita, M. javanica, Pratylenchus brachyurus,* and *Radopholus similis*. Insect pests include the following species, mainly in Sri Lanka and the East Indies: *Araecerus fasciculatus* (the coffee bean weevil, attacks dry stored seeds), *Cerococcus indicus,* mealy bugs, *Xyloborus* spp. (various borers),

Tephrosia candida (Roxb.) DC.

grasshoppers, and caterpillars. Mites can also cause problems.

REFERENCES:

Ceylon Department of Agriculture, 1929, Green manuring with particular reference to coconuts, *Ceylon Dep. Agric., Leaflet* **57**; *Trop. Agric. (Ceylon)* **73**:144–145.

Roark, R. C., 1937, *Tephrosia,* as an insecticide—A review of the literature, *U.S. Dep. Agric., Bur. Ent. Pl. Quarant.* **E-402**:1–165.

CONTRIBUTORS: J. A. Duke, C. F. Reed.

Tephrosia vogelii Hook. f.

FAMILY: Fabaceae

COMMON NAMES: Vogel tephrosia, Fishbean, Fish-poison bean

SYNONYM: *Tephrosia periculosa* Baker

USES: Widely cultivated and used in East Africa as a fish poison, molluscicide and insecticide. Leaves are said to be effective as an insecticide against lice, fleas, and ticks and strewn leaves are said to repel fleas. The unboiled leaf is lethal to snails. Plants used for cover crops for coconut, coffee, rubber, and tea in tropical countries. Plants used to fix nitrogen in tropical agriculture. No animal in East Africa eats the seeds, and fowl are said to starve to death rather than eat them. In many areas plants used for ornamental, hedge, fence plant, or wind break.

FOLK MEDICINE: Roots used as a purgative; weak infusion of leaves used internally as an anthelmintic. Said to be used for bactericide, ecbolic, emetic, pediculicide, piscicide, purgative, repellent, schistosomiasis, skin sores (lf).

CHEMISTRY: Leaves and seeds are high in rotenone. Oil from fresh seeds closely resembles that from clover or peanut oil. Leaves contain tephrosin and deguelin; 1 part in 50 million parts water kills fish in 105 min. Salt water fish are more resistant than freshwater fish to the tephrosin. On a dry basis, the rotenoid content of the leaves is 0.65–4.25%, the stem 0.40–0.90%, the root 0.30–0.45%, and the seed 0.90–1.40%.

DESCRIPTION: A soft woody branching herb or short-lived shrub, 1–4 m tall: stems tomentose with both short and long white or fulvous hairs; leaves compound, the leaf-rachis 10–25 cm long, including a petiole 1–3 cm long, prolonged 2–7 mm beyond the lateral leaflets; stipules lanceolate, 1–2 cm long, tomentose, very caducous; leaflets 11–29, elliptic-oblong to elliptic-lanceolate, up to 5 cm long, 2 cm broad, silky tomentose above and more densely tomentose beneath; flowers white or violet, in dense terminal or axillary pseudoracemes, peduncles rusty tomentose, at least 2.5 mm in diameter, stout and as long as the pseudoracemes; bracts ovate-acuminate, up to 12 mm long, 9 mm broad; pedicels up to 2.3 cm long; calyx brown tomentose, tube about 4 mm long; 2 upper lobes oblong and united almost to tip, ca. 6 mm long, rounded at apex; lateral lobes oblong, ca. 7 mm long, 4 mm broad, rounded at apex; lower lobe narrow, acute, ca. 10 mm long; standard white and silky, about 23 mm long and 25 mm broad; keel pubescent at margins; upper filament lightly attached and strongly dilated, anthers 1.7 mm long; style about 10 mm long, short-pubescent, incurved at tip; pod straight, densely white to rusty tomentose, about 11 cm long, 13 mm broad and 4 mm thick, with a straight or slightly curved beak; seeds 10–20, black,

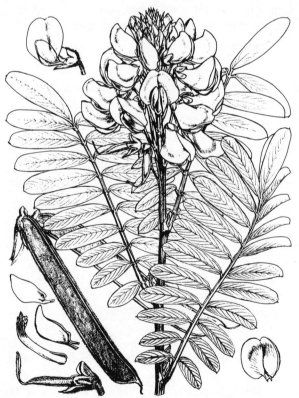

Figure 112. *Tephrosia vogelii* Hook. f.

smooth, oval, obliquely transverse, ca. 7 mm long, 3 mm broad; hilum at side of narrower end, with a well-marked annular aril. Fl. and Fr. year round.

GERMPLASM: In East Africa, the white-flowered form predominates; in West Africa, the purple-flowered form. Assigned to the African Center of Diversity, vogel tephrosia or cvs thereof is reported to exhibit tolerance to drought, fire, grazing, insects, low pH, poor soil, and wind. ($2n = 22$.)

DISTRIBUTION: Native throughout West Africa (Congo, Rwanda, Burundi) east to Sudan, Ethiopia, Malawi, Mozambique, Rhodesia, Zambia, and Angola. Introduced and cultivated also in India, Assam, Java, and other tropical countries.

ECOLOGY: Found in grasslands, forest margins, waste ground and in old cultivations. Thrives in poor soils, from sea level to 2,100-m altitudes, where rainfall is over 11 dm/yr. Burning the steppes in East Africa does not harm this plant as it has very deep root system. Ranging from Warm Temperate Moist through Tropical Dry to Wet Forest Life Zone, vogel tephrosia is reported to tolerate annual precipitation of 8.7–26.7 dm (mean of 7 cases = 14.5), annual mean temperature of 12.5°–26.2°C (mean of 7 cases = 19.8), and pH of 5.0–6.5 (mean of 5 cases = 5.6).

CULTIVATION: Propagated by seed, often cultivated along rivers, or in small patches in the native villages for handy use as a fish poison. Requires no attention after sowing seed.

HARVESTING: Leaves and small branches are gathered as they are needed, and the leaves (Kassa) are macerated in water or beaten to a pulp, and then thrown into the water. Men then enter the water and splash about, and in about 10 min fish begin to appear on the surface and are collected in baskets or by hand. The skin of the men who enter the water into which kassa has been thrown, is affected and becomes tough like bark. In southern Africa, natives throw pounded root into the water to stupefy the fish. Fish so treated are not poisonous to eat after cooking.

YIELDS AND ECONOMICS: In Tephrosia the leaf crop can be harvested for rotenone in only 6 months, whereas in Derris and Lonchocarpus, the land is tied up for 2–3 yr.

BIOTIC FACTORS: Following fungi have been reported to infect this plant: *Corticium salmonicolor, Fomes noxius, Nematospora coryli, Ravenelia tephrosiicola, Rhizoctonia bataticola, Rosellinia bumodes,* and *Uredo tephrosiicola*. Nematodes isolated from this plant include the following species: *Meloidogyne arenaria, M. brevicauda, M. incognita acrita, M. javanica, Pratylenchus coffeae, P. pratensis,* and *Scutellonema brachyurum*. Insect pests, mainly in Java, include: *Araecerus fasciculatus* (coffee bean weevil, attacks dry seed), *Xyleborus* spp. (borers), *Pseudococcus virgata,* and other scale insects and *Helopeltis* spp.

CONTRIBUTORS: J. A. Duke, C. F. Reed.

Tetragonolobus purpureus Moench

FAMILY: Fabaceae

COMMON NAMES: Winged pea, Asparagus pea

SYNONYMS: *Lotus tetragonolobus* L.
Tetragonolobus edulis Link

USES: Winged pea is grown for its edible pods that are picked young, cooked whole, and served with melted butter. Seeds are roasted and used as substitute for coffee. Sometimes grown as a culinary herb.

FOLK MEDICINE: No data available.

CHEMISTRY: No data available.

DESCRIPTION: Annual trailer, soft hairy, spreading, stems erect, procumbent or suberect, 10–40 cm long, succulent, sulcate, pubescent; leaves trifoliolate, fleshy, slightly pubescent on both surfaces, petiole to 1.2 cm long; leaflets obovate to oval-rhombic, acuminate, cuneately narrowed, 4–5 cm long, 3 cm broad; stipules ovate, acute, to 1 cm long; peduncles shorter than or equal to leaves, densely pubescent; bracts trifoliolate with obovate segments; flowers solitary or paired, crimson or purplish-red, corolla 1.5–2.1 cm long; calyx-lobes slender, about one-half length of corolla, divided to below the middle, densely pubescent, the teeth acutely lanceolate; pods 4-sided, 4-winged, wings broad, undulating, 2–4 mm wide, pods 3–9 cm long, 6–8 wide, fleshy when immature, becoming

Tetragonolobus purpureus Moench

Figure 113. *Tetragonolobus purpureus* Moench

black; seeds smooth, yellow-brown, spherical-oblong, 0.3 cm crosswise. Fl. Mar.–June; Fr. early summer.

GERMPLASM: Assigned to the Mediterranean Center of Diversity, winged pea or cvs thereof is reported to exhibit tolerance to slope. ($2n = 14$.)

DISTRIBUTION: Native to the Mediterranean region of southern Europe, extending to 47° in the Ukraine through the Balkans, Caucasus, Near East, and Transcaucasus. Becoming widely cultivated in warm temperate regions of the world. Widely cultivated in Europe, and naturalized.

ECOLOGY: Winged pea requires an exposure similar to that for snap and field beans, with climate and rainfall comparable to that of the Mediterranean region. Ranging from Cool Temperate Wet through Moist Forest Life Zone, winged pea is reported to tolerate annual precipitation of 5–6 dm, annual mean temperature of 7°–8°C and pH of 6.0–7.0.

CULTIVATION: Propagated from seed sown in drills in April northward, earlier in warmer regions. Plants require no special care except watering during drought.

HARVESTING: Immature pods are harvested before the beans begin to develop.

YIELDS AND ECONOMICS: No production data available; very slight in the United States, mainly grown in home gardens.

BIOTIC FACTORS: The fungi *Uromyces anthyllidis* and *U. loti* are known to attack asparagus pea.

CONTRIBUTORS: J. A. Duke, C. F. Reed.

Trifolium alexandrinum L.

FAMILY: Fabaceae
COMMON NAMES: Berseem clover, Egyptian clover

USES: Extensively cultivated as a forage plant in eastern Mediterranean regions and as a winter annual in Florida and other subtropical areas, berseem clover can be green-chopped, grazed or ensiled. Production, palatability and longevity are considered excellent.

FOLK MEDICINE: No data available.

CHEMISTRY: Seeds of the genus *Trifolium* are reported to contain trypsin inhibitors and chymotrypsin inhibitors. The genus *Trifolium* is reported to contain histamine, hydrocyanic acid, malonic acid, maltose, quercitin, and saponin. Tyramine occurs in berseem. Studying 10 cvs of berseem, Gupta and Pradham (1974) found that DM was 10.5–37.7%, crude protein, 13.6–25.9%; neutral detergent fiber, 40.4–54.6; acid detergent fiber, 30.1–43.2; hemicellulose 21.3–28.8; lignin, 8.0–15.2; silica, 0.4–2.1. *In vitro* DM digestibility was 62.2–74.5% and *in vitro* cell-wall digestibility was 28.4–45.8%. Data from

Trifolium alexandrinum L.

Figure 114. *Trifolium alexandrinum* L.

Miller (1958) suggest that the moisture content of the hay ranged from 8.3 to 9.9%. On a moisture-free basis, hay averaged 15.0% crude protein (12.6–17.9; $n = 20$), 2.9% fat (1.2–3.4; $n = 20$), 25.0% crude fiber (21.2–26.9; $n = 20$), 12.7% ash (11.1–15.1; $n = 20$), Ca analyses were 1.47–3.56%; P, 0.28–0.32%; K, 2.39%; and Mg, 0.40%.

DESCRIPTION: Sparingly appressed pubescent annual legume; stems 3–6 dm tall, branching, glabrous to glabrate; petioles of lower leaves about equaling the leaflets, those of the upper leaves very short; leaflets oblong to broadly elliptic or oblong-lanceolate, obtuse, 1.5–5 cm long, rather shallowly denticulate; stipules lanceolate, the free part subulate to setaceous, villous to bristly-ciliate heads short-peduncled to almost sessile, ovoid, becoming cuneate-oblong, 14–20 mm long, 15 mm thick; bracts linear, equaling to somewhat exceeding the calyx-tube, caducous; calyx sessile, the tube obconic, appressed-hirsute, 2–3 mm long in fruit, the triangular-subulate teeth hirsute, finely spreading, the lower 3–4 mm long, exceeding the tube; corolla yellowish, twice as long as the calyx; pod oblong-ovoid, included, 1-seeded; seed similar to that of red clover but larger, less distinctly lobed. Fl. April–May; Fr. June (NH). Seeds 441,000/kg; wt. kg/hl = 77.

GERMPLASM: Berseem cvs are characterized by the form of stem-branching: (1) basal and profuse, (2) basal and along the main stem, and (3) apical along the upper part of stem. Once cut, the latter type does not renew growth. The best known cvs are 'Fahl' (var *alexandrinum*), a spring form unable to regenerate after harvesting and therefore grown from seeds only; and 'Mescavi' (var *serotinum* Zoh. & Lern.), an early summer crop that regenerates readily after each of 4–6 harvests. Multicut domestic cvs 'Nile' and "Hustler" are adapted to Florida. 'Saidi' and 'Kahdrawi' have also been tested in Florida. Several closely related species in the Near East can contribute to this gene pool. *T. berytheum* Boiss. & Bl., centered in Israel, Lebanon, and Syria, is fully fertile and is interconnected with berseem through a series of intermediate forms. It may represent the wild progenitor of berseem. Other "species" interfertile with berseem include *T. apertum* Bobr., *T. meironensis* Zoh. & Lern., and *T. salmoneum* Mout. Assigned to the Mediterranean Center of Diversity, berseem clover or cvs thereof is reported to exhibit tolerance to high pH and virus. ($2n = 16$.)

DISTRIBUTION: Native to the Mediterranean region, Near East, and India (because of its importance as a feed crop in Egypt it is called Egyptian clover). Introduced and cultivated in tropical and subtropical regions, often escaping from cultivation. Introduced in California and southern United States, especially Florida.

ECOLOGY: Berseem clover is the least winter-hardy of the cultivated clovers. Adapted in the US to the warmest Gulf States and southwest areas, where temperature seldom falls below freezing. In such regions, the advantage of berseem over other clovers and alfalfa is its fast winter growth, providing additional feed in January, February, and March. An average temperature of about 17°C is required for vigorous growth. Cold temperatures in northern Florida limit its use there. Mature plants can withstand much lower temperatures than seedlings. Berseem clover probably should not be sown where temperatures reach −6°C or below several times during the winter. Some cvs are as tolerant to drought as alfalfa ('Mescavi' is not). Berseem can tolerate more soil moisture than alfalfa. Ranging from Boreal Wet through Subtropical Moist Forest Life Zone, berseem clover is reported to tolerate annual precipitation of 3.8–16.6 dm (mean of 17 cases = 8.7), annual mean temperature of 7.0°–26.7°C (mean 17 cases = 16), and pH of 4.9–7.8 (mean of 15 cases = 6.8).

CULTIVATION: Berseem clover has germinated well and grown faster than white clover when broadcast as a mixture in pangolagrass 10–20 cm tall. However, stands and plant vigor were best in a clean-fallow seedbed. Seed can be planted with a drill or broadcast. Berseem clover can germinate when soil surfaces are moderately dry and from a greater depth than white clover. Seeds should be planted 1.3–2.5 cm deep. When broadcast, seed should be disked into the soil, and the area rolled immediately after disking to conserve moisture. Seeds germinate in about 5 days. Seeding rates vary from 17–22 kg/ha, planted from the first of October to mid November (in southern Florida). Earlier sowings result in earlier forage production and about equal total crop yields. Growth is good from January or February plantings, but total production is reduced because berseem cannot tolerate the warm weather in May or June. For a first harvest or grazing in late December or early January, seed should be planted about October 15. Berseem clover should be inoculated; commercial white clover inoculant is effective. Molasses or some other sticking agent is used to assure that the inoculant sticks to the seed. Soil fertility and liming needs of berseem have not been precisely determined. On Florida flatwoods soils, growth was satisfactory with a liming program similar to that for white clover, i.e., 2.5–5 MT/ha of liming materials. On virgin flatwoods soils about 700–800 kg/ha of 0–12–12 plus 0.5% CuO, MnO, ZnO, and B_2O_3 or equivalent fertilizer should be applied and disked in prior to drilling the seed. Sometimes the fertilizer and seed are broadcast separately and disked in lightly. About 400 kg/ha of 0–10–20 should be applied after the second cutting or after the first cutting if there have been heavy rains. If soil tests indicate high phosphorus levels, muriate of potash can be used for the second application. When berseem is used for grazing, subsequent fertilization would depend upon climatic conditions, grazing intensity, and plant stand. Minor elements, as copper and boron, may be a problem and should be added as determined by soil test. Berseem clover can be seeded in mixture with white clover to newly planted pasture to boost first-year yields. This does not prevent seed production of the white clover, and provides additional nitrogen for summer grass growth. It can also be successfully grown with oats and rye. In areas where corn or sorghums are seeded in the early spring, an early fall planting of 'Fahl' may fill the feed gap in the winter, and still permit time to plant corn on the same land.

HARVESTING: When planted in October or November in southern Florida, the multicut cvs of berseem clover generally can be harvested four times with the last cutting in April or May. Initial cutting or grazing should begin when new shoots arising from the plant crown are about 2.5 cm tall. Depending upon planting date, plant population, and other factors, the clover will have attained a height of 25–37.5 cm. Removal of the plant tops at this time permits rapid growth of new shoots and increased DM production. The plants will be somewhat higher when ready for subsequent cuttings. Delay in cutting does not appear to injure regrowth but may reduce total production. Tests indicate that stubble is better left 7.5 cm tall than 3 cm.

YIELDS AND ECONOMICS: About 30% more DM is obtained from 'Fahl' berseem than from sweet clover or alfalfa at the first cutting. More early forage and about 50% more total forage were obtained from 'Mescavi' than from alfalfa cvs. Early berseem clover yields are superior to those of white clover, alfalfa, or sweet clover. For the first two harvests, dry matter contents of multicut berseem generally are about 10–12% and fresh weight yield of forage are 25–37.5 MT/ha. Crude protein content of 'Mescavi' berseem is 18.3–26.6%; of 'Fahl', about 16.6%. Berseem is reported to have about one-half the crude fiber and more carbohydrate and fat than alfalfa. Berseem produces oven-dry yields of 3–7 MT/ha, more winter forage than other legumes if not damaged by low temperatures. A very important legume forage crop in the Mediterranean region, Near East, east to Pakistan and India, and south into Egypt. Also important as a winter annual forage crop in southern United States.

BIOTIC FACTORS: Fungi reported on berseem clovers include: *Cercospora cruenta, C. zebrina, Fusarium oxysporum, F. solani, Katatiella caulivora, Pellicularia filamentosa, Peronospora trifoliorum,*

Pleospora herbarum, Rhabdospora alexandrina, Sclerotinia sclerotiorum, Sphaerulina maroceana, and *Stagonospora recendens.* The bacteria *Bacillus megatharium* and *Pseudomonas syringae;* and the cucumber mosaic virus *(Marmor cucurmeris)* also affect berseem. Plants may be parasitized by *Cuscuta arvensis* and *Orobanche muteli.* Nematodes isolated from berseem clover include: *Ditylenchus dipsaci, Heterodera glycines, Meloidogyne hapla, M. incognita, M. i. acrita, M. javanica, Pratylenchus brachyurus, P. minyus, P. penetrans, P. pratensis, P. vulnus,* and *Trichodorus teres.*

REFERENCES:

Gupta, P. C., and Pradhan, K., 1974, Studies on the nutritive value of forages. II. Berseem *(Trifolium alexandrinum),* Haryana Agric. Univ. J. Res. **4**(1):75–81.

Kaddah, M. T., 1962, Tolerance of berseem clover to salt, Agron. J. **54**:421–425.

Kretschmer, A. E., Jr., 1964, Berseem clover, a new winter annual for Florida, Univ. Fla., Fla. Agric. Exp. Stn., Circ. **S-163**:1–16, illus.

CONTRIBUTORS: J. A. Duke, A. E. Kretschmer, Jr., C. F. Reed, J. K. P. Weder, D. Zohary.

Trifolium ambiguum Bieb.

FAMILY: Fabaceae

COMMON NAMES: Kura clover, Pellett clover, Honeyclover, Caucasian clover

SYNONYM: *Trifolium vaillantii* Bieb. ex Fisch.

USES: Cultivated for its abundance of easily accessible nectar of high sugar content, kura is a valuable plant for honey production. Plant palatable to livestock. Of interest to forage researchers because of its rhizomatous nature, resistance to pests, winter-hardiness, persistence, and drought tolerance.

FOLK MEDICINE: No data available.

CHEMISTRY: Green forage of kura clover is reported to contain: 79.5% moisture, 4.1% protein, 0.8% fat, 7.4% fiber, 2.0% ash. Hay of honeyclover contains 15.5% moisture. 16.7% protein, 2.8% fat, and 30.3% fiber.

DESCRIPTION: Long-lived perennial legume, spreading vegetatively by underground rootstocks; stems creeping, ascending, 1–4 dm long, glabrate or sparingly pubescent above; leaves all petiolate;

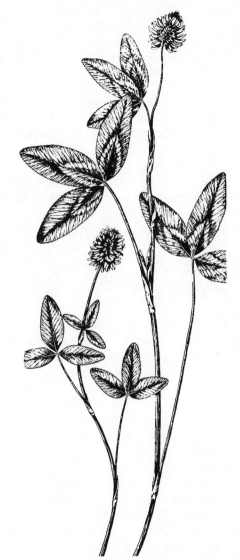

Figure 115. *Trifolium ambiguum* Bieb.

leaflets broadly elliptic to elliptic-lanceolate, obtuse to shallowly emarginate, 1–8 cm long, 0.5–5 cm broad, setose-dentate; stipules pale, marcescent, ovate, acuminate; peduncles moderately long, occasionally very long; heads lateral and terminal, subglobose or ovate, becoming oblong in fruit; bracts linear-lanceolate, much exceeding the pedicels, membranous with a prominent excurrent midrib; pedicels one-fourth to one-half the length of the calyx-tube, sparingly pilose, reflexed in fruit; flowers 10–12 mm long; calyx glabrous except for the sparsely hairy base and rim of the tube, the tube often more or less plicate between the nerves, 3 mm long, the teeth lanceolate-subulate with membranous margins, widely spreading to reflexed in

Trifolium ambiguum Bieb.

fruit, subequal, 2.5–3 mm long; corolla white, becoming reddish in age, about twice the length of the calyx; pod oblong, glabrous, mostly 2-seeded; seeds reniform, dull-yellowish or reddish-brown. Fl. June–July; Fr. July–August.

GERMPLASM: In contrast to several other species, kura is resistant to seven viruses (AMV, BYMV, CYMV, CYVV, PSV, PCVMV, and WCMV) and is a potential source of resistance, if it can be hybridized with related species such as *T. hybridum*. Assigned to the Near Eastern Center of Diversity, kura clover or cvs thereof is reported to exhibit tolerance to disease, drought, frost, insects, low pH, slope, and ultraviolet. Survival was over 80% in 'CPI 2264,' 'CPI 6884,' and 'CPI 20871,' following inundation up to 15 min in early spring. Diploid, tetraploid, and hexaploid cvs have been developed in Australia. ($2n = 16, 32, 48$.)

DISTRIBUTION: Native to Asia Minor and southeastern European countries. Widely distributed and more commonly found in mountain meadows of the Caucasus. Introduced in the United States, especially in and around Iowa.

ECOLOGY: Kura clover grows under a wide range of climatic conditions, but is better adapted to the northern humid regions than to southern areas. Kura persists during drought, but it should not be considered as a dryland legume. It shows considerable winterhardiness and has persisted in Iowa, Wisconsin, and other northern states, and in Canada for many years. In nature, found in steppe depressions, forest margins, and glades, mountain meadows up to the subalpine regions, and even in mountain alpine regions, up to 3,000-m elevations. Ranging from Cool Temperate Steppe through Wet Forest Life Zone, kura clover is reported to tolerate annual precipitation of 4.9–11.6 dm (mean of 3 cases = 9.3), annual mean temperature of 8.4°–12.5°C (mean of 3 cases = 10.2), and pH of 4.5–7.3 (mean of 3 cases = 5.8). Frequent on noncalcareous clays and clay loams. In Armenia, populations grow in alpine situations with frost-free period of only 60–90 days. Diploids generally seem more cold hardy than hexaploids.

CULTIVATION: Little is known about the value or culture of kura clover as a field crop. While small seed samples and rootstock cuttings have been widely distributed, the plantings have been made mostly in short rows in plant nurseries or gardens. Germination and seedling growth are slower than such legumes as red clover and alfalfa. Stands can be established by planting pieces of rootstocks 15–30 cm long with one to several nodes. Plants spread vegetatively by underground rootstocks, but final stand density depends on the fertility and moisture of the soil. Plants usually do not flower the first year. From established plants in middle latitudes flower stems emerge in May, and reach maximum bloom in June and July. However, flower heads have developed as late as September (Beltsville, MD). Seeds have hard coats and must be scarified. Seeds must be inoculated with the special nitrogen-fixing bacteria for kura. The inoculant, isolated from soil from Turkey, is now available.

HARVESTING: Crop used for grazing in northern United States and southern Canada. This clover is still being tested to determine harvest practices and grazing management for highest yields.

YIELDS AND ECONOMICS: This clover is being tested for use in the United States as a forage crop. Early reports from trials in Australia suggest that kura clover is less productive than white clover.

BIOTIC FACTORS: This is a self-incompatible clover that is cross-pollinated by bees, and yields of seed are large when bees are placed near fields. Plants from seed indicate that growth and vegetative characteristics vary widely.

REFERENCES:
Bryant, W. G., 1974, Caucasian clover (*Trifolium ambiguum* Bieb.): A review, *J. Aust. Inst. Agric. Sci.* **40**(1):11–19.
Hollowell, E. A., 1955, Kura clover, *U.S. Dep. Agric., Field Crops Notes*, 3 pp., Beltsville, MD.
Townsend, C. E., 1970, Phenotypic diversity for agronomic characters and frequency of self-compatible plants in *Trifolium ambiguum*, *Can. J. Plant Sci.* **50**:331–338.

CONTRIBUTORS: J. A. Duke, C. F. Reed, M. B. Forde, J. K. P. Weder.

Trifolium fragiferum L.
FAMILY: Fabaceae
COMMON NAME: Strawberry clover

Trifolium fragiferum L.

SYNONYMS: *Trifolium ampullescens* Gilib.
Trifolium bonanni Presl.
Trifolium congestum Link
Trifolium neglectum Fisch., Mey, & Ave-Lall.

USES: A European perennial clover, becoming important in lawns and permanent pastures in New Zealand and Australia. Cultivated as an important pasture legume for saline and alkaline soils in western United States. Used also for green manure, this seldom grows tall enough to be used for hay. Extremely valuable for large irrigated areas where drainage is a limiting factor for crop production. Animals and poultry have grazed this clover with good results. Very palatable and as rich in animal feed units as white clover. A good honey plant.

FOLK MEDICINE: No data available.

CHEMISTRY: No data available.

DESCRIPTION: Pubescent perennial legume; stems creeping, branching, 1–3 cm long, rooting at nodes; petioles long, pilose; leaflets broadly elliptic to narrowly obovate, obtuse or emarginate, often mucronate, 5–15 mm long, 4–12 mm broad, minutely toothed, glabrous, the nerves parallel, thickened and conspicuous at the margin; stipules lanceolate-acuminate, dilated and white-chartaceous toward the base; peduncles scapose, pilose, usually curved, ascending and exceeding the subtending leaves; heads densely many-flowered, ovate to globose, becoming 1.2–2 cm thick in fruit; involucre composed of lanceolate, often irregularly divided, bracts about equaling the calyx; flowers crowded, subsessile, the outer subtended by bracteoles as long as the calyx; calyx bilabiate, the upper side villous, becoming greatly inflated and coarsely reticulate in fruit, the spreading upper teeth shorter than the tube, projecting from the fruiting head as rigid bristles, the approximate lower teeth equaling the tube; corolla white to pinkish, longer than the calyx, standard about 6 mm long, strongly nerved; pods obliquely ovate, included; seeds ovoid-truncate, spotted with light brown. Fl. May–June; Fr. June–July. Seeds 661,500/kg; wt. kg/hl = 77.

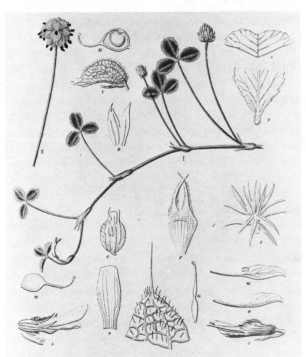

Figure 116. *Trifolium fragiferum* L.

GERMPLASM: Most of the named cvs originated in Australia. Limited trials indicate that Australian cvs are better adapted west of the Cascade Range and the Sierra Nevada than in the region between these mountains and the Rocky Mountains. 'Palestine' is larger and more productive than other Australian cvs tested, but it lacks winterhardiness. 'Salina' was developed in California from selections of 'Palestine,' and is better adapted to California conditions. Most seed available is harvested from strains naturalized in the region between the Cascade Range and Sierra Nevada and the Rocky Mountains, and these strains are similar in growth characteristics. Assigned to the Eurosiberian and Mediterranean Centers of Diversity, strawberry clover or cvs thereof is reported to exhibit tolerance to alkali, grazing, high pH, salt, virus, waterlogging, and weeds. ($2n = 16$.)

DISTRIBUTION: Native to South and Central Europe, north to southern Scandinavia, and east to Asia Minor, north to southern Russia, Lapland to Crimea, Caucasus, eastern Transcaucasus and Western Siberia, and naturally spread in this area. Introduced to New Zealand and Australia, later to United States, where it spread from cultivated fields throughout northern United States.

ECOLOGY: Found in wet meadows, in river valleys, along banks of brooks and in bogs. Adapted to wet

Trifolium fragiferum L.

saline and alkaline soils of western United States. Adapted to wide range of soil conditions and temperature, but adaptation to temperature varies with cv. Very tolerant to wet soils, and to soils with levels of salt that inhibit the growth of most other crop plants. This clover lives under relatively dry conditions and survives short droughts, but does not grow well. It survives flooding for 1–2 months. Some strains adapted to cool climates in northern states, and other strains, to milder climates farther south. Makes good growth in association with saltgrass and sedges in locations where salt concentrations are more than 1%. In established stands, plants can survive over 3% salt for long periods. After salt concentration is reduced by flooding, plants renew growth. Salt tolerance of seedling plants is greater under cool than warm temperatures. Ranging from Cool Temperate Steppe to Wet through Warm Temperate Thorn to Moist Forest Life Zones, strawberry clover is reported to tolerate annual precipitation of 4.4–11.6 dm (mean of 20 cases = 7.0), annual mean temperature of 6.6°–21.8°C (mean of 20 cases = 11.9), and pH of 5.3–8.2 (mean of 19 cases = 6.8).

CULTIVATION: If seedbeds are prepared in fall, early spring planting allows this clover to get established ahead of competing vegetation. Should be plowed or disked thoroughly, then harrowed to level and firm the soil. If the soil is too wet for tilling, vegetation should be mowed and removed. On tilled seedbeds, scarified seed should be sown early in spring when soil is moist. On ground that has not been prepared except for mowing and removing vegetation, either scarified seed should be sown in early spring or unhulled seed should be sown in late winter. Seed may be broadcast or very shallowly drilled. Seedcoats of strawberry clover are hard, and even after hulling, as much as 40–75% may remain hard. Except when unhulled seed is sown in early winter on untilled ground, seed should be scarified. Seeding rate is 2–5 kg/ha, but good stands have been established with 2 kg/ha. A thick stand the first year requires a higher rate. For a good stand more seed is required on poorly prepared than on well-prepared soil. Strawberry clover spreads rapidly and originally thin stands may become thick by the end of the second year. On normal soils, strawberry clover may be spring seeded with a companion grain crop, assuming an ample supply of moisture throughout the season. On saline soils, barley is the only salt-tolerant small grain that can be used as a companion crop. On prepared seedbed, other vegetation usually does not crowd out the seedlings and prevent their establishment. When the seed is broadcast without seedbed preparation, rushes and sedges are apt to crowd seedlings and should be mowed. After seedlings are well established, crop should be grazed. Strawberry clover withstands grazing and trampling better than competing vegetation; it spreads and becomes better established as other plants decline. When strawberry clover is planted with grasses or other legumes in normal soils, it may be grazed at the same time as other pasture plants.

HARVESTING: When used for pasture, strawberry clover, like low-growing forms of white clover, survives under close grazing. It may be more productive if grazed moderately but close grazing reduces less desirable plants and encourages clover spread, and development of good pasture turf. It can be grazed continuously from early spring until late fall without affecting stand. Rotational grazing might favor greater production. Salt content of soil may increase to the point where it inhibits growth of the clover, and pastures so affected may look overgrazed. If closely grazed in late fall, part of the stand may die during winter. Strawberry clover is particularly valuable as a green manure crop on soils where salinity prevents the growth of other legumes. It seldom grows well enough to harvest for hay. When grown for seed, crops may be grazed until the first of June. The growth made after animals are removed is followed by blooming and seed formation. Early season grazing reduces growth of weeds, which often reduce seed yields and handicap harvest. Close grazing up to blooming time does not prevent seed production, but the heads form close to the soil and cannot be machine-harvested. For seed harvest, grazing should be stopped early enough in the season to allow new growth before blooming starts. When clover is grown in soils that are strongly saline preceding and during the blooming period, the heads are short stalked. Seed crop should be cut when most seed envelopes are light brown. If seed crop is cut too early, seed will be shrunken and immature. Because ripe heads of strawberry clover shatter readily, crop should be cut and handled when the heads are slightly damp and seed envelopes are tough and seed shatter less. Crop may be harvested with a mower and later picked up from the swath or windrow. The use of a closely set windrowing attachment or buncher on the mowing machine reduces shattering by reducing the number of times the crop must be handled. Harvesting with vacuum machines has been successful in some areas where the soil surface is relatively smooth. Seed crop may be cured in windrows or stacks, and threshed from the windrows or stacks by clover hullers or combines equipped with hulling attachments.

YIELDS AND ECONOMICS: Seed yields are 40–300 kg/ha and average 100 kg/ha. For pasturage, one hectare of strawberry clover carries 2–5 cows through the entire growing season, if grazing conditions are favorable. Because of its tolerance to saline soils, this clover is extensively cultivated on land in western United States that otherwise would be wasteland. Most seed available from seed trade is harvested from strains naturalized in western United States.

BIOTIC FACTORS: Flowers are visited by honeybees, and this clover is considered a good honey plant. Following fungi have been reported on strawberry clover: *Ascochyta trifolii-alpestris, Dothidella trifolii, Erysiphe communis* and forma *trifolii, Phyllachora trifolii, Pleospora herbarum, Polythrincium trifolii, Pseudopeziza trifolii,* and *Pseudoplea trifolii, Uromyces flectens, U. nerviphilus, U. trifolii,* and *U. trifolii-repentis*. It is also attacked by the bean local chlorosis virus and lucerne mosaic virus. Nematodes isolated from strawberry clover include: *Ditylenchus dipsaci, Heterodera galeopsidis, H. trifolii, Meloidogyne javanica,* and *Pratylenchus penetrans*.

REFERENCE:
Hollowell, E. A., 1960, Strawberry Clover: A legume for the West, *U.S. Dep. Agric., Leaflet* **464**.

CONTRIBUTORS: M. B. Forde, J. A. Duke, P. Gibson, C. F. Reed, R. R. Smith.

Trifolium hirtum All.

FAMILY: Fabaceae

COMMON NAME: Rose clover

SYNONYMS: *Trifolium hispidum* Desf.
Trifolium oxypetasum Heldr. & Sart. ex Nym.
Trifolium pictum Roth

USES: Cultivated and highly adaptive for range use, especially in California, but used more extensively in southern Europe and Russia. Plants highly palatable even when dry. Planted on denuded land surfaces adjacent to highways and freeways, on and about dam sites and in chaparral after fires.

FOLK MEDICINE: No data available.

CHEMISTRY: No data available.

DESCRIPTION: Very villous annual legume; stems 1–4 dm tall, erect or ascending, the branches widely spreading; leaves all petioled or the uppermost sometimes sessile; leaflets obovate or rarely oblong, strongly cuneate, 0.8–2 cm long, the upper shallowly denticulate, the nerves prominent, outwardly curved; stipules narrow, abruptly contracted into a long setose tip much longer than the free blade, the united basal portion pale-chartaceous with green or purple ribs, villous, the uppermost stipule broadly ovate-acuminate; heads terminal, solitary, globose, closely subtended by the upper-

Figure 117. *Trifolium hirtum* All.

Trifolium hirtum All.

most leaf or pair of leaves, their dilated stipules forming an involucre; calyx-tube obconic, densely lustrous-villous, about one-half the length of the subequal, ciliate, setaceous teeth; corolla purplish-red, exceeding the calyx; standard lanceolate-acuminate; lower half of the style united with the stamen tube, pods ovate, 2-valved; seeds rounded-ellipsoid, yellowish, cotyledon margins visible. Fl. May; Fr. June. Seeds 308,700/kg., wt. kg/hl = 77.

GERMPLASM: Assigned to the Mediterranean and Near Eastern Centers of Diversity, rose clover or cvs thereof is reported to exhibit tolerance to grazing, high pH, low pH, poor soil, and slope. ($2n = 10$.)

DISTRIBUTION: Native to the western and eastern Mediterranean regions, Balkans, Asia Minor, Crimea, Caucasus, and Transcaucasus. Introduced to range areas of California.

ECOLOGY: Found native on dry slopes of low mountains, on scrublands and in open woods. Adapted to a wide range of soils, but not to poorly drained soils nor to areas receiving less that 2.5 dm rainfall annually, or above 1,000 m in elevation. Adapted to lower southeastern United States, but not a major pasture crop in that area. Ranging from Warm Temperate Thorn to Dry through Subtropical Dry to Moist Forest Life Zones, rose clover is reported to tolerate annual precipitation of 4.4–13.6 dm (mean of 6 cases = 7.5), annual mean temperature of 14.1°–19.7°C (mean of 6 cases = 17.4), and pH of 5.5–8.2 (mean of 6 cases = 6.8).

CULTIVATION: Seed usually broadcast at rate of 17–22 kg/ha over unprepared land, as along highways, on chaparral or on denuded land. Sometimes seed lightly disked into soil. Requires little to no further attention. Fertilizer requirements negligible as this clover can adapt to most soil mineral contents.

HARVESTING: With proper grazing management, this clover reseeds itself and contributes greatly to the grass–legume pasture of rangeland. Summer grazing of accumulated dry forage shatters the hard seed and animals trample them into the soil.

YIELDS AND ECONOMICS: Yield data are not available, as most stands are self-seeding, after once established. Seed often collected from such stands as needed for other stands. Forage stands are often on rangeland or along highways where they are either eaten by grazing animals or left for soil improvement. No crop is actually harvested. Used more extensively in southern Europe and Russia than in United States.

BIOTIC FACTORS: Except for the nematodes *Meloidogyne incognita acrita* and *M. javanica*, no serious pests or diseases have been reported.

REFERENCES:
Raguse, C. A., Meake, J. W., and Summer, D. C., 1974, Developmental morphology of seedling subterranean and rose clover leaves, *Crop Sci.* **14**(2):333–334.
Thomas, J. H., 1971, *Trifolium hirtum* L. (Fabaceae) in California, *Madrono* **21**:258.
Williams, W. A., Love, R. M., and Berry, L. J., 1957, Production of range clovers, *Calif. Agric. Exp. Stn., Circ.* **458**.

CONTRIBUTORS: J. A. Duke, C. F. Reed.

Trifolium hybridum L.

FAMILY: Fabaceae

COMMON NAMES: Alsike clover, Swedish or Hybrid clover

SYNONYMS: *Trifolium fistulosum* Gilib.
Trifolium michelianum Gaud.

USES: Alsike clover is mainly cultivated for forage and hay. Used mostly in mixtures with red clover, timothy or other grasses, improving the hay and increasing the yield. It has certain advantages over red clover: stands up to excessive moisture in soil, does not produce clover sickness, does not blacken in drying for hay, gives consistent seed yields, and has comparable quality. It is recommended for creek bottoms, wet natural meadows and swales and volunteers and spreads in such places.

FOLK MEDICINE: No data available.

CHEMISTRY: Full-bloom alsike hay contained: 89.0–94.2% DM which analyzed on a zero-moisture basis: 13.5–21.0% crude protein (avg. of 11 = 15.1), 1.4–4.3% fat (avg. of 11 = 35), 24.6–33.4% crude fiber (avg. of 11 = 30.3), 7.7–15.2% ash (avg. = 9.1) and 34.8–43.7% N-free extract (avg. 42.0). Green roughage of full-bloom alsike contains 23.6% DM, 1.3% Ca, 0.29% P, 2.6% K, 0.05% Fe, and

Trifolium hybridum L.

Figure 118. *Trifolium hybridum* L.

about 120 ppm Mn. With 22.6% DM, there is 0.22% S, 0.45% Na, 0.77% Cl, and ca. 60 ppm Zn, 10 ppm thiamine, and 2 mg riboflavin g (Miller, 1958). It can cause *trifoliosis* or bighead, a photosensitized dermatosis. It is thought to contain a volatile oil, salicylic acid, and several glucosides, but no alkaloids. Seeds are reported to contain trypsin inhibitors. The isoflavin phytoalexin isosativan has been reported from this species.

DESCRIPTION: Glabrous, short-crowned perennial legume, often treated agriculturally as a biennial; stems firm, usually hollow, erect or ascending from a procumbent base, 2–8 dm tall, rarely up to 1.6 m under favorable conditions, sometimes branching; leaves, except the uppermost, long-petioled; leaflets ovate or oval to cuneate-obovate, obtuse to emarginate, 1–6 cm long, 1–3 cm broad, dull green, finely setose-toothed, many-nerved; stipules broadly oblong–lanceolate, the tip usually long-attenuate; peduncles axillary, longer than the leaves; heads globose, becoming semiglobose, dense, 30- to 50-flowered, 2–3.5 cm in diameter; bracts subulate, much shorter than the pedicels; pedicels unequal, those of the inner flowers becoming twice the length of the calyx-tube; flowers 6–11 mm long, becoming reflexed in age; calyx about equaling the pedicels of the outer flowers, 3–4.5 mm long, the upper teeth slightly longer than the lower and somewhat exceeding the campanulate tube; corolla pink and white or roseate, becoming dull brown; pods stipitate, 2- to 4-seeded; seeds ovoid-truncate, dull green to nearly black. Fl. and Fr. May–Sept. Seeds 1,543,500/kg; wt. kg/hl = 77.

GERMPLASM: Relatively few cvs have been developed, mainly because of the low degree of variation in this clover. 'Aurora,' developed in Canada, is outstanding in hardiness and seed yield. 'Tetra,' developed in Sweden, a tetraploid cv, may be more persistent and higher-yielding than diploids. Assigned to the Eurosiberian Center of Diversity, alsike clover or cvs thereof is reported to exhibit tolerance to alkali, aluminum, anthracnose, disease, frost, fungus, grazing, heavy soil, mycobacteria, salt, slope, virus, waterlogging, and weeds. ($2n = 16$.)

DISTRIBUTION: Native to northern Europe, but spread naturally throughout Europe to Central Asia, Asia Minor. Introduced and freely escaped throughout temperate regions of the world.

ECOLOGY: Alsike clover grows well in northern latitudes and at high altitudes. Thrives best in a cool, moist climate, and well-adapted to low, wet, fertile land. A rather heavy silt or clay soil with plenty of moisture is recommended. Thrives on good loams, but usually does poorly on dry, sandy or gravelly soils. Alsike also endures overflow that would kill most crops. It has been known to survive a year and make heavy growth in water-soaked and water-covered soil. It withstands severe winters better than red clover and thrives best where the summers are cool. In the South, it is less successful, and only thrives where abundant moisture enables it to overcome the injurious effects of warm summers. Grows on soils too acid for red clover, but also tolerates more alkalinity than most clovers. Alsike is reported to range from Boreal Moist to Wet through Tropical Moist Forest Life Zones, to tolerate annual precipitation of 3.5–17.6 dm (mean of 42 cases = 8.2), annual mean temperature of 4.9°–19.7°C (mean of 42 cases = 8.4), and pH of 4.8–7.5 (mean of 41 cases = 6.3).

CULTIVATION: Alsike clover is seldom cultivated alone. In northern United States, seed is usually sown in early spring; farther south, as in New Jersey, it may be sown in late summer, following early potatoes or cabbage in July or August. May

Trifolium hybridum L.

also be sown in corn at the last working, or seeded with buckwheat in July as in Michigan. If a stand is not secured from spring seeding, the stubble field may be disked and clover seeded in late July or August. In the South where alsike clover is used it is commonly seeded in the fall. Seeding on winter grain is usually done as it is with red clover; seed is broadcast when the ground is checked by frosts, or else on a late snowfall, so that the seed may be carried into the soil with the thawing of the snow. When seeded with spring-sown grain, seedbed is in good shape and clover can be most advantageously drilled. Usual rate of seeding is 4.5–7 kg/ha when drilled, about 7–9 kg/ha when broadcast. Alsike clover is usually seeded with red clover or timothy, or with both. A common practice is to sow timothy in the fall and to sow alsike alone or alsike and red clover in the spring. The proportion of each varies with the intended use of the total crop: 1 part alsike clover, 1–2 parts red clover and 2–3 parts timothy, planted at 8–12 kg/ha. Alsike clover must be inoculated with the appropriate nodule-forming bacteria. As the bacteria used for alsike clover are so widely distributed, soil from a field on which white, red, alsike or crimson clover has been grown can be scattered over the seed, or pure culture of the proper bacteria may be applied to the seed before planting.

HARVESTING: Alsike clover makes good hay; with timothy, it makes an exceptionally good hay. Alsike clover cures to the same color as timothy, and the timothy helps keep the clover from packing when cut, so that it is more airy in the swath and cures better and quicker than pure alsike. With good weather such mixed hay is ready to rake and stack about a day after cutting. On rich soil alsike planted alone tends to grow rank and at harvest is so dense that it molds before drying thoroughly. Alsike is best harvested for hay when in full bloom. It blooms irregularly, and when the field as a whole is in full bloom, there may be many heads with ripe seed. Fortunately, the fine stems of alsike do not harden as quickly as red clover stems. They keep on growing and blooming as the lower flower heads mature. Consequently good hay may be made within a range of several days. Both good hay and a fair seed crop may then be secured. It is not desirable to allow alsike clover to get too ripe. Ripe seed are said to cause slobbering of horses fed on the hay. When seeded with red clover the hay will be at its best if cut when the red clover is in full bloom. Alsike, when grown with red clover, improves the quality of the hay, thickens up the lower parts of the stand and adds materially to the yield, which is said to be greater than that of either variety alone. Alsike may also be grown in mixtures with sweetclover, redtop, Kentucky bluegrass, alfalfa, brome-grass, meadow fescue, ryegrass, orchardgrass, tall meadow oatgrass, and white clover. Alsike clover seed is produced mainly throughout the northern United States as far west as Kansas and the Dakotas and into Idaho. Seed is taken from the first crop which ripens in July or later, depending on the season, latitude and method of handling. Sometimes fields are pastured in the early part of the season, from late May until early July. This delays maturity and the seed crop is ready to cut in August. Cutting is best done with a self-rake reaper and the cut clover is left in small bunches. Some farmers take off every other rake. It is usually cut when three-fourths of the heads are ripe. If the growth is rank and much tangled, the clover is cut with a mower in early morning when the dew is on. A buncher attachment or swather may be used. Alsike clover shatters readily and great care should be taken in handling it. As soon as dry, the clover should be stacked unless threshed direct from the field. Alsike clover straw makes fair feed, even when the crop is not cut until it is too old for good hay. Straw serves as a winter feed for sheep, colts, and young cattle.

YIELDS AND ECONOMICS: Seed yields have been highest in Idaho and Oregon and average 300–375 kg/ha; in other areas yields average 160–200 kg/ha, and occasionally 500–1,000 kg/ha in special seed-producing areas. Alsike is an important legume in the agriculture of Europe, Canada, and northern United States, and in other areas of adaptation. Especially important as a hay crop when grown with red clover, timothy and other grasses or legumes. Annual production of alsike clover seed is about 11,000 MT with an average yield of 163 kg/ha.

BIOTIC FACTORS: Alsike clover is considered resistant to both northern and southern anthracnose. The following fungi have been reported as causing diseases in this clover: *Alternaria tenuis, Cercospora zebrina, Chaetomium globosum, Cymadothea trifolii, Colletotrichum destructivum, C. trifolii, Corticium vagum, Erysiphe martii, E. polygoni, Fusarium acuminatum, F. avenaceum, F. equiseti,*

F. moniliforme, F. oxysporum, F. solani, F. poae, Pellicularia filamentosa, Peronospora trifolii-hybridi, P. trifoliorum, Phyllachora herbarum, Phymatotrichum omnivorum, Pleospora herbarum, Pseudopeziza trifolii, Pseudoplea trifolii (Sphaerulina trifolii), Pythium debaryanum, Sclerotinia sclerotiorum, S. trifoliorum, Sclerotium rolfsii, Stemphylium sarcinaeforme, Stagonospora meliloti, Thielavia basicola, Typhula trifolii, Uromyces trifolii, U. trifolii-hybridi, and *U. trifolii-repentis*. Alsike clover is also attacked by the bacteria, *Bacillus lathyri, Phytomonas cerasi,* and *Pseudomonas syringae*. The following viruses are known to attack alsike clover: alfalfa mosaic, alsike clover mosaic virus 1 and 2, bean chlorotic ringspot, bean mosaic, rosetten krankheit (Phaseolus virus 2), stolbur virus, white clover mosaic, yellow bean mosaic, red clover vein mosaic *(Marmor trifolii)*. Nematodes isolated from alsike clover include the following species: *Ditylenchus destructor, D. dipsaci, Helicotylenchus microlobus, Heterodera glycines, H. trifolii, Longidorus menthasolanus, Meloidogyne hapla, M. incognita acrita, M. javanica, Pratylenchus penetrans, P. pratensis, Trichodorus christiei,* and *Tylenchorhynchus maximus*.

REFERENCES:
Hermann, F. J., 1966, *Notes on Western Range Forbs: Cruciferae to Compositae,* pp. 150–151, illus., Washington, D.C.
Pieters, A. J., 1947, Alsike clover, *U.S. Dep. Agric., Farmers Bull.* **1151**:1–18, illus.

CONTRIBUTORS: J. A. Duke, C. F. Reed, J. K. P. Weder.

Trifolium incarnatum L.

FAMILY: Fabaceae

COMMON NAMES: Crimson clover, Scarlet clover

SYNONYMS: *Trifolium incarnatum* var *elatius* Gibelli & Belli
Trifolium molineri Balb.
Trifolium noeanum Reichb. ex Mert. & Koch
Trifolium spicatum Perret ex Colla
Trifolium stramineum Presl

USES: Crimson clover is the most important winter annual legume for southern United States. Besides being an excellent pasture and hay crop, it covers the soil and protects it against washing and leaching

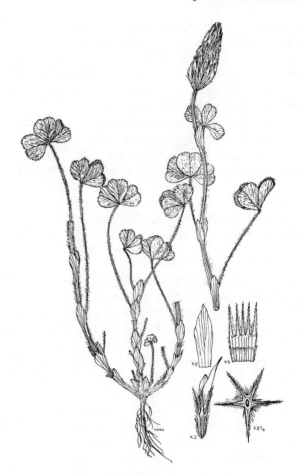

Figure 119. *Trifolium incarnatum* L.

during the fall, winter, and spring. It may be turned under for green manure in the spring. It produces large quantities of seed which is harvested and sown without the use of expensive machinery. All clovers when grown in thick stands are useful for soil improvement.

FOLK MEDICINE: No data available.

CHEMISTRY: On a moisture free basis, the average of 39 analyses was: 16.7% crude protein, 3.1% ether extract, 28.3% crude fiber, 9.3% ash, and 42.6% N-free extract. Ca analyses averaged 1.6%, P, 0.35%, K, 2.41%, Mg, 0.41%, Fe, 0.70%, and Mn 330 ppm (Miller, 1958).

DESCRIPTION: Stout, soft-pubescent annual legume; stems erect, usually simple, occasionally with elongate branches from the base, 2–6 dm long; petioles of lower leaves and median leaves very long, those of upper leaves very short; leaflets broadly cuneate-

Trifolium incarnatum L.

obovate, usually emarginate, 1.5–3 cm long, and nearly as broad, strigose, more or less denticulate above; stipules large, the free part short, ovate-obtuse, toothed; heads terminal, solitary, long-peduncled, oblong-conic, becoming cylindrical, 2–7 cm long, 1–2.5 cm thick; calyx fulvous-villous, the tube cylindric to campanulate, strongly 10-ribbed, the teeth linear-setaceous, erect, stiff, spreading, in fruit, subequal, more or less exceeding the tube; corolla scarlet or dark red, rarely pink or white, exceeding the calyx-teeth; pods ovoid; seeds narrowly ovoid, large, greenish-yellow, becoming reddish, moderately lustrous. Fl. spring; Fr. late spring and early summer to July. Seeds 308–700/kg. Wt. kg/hl = 77.

GERMPLASM: More than half of the domestic seed produced now is of the reseeding type. Fall volunteer stands are possible because the hard seed coat delays germination from late spring, when seed shatter, until fall. Common crimson clover is not a reseeding type. Five reseeding cvs are widely used: 'Dixie,' 'Auburn,' 'Autauga,' 'Chief,' and 'Talladega,' the first 3 are early cvs, their seed matures about a week before the other 2, and before 'Common.' 'Frontier,' a soft-seeded type, developed in Mississippi and 'Tibbee,' a reseeding cv, have large seed, superior seedling vigor, greater fall and winter growth, equal or superior forage and seed yields, and early maturity. Assigned to the Mediterranean and Eurosiberian Centers of Diversity, crimson clover or cultivars thereof is reported to exhibit tolerance to grazing, high pH, heavy soil, nematodes, sand, slope, virus, and weeds. ($2n = 14, 16$.)

DISTRIBUTION: Native to Atlantic and southern Europe, Caucasus, and Transcaucasus. Cultivated and introduced in many temperate areas; especially useful in southern United States.

ECOLOGY: Crimson clover does well in the cool, humid climate in most of southern United States in winter. May be grown as a summer annual in northern Maine, northern Michigan, and other locations where the weather remains cool and moist. Crop thrives on both sandy and clay soils and is tolerant of ordinary soil acidity. Does not thrive on heavy, poorly drained soils that remain wet and cold during the winter. On very poor soils, stands are difficult to obtain and growth is stunted. Ranging from Boreal Wet through Subtropical Moist Forest Life Zone, crimson clover is reported to tolerate annual precipitation of 3.1–16.3 dm (mean of 30 cases = 9.2), annual mean temperature of 5.9°–21.3°C (mean of 30 cases = 12.9) and pH of 4.8–8.2 (mean of 26 cases = 6.5).

CULTIVATION: A firm seedbed is best for crimson clover. If the seedbed is loose, roots grow into air pockets between soil particles, dry out, and die. It is difficult to make a seedbed between rows of cultivated crops. Row crops also shade the clover seedlings and compete for moisture. If seeded in wider rows and more thinly than normal, the clover establishes better. Between rows of other crops, seed is usually broadcast on the soil surface and covered by cultivating or harrowing. More complete stands are obtained by lightly stirring the soil surface with a harrow, and then drilling the seed. Seeding following a grain crop is a surer method of establishing a stand than planting between the rows of the cultivated crops. For a permanent pasture, ground is prepared by closely grazing or clipping the grass. Heavy turf is thoroughly disked or burned before seeding. Crimson clover may be seeded alone or in combination with winter grains or ryegrass. Seed may be broadcast or drilled, which gives more uniform stands. Should not be planted more than 1.3 cm deep in clay soil, or more than 2 cm in sandy soil. Under ordinary conditions 12–15 kg/ha of hulled seed gives a good stand. Thick stands with plants 2.5 cm apart produce much greater fall and winter growth than thin stands. Seed sown from mid-July until November; best time depends on location. The farther north the crop is planted, the more important it is to establish the stand early. Seeding either immediately before or after heavy rains increases the chance of a good stand. Spring planting is not recommended as it usually results in short, stunted growth followed by poor flowering and low seed yield. When crimson clover is grown with companion crops, as rye, vetch, ryegrass or fall-sown grain, the crimson clover usually is seeded at two-thirds the normal rate and the companion crop at one-third to one-half its normal rate. Crimson clover and its companion crop should be seeded at the same time. Since the companion crop is usually seeded at a greater depth than the crimson clover, two seeding operations may be necessary. Reseeding cvs have given good results when planted with Johnsongrass, bermudagrass or other summer-growing perennial grasses. Crimson clover requires phosphate and potash fertilizer on most soils. Soils should be tested first. Common rates used on deficient soils range from 200–400 kg/ha of a 0 : 20 : 20 fertilizer. A large single application of these two is sometimes sufficient to produce two crops, but annual fall applications on most soils are recommended to maintain high yields. On some soils boron may improve growth and increase seed yields. Liming to reduce soil acidity is widely practiced in most of eastern United States.

HARVESTING: Crimson clover is used for many purposes. It produces abundant green forage in fall and spring. It also furnishes grazing during winter months if it is planted early and makes good fall growth. This clover combined with small grains or ryegrass has been widely used for winter grazing. Animals grazed on crimson clover seldom bloat. Quality of hay is highest if the crop is cut at the early bloom stage; the yield is highest if cut at full bloom. Hay is easily cured either in the swath or in the windrow. Fewer leaves are lost and less bleaching occurs in windrowed hay. This clover may be made into silage by the methods used for other legumes. Also crimson clover is ideal for green manure. For best results, it is plowed under 2–3 weeks before a succeeding crop is to be planted, giving sufficient time for decomposition. In fields in which row crops are to be planted, strips are plowed under for green manure and the row crop is later planted in the plowed strips. Clover between the plowed strips matures and seed may be harvested by hand from the clover between the row crops, and the remaining stubble may be allowed to mat and serve as a mulch. Or, the crop may be left unharvested and permitted to form a mulch. In orchards, crimson clover is often allowed to mature, then disked into the soil. Reseeding cvs volunteer good stands in fall, year after year, if properly managed. In rotation with row crops, crimson clover volunteers if the seed is allowed to mature in the spring before the seedbed is prepared for the cultivated crop. Stands have been good following sorghum or late-planted corn. Grazing or clipping during late winter or early spring may slightly reduce seed yields. However, if some growth is not grazed off and if spring growing conditions are favorable, growth may be so rank that it will lodge and result in low seed yields or failure. Crop may be grazed in late winter or early spring and animals removed 4–6 weeks before crop is due to be in full bloom. After animals are removed, the clover is clipped. This causes more uniform flowering and seed ripening. Hand harvesting may be practical for small quantities of seed. Mature seed heads shatter readily and are easily stripped by hand or with a homemade stripper. Seed may be mechanically harvested in three ways: combined direct from standing plant, cut with mower and left in swath or windrowed to dry and then picked up and threshed with the combine, or cut with a mower and left in the swath or windrowed to dry out, then hauled to a stationary huller or thresher. For minimum shattering crop should be cut and windrowed when heads are damp and tough. Most of hulls should be light brown if seed is stripped or cut with a mower. Hulls should be dark brown for direct combining. In humid areas harvested seed usually is dried to lower moisture content to a safe level for storing. Seed may be thinly spread under shelter and turned until dry enough to store, or may be dried under hot-air driers.

YIELDS AND ECONOMICS: Yields of hay of crimson clover average 4.5–5 MT/ha, yields up to 7.5 MT/ha are possible on very good soils with excellent growing conditions. Yields of 340–410 kg/ha of seed are common. Seed yields usually are higher on soils of medium fertility than on rich soils.

BIOTIC FACTORS: Florets are self-fertile, but not self-tripping. Bees, visiting the flowers, increase the number of seeds per head by tripping the florets and transferring the pollen. For each hectare, 2–3 strong colonies of honeybees should be placed at the edge of the field at blooming time. The following fungi have been reported as causing diseases in crimson clover: *Cercospora zebrina, Colletotrichum trifolii, Cymadothea trifolii, Erysiphe martii, E. polygoni, Kabatiella caulivora, Peronospora trifoliorum, P. viciae, Phyllachora trifolii, Pleospora herbarum, Polythrincium trifolii, Pythium debaryanum, P. irregulare, Sclerotinia minor, S. sclerotiorum, S. trifoliorum, Sclerotium rolfsii, Stemphylium sarciniforme, Uromyces trifolii*, and its vars. Also damaged by the bacterium *Pseudomonas syringae*. Disease-causing viruses isolated from crimson clover include: alfalfa mosaic, alsike clover mosaic, bean local chlorosis, bean mosaic, broadbean mottle, New Jersey potato yellow dwarf, pea enation mosaic, pea mosaic, red clover vein mosaic

Trifolium incarnatum L.

(Marmor trifolii), Rosetter Krankheit *(Phaseolus virus* 2), woundtumor virus *(Aureogenus magnijena)*, yellow virus of garden pea. Nematodes isolated from crimson clover include the following species: *Belonolaimus gracilis, Ditylenchus dipsaci, Heterodera glycines, H. radicicola, H. schachtii, H. s.* var *trifolii, H. trifolii, Meloidogyne arenaria, M. hapla, M. incognita, M. i. acrita, M. javanica, Paratylenchus projectus, Pratylenchus brachyurus, P. penetrans, Rotylenchulus reniformis, Trichodorus christiei, Tylenchorhynchus claytoni,* and *Tylenchus dipsaci.*

REFERENCES:
U.S. Department of Agriculture, 1971, Growing crimson clover, *U.S. Dep. Agric., Leaflets* **842**:1–10, illus.
Lee, W. O., 1964, Chemical control of weeds in crimson clover grown for seed production, *U.S. Dep. Agric., Tech. Bull.* **1302**:1–21, illus.

CONTRIBUTORS: J. A. Duke, C. F. Reed.

Trifolium medium L.

FAMILY: Fabaceae

COMMON NAME: Zigzag clover

USES: For hay, pasture, and erosion control.

FOLK MEDICINE: No data available.

CHEMISTRY: Not known to contain toxic compounds. Similar to other cultivated forage legumes in percentage of crude protein, phosphorus, potassium, calcium, and magnesium.

DESCRIPTION: A long-lived, rhizomatous perennial, up to 0.8 m tall; stems flexuose, ascending; leaflets elliptic-oblong to oval, obtuse to acute, 1.5–5.5 cm long, 0.5–1.5 cm wide, margins finely ciliate; petioles similar to leaflets in length; stipules pale, lanceolate, ciliate, the free part lanceolate-subulate, usually shorter than the petiole; peduncle very short at first, then some elongation; heads mostly terminal, usually solitary, subglobose to ovoid, 2–4 cm long; flowers 13–18 mm long, light-purple; calyx tube cylindrical, rounded at base, with ring of stiff hairs at inner margin of throat; corolla about twice as long as the calyx; pod ovoid to obovoid; seeds triangular-ovoid, yellow to brown.

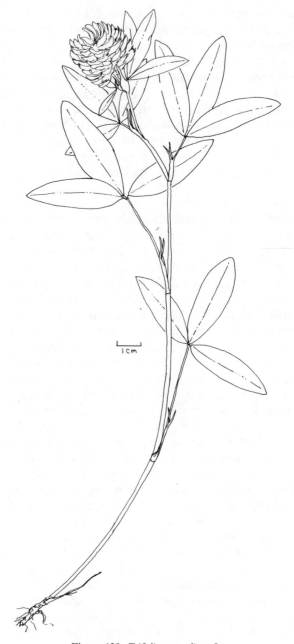

Figure 120. *Trifolium medium* L.

GERMPLASM: Named cultivars have not been developed. One germplasm pool was developed and released by the USDA and the Colorado Agricultural Experiment Station. Germplasm is available from researchers in Canada, Europe, and the United States from native stands in Europe and Asia; and from naturalized stands in Canada and the United States. There is considerable variability within the species for growth habit and it should be possible to select types adapted to grazing or to hay production. From the Eurosiberian Center of Diversity, zigzag clover is reported to tolerate frost and shade.

DISTRIBUTION: Grows throughout Eurasia and is native to the area. A few naturalized stands occur in eastern Canada and the northeastern United States.

ECOLOGY: In its native habitat it is common in both damp and dry meadows, forested areas, and along roadsides and stream banks. Appears to be less sensitive to shade than most cultivated clovers. It prefers permeable soils that may be somewhat acid, but avoids purely calcareous soils. Very winter-hardy. Cross-pollinated and highly self-incompatible. The few data available suggest that zigzag clover is at home where annual precipitation is 5.2–6.6 dm (mean of 2 cases = 5.9), annual mean temperature is 8.0°–8.9°C (mean of 2 cases = 8.2), and pH is 6.0–7.0 (mean of 2 cases = 6.5).

CULTIVATION: Propagated from seeds but can be propagated from rhizomes. It is seldom grown alone and is generally found as an admixture of red clover, *Trifolium pratense* L. Spreads rapidly and under favorable conditions some plants will spread over 1 m within 2 years after transplanting. Low seed production has contributed to limited agricultural interest in the species.

YIELDS AND ECONOMICS: When harvested for hay under irrigation, zigzag clover yielded 50% less than red clover and alfalfa. Certain clonal progenies, however, have produced substantially higher yields. It is particularly slow to recover following harvest and hence is not a suitable component in a mixture with orchardgrass, *Dactylis glomerata* L. As with many long-lived perennial species, seedling vigor is poor and establishment difficult. Once established, however, persistence is excellent.

REFERENCES:
Kownacka, M., 1958, Preliminary observations on zig-zag clover, *Rocz. Nauk Roln., Ser. F* **72**(3):1–6.
Robertson, R. W., and Armstrong, J. M., 1964, Factors affecting seed production in Trifolium medium, *Can. J. Plant Sci.* **44**:337–343.
Townsend, C. E., 1971, Registration of C-1 zigzag clover germplasm, *Crop Sci.* **11**:139.
Townsend, C. E., Dotzenko, A. D., Storer, K. R., and Edlin, F. E., 1967, Response of zigzag clover genotypes to management practices, *Can. J. Plant Sci.* **48**:273–279.

CONTRIBUTORS: J. A. Duke, C. E. Townsend.

Trifolium nigrescens Viv.

FAMILY: Fabaceae
COMMON NAME: Ball clover

USES: A cool-season diploid nonstoloniferous clover, cultivated as a winter annual legume, useful for both pasture and soil improvement. It has excellent reseeding ability, and is used in permanent grass sods in southeastern United States.

FOLK MEDICINE: No data available.

CHEMISTRY: Crude protein ranged from 17.6 to 26.1%, averaging about 22%. Highly succulent, ball

Figure 121. *Trifolium nigrescens* Viv.

Trifolium nigrescens Viv.

clover may have lower DM content than crimson clover.

DESCRIPTION: Glabrous annual legume, usually branching from the base; stems decumbent to ascending, 1–5 dm long, not rooting at nodes; petioles of lower leaves exceeding the leaflets, those of the uppermost much shorter; leaflets obovate from a cuneate base, truncate to emarginate, 10–30 mm long, 7–30 mm broad, sharply setose-toothed above, entire near the base, prominently nerved; stipules membranous, triangular-lanceolate, abruptly acuminate to a setose tip; peduncles axillary, exceeding the leaves; heads many but loosely flowered, globose to semiglobose, 12–20 mm in diameter; bracts small, triangular-subulate, strongly keeled; pedicels in anthesis equaling the calyx-tube, the inner greatly elongating in fruit; flowers 7–10 mm long, fragrant, white to yellowish-white, becoming sordid-yellow and strongly reflexed on drying; calyx glabrous, 5-ribbed with prominent intermediate nerves, somewhat 2-lipped, the teeth triangular-lanceolate, recurved after anthesis, unequal, the upper approximate, longer than the lower and about equaling the tube; corolla 5–8 mm long, 2 to almost 3 times the length of the calyx; pods linear, exserted, 3- to 4-seeded; seeds ovoid, dark brown. Seeds 2,205,300/kg: wt. kg/hl = 77.

GERMPLASM: Assigned to the Near East and Eurosiberian Centers of Diversity, ball clover or cvs thereof is reported to exhibit tolerance to grazing, high pH, and slope. ($2n = 16, 32$.)

DISTRIBUTION: Native to Turkey and the Near East. Introduced to the United States especially Alabama and the southeastern states.

ECOLOGY: Ball clover succeeds on various soil types, but is best adapted to loam or clay soils. It may suffer more during hot, dry spring weather than crimson clover, especially on droughty soils. When soil moisture is adequate, ball clover furnishes large quantities of high-quality grazing to bridge the gap between crimson clover and summer crops. Ball does not provide much early forage; in Alabama, farmers graze cattle on ball from winter through spring (Jan.–Feb. through May). Ranging from Warm Temperature Dry to Moist through Subtropical Moist Forest Life Zone, ball clover is reported to tolerate annual precipitation of 6.2–13.3 dm (mean of 4 cases = 11.0), annual mean temperature of 12.5°–21.3°C (mean of 4 cases = 16.4), and pH of 5.6–6.5 (mean of 4 cases = 5.9).

CULTIVATION: Seed broadcast during early fall. Stands may be spotty in first year but natural reseeding usually produces good stands by the second year. Seed rates are 2–4 kg/ha, depending on the area. Stand failures or poor stands on permanent grass sods of coastal bermudagrass or Pensacola bahiagrass can be expected when a dense growth of ungrazed grass is left in the fall. Such sod should be mowed for good ball clover stands. Farmers maintain volunteering stands of ball in common bermudagrass, coastal bermudagrass or bahiagrass pastures. Even after late, heavy grazing, ball produces a seed crop because seedheads are close to the ground. Ball clover does well when planted on cultivated land with Abruzzi rye. This forage mixture gives longer, more uniform production than rye and crimson clover. Both clovers have a place in a well-balanced forage program—crimson for early- and ball for late-spring use. Adequate lime, phosphate and potash are essential for good growth. Like other clovers, ball does not grow in extremely acid soils. Lime and fertilizers should be applied in accordance with soil analyses.

HARVESTING: Over 60% of combine-harvested seed have hard coats. High late-summer and early-fall temperatures inhibit germination, even when seed is scarified. Consequently, most seeds germinate during late fall when temperatures are more favorable for growth. Ball tolerates grazing, and reseeds well, even under heavy grazing, producing most of its growth about a month later in spring than crimson clover. Ball can be grazed until midspring and still set a good seed crop. Seed is harvested by direct combining. Rain during harvesting can severely cut seed yields.

YIELDS AND ECONOMICS: Ball is an excellent seed producer, yielding 200–600 kg/ha of seed. Dry forage yields are 2.5–7.5 MT/ha. Year to year production varies mainly because ball makes most of its total growth during the late winter and spring months. Florida Ball seeded in Oct. yielded only ca. 2 MT/ha by the following April, seeded in Nov. yielded only ca. 2½ MT/ha by the following May. It does not produce as well in Florida as white clover, but its hard seededness results in good self-regeneration. 'Giant Ball' appears to be much more vigorous. At present the largest use of ball clover is in Alabama and southeastern United States, where it is used mainly for winter pasture and for soil improvement.

BIOTIC FACTORS: The fungus *Pseudopeziza trifolii* has been reported on ball clover; also the nematode *Meloidogyne incognita acrita* has been used in resistance studies for the southeastern states. Clover head weevil *(Hypera meles)* may be a serious pest, reducing seed yields.

REFERENCES:
Hoveland, C. S., 1960, Ball clover, *Auburn Univ. Agric. Exp. Stn., Leaflet* **64**.
Hoveland, C. S., 1962, Ball clover is rolling, *Crops Soils* **15**(2):8.
Kretschmer, A. E., 1966, Production and adaptability of Trifolium sp. in South Florida, *Proc. Soil Crop Sci. Soc. Fla.* **26**:81–93.

CONTRIBUTORS: J. A. Duke, C. F. Reed.

Trifolium pratense L.

FAMILY: Fabaceae

COMMON NAMES: Red clover, Peavine clover, Cowgrass

SYNONYMS: *Trifolium alpicola* Hegetschw. & Heer
Trifolium brachystylos Knaf
Trifolium carpaticum Porc.
Trifolium expansum Waldst. & Kit.
Trifolium nivale Sieb. ex Koch
Trifolium silvestre Ducomm.

USES: Extensively grown for pasturage, hay, and green manure. As a forage crop, it is excellent for livestock and poultry.

FOLK MEDICINE: Said to be used for: alterative, antiscrofulous, antispasmodic, aperient, athlete's foot, constipation, detergent, depurative, diuretic, expectorant, gall-bladder, gout, liver, rheumatism, sedative, skin, sores, and tonic. An extract of the flowers is used for cancerous ulcers, corns, etc. American Indians are said to have used the herb for burns and sore eyes.

CHEMISTRY: Seeds are reported to contain trypsin inhibitors and chymotrypsin inhibitors. Green forage of red clover is reported to contain: 81% moisture, 4.0% protein, 0.7% fat, 2.6% fiber, 2.0%

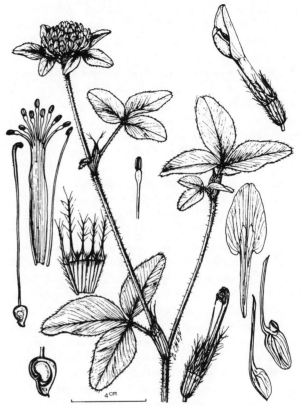

Figure 122. *Trifolium pratense* L.

ash. Hay of red clover contains 12.0% moisture, 11.8% protein, 2.6% fat, 27.2% fiber, and 6.4% ash. On the basis of more than 500 analyses, Miller (1958) reported the hay contained on a moisture-free basis: 8.3–24.7% protein (avg. 14.9%), 1.0–6.6% fat (avg. 2.9%), 12.5–39.3% crude fiber (avg. 30.1%), 3.1–14.0% ash (avg. 7.9), and 33.4–59.1% N-free extract (avg. 44.2). For green red clover forage he reported 12.4–34.8% protein (avg. 18.2), 3.2–5.9% fat (avg. 4.0%), 12.7–30.8% crude fiber (avg. 24.2), 7.0–13.6% ash (avg. 8.8), and 37.1–49.7% N-free extract (avg. 44.8%). The hay (dry matter averaging 87.7%) contained 0.97–2.29% Ca (avg. 1.61), 0.09–0.45% P (avg. 0.22), 0.57–2.67% K (avg. 17.6%), 0.24–0.81% Mg (avg. 0.45%), 0.001–0.185% Fe (avg. 0.013%), 9.9–17.6 ppm Cu (avg. 11.2 ppm), and 24.9–120.8 ppm Mn (avg. 65.6). The green forage contained 0.58–3.21% Ca (avg. 1.76), 0.24–0.53% P (avg. 0.29), 1.49–2.94% K (avg. 2.10%), 0.36–0.57% Mg (avg. 0.45), 0.016–0.032% Fe (avg. 0.03), 7.3–10.3 ppm Cu (avg. 8.8 ppm), 121–464 ppm Mn (avg. 159 ppm). The leaf-protein concentrate (59% protein) contains 6.4% arginine, 2.5% histidine, 5.4% thre-

Trifolium pratense L.

onine, 1.7% tryptophan, 9.5% leucine, 5.3% isoleucine, 1.7% methionine, 6.8% lysine, 6.1% phenylalanine, and 6.8% valine. Estrogenic disorders have been reported in cattle grazing largely on red clover, apparently due to activity of the isoflavones-formononetin, biochanin A, and to some small extent daidzein and genistein. The flowers contain a number of phenolic compounds: daidzein, genistein, isotrifolin, isorhamnetin, pratol, pratensol, trifolin, and an antifungal compound trifolirhizin. They also contain coumaric acid, hentriacontane, heptacosane, myricyl alcohol, and β-sitosterol. On a dry basis flowers yield 0.028% of an oil containing furfural.

DESCRIPTION: Perennial, sometimes biennial, legume, sparingly pilose to glabrous, sometimes densely pilose (forma *pilosum* (Griseb.) Hayek); stems erect or ascending, 1–5 cm long; leaves of basal rosette all long-petioled, those of stem moderately long-petioled to nearly sessile; leaflets oval or elliptic to cuneate-obovate, 1–3 cm long, 0.5–1.5 cm broad, subentire; stipules oblong-oval to oval-triangular, the free part broadly triangular, abruptly tapering to an erect setaceous tip; peduncles short or absent; heads mostly terminal, sessile, short-peduncled, usually closely subtended by the stipules of the upper pair of leaves, dense, subglobose to ovoid, 1.2–3 cm long; flowers sessile, 10–15 mm long, rosy purple to creamy-white (forma *leucochraceum* Aschers. & Prantl), erect; calyx-tube campanulate, narrower at base, 10-nerved, pubescent including the inner margin of the throat, the teeth filiform from a triangular base, sparsely hirsute, porrect, the upper about equaling the tube, the lower almost twice as long; corolla about twice the length of the calyx; pods oblong-ovoid, circumscissile; seeds ovoid, asymmetrical, yellowish to purplish. F. May–Sept. Seeds 606,375/kg; wt. kg/hl = 77.

GERMPLASM: Red clover cvs are of two types, early-flowering cvs that produce at least two cuttings per season, and late-flowering cvs that produce one cutting per season. Early-flowering cvs, also called medium or double-cut, are the most common type grown in the United States. Late-flowering cvs, called mammoth or single-cut, are grown where the growing season is short, as in high elevations or latitudes. Numerous cvs of both types are available. Most pubescent American cvs have excellent resistance to the potato leafhopper. Assigned to the Eurosiberian Center of Diversity, red clover or cvs thereof is reported to exhibit tolerance to aluminum, disease, frost, grazing, hydrogen fluoride, high pH, low pH, mines, mildew, mycobacteria, rust, slope, virus, waterlogging, and weeds. ($2n = 14, 28$.)

DISTRIBUTION: Native to southwestern Europe and Asia Minor, found also in north Atlantic and central Europe, the Mediterranean region, Balkans, Iran, India, Himalayas, Russia from Arctic south to eastern Siberia, Caucasus, and the Far East. Widely introduced and cultivated.

ECOLOGY: Native on wet to dry meadows, open forests, forest margins, field borders and paths. Grows best on well-drained loam soil, but also adapted to wetter soils. Most soils that produce good crops of corn, tobacco or small grains also produce a good crop of red clover. Some of these soils may need lime or fertilizer, or both. Red clover is most productive on soil with pH 6.6–7.6, but is better adapted than alfalfa at pH 5.7–6.5. It also needs P and K to produce good yields; amounts needed can be determined by soil tests. Ranging from Boreal Moist to Wet through Subtropical Moist Forest Life Zone, red clover is reported to tolerate annual precipitation of 3.1–19.1 dm (mean of 91 cases = 8.6), annual mean temperature of 4.9°–20.3°C (mean of 91 cases = 10.6), and pH of 4.5–8.2 (mean of 84 cases = 6.3).

CULTIVATION: In northeastern United States and Canada, and at higher elevation in southeastern and western United States, red clover grows as a biennial or short-lived perennial; at lower elevations in southeastern United States, it grows as a winter annual, and at lower elevation in western United States and Canada, it grows under irrigation as a biennial. Most red clover is spring-seeded in a crop of fall- or spring-sown small grain. In the early spring the soil alternately freezes and thaws, thus covering the seed with soil. The small grain holds weeds in check while the clover is getting started. Seeds are also drilled in, depending on circumstances. At lower elevations in southeastern and western United States, red clover is sown from Oct. on into Jan. In these areas it is most frequently sown without a companion crop. In southeastern United States, late-summer seedings can be successful if they are made on a seedbed that has been fallowed to prevent weed growth. Grass is extensively seeded with red clover. Clover–grass mixture is preferred to clover alone in the southeast. *In vitro* experiments show that some lines of red clover perform well when intercropped with ryegrass *(Lolium multiflorum)*. Clover–grass hay cures more rapidly than pure clover hay, and it produces more hay per hectare. Animals are less likely to bloat on clover–grass pasture than on pure clover. Timothy is frequently mixed with red clover. It has a high yield, and it is ready to cut for hay when the red clover is ready. Grass is sown in the early fall in the small-grain crop; red clover is sown in the small grain-grass in the spring. When the grain crop is harvested, straw and stubble are removed from the field as they smother the clover and favor clover disease development. Clover-hay yields from fields where the straw and stubble have been left are only about one-half as large as the yields from fields where they have been removed immediately after combining. Small-grain companion crops compete with red clover for mineral nutrients, moisture, and light. This intense competition can be reduced by grazing or clipping the small grain in late winter or early spring, just before its stems begin to grow. Grazing or clipping after clover stems have begun to joint reduces small-grain yield.

HARVESTING: The first year, graze or mow the clover 4 to 6 weeks before the first frost in the fall. If the stand is mowed, remove the clippings unless the total amount is quite small. The first crop of red clover, harvested early the second year is almost always harvested for hay or silage. In early bloom, red clover is leafy and produces its largest yield of protein per hectare. Red clover should be cut shortly (5-15 days) after the very first bloom appears in the stand. When grown with a grass, the stand is cut when the clover is ready rather than when the grass is ready. Usually the second crop of red clover is pastured, harvested for seed, or grown for soil improvement and green manure. For hay, it should be cut in early bloom. A medium stand produces two to four crops of hay in the harvest year. Mammoth clover produces one crop. The cut crop is allowed to wilt in swath and then raked into small, loose windrows. Modern technology uses a crimper-swather. Hay cures about as rapidly in the windrows as in the swath, and fewer leaves are lost in baling. Energy permitting, it can also be forced-air dried, which preserves the green color, lessens leaf shattering and practically eliminates spoilage. Red clover and red clover–grass mixtures are frequently ensilaged. These crops make good ensilage if they are wilted slightly before ensiled, or if carbohydrate or chemical preservatives are added as they are ensiled. Red clover is one of the best legume pasture plants for livestock and poultry. Red clover and red clover–grass mixture pastures can be grazed or cut green and fed to livestock and poultry. Red clover is also one of the better legumes for renovating old pastures. Old pasture should be clipped or grazed closely. Sod is chopped with a disk or harrow before red clover is sown. In addition to its principal value as livestock feed, red clover may be turned under as green manure which improves the physical properties of the soil and increases the yield of the following cultivated crop. Many crop rotations are possible for red clover, the oldest is a 3-yr rotation of corn, oats, or wheat and red clover. Other common rotations are: corn, soybeans, small grain, red clover; corn, small grain, red clover; rice, red clover; sugar beets, small grain, red clover; tobacco, rye, or wheat, red clover–grass, grass, grass; potatoes, small grain, red clover. For seed production, the first crop of the second-year stand is usually harvested for hay or silage, the second crop may be harvested for seed. In eastern United States, most red clover is pollinated by native bumblebees. Where native pollination is inadequate, 2–8 colonies of honey bees per hectare are recommended. Seed yields are best when bees are abundant and soil fertility and moisture are adequate to promote good growth, and when the weather is warm and clear during the flowering period. The seed crop is harvested when most of the seed heads are black, usually 25–30 days after full bloom. For seed, the crop is mowed and cured in the swath or in small windrows. During rainy weather, the mowed crop cures better in swaths than in windrows. Windrowing is better during clear, warm weather because it reduces harvesting losses. Handling should be minimum; each handling causes shattering losses. The swathed or windrowed crop is harvested with a combine with a pickup attachment. Combine should be operated carefully to do a good harvesting job and to reduce harvesting losses. Seed yields may be best from direct combining following paraquat treatment. Artificial drying or drying by spreading seed thinly on

Trifolium pratense L.

a floor maintains the quality of the seed. Seed should be turned every few days until completely dry. Rough cleaning immediately after combining reduces the drying time and maintains seed quality.

YIELDS AND ECONOMICS: Under favorable conditions, seed yields average about 70–100 kg/ha, but under irrigation in western United States may reach 600–800 kg/ha. Several cvs of red clover, as 'Arlington,' 'Kenland,' 'Kenstar,' and 'Lakeland,' where adapted, produce higher yields of forage and are more persistent than common red clover. Depending upon stage at which red clover is cut for forage, hay or silage, yields vary from 1 to 9 MT/ha. Red clover is one of the most widely grown of the true clovers and forms one of the most important hay crops in temperate regions, especially when combined with grasses. Extensively grown and used for pasture and soil improvement throughout the United States. In 1970, about 24,000 MT of red clover seed were produced on ca. 200,000 ha, mainly in California, Iowa, Oregon, Kansas, Virginia, and Washington.

BIOTIC FACTORS: Red clover is attacked by many fungi, some of which may cause serious losses. The following fungi have been reported on red clover: *Alternaria tenuis, Ampelomyces quisqualis, Ascochyta trifolii, Botrytis anthopila, B. cinerea, Brachysporum trifolii, Cercospora zebrina* (summer black stem), *Chaetomium cochliodes, Colletotrichum destructivum, C. trifolii* (southern anthracnose), *Corticium solani, Cylindrocladium scoparium, Cymadothea trifolii, Didymella trifolii, Didymium sturgisii, Erysiphe communis* f. *trifolii, E. martii, E. polygoni* (powdery mildew), *Fusarium acuminatum, F. avenaceum, F. equiseti, F. gramineaum, F. oxysporum* (root rot), *F. poae, F. roseum, F. solani, Kabatiella caulivora* (northern anthracnose), *Leptosphaerulina americana, L. briosiana, L. trifolii, Metasphaeria boucera, Microsphaeria alni, Mycosphaerella carinthiaca, Oidium erysiphoides, Ophiobolus collapsus, O. graminis, Peronospora pratensis, P. trifoliorum, Phoma trifolii* (spring black stem), *Phyllachora trifolii, Phymatotrichum omnivorum, Phytophthora cactorum, Phyllosticta trifolii, Plenodomus melioti, Pleospora herbarum, Polythrincium trifolii, Pseudopeziza trifolii, Pseudoplea medicaginis, P. trifolii* (pepper spot), *Pyrenopeziza jonesii, Pythium debaryanum, Rhizoctonia crocorum, Rh. leguminicola* (black patch), *Rh. solani, Rh. violacea, Sclerotinia kitajimana, S. sclerotiorum, S. spermophila, S. trifoliorum* (crown rot), *Septoria compta, Sporonema phacidioides, S. trifolii, Sphaerulina trifolii, Stagonospora compta, S. meliloti, S. recdens, Stemphylium sarcinaeforme* (target spot), *S. botryosum, Sclerotium delphinii, S. rolfsii, S. baticola, Stictus pustulata, Thielayiopsis basicola, Thyrospora sarcinaeformis, Uromyces fallens, U. minor, U. nerviphilus, U. trifolii, U. trifolii-repentis, Vermicularis dematium, Verticillium dichotomum,* and *Volutella fusarioides.* Bacteria causing diseases in red clover include: *Bacillus lathryi* (red clover streak), *Pseudomanes radiciperda,* and *Ps. syringae.* Parasitic on red clover are *Cuscuta epithymum* and *C. pentagona.* Viruses causing diseases in red clover include the following: bean yellow mosaic (BYMV), red clover vein mosaic (*Marmor trifolii*), clover mosaic, Pisum virus 2, lucerne mosaic, Trifolium virus 1, common pea mosaic, clover phyllody virus, rugose leaf curl, tobacco mosaic, white clover mosaic, and potato calico (*Marmor medicaginis* var *solani*). The most promising method for control of many red clover diseases is development of resistant cvs. Some progress has been made in developing cvs resistant to northern and southern anthracnose and powdery mildew. For control methods, local agents should be consulted. Nematodes isolated from red clover include: *Acrobeles ciliatus, Acrobeloides emarginatus, Aphelenchoides ritzemabosi, Aphelenchus avenae, Boleodorus thylactus, Cephalobus* spp., *Chiloplacus* spp., *Criconemoides curvatus, C. lobatum, C. rusticus, Ditylenchus destructor, D. dipsaci, Eucephalobus* spp., *Helicotylenchus cairnsi, H. canadensis, H. digonicus, H. dihystera, H. erthrinae, H. microlobus, H. multicinctus, Heterodera avenae, H. cruciferae, H. galeopsidis, H. glycines, H. goettingiana, H. lespedezae, H. major, H. schachtii, H. trifolii, Hoplolaimus galeatus, H. tylenchiformis, Longidorus elongatus, L. maximus, Meloidogyne arenaria, M. artiellia, M. hapla, M. incognita, M. i. acrita, M. javanica, Neotylenchus* spp., *Paratylenchus aciculus, P. brevihastus, P. hamatus, P. projectus, P. sarissus, P. tenuicaudatus, Pratylenchus brachyurus, P. coffeae, P. neglectus, P. penetrans, P. pratensis, P. scribneri, P. tumidiceps, Psilenchus hilarulus, Pungentus pungens, Radopholus similis, Rotylenchulus reniformis, Rotylenchus erythrinae, R. goodeyi, R. robustus, Scutellonema brachyurum, Trichodorus christiei,*

Tylencholaimus mirabilis, Tylenchorhynchus agri, T. brevidens, T. claytoni, T. dubius, T. macrurus, T. martini, T. maximus, T. parvus, Tylenchus costatus, T. davainii, T. devastrix, and *Xiphinema americanum*. The most destructive insects reported on red clover include the following: clover root borer (*Hylastinus obscurus*), clover root curculio (*Sitona hispidulus*), clover seed chalcid (*Bruchophagus platypterus*), lesser clover leaf weevil (*Hypera nigrirostris*), potato leafhopper (*Empoasca fabae*), yellow clover aphid (*Therioaphis trifolii*), meadow spittlebug (*Philaenus spumarius*), clover seed midge (*Dasineura leguminicola*) clover leafhopper (*Aceratagallia sanguinolenta*), and pea aphid (*Acyrthosiphon pisum*).

REFERENCES:

Taylor, N. L., 1973, Red clover and alsike clover, in: *Forages* (M. E. Heath, D. S. Metcalfe, and R. F. Barnes, eds.), 3rd ed., pp. 148–158, Iowa State University Press, Ames, 755 pp.

U.S. Department of Agriculture, 1954, The clover seed midge (*Dasyneura leguminicola*). How to control it, *U.S. Dep. Agric., Leaflet* **379**.

U.S. Department of Agriculture, 1968, Growing red clover, *U.S. Dep. Agric., Leaflet* **531**:1–8, illus.

CONTRIBUTORS: J. A. Duke, M. B. Forde, C. F. Reed, R. R. Smith, N. L. Taylor, J. K. P. Weder.

Trifolium repens L.

FAMILY: Fabaceae
COMMON NAMES: White clover, Ladino clover
SYNONYM: *Trifolium nigrescens* Schur, not Viv.

USES: White clover is one of the most nutritious and widely distributed forage legumes of the world. Its value cannot be estimated accurately because of its volunteering characteristic and the seemingly spontaneous development of stands in pastures, lawns, turfs, orchards, and other areas, if cultural practices permit its growth. About half the 45 million hectares of humid or irrigated pastureland in the United States is estimated to have varying amounts of white clover. Extensively and effectively used in mixtures or alone, white clover provides a highly nutritive feed as pasture, hay, and silage for livestock and poultry, usually grown in mixtures with grasses for grazing, or alone for swine or poultry. As a cover crop, after seedlings are well established, it stabilizes soil and reduces erosion.

Figure 123. *Trifolium repens* L.

FOLK MEDICINE: Said to be antirheumatic, antiscrofulous, depurative, detergent, tonic, and prophylactic for mumps. A tincture of the leaves is applied as an ointment in gout.

CHEMISTRY: On the basis of more than 50 analyses of white clover hay Miller (1958) reported DM content of 86.6–93.8% and, on a moisture-free basis 13.1–32.4% crude protein (avg. 22.0%), 1.5–4.7% fat (avg. 2.6), 12.8–37.1% crude fiber (avg. 23.3), 4.7–14.8% ash (avg. 10.1), and 37.8–46.5% N-free extract (avg. 42.0). Green white clover roughage contained 11.7–45.4% DM (avg. 18.5) and, on a moisture-free basis, 13.7–32.0% crude protein (avg. 27.8), 1.7–5.7% fat (avg. 3.2), 11.7–27.4% crude fiber (avg. 15.6), 7.9–30.2% ash (avg. 12.1), and 27.6–48.7% N-free extract (avg. 41.3). The hay contained 0.9–1.9% Ca (avg. 1.2), 0.2–0.5% P (avg. 0.4), 1.8–2.8% K (avg. 2.0), 0.2–0.6% Mg (avg. 0.5), 0.05–0.15% Fe (avg. 0.11), 11.9 ppm Cu, and 142.3 ppm Mn. The green roughage contained (moisture-free): 0.6–1.9% Ca (avg. 1.5), 0.3–0.6%

Trifolium repens L.

P (avg. 0.5), 1.6–3.4% K (avg. 2.2), 0.2–0.5% Mg (avg. 0.3), 0.02–0.04% Fe (avg. 0.03), and 370 ppm Mn. In addition, Ladino hay contained 0.18–0.29% S (avg. 0.23), 0.25–0.37% Cl (avg. 0.31), 0.16 ppm Co, 16.9 ppm Zn, and 300 ppb I, 4.2 ppm thiamine, 16.9 ppm riboflavin, 1.1 ppm pantothenic acid, and 160 ppm carotene. On the basis at 10 analyses, Ladino hay contained (moisture-free basis) 1.0–1.2% arginine (avg. 1.1), 0.4–0.6% histidine (avg. 0.5), 1.1–1.3% isoleucine, 1.9–2.3% leucine, 1.1–1.3% lysine, 1.1–1.3% phenylalanine, and 1.2–1.4% threonine. Another analysis (WOI) indicates (g/16 g N): 4.3 arginine, 6.2 alanine, 14.6 aspartic acid, 1.6 cystine, 6.8 glutamic acid, 3.9 glycine, 1.9 histidine, 4.8 leucine, 8.4 isoleucine, 4.7 lysine, 1.3 methionine, 4.7 phenylalanine, 6.0 proline, 3.9 serine, 5.3 threonine, 1.8 tryptophan, 2.9 tyrosine, and 5.3 valine. Seeds are reported to contain trypsin inhibitors and chymotrypsin inhibitors. The species is polymorphic for cyanogenic glucosides. Leaves and flowers of certain cyanogenic phenotypes contain a glucoside which releases cyanide on contact with the enzyme linamarase. This can be induced by frosting, droughting, or chewing. The level of cyanide in white clover pastures might be as high as 350 ppm. White clover juice may contribute to bloat in ruminants. All cvs have a low estrogen content (usually < 6.3% for mononetin) and ewes grazing white clover pastures are considered safe from clover disease (W. Foulds, personal communication, 1978). The dominant estrogen is coumestrol. Late in the season diseased leaves may be more estrogenic than healthy leaves. Leaves do not contain tannins but flowers contain up to 1% prodelphinidins.

DESCRIPTION: A long-lived perennial, in warmer areas a winter annual and often used as such, in cooler areas acts as a biennial; (primary axis seldom persists more than two years; plants persist vegetatively by forming new growth centers that arise from rooted stolons and their apical meristems; under adverse conditions behaves as a winter annual; often treated as a winter annual in southern United States); plants shallow-rooted, spreading by stolons that root at the nodes; stolons glabrous, solid, creeping, 1–4 dm long, leaves long-petioled; leaflets cuneate-obovate to broadly oblong, emarginate to obtuse, 1–3 cm long, 0.75–2.5 cm broad, solid dull green or occasionally marked with a white "V" and sometimes with dark red flecks, denticulate; stipules pale except for the conspicuous nerves, abruptly narrowed to a point; peduncles scapiform, usually very elongate; heads subglobose, becoming semiglobose in fruit, rather loose, 1.5–2 cm in diameter; flowers 20–40 (100); bracts membranous, at first equaling the pedicels but much exceeded by them as the latter elongate; pedicels often sparsely puberulent, elongating and recurving in age, the inner longer than the outer; flowers 6–10 mm long, fragrant; calyx-teeth unequal, the upper slightly shorter than the tube, the lower two-thirds its length; corolla white or rose-tinged, becoming brown, 8 mm long, twice the length of the calyx; pods sessile, linear, exserted, usually 3- to 4-seeded, with 75–100 seeds per head; seeds ovoid-truncate, ca. 1.5 mm long and wide, surface smooth, usually entirely yellowish, sometimes reddish. Fl. and Fr. nearly year round in some areas; spring to fall in northern areas. Tetraploid. Seeds 1,764,000/kg; wt. kg/hl = 77.

GERMPLASM: For descriptive purposes cvs are arbitrarily classified primarily on size into three types: small, intermediate, and large. Seed of cv and types are indistinguishable, and noncertified seed often may contain more than one type. Cv names of the small type frequently contain the words "Wild White," e.g., 'Kent Wild White.' Plants of the small type may selectively persist where, because of heavy grazing or other reasons, the plants remain short. Because low growth restricts forage yields, cvs of the small type are seldom planted in improved pastures. Intermediate white clover is intermediate in size between the small and large types. In general, plants of the presently used cvs of the intermediate type flower earlier and more profusely than do plants of the large type. Profuse flowering usually produces ample seed for reseeding even under close grazing. The term "common" when applied to white clover seed usually implies that the seed were harvested from a local ecotype or the cv name is not known. Most white clovers so designated are of the intermediate type. Louisiana White and Louisiana S-1, a five clone synthetic cv developed from the former are well known U.S. cvs of the intermediate type. Foreign cvs of the intermediate type include from the U.K. 'Aberystwyth S100' and 'Kersey,' and from New Zealand 'Grasslands Huia.' The large type was introduced from Italy into the United States as "Ladino" early in the 1900s. Until the early 1950s seed derived from the Italian ecotype were designated Ladino and were the only U.S. source of the large type. Consequently the term Ladino had meant both cv and type. In the early 1950s 'Pilgrim' the first large type cv developed in the United States was released. 'Merit,' 'Regal,' and 'Tillman,' also cvs of the large type developed in the US from the original Ladino and new plant introductions, soon followed. 'Espanso' was developed in Italy. It and many other foreign varieties and strains have been introduced into the United States for testing and use in breeding improved cvs. Plants of the small, intermediate, and large types are cross compatible, and the F_1 plants are intermediate to their parents. Seed production fields must be isolated from other white clover to maintain cv purity. New Zealanders recognize four major cv groups: (1) small leaved, prostrate "wild white" types, very persistent in low-fertility hill pastures; (2) early flowering, short lived, common commercial or "Dutch white" types; (3) persistent pasture types intermediate in leaf size and flowering date, e.g., New Zealand and Danish improved pasture cvs; and (4) tall, large-leaved, late-flowering ladino types. 'Louisiana' white clover flowers profusely throughout the South and seed production is ample for a volunteer crop even under close grazing. 'Louisiana S-1' is a synthetic cv, developed from five clones of 'Louisiana White' for increased heat and drought tolerance. 'Regal,' a synthetic variety from 5 clones, developed in Alabama, for persistency, vigor and sparse blooming under Alabama conditions. 'Merit' is a synthetic cv of 30 clones selected from Ladino, showing superiority in forage yield, winter survival, and summer drought resistance. 'Pilgrim' is a synthetic cv of 21 clones of Ladino. 'Espanso,' developed in Italy, has improved winter and drought tolerance together with good persistence. 'Nordic,' developed in Germany from a cross between 'Ladino' and 'Morso,' a Danish cv, is being grown in the United States for seed export. Many foreign cvs and strains have been introduced to the United States for testing and as additional superior germplasm for hybridizing. In Canada, Russia's 'ML 48-65' is winter hardy and has outyielded all *repens* introduced since 1953. The similarity of karyotypes of *T. nigrescens, T. occidentale, T. petrisavii,* and *T. repens* suggests close relationships. White clover has been crossed with *T. nigrescens, T. occidentale,* and *T. uniflorum.* It is tetraploid, perhaps an amphidiploid derived from the annual self-incompatible *T. nigrescens* and the perennial self-compatible *T. occidentale.* Perhaps it is an autotetraploid of *T. occidentale.* Assigned to the Mediterranean, Near Eastern, and Eurosiberian Centers of Diversity, white clover or cvs thereof is reported to exhibit tolerance to aluminum, disease, drought, frost, grazing, heat, heavy soil, herbicides, high pH, low pH, mines, mycobacteria, nematodes, poor soil, salt, smog, virus, weeds, and wilt. ($2n = 32, 48, 64.$)

DISTRIBUTION: White clover probably originated in the eastern Mediterranean region of Asia Minor. Its rapid spread to other continents was apparently associated with early colonization and the presence of domesticated grazing animals.

ECOLOGY: White clover grows nearly anywhere, and is found naturally in dry meadows, mud flats, rarely on saline meadows, wood margins, open woods, banks of rivers and brooks, plains, semi-desert regions, mountains up to the subalpine meadows, often a weed at roadsides, near dwellings, and in barrens. Thrives best under cool, moist growing conditions, in soils with plenty of lime, potash, and P. Tolerates poor conditions better than most clovers. Grows better in clay and loam soils with sufficient moisture than on sandy soils which may be droughty and contain less minerals. Responses of white clover to climatic and agronomic conditions vary greatly. Cell suspension cultures derived from callus of germinating, seeds of 'Regal Ladino' were selected for tolerance to 2,4-D, 2,4-DB, and 2,4,5-T. Tolerance increased with selection, and was transmitted through succeeding asexually propagated cell generations. Cells selected for tolerance to one of the 3 chemicals showed tolerance to the others (Oswald et al., 1977). Ranging from Boreal Moist to Wet through Subtropical Dry to Rain Forest Life Zones, white clover is reported to tolerate annual precipitation of 3.1–19.1 dm (mean of 101 cases = 8.6), annual mean temperature of 4.3°–21.8°C (mean of 101 cases = 11.6), and pH of 4.5–8.2 (mean of 89 cases = 6.3).

CULTIVATION: Practically all white clover seeded for pasturage, hay or cover in northern and western United States is 'Ladino' or other large types. Many unimproved pastures or night pastures with close continuous grazing are 'Common White.' Seeding rates and mixtures are about the same for

Trifolium repens L.

both. Lime and fertilizers may be applied before seedbed preparation or, better yet, worked into the soil surface of the seedbed. Seedbed should be firm, smooth, and free of weeds. Seed may be surface broadcast, drilled, or planted by a combination corrugated roller-seeder. Since white clover seeds are very small, they must be planted at or near soil surface not more than 0.4 cm deep. Depending on the mixture used or if seeded alone, the rate of seeding varies from 0.45 to 5 kg/ha. Seed should be inoculated with a white clover culture before planting. Except for swine and poultry pastures and seed crops, which are seeded alone, most white clover is seeded with one or more grasses—orchard, timothy, brome, tall fescue, or the legumes red clover and alfalfa. Kentucky bluegrass should be avoided because it is highly competitive with white clover. Spring seeding should be early, alone or with a companion grain crop. Late-spring seedings are less successful. Frequently the companion grain is seeded at one-half the normal rate to reduce its competition for light, moisture, and nutrients. If the legume-grass seeding is of first importance, the grain crop may be grazed, starting when it is ca. 30 cm tall, or cut for hay or silage when the heads are in the soft dough stage. It is not advisable to practice close grazing or let animals trample young seedlings when the soil is wet. Young clover may be grazed when growth permits. Weeds may be clipped or sprayed for control. If harvested for grain, the cereal stubble should be clipped and straw removed as soon as possible. Summer seedings are successful under humid, cool conditions. They germinate when other farm operations are few and more time can be spent preparing a good seedbed. Seeding must be sufficiently early for plants to become well established before winter. Small plants are readily heaved by freezing and thawing. Essential to stand establishment and good growth, P and K may be supplied in various commercial fertilizers, Ca as agricultural ground limestone. Where Ca and Mg are deficient, they may be supplied by dolomite limestone. Use of superphosphate as a source of P also supplies S. Minor elements must be applied for the highest yields of forage and seed in some areas. Of the major elements needed, 300–600 kg/ha of 20% superphosphate and 60–250 kg/ha of muriate of potash, or their equivalents are sufficient. Finely ground limestone (2.5–10 MT/ha) is generally applied when needed. Although white clover will grow in slightly to medium-acid soils, higher yields, longer-lived stands and better inoculation are obtained at pH 6–7. Nitrogen fixation is optimal at pH 6.5. Well-inoculated plants do not benefit from N applications. In cool regions where white clover has recently grown, enough bacteria are generally present in the soil for good inoculation. In warmer regions, every seeding should be inoculated. High soil temperatures over several months may seriously reduce the number of bacteria at or near the soil surface.

HARVESTING: With good management forage may be either grazed or cut at intervals of 15–30 days during spring and summer. White clover has been used in green feeding practices, but the heavy machinery necessary for this method of utilization severely injures the plants, making them susceptible to diseases and early loss of stands. Where winter conditions are severe, white clover should not be closely grazed or frequently cut in the fall. In these areas, white clover may be top-dressed with barnyard manure soon after the soil surface becomes frozen in late fall. Grown in association with other legumes or grasses for hay or silage, white clover should be harvested early even though some yield is sacrificed. Delayed harvesting favors the development of leaf diseases, reduces light reaching the clover, and consequently is detrimental to rapid recovery and growth of the clover. Clover growth is dependent on the amount of functional leaf surface left after grazing or cutting. Forage should not be cut or grazed shorter than 5 cm. Rotational grazing is more favorable for clover than continuous grazing. White clover has been successfully used under daily ration grazing, but excessive trampling on fleshy stolons should be avoided. The seed crop should be harvested when most seed heads are light brown, about 25–30 days after full bloom. Seed crop is cut with a mower and cured either in the swath or in small windrows and is picked up with combines equipped with pickup attachments. During rainy weather curing in the swath facilitates combining, whereas during clear warm weather, windrowing reduces harvesting losses. Cut crop should be handled as few times as necessary, since each handling causes shattering losses. Under humid conditions, artificial drying or spreading the seed thinly under cover for complete drying maintains seed quality. With the latter method, the seed should be turned every few days until dry.

YIELDS AND ECONOMICS: Seed yields are 30–200 kg/ha in humid areas and 150–600 kg/ha in western areas, depending on cv and environmental conditions. Yields of forage vary greatly depending on the number of cuttings per season, or the time allowed for grazing of animals on a stand. White clover is widely distributed throughout the world and is one of the most widely used forage legumes. In the United States between 1,500 and 2,500 MT of seed are sown annually. Principal U.S. seed producing areas are: (1) Idaho, Washington, Oregon, and California; (2) Mississippi and Louisiana; (3) Wisconsin and northern Corn Belt States. Annual value of seed crop in United States has reached $5,500,000.

BIOTIC FACTORS: An abundance of bees for cross-pollination and dry, warm weather are necessary for high seed yields. Bees work white clover flowers for both nectar and pollen. The honeybee population can be controlled readily by moving colonies in and out of seed fields. Prolonged cloudy, rainy weather reduces flowering and retards bee visits. For this reason, most seed is produced where clear, bright weather prevails and favors flowering and seed development, facilitates harvesting, and reduces harvesting losses. Several pathogens attack white clover, reducing its quality, yield, and persistence. Their prevalence depends on many factors, including presence of susceptible plants, favorable temperature, and moisture conditions, age of plants, and for some viruses, the presence of suitable vectors. Fungi infecting white clover include: *Ascochyta trifolii, A. volkartii, Badhamia utricularis, Blennoria trifolii, Brachysporum trifolii, Cercospora zebrina* (leaf and stem spot), *Colletotrichum destructivum, C. graminicola, C. trifolii, Curvularia trifolii* (leaf spot), *Cylindrocladium scoparium, Cymadothea trifolii* (sooty blotch), *Dothidella trifolii, Erysiphe communis* forma *trifolii, E. martii, E. polygoni, Fusarium arthrosporiodes, F. avenaceum, F. moniliforme, F. oxysporum, Gloeosporium caulivorum, G. trifolium, Kabatiella caulivora, Macrophomina phaseoli* (charcoal rot), *Leptosphaerulina australis, L. trifolii* (pepper spot), *Olpidium trifolii, Pellicularia filamentosa, Peronospora repentis, P. trifolii-minoris, P. trifolii-repentis, P. trifoliorum, Phyllachora trifolii, Phyllosticta trifolii, Phymatotrichum omnivorum, Physarum cinereum, Pleospora herbarum, Polythrincium trifolii, Pseudopeziza trifolii, Pseudoplea trifolii, Pythium debaryanum, Rhizoctonia solani, Sclerotinia kitajimana, S. sclerotiorum, S. spermophila, S. trifoliorum* (stolon and root injury), *Sclerotium rolfsii, Sphaerulina subglacialis, S. trifolii, Spongospora subterranea, Stagonospora compta, S. meliloti, S. trifolii, Stemphylium sarcinaeforme, Thielaviopsis basicola, Thyrospora sarcinaeformis, Uromyces flectens, U. nerviphilus, U. striatis, U. trifolii, U. trifolii* var *trifolii-repentis, Urophylyctis trifolii,* and *Thanatephorus cucumeris.* White clover is also attacked by the bacterium *Pseudomonas syringae,* and the following viruses: alfalfa mosaic, alfalfa mosaic virus from soybean, aster yellows (Phyllody), bean chlorotic ringspot, bean yellow mosaic, bean local chlorosis, bean mosaic, bigbud or bluetop, lucerne mosaic, white clover mosaic, Medicago virus 2, virus necrosis (*Marmor repens, M. efficiens*), potato calico (*Marmor medicaginis* var *solani*), red clover vein mosaic (*Marmor trifolii*). Nematodes isolated from white clover include the following species: *Aphelenchoides fragariae, Belonolaimus gracilis, Ditylenchus destructor, D. dipsaci, Helicotylenchus dihystera, Hemicycliophora similis, Heterodera estonica, H. glycines, H. lespedezae, H. schachtii* var *galeopsidis* and var *trifolii, Longidorus menthasolanus, Meloidogyne arenaria, M. hapla, M. incognita acrita, M. javanica, M. marioni, Paratylenchus projectus, P. tenuicaudatus, Pratylenchus brachyurus, P. pentrans, P. scribneri, P. minyus, Pseudhalenchus anchilisposomus, Psilenchus duplexus, Ps. hilarulus, Ps. magnidens, Radopholus similis, Rotylenchus erythrinae, R. goodeyi, R. reniformis, Scutellonema brachyurum, Trichodorus minor, Tylenchorhynchus claytoni, T. dubius, T. maximus, T. striatus, Tylenchus discrepans, T. exiguus, T. leptosoma, T. thornei,* and *Xiphinema radicicola.* Many species of insects attack and damage white clover by preventing or reducing plant growth, destroying plant tissue, or damaging the seed. The most common ones include: grasshoppers, clover leaf weevil (*Hypera punctata,* potato leafhopper (*Empoasca fabae*), meadow spittlebug (*Philaenus spumarius*), garden fleahopper (*Halticus bractatus*), spider mites, clover root curculio (*Sitona*

Trifolium repens L.

hispidula), green june beetle (*Cotinis nitida*), lesser clover leaf weevil (*Hypera nigrirostris*), clover head weevil (*Hypera meles*), clover seed weevil (*Microtrogus picirostris*), white clover flower midge (*Dasineura gentneri*), Lygus bugs, clover aphid (*Nearctaphis bakeri*), green cloverworm (*Plathypena scabra*), cutworms, lesser cornstalk borer (*Elasmopalpus lignosellus*), alfalfa weevil (*Hypera postica*), blister beetles (*Epicauta* spp.), and the fall armyworm (*Spodoptera frugiperda*).

REFERENCES:
Foulds, W., 1977, The physiological response to moisture supply of cyanogenic and acyanogenic phenotypes of *Trifolium repens* L. and *Lotus corniculatus* L., *Heredity* **39**(2):219–234.
Foulds, W., and Young, L., 1977, Effect of frosting, moisture, stress and potassium cyanide on the metabolism of cyanogenic and acyanogenic phenotypes of Lotus corniculatus L. and Trifolium repens L., *Heredity* **38**(1):19–24.
Gibson, P. B., and Hollowell, E. A., 1966, White clover, *U.S. Dep. Agric., Agric. Handb.* **314**:1–33, illus.
Leffel, R. C., and Gibson, P. B., 1973, White clover, in: *Forages* (M. E. Heath, D. S. Metcalfe, and R. F. Barnes, eds.), 3rd ed., pp. 167–176, illus., Iowa State University Press, Ames, 755 pp.
Marble, V. L., *et al.*, 1970, Ladino clover seed production in California. *Calif. Agric. Exp. Stn. Ext. Serv., Circ.* **554**:1–36, illus.
Oswald, T. H., Smith, A. E., and Phillips, D. V., 1977, Herbicide tolerance developed in cell suspension cultures of perennial white clover, *Can. J. Bot.* **55**(10):1351–1358.
Stewart, I., and Bear, F. E., 1951, Ladino clover, its mineral requirements and chemical composition, *N.J. Agric. Exp. Stn., Rutgers Univ., Bull.* **759**:1–32, illus., New Brunswick, N.J.

CONTRIBUTORS: J. A. Duke, M. B. Forde, W. Foulds, P. B. Gibson, C. F. Reed, J. K. P. Weder.

Trifolium resupinatum L.

FAMILY: Fabaceae

COMMON NAMES: Persian clover, Shaftal, Birdseye clover, Reversed clover

SYNONYMS: *Trifolium bicorne* Forsk.
Trifolium clusii Gren. & Godr.
Trifolium suaveolens Willd.

USES: Persian clover provides nutritious winter forage crop, and is relished by all classes of livestock and poultry. Also used for hay, silage, green manure, and seed. Produces feed in late winter and early spring, when southern grasses are dormant, and extends the grazing season. Sometimes grown as ornamental.

Figure 124. *Trifolium resupinatum* L.

FOLK MEDICINE: No data available.

CHEMISTRY: The hay, with 10% moisture, contains (moisture free basis): 16.3% crude protein, 1.8% fat, 30.8% crude fiber, 10.3% ash, and 40.8% N-free extract. The leaves contain 23.1% crude protein, 2.2% fat, 19.4% crude fiber, 10.6% ash, and 44.7% N-free extract. The stems contain: 10.3% crude protein, 1.3% fat, 39.8% crude fiber, 10.7% ash, and 37.9% N-free extract. The hay contains 1.66% Ca and 0.23% P. Green leaves are said to contain 4,000 IU/100 g vitamin A and 163 mg/100 g vitamin C. The essential amino acids in the leaf proteins are (g/16 g N): 9.7 arginine, 1.9 histidine,

1.9 methionine, 1.8 tryptophan, and 5.4 total lysine (5.0 available).

DESCRIPTION: Glabrous, often coarse, annual or biennial legume; stems ascending, diffuse or decumbent, 1–4 dm tall, branching; lower and middle leaves long-petioled, the uppermost subsessile; leaflets cuneate-ovate to cuneate-obovate, acute to obtuse or truncate, denticulate, strongly nerved beneath; stipules lanceolate, long-acuminate, the base chartaceous-dilated; upper peduncles mostly equal or shorter than the leaves, the lower considerably longer; heads soon becoming globose, 1.5–2 cm in diameter in fruit; bracts minute, membranous, truncate, forming a very short, toothed and often divided, involucre; flowers small, 5–6 mm long, subsessile, resupinate; calyx bilabiate, the upper lip of 2 bristlelike divaricate teeth; corolla roseate to purplish, 2–3 times the length of the calyx; pods orbicular to ovate, included; seeds ovoid-truncate, greenish-brown to greenish-black, lustrous. Fl. April; Fr. May. Seeds 1,488,375/kg; wt. kg/hl = 77. Seeds enclosed in balloonlike capsules that shatter easily, float on water or are readily blown about by wind.

GERMPLASM: Several cvs have been developed. 'Abon,' developed in Texas, allows earlier grazing in fall and later grazing in spring, lodges less, and shatters less than other cvs. It can be harvested for seed by combine standing or from the swath, survives −12°C, has potential in Gulf Coast, and reseeds itself freely. Assigned to the Mediterranean and Near Eastern and Eurosiberian Center of Diversity, Persian clover or cvs thereof is reported to exhibit tolerance to drought, frost, high pH, heavy soil, limestone, sand, and weeds. ($2n$ = 14, 16, 32.)

DISTRIBUTION: Native to southern Asia Minor and Iran and Mediterranean countries west to Greece and Egypt. Early introduced to West Pakistan and Punjab, and later to England and the United States. Now widely cultivated in temperate countries.

ECOLOGY: Occurs in wet meadows near brooks, seaside sands, and sometimes cultivated in oases. Best adapted to wet heavy soils in low-lying areas, but does not stand waterlogging for more than a few days at a time. Grows successfully on medium to slightly acid soils; on strongly acid soils, 2.5–5 MT/ha of limestone may be applied for best results. Adapted to climatic conditions in southeastern United States and along the Pacific Coast. Ranging from Boreal Wet through Subtropical Thorn to Moist Forest Life Zone, Persian clover is reported to tolerate annual precipitation of 2.8–13.6 dm (mean of 20 cases = 7.8), annual mean temperature of 6.3°–20.3°C (mean of 20 cases = 12.7), and pH of 4.8–8.2 (mean of 19 cases = 6.7).

CULTIVATION: Persian clover usually is seeded on a grass turf. Stands are best on bermudagrass or dallisgrass turfs, but stands also have been good on carpetgrass and bahiagrass. Turfs of bermudagrass or dallisgrass are prepared for seeding Persian clover by either clipping or grazing the grass closely. Turfs of carpetgrass and bahiagrass are dense and are more difficult to prepare. They should be disked thoroughly, even to extent of turning over slices of turf. Persian clover also can be seeded successfully on cultivated land. A firm seedbed is essential, and soil should be rolled or dragged before seeding. Soil between cotton or corn rows, which often is compacted or crusted, should be loosened before seeding with Persian clover. In India, Persian clover is interplanted with *Melilotus indica*. Where seeded alone, seeding rates are 12–15 kg/ha drilled, 15–18 kg/ha broadcast; where seeded in grass, seeding rate is 5–7 kg/ha. Persian clover does best on alkaline soil where pH is about 6.0. For best results on acid soils, 2.5–5 MT/ha of finely ground limestone is applied in midsummer. Persian clover may need mineral fertilizers. Phosphates vary in soils in southern United States; on more fertile soils about 200 kg/ha of 20% superphosphate applied at time of sowing gives excellent growth; on less fertile soils up to 500 kg/ha may be necessary for initial application, with 100–200 kg/ha supplements in subsequent years. When potash is deficient, 100–200 kg/ha of muriate of potash may be applied. Fertilizers may be broadcast or drilled. Concentrated fertilizers can be applied over large areas by airplanes. Inoculation may be necessary the first and second years but is not necessary if the Persian clover has grown successfully in the same field for 2 yr.

HARVESTING: Properly cured hay is highly nutritive and relished by livestock. For hay, it should be cut when plants are one-fourth to full bloom. Yield of hay is largest in full-bloom stage, but quality is higher from earlier cuttings. Hay is mowed, allowed to wilt in the swath and then windrowed for final curing. Crop cut in full bloom, does not reseed. When used for green manure on heavy low-lying soils in s US, the crop usually is grazed lightly during late winter and early spring, and then, if allowed to approach maturity before being turned under, yields up to 37.5 MT/ha WM and sufficient seed for volunteer stands for 2 yr. In this practice, corn or sorghum is planted later than normal during the year so the clover can replenish seed for volunteer stands. When harvested for seed and followed by a summer-growing cultivated crop, sufficient seed shatters to ensure a thick stand in the fall. Seed crop is cut when most seed capsules have turned light brown. Cutting by mower equipped with lifter guards. In thick stand, or when crop is lodged or entangled, a heavy short weed bar without guards is used. Curing in the windrow is recommended, rolling the seed heads inward to reduce shattering. Crop threshed either by grain separators equipped with hulling attachments, by combine used as stationary machines, or with attachments to pick up crop from windrows.

YIELDS AND ECONOMICS: Yields of green manure may be up to 37.5 MT/ha. Hay yields are 2.5–5 MT/ha, depending on fertilizer rate and method of handling. Light winter grazing is possible, and heavy grazing should start in early in spring; heavy grazing in late spring is not recommended. Seed yields are 150–300 kg/ha, but have reached 600 kg/ha with sufficient shattered to reseed.

BIOTIC FACTORS: Persian clover is self-pollinating and self-fertile. Seed yields are 9-fold when bees are provided near fields. Blister beetles (Meloidae) are pests in seed fields, especially in May as they eat the tender leaves and ruin the crop for seed. Fungi reported to attack Persian clover include: *Cymadothea trifolii*, *Dothidella trifolii*, *Cercospora zebrina*, *Peronospora trifolii-repentis*, *Phyllachora trifolii*, *Physoderma trifolii*, *Thielaviopsis basicola*, *Uromyces flectens*, *U. minor*, *U. striatus*, *U. trifolii*, and *U. trifolii-repentis*. Nematodes isolated from this clover include: *Ditylenchus dipsaci*, *Heterodera avenae*, *H. glycines*, *H. schachtii*, *Meloidogyne incognita acrita*, and *M. javanica*.

REFERENCES:
Relwani, L. L. 1973. Shaftal (*Trifolium resupinatum*) or (*Trifolium suaveolens*), Indian Dairyman **25**(3):117–120.

Texas A & M University, 1964, Abon, Persianclover, *Leaflet* **618**:1–4, illus., College Station, Texas.
U. S. Department of Agriculture, 1960, Persian clover—A legume for the South, *U.S. Dep. Agric., Leaflet* **484**.
CONTRIBUTORS: J. A. Duke, C. F. Reed.

Trifolium subterraneum L.

FAMILY: Fabaceae
COMMON NAMES: Subclover, Subterranean clover
SYNONYMS: *Trifolium oxaloides* Bunge ex Nyman
Trifolium subterraneum var *oxaloides* Rouy & Foucaud

USES: A decumbent, large-seeded, winter annual legume, relished by all livestock; best used for pasture when grown in mixtures with adapted companion grasses. Also suited for silage and hay. Mat of creeping stems is good for erosion control.

FOLK MEDICINE: No data available.

Figure 125. *Trifolium subterraneum* L.

CHEMISTRY: Formononetin contents of 'Yarloop' may be 200 times as great as that of other cvs, and may exceed the level suggested as a cause of infertility in ewes. Many cvs have oestrogenic potential.

DESCRIPTION: More or less softly pubescent annual legume; stems slender, procumbent to decumbent, 0.5–4 (rarely to 8) dm long, forming circular clumps; leaves all long-petioled; leaflets broadly obcordate, entire except for the shallowly crenate apex; stipules ovate to oblong-ovate, the lowermost acuminate, otherwise acute to obtuse; peduncles axillary, reflexing, elongating, and burying the heads in the soil after anthesis; inflorescence a few-flowered fascicle becoming a globose head in fruit; fertile flowers 2–5, whitish, striped with rose, 12–14 mm long; sterile flowers developing after anthesis, numerous, apetalous, finally completely enclosing the pods; calyx of fertile flowers nerveless, the tube glabrous, the subequal flexous teeth ciliate, equaling the tube, at first setaceous, becoming stout; corolla about twice the length of the calyx; sterile flowers with calyx-teeth rather unequal, narrow, irregularly bent, slightly exserted, 1-seeded; seeds large, broadly ellipsoid, purplish-black. Fl. April-May; Fr. May–June. Seeds 143,325/kg; wt. kg/hl = 77.

GERMPLASM: More than 40 strains of subclover have been recognized, and grouped as early, midseason, and late. Most cvs were developed in Australia. Foliage color varies from dark to light green with various markings. Early introductions were adapted only to southern California; later cvs are adapted to the Pacific Northwest and to some extent to southern United States. In the Pacific Northwest, 'Mt. Barker,' a midseason strain, and 'Tallarook,' a late strain, are well adapted and are used almost exclusively. Both cvs are most satisfactory for forage purposes, especially for pasture, in that they bury a high percentage of the seed which assures self-perpetuation even when heavily grazed. Other cvs in common use are: 'Yarloop,' 'Dwalganup,' and 'Bacchus Marsh.' 'Nangeela,' a late maturing cv is also used extensively. *T. brachycalycinum*, *T. subterraneum*, and *T. yanninicum* are very closely related Mediterranean species. Although F_1 hybrids show some chromosome divergence, some think that they should be considered as components of a single polymorphic species, not as fully diverged species. The gene pools of these species used in current breeding programs are extremely mixed. *T. israeliticum* D. Zoh. & Katzn with $2n = 12$ is not associated with the cultivated ensemble now used in Australia and elsewhere. Genetically, it is remote from subclover and cannot be crossed with it. Assigned to the Mediterranean and Australian Centers of Diversity, subclover or cvs thereof is reported to exhibit tolerance to aluminum, disease, grazing, high pH, low pH, and virus. ($2n = 16$.)

DISTRIBUTION: Native of Mediterranean Europe, Asia, Africa, and southern England. Introduced and cultivated in Australia and New Zealand. Later, introduced to the United States from Australia. Often in coastal areas.

ECOLOGY: Found native in low grass meadows and scrubland. Adapted to climates having relatively warm moist winter and dry summer. Grows especially well on light soils, but is not tolerant of poorly drained areas. Makes good growth and reseeds on practically all well-drained soils that will grow white or ladino clover. Well-established seedlings can survive temperatures as low as −15°C. Adapted to areas with more than 400 mm annual rainfall below 1,000 m in elevation. Especially tolerant to acid soils. Ranging from Cool Temperate Steppe to Wet through Subtropical Dry to Moist Forest Life Zones, subclover is reported to tolerate annual precipitation of 3.8–16.3 dm (mean of 34 cases = 9.0), annual mean temperature of 5.9°–21.3°C (mean of 34 cases = 14.4), and pH of 4.5–8.2 (mean of 31 cases = 6.1).

CULTIVATION: Subclover should be sown on clean seedbed, well-prepared, level and firm because close cutting is essential in harvesting a seed crop. If a field is to remain in seed production for more than 1 yr, it should be reworked each fall in preparation for the next seed crop. For fall planting for forage, looser, rougher seedbeds are satisfactory. The ash remaining after a fall burn provides good coverage for fall-sown seed. In Mississippi, seed is planted in fall, Sept. 1–Oct. 15. Earlier plantings produce the earliest forage. Initial growth is slower than crimson clover but somewhat faster than white clover. Once established, it makes a dense growth. For forage subclover should always be planted in mixtures with adapted companion

Trifolium subterraneum L.

grasses. Subclover is suited for growth with grasses which give winter protection to seedlings. In this association, subclover stimulates the grasses by improving soil fertility. Clover can be established on sod by broadcasting the seed in the fall without seedbed preparation. In a better method of planting without seedbed preparation, threshed subclover straw is scattered over the land and trampled into the soil by livestock. Mixtures of clover and grasses are best established by planting them together. For seed production subclover is broadcast at 22–28 kg/ha. Yields of seed are highest when grown alone. Subclover seed should be inoculated with rhizobia bacteria: 32–37 nodules per plant were produced on 'Mt. Barker' with *Trifolium* inoculum, special subclover inoculum or soil from subclover fields. Plants reseed well but are unable to withstand hot summers, even in the Northwest. Fertilizer requirements about the same as for white clover—phosphate, potash, and lime. General recommendation for western Oregon is a yearly application of superphosphate at 200–300 kg/ha, applied in fall. Lime applied at rate of 2.5–5 MT/ha may be beneficial on more acid, hilly soils.

HARVESTING: Season for green forage on West Coast is usually from late March to mid-July. After plants mature, unburied seed heads are eaten eagerly by sheep and cattle. During dry years, when most seed may mature unburied, careful pasturing is advisable to avoid heavy consumption of heads and consequent reduction of seed for the following year's crop. Permanent pastures have persisted for 25 years in southern United States. Subclover seed is ready to harvest when the plants are dead and thoroughly dry, usually in late July or in August, depending on the strain and the season. A heavy crop is harvested with a tractor-driven, power take-off mower equipped with a lespedeza cutter bar and windrowed. The bar clips closer to the soil surface than the standard mower bar. Where the mat of clover vines is especially dense and heavy, pea lifter guards that attach over the upper surface of the bar should be used. Lifter guards attached underneath the bar cause higher clipping than is desirable. A harvesting method that requires no special equipment consists of mowing above the seed heads with a standard mower and windrower to clip off weeds, grasses and excess straw, which are removed and may be used for feed. Field is then raked with a strong hay rake with heavy teeth. The runners and heads thus pulled up are dropped into windrows. This method is successful only when the clover plants are thoroughly dry and break off easily at the surface of the soil. After removal of the seed crop, some heads always remain and can be gathered by vacuum. The seed crop is usually threshed with standard stationary threshers, but can be threshed directly from the windrows with combine threshers. The toothed cylinder machine is superior to the rub-bar type for this purpose. Warm, dry weather is essential for efficient threshing. Rethreshing of the straw is often worthwhile as the seeds are rather difficult to separate from the heads and pods.

YIELDS AND ECONOMICS: Forage is harvested in later winter or spring as growth warrants. Yields of 3,005 kg/ha of 'Mt. Barker' and 4,197 kg/ha of 'Tallarook' have been obtained. In India yields were reported of 4,185 kg/ha WM for 'Yarloop' and 19,800 kg/ha WM for 'Bacchus marsh' (in 2 cuttings). Forage is harvested 4–8 times each spring. Yields with perennial ryegrass were for silage, 19.4–21.3 MT/ha and for hay, 5.6–5.9 MT/ha. A valuable winter annual legume, useful for forage and pasture in Eurasia, Australia, New Zealand, and western and southern United States.

BIOTIC FACTORS: Following fungi have been reported on subclover plants: *Cercospora medicaginis, C. zebrina, Erysiphe communis, E. polygoni, Kabatiella caulivora, Pellicularia filamentosa, Physoderma trifolii, Pleospora herbarum, Pseudopeziza trifolii, Pseudoplea trifolii, Rhizoctonia solani, Sclerotinia homeocarpa, S. trifoliorum, Sclerotium baticola, Uromyces trifolii,* and var *subterranei.* Stem rot is the most serious disease of subclover; powdery mildew and leaf rust are less destructive. Subclover may be parasitized by *Cuscuta epithymum*. Several viruses attack subclover: alfalfa mosaic virus isolated from soybean, bean yellow mosaic, big-bud or blue-top, broad-bean mottle, subterranean clover stunt, Pisum virus 2, pea mosaic, pea mosaic 2 *(Marmor leguminosarum)*, soybean stunt, and yellow bean mosaic. Nematodes isolated from subclover include: *Aphelenchoides parietinus, Heliotylenchus dihystera, Meloidogyne exigua, M. incognita, M. hapla, M. javanica, Monhystera stagnalis, Pratylenchus zeae, P. pratensis, Rotylenchus brevicaudatus.*

REFERENCES:
Coats, R. E., and Johnson, C. M., 1959, Subclover, a satisfac-

tory reseeding winter annual legume, *Miss. State Univ., Agric. Exp. Stn., Inf. Sheet* **647**:1–2.

Cocks, P. S., 1973, The influence of temperature and density on the growth of communities of subterranean clover (*Trifolium subterraneum* L. cv. Mount Barker), *Aust. J. Agric. Res.* **24**(4):479–495.

Katznelson, J., 1974, The subterranean clovers of *Trifolium* subsect. *Calycomorphua* Katzn. *Trifolium subterraneum* L. sensu latu, *Israel J. Bot.* **23**:69–108.

Raguse, C. A., Menke, J. W., and Sumner, D. C., 1974, Developmental morphology of seedling subterannean and rose clover leaves, *Crop Sci.* **14**(2):333–334.

Rampton, H. H., 1952, Growing subclover in Oregon, *Agric. Exp. Stn., Ore. State Coll., Stn. Bull.* **432**:1–12, illus., Corvallis.

Read, J. W., 1973, Comparison of introduced lines of *Trifolium subterraneum* subsp. *yanninicum* with cultivars of *T. subterraneum* 2, yield and formononetin concentration under irrigation at Leeton, New South Wales, *Fld. Sta. Rec. Div. Pl. Ind., CSIRO* **12**:5–9.

CONTRIBUTORS: J. A. Duke, M. B. Forde, C. F. Reed, D. Zohary.

Trifolium variegatum Nutt.

FAMILY: Fabaceae

COMMON NAME: White tip clover

SYNONYMS: *Trifolium melananthum* Hook. & Arn.
Trifolium pauciflorum Nutt.
Trifolium rostratum Greene
Trifolium trilobatum Jeps.

USES: Produces thick stands and good growth under varying conditions. Relished by all livestock. Food for wildlife, as valley quail and California quail.

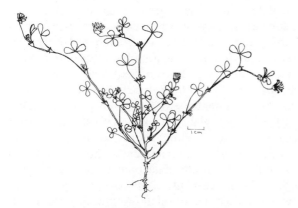

Figure 126. *Trifolium variegatum* Nutt.

FOLK MEDICINE: No data available.

CHEMISTRY: No data available.

DESCRIPTION: A variable annual legume; stems glabrous, up to 60 cm long, thick and fistulous or slender, ascending or decumbent, several from the base and freely branching; stipules ovate, laciniately toothed; leaflets usually obovate, sometimes oblong-oblanceolate, very small to large, 0.4–1.5 cm long; peduncles slender; heads irregularly subglobose, 0.7–1.3 cm broad, few- to many-flowered; involucre much smaller, 4- to 12-lobed, the lobes 3- to 7-toothed; flowers small, corolla purple, white-tipped, or purple throughout; calyx 5- to 20-nerved, its teeth subulate-setaceous, often purple, simple or one tooth bifid; pods stalked in the calyx, 2-seeded.

GERMPLASM: Several species, varieties and cvs have been described from the variations within this species. Var *pauciflorum* McDer. is a dwarf caespitose plant, or with stems short and slender, 2.5–3.0 cm long; leaflets very small; heads 0.4–0.6 cm long, of 1–7 small purplish flowers; involucre 1–4-lobed, the lobes 3–5-toothed; a montane var. occurring at 700–3,000-m elevations, Coastal Ranges, Sierra Nevada, California. Var *trilobatum* Jeps. is a very small legume; stems slender, 10–20 cm tall, sparsely branched at base; margin of stipules laciniate; petioles slender, 3–5 cm long; leaflets lanceolate, acute at each end or often trilobate at apex; heads on long slender peduncles; lobes of involucre deeply and laciniately toothed; flowers long, corolla dark purple, cream-colored at tips; calyx-teeth slender, acute, usually purple-tinted; pods 2-seeded; Marysville Buttes, California. Var *melananthum* Greene, has leaflets large, oblanceolate to oblong or obovate, obtuse, 1.3–2.5 cm long; heads 2–3 cm broad, large-flowered, showy; involucre small; calyx-teeth pungent and purple-tipped; common in low moist places, attaining a luxuriant growth; coastal Ranges and Great Valley, California. Assigned to the North American Center of Diversity, white tip clover or cvs thereof is reported to exhibit tolerance to waterlogging. ($2n = 16$.)

DISTRIBUTION: Widely distributed and common throughout the cismontane regions of California, and ranging from Baja California, Mexico, north to British Columbia, east to Montana, Wyoming, and Arizona.

Trifolium variegatum Nutt.

ECOLOGY: Occurs in moist habitats, usually below 2,600 m in elevations. Adapted to flooded mountain meadows in western United States and western Canada. Ranging from Cool Temperate Moist through Warm Temperate Moist Forest Life Zone, white tip clover is reported to tolerate annual precipitation of 5.6–12.8 dm (mean of 2 cases = 9.2), annual mean temperature of 8.9°–17.6°C (mean of 2 cases = 13.3), and pH of 6.3–6.5 (mean of 2 cases = 6.4).

CULTIVATION: Rarely if ever actually cultivated as a crop. Where luxuriant, it is sometimes grazed. Spread by seed.

HARVESTING: Grazed in summer where adapted and growing in quantity. A preferred food for quail. Relished by all stock.

YIELDS AND ECONOMICS: A valuable annual legume for local grazing during summer, especially on Coastal and cismontane regions of western North America.

BIOTIC FACTORS: White tip clover is known to be attacked by the following fungi: *Erysiphe communis, E. polygoni, Uromyces minor,* and *U. oblongus.*

CONTRIBUTORS: J. A. Duke, C. F. Reed.

Figure 127. *Trifolium vesiculosum* Savi

Trifolium vesiculosum Savi

FAMILY: Fabaceae
COMMON NAME: Arrowleaf clover
SYNONYM: *Trifolium turgidum* Bieb.

USES: Cultivated in southern United States and southern Europe for high forage yields, especially useful in late winter and early spring. Readily grazed by cattle.

FOLK MEDICINE: No data available.

CHEMISTRY: Digestible dry matter (DDM), when cut for forage at early bloom stage, is about 84% (cf. 78% for clover).

DESCRIPTION: Annual legume; stems 15–40 cm long, glabrous, angled, often reddish, simple or branched; stipules linear-lanceolate, scarious, prominently nerved, the free portion long-acuminate; basal petioles long, the upper ones shorter, subopposite; leaflets petiolulate, glabrous, lanceolate, 1.5–4 cm long, 0.5–1.5 cm broad, prominently veined beneath, acutely serrate-dentate, usually apiculate with a pale arrow-shaped spot above; peduncles surpassing terminal leaves; heads spuriously terminal at ends of branches, solitary, at first broadly conical, finally oblong to ovoid-cylindric, 3–5 cm long, 2.5–3.5 cm broad; flowers numerous, 15–18 mm long; bracteoles scarious, lanceolate, acuminate, exceeding calyx-tube; calyx 8–10 mm long, glabrous, at first fusiform, in fruit regularly inflated; subpyriform, distinctly 20–30-nerved, narrow, subulate-acuminate, erect or in fruit reflexed; corolla pale pink or almost purple, coriaceous-marcescent, about twice length of calyx; standard free, obsoletely clawed, oblong-lanceolate, attenuate at summit, striate by numerous longitudinal nerves; wings and keel slightly shorter, also striate, narrow, acuminate; ovary sessile, 2- or 3-ovuled, with a long style; pods scarious, containing 2 or 3 seeds. Fl. and Fr. June–July. Seeds 880,000/kg.

GERMPLASM: There are three main cvs in the United States. 'Yuchi,' a selection of arrow-leaf clover introduced from Italy and named for an Indian tribe in east-central Alabama is especially adapted to well-drained soils in Alabama, maturing about 2 months later than crimson clover. 'Amclo,' developed in Georgia, matures 2–4 weeks earlier in spring than 'Yuchi,' with a protein content of 11.88%. 'Meechee,' developed in Mississippi, is somewhat later maturing than 'Yuchi' and is more cold hardy. Assigned to the Mediterranean Center of Diversity, arrowleaf clover or cvs thereof is reported to exhibit tolerance to drought, frost, grazing, heavy soil, nematodes, and slope. ($2n = 16$.)

DISTRIBUTION: Native to the western and eastern Mediterranean regions, Balkan Peninsula, Greece, Crimea, western Caucasus, and southern Russia. Introduced and widely cultivated in southern United States from Georgia, Alabama, and Mississippi to Texas, north to Arkansas, South Carolina, and Tennessee.

ECOLOGY: Usually found on dry grassy slopes. Fares well on silty loams, but grows on various types of clay soils. Requires a warm temperate climate, is drought resistant; some cvs show considerable cold resistance. More than 14 hr of light is required for flowering under warm night temperatures. Cool nights delay flowering. Produces best at ca. pH 6.0; not adapted to lime soils of pH 7.5 or higher or acid soils as low as 5.0 pH. Requires a well-drained soil, making little growth on poorly drained soils. Ranging from Cool Temperate Moist to Subtropical Moist Forest Life Zone, arrowleaf clover is reported to tolerate annual precipitation of 6.1–13.6 dm (mean of 3 cases = 10.4), annual mean temperature of 11.1°–19.7°C (mean of 3 cases = 14.4), and pH of 5.5–8.2 (mean of 3 cases = 6.4).

CULTIVATION: Seed-coats of this clover are unusually hard and require more scarification than most other reseeding winter legumes. Mechanical scarification of 'Yuchi' arrowleaf by commercial scarifiers reduces the hard seed content from ca. 70% to ca. 25%. It is essential that scarified seed be planted to ensure a stand the first year. Nonscarified seed may lie in the soil for several years before germinating. Good stands of 'Yuchi' have been obtained in fields where no natural reseeding has been permitted for 4 yr. Arrowleaf clover germinates more rapidly at 6°C than crimson clover. The ability of arrowleaf to germinate well at low temperatures is a valuable adaptive feature enabling it to produce new plants throughout the winter. On prepared land, 'Yuchi' can be planted alone, with rye, or with rye and ryegrass for a high-quality, long-season pasture. Planting in mid-September on prepared land, yields forage from Jan. to May of Ca. 6.5 MT/ha cutting every 4–6 weeks during that period. Overseeding of bahia and bermudagrass sods with winter annual clovers can extend the grazing season by 4–8 weeks and provide higher quality forage. Good natural reseeding by the clover is desirable. Reseeding of clovers is generally more uncertain on bahiagrass sod than on bermudagrass. Satisfactory seeding rates are about 4 kg/ha drilled in rows 90 cm apart, or 10 kg/ha broadcast. Both of these seeding rates have produced complete cover. Row seeding may be best for seed production where weed control is necessary. In Georgia and vicinity should be seeded between Oct. 15 and Nov. 20 for either seed or forage. Earlier seeding does not increase fall growth. 'Yuchi' and other cvs are more sensitive than crimson clover to low soil phosphorus, showing poorer growth than crimson. When no phosphorus was added to soil, 'Yuchi' and ball clover plants died in the seedling stage, whereas crimson survived. About 900 kg/ha of 0–14–14 fertilizer is disked in at time of seedbed preparation. 'Yuchi' and other arrowleafs must be inoculated. A standard clover inoculant seems to be sufficient, although plants are slow to form the nitrogen-fixing nodules.

HARVESTING: Continuous grazing after plants are 15 cm tall permits a long productive season. Arrowleaf clover has considerable tolerance to animal trampling. Clover stands and growth are good even when grazed during the wet winter months. If it appears that there will be surplus forage during the peak growing period of April and May, grazing animals can be confined to a part of the pasture in early April to permit cutting the ungrazed portion in May for hay, or combining seed in June. Since growth is heavy, the use of a crimper or stem crusher speeds up field drying. In southern United

Trifolium vesiculosum Savi

States, 'Yuchi' begins blooming in late May or early June, and matures seed in late June or July. Seeds are borne in clustered pods produced at tips of stems that remain erect if plants have not made too much vegetative growth. Seed harvested by combine. Potential seed fields should be grazed during winter until early April. Regrowth after grazing in a seed field, even under pecans *(Carya)*, is very productive. Thin stands of clover or late planted clover may give little or no grazing, but often produce the most seed. Seed harvest should be started in late June when weather is usually sunny, dry, and hot. In wet months, seed harvest is difficult and sometimes impossible. 'Yuchi' continues to flower and the stems may remain green much of July, even though there is a heavy crop of mature seed heads. Plant material can be dried by cutting and swathing, but some seed may be lost. Dessicants are more effective and permit direct combining of the dry plants 2–5 days after application.

YIELDS AND ECONOMICS: Forage yields average 7–12 MT/ha depending on area and cv. In Oklahoma trials, e.g., 'Meeche' and 'Yuchi' yielded best, ca. 5 MT/ha. Soils greatly influence the total yield, silty loam is most productive. Seed yields average 100–500 kg/ha. The highly productive winter annual clover is well adapted to well-drained soils through most of southeastern United States and is being increasingly grown.

BIOTIC FACTORS: Bees are essential for pollination. If native bees are scarce, two or three colonies per hectare should be placed in the seed field. All dodder *(Cuscuta* sp.) should be removed. Dodder seed are extremely difficult to separate from 'Yuchi' seed since both are of similar size and have rough seed coats. No serious fungal diseases have been encountered with 'Yuchi,' although it is susceptible to crown and stem rot *(Sclerotinia trifoliorum)*, and stand losses may occur during warm, wet, winter periods. Problem more serious when forage is heavy. Grazing the clover to remove the surplus growth and permit light to penetrate the sward reduces disease losses. 'Yuchi' is remarkably resistant to insects that attack many other legumes. Clover head weevil *(Hypera meles)* causes severe seed losses in crimson clover, but has little effect on 'Yuchi.' Alfalfa weevil *(Hypera postica)* causes little or no damage to 'Yuchi,' as compared to severe damage to alfalfa.

REFERENCES:

Ahlich, V. E., and Byrd, M., 1966, Meechee—A new variety of arrowleaf clover, *Miss. State Univ., Agric. Exp. Stn., Inf. Sheet* **948**:1–2, illus.

Ball, D. M., Hoveland, C. S., and Buchanan, G. A., 1974, Flower and seed production in Yuchi arrowleaf clover, *Agron. J.* **66**(4):581–583.

Bates, R. P., undated, 1974–75 forage yields from ryegrass and legume varieties and strains, *Samuel Roberts Noble Foundation Publ.* **R-152**: 2 pp.

Beaty, E. R., and Powell, J. D., 1969, Forage production of amclo and crimson clover on Pensacola bahiagrass sods, *J. Range Manage.* **22**:36–39.

Beaty, E. R., Powell, J. D., and McCreery, R. A., 1963, Amclo—A high-yielding winter clover, *Univ. Ga. Coll. Agric. Exp. Stn., Circ.*, n.s. **35**:1–11, illus., Athens.

Hoveland, C. S., *et al.*, 1969, Yuchi arrowleaf clover, *Auburn Univ., Agric. Exp. Stn., Bull.* **396**:1–27, illus.

Hoveland, C. S., *et al.*, 1970, Management effects on forage production and digestibility of Yuchi arrowleaf clover (*Trifolium vesiculosum* Savi), *Agron. J.* **62**.

CONTRIBUTORS: J. A. Duke, C. F. Reed.

Trigonella foenum-graecum L.

FAMILY: Fabaceae

COMMON NAME: Fenugreek

USES: Fenugreek is widely cultivated as a condiment crop; seeds, containing coumarin, are used in soups and curries. Plant is used to make horse hair shiny. Harem women eat roasted fenugreek seed to attain buxomness. Powdered seeds are used locally for a yellow dye. In India, the crop is grown for forage. Mixed with cottonseed, the seed increases the flow of milk in cows, but imparts the fenugreek aroma to milk. Plant serves as a potherb; used to flavor imitation maple syrup. In North Africa it is mixed with breadstuffs. In Greece, raw or boiled seeds are eaten with honey. In Switzerland, the seeds are used to flavor cheese. Fenugreek contains ca. 5% oil, extracted by ether. Oil, with a strong celery odor, is used in butterscotch, cheese, licorice, pickle, rum, syrup, and vanilla flavors, and may be of interest to the perfume industry. Used also in cosmetics and hair preparations. In Punjab, dried plants are added to stored grains as insect repellents. As a legume, it is also considered a good soil renovator. Recently fenugreek has proved to be a useful source of the drug diosgenin, used in the synthesis of hormones. As with other chemurgic

Trigonella foenum-graecum L.

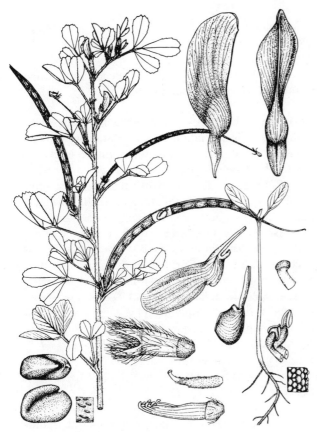

Figure 128. *Trigonella foenum-graecum* L.

crops, a large percentage of the crop must be thrown into the pot to extract a small percentage (1–2% on a dry-weight basis) of pharmaceutical (diosgenin). While the "pot is boiling," proteins, fixed oils, oleoresins, e.g., coumarin, mucilages and/or gums might be extracted. Organic residues might be used for biomass fuels or manures, inorganic residues for "inorganic" chemical fertilizers. The husk of the seed might be removed for its mucilage, with the remainder partitioned into oil, sapogenin, and protein-rich fractions.

FOLK MEDICINE: Mucilaginous seeds regarded as having tonic, emollient and vermifugal properties, and used for oral ulcers, chapped lips, and stomach irritation. Indian women believe the seeds promote lactation. Seeds carminative, tonic, used for diarrhea, dyspepsia, rheumatic conditions. Crushed leaves taken internally for indigestion. Said to be used for aphrodisiac, astringent, bronchitis, demulcent, diuretic, emmenagogue, emollient, expectorant, fever, fistula (sd), glands (sd), gout (sd), lactogogue, neuralgia (sd), restorative, sciatica (sd), skin (sd), sore (sd), sore throat, stomach (sd), tonic (sd), tuberculosis, tumor (sd), and wounds (sd).

CHEMISTRY: Reporting on a new C_{27}-steroidal sapogenin peptide ester, fenugreekine, Ghosal *et al.* (1974) describe its pharmacological activities as cardiotonic, hypoglycaemic, diuretic, antiphlogistic, and antihypertensive, which would account for the reported therapeutic uses of the seed extracts in the Indian system of medicine. Further fenugreekine inhibited ca. 80% of vaccinia virus replication, consistent with the reported uses of the seed extract as a prophylactic for chickenpox and smallpox. Seeds are reported to contain 6.2% moisture, 23.2% protein. 8.0% fat, 9.8% fiber, 26.3% mucilaginous material, and 4.3% ash. Seeds are also reported to contain trypsin inhibitors and chymotrypsin inhibitors. Whole grain is reported to contain (per 100 g edible portion): 369 calories, 7.8% moisture, 28.2 g protein, 5.9 g fat, 54.5 g total carbohydrate, 8.0 g fiber, 3.6 g ash, 220 mg Ca, 358 mg P, 24.2 mg Fe, 55 μg β-carotene equivalent, 0.32 mg thiamine, 0.30 mg riboflavin, 1.5 mg niacin, and 274 mg tryptophane. The protein is characterized by low levels of S-amino acids and high level of lysine and tryptophan. The amino acids values for the seed proteins are (% of protein): 8.0 lysine, 1.1 histidine, 8.0 arginine, 3.0 tyrosine, 9.0 aspartic acid, 9.0 glutamic acid, 6.0 serine, 9.5 glycine, 5.0 threonine, 5.9 alanine, 1.0 phenylalanine, 11.0 leucines, 1.0 proline, and 6.0 valine + methionine. The component fatty acids of the seeds are (% of total acids): 9.6 palmitic, 4.9 stearic, 2.0 arachidic, 0.9 behenic, 35.1 oleic, 33.7 linoleic, and 13.8 linolenic acid. Flour contains per 100 g 375 calories, 9.9% moisture, 25.5 g protein, 8.4 g fat, 53.1 g total carbohydrate, 7.1 g fiber, 3.1 g ash, 213 mg Ca, 270 mg P, 32.4 mg Fe, 0.06? mg thiamine, 0.05? mg riboflavin, and 1.5 mg niacin. Raw leaves contain: 35 calories, 87.6% moisture, 4.6 g protein, 0.2 g fat, 6.2 total carbohydrate, 1.4 g fiber, 1.4 g ash, 150 mg Ca, 48 mg P/100 g. Another analysis (from India) gave 86.1% moisture, 4.4% protein, 0.9% fat, 1.1% fiber, 6.0% other carbohydrates, and ash 1.5%. The mineral components were (mg/100 g): 395 Ca, 67 Mg, 51 P, 16.5 Fe, 76.1 Na, 31.0 K, 0.26 Cu, 167.0 S, and 165.0 Cl. Carotene content was 2.34, thiamine 0.04, riboflavin 0.31, nicotinic acid 0.8, and vitamin C 52.0 mg/100 g edible portion. Rao and Sriramulu (1977) report that vegetative

Trigonella foenum-graecum L.

material contains 36 ppm protocatechuic acid, 80 ppm chlorogenic acid, 227 ppm *p*-hydroxybenzoic acid, 302 ppm vanillic acid, 111 ppm rutin, 27 ppm caffeic acid, 53 ppm cis *p*-coumaric acid, 80 ppm trans-*p*-coumaric acid, 111 ppm trans ferulic acid. Reproductive material contains: 60 ppm protocatechuic acid, 90 ppm chlorogenic acid, 260 ppm p-hydroxybenzoic acid, 360 ppm vanillic acid, 100 ppm rutin, 60 ppm caffeic acid, 105 ppm cis-*p*-coumaric acid, 90 ppm trans-*p*-coumaric acid, 80 ppm trans ferulic acid, and 320 ppm cis-ferulic acid. Seeds of the genus *Trigonella* are reported to contain choline, malonic acid, and trigonelline.

DESCRIPTION: Annual herb; stems erect, rounded, sparsely pubescent, to 60 cm tall, loosely branched, except when growing in thick stands when few branches form, or unbranched; plants and seeds have a very strong distinctive odor; leaves trifoliolate with 2 light green stipules; leaflets obovate to oblong-lanceolate, thick, toothed near the apex, 20–50 mm long, 10–15 mm wide, those on lower stem often broader; flowers solitary or paired, axillary, subsessile; calyx 6–8 mm long, teeth about as long as the tube; corolla yellowish-white, 12–18 mm long, tinged with violet at the base; legume a slender pointed pod, 7.5–15 cm long (excluding beak), 4–6 mm wide, erect or patent, linear, somewhat curved, glabrous or glabrescent, with longitudinal veins, the beak persistent, 2–3 mm long; seeds 10–20 per pod, oblong, ca. 5 × 3 mm, quadrangular, somewhat compressed, yellow or pale brown, finely tuberculate. Fl. April–May.

GERMPLASM: Breeding programs have been established to determine whether yields of the important steroid, diosgenin, that is used for birth control, could be increased. One spontaneous mutant of 'IC74' is trailing and green (conditioned by a single dominant gene), flowers 30 days earlier and has larger seed (Singh and Singh, 1974). Banjai (1973) reports studies of 29 "ecotypes." Assigned to the Mediterranean and Near East Centers of Diversity, fenugreek is reported to tolerate disease, drought, high pH, insects, poor soils, and salt. ($2n = 16$.)

DISTRIBUTION: Fenugreek is native to the Mediterranean region and Near East, and is cultivated there, in Ethiopia and in India. In the Old World it is cultivated mainly for the seeds, but in India the young plants are used as a potherb. In North Africa, crop is sometimes cut for hay, and has been cultivated around the Saharan oases since very early times. In the United States, it is successfully grown in the fog belt of southern California.

ECOLOGY: Fenugreek is suitable for areas with moderate or low rainfall. A temperate cool growing season without extremes of temperature is favorable for best development. Fenugreek can tolerate 10°–15° of frost. In areas with mild winters, fenugreek is best sown in fall to mature in spring. It does well on a good, well-drained, deep loamy soil, but also grows fairly well on gravelly or sandy soils. Not adaptable to heavy clay or soil that becomes hard, it is fairly tolerant of salt. It does well without fertilizers, but lime should be added to acid soils. Ranging from Cool Temperate Steppe to Wet through Tropical Very Dry Forest Life Zone, fenugreek is reported to tolerate annual precipitation of 3.8–15.3 dm (mean of 18 cases = 7.1), annual mean temperature of 7.8°–27.5°C (mean of 18 cases = 15.9), and pH of 5.5–8.2 (mean of 15 cases = 7.3).

CULTIVATION: Seed may be sown in the fall in mild climate or in the spring further north and germinate in 4–5 days. Seed should be sown in close drills ca. 7.5 cm apart, the rows ca. 45 cm apart, or broadcast at ca. 22.5 kg/ha. Deep plowing and thorough harrowing are essential. Clean cultivation either mechanical or manual is necessary. In southern India fenugreek is intercropped with coriander, gingelly or Bengal gram. For seed production, potash and phosphoric acid have been recommended; for forage, nitrogenous manures. When grown as an irrigated crop, seeds are broadcast rather thickly onto beds at the rate of 25–30 kg/ha, and then stirred into the soil. Irrigation should start immediately after sowing and continue when necessary.

HARVESTING: Fenugreek ripens ca. 3–5 months after planting. It can be cut and handled with ordinary farm machinery. As a garden crop, it requires same conditions as garden peas. Plants are uprooted and dried a few days. Then the seeds are threshed, winnowed, further dried, and stored, or prepared for shipment. Pods open slowly, so that crop is easily handled for seed production. Seed retain their viability for many years. In India, where grown for forage, Oct. plantings are harvested in Feb.–Mar., while Jan. plantings are harvested in April.

YIELDS AND ECONOMICS: Studying 24 genotypes, Paroda and Karwasra reported forage yields of about 150–275 kg/ha with the mean ca. 200, while the Wealth of India reports 9–10 MT/ha green forage. Studying 29 ecotypes, Banyai (1973) reported seed yields of 500–3,320 kg/ha and that yields of 1,800 kg were economically viable. Diosgenin content ranged from 0.4% to 1.2%. Experimental seed yields of 3,700 kg/ha were reported from Bath, England and 1,000 kg from Morocco. Principal exporters are India, France, Lebanon, Egypt, and Argentina. Morocco seed commanded about 44¢/kg f.o.b. in the United States in 1975.

BIOTIC FACTORS: Fenugreek is reportedly attacked by: *Entyloma trigonellae; Alternaria tenuis; Cercospora traversiana; Cladosporium cladosporioides; Curvularia penniseti; Erysiphe polygoni; Leveillula taurica; Oidiopsis taurica; Peronospora trigonellae; Pythium debaryanum; Uromyces anthyllides; U. trigonellae;* and *Xanthomonas alfalfae.* In India, the plant is often infested with the mite *Tetranychus cucurbitae,* resulting in no or poor fruit set. Aphids, *Aphis craccivora,* and *Myzus persicae* are also problematic. Viruses: pea mosaic virus strain 2A; bean yellow mosaic; cucumber mosaic; red clover vein mosaic; subterranean clover mosaic; tobacco ring spot; Wisconsin pea streak. *Meloidogyne* spp. attack the roots. Flowering Plants: *Orobanche indica* Ham.

REFERENCES:
Banyai, L., 1973, Botanical and qualitative studies on ecotypes of fenugreek (*Trigonella foenum-graecum* L.), *Agrobotanika* **15**:175–187.
Ghosal, G., Srivastava, R. S., Chatterjee, D. C., and Dutta, S. K., 1974, Fenugreekine, a new steroidal sapogenin-peptide ester of *Trigonella foenum-graecum, Phytochemistry* **13**:2247–2251.
Paroda, R. S., and Karwasra, R. R., 1975, Prediction through genotype environment interactions in fenugreek, *Forage Res.* **1**(1):31–39.
Rao, P. G., and Sriramuly, M., 1977, Physiological characterization of a spice (*Coriandrum sativum*) and a condiment (*Trigonella foenum-graecum*) during vegetative and reproductive stages, *Curr. Sci.* **46**(17):615–616.
Singh, D., and Singh, A., 1974, A green trailing mutant of *Trigonella foenum-graecum* L. (Metha), *Crop Improvement* **1**(1/2):98–100.

CONTRIBUTORS: J. A. Duke, C. F. Reed, J. K. P. Weder.

Vicia benghalensis L.

FAMILY: Fabaceae
COMMON NAME: Purple vetch
SYNONYM: *Vicia atropurpurea* Desf.

USES: One of the least hardy of the cultivated vetches, purple vetch is restricted in its range of usefulness. In milder parts of California it is winter hardy. In the U.S. Cotton Belt, it cannot stand the

Figure 129. *Vicia benghalensis* L.

fluctuating winter temperatures. Plants winterkill severely in that region, except in southern Georgia and Alabama. All cultivated vetches make good hay, silage, pasture, green manure, and winter cover crops. For pasture, vetches are grown alone or in mixtures with small grains or ryegrass, thus providing winter and early spring grazing. Vetches compare in feeding value to clover, alfalfa and other common legumes. Protein content of vetch hay is 10–20% depending on stage of development of crop when cut. Seed used as one of the ingredients of ground poultry feed. All cultivated vetches make good hay, silage, and green manure. Some are also used for cover crops.

FOLK MEDICINE: No data available.

CHEMISTRY: The genus *Vicia* is reported to contain choline, guanidine, hydrocyanic acid, lupine, physostigmine, quercitrin, shikimic acid, and xanthine. Seeds of the genus *Vicia* are reported to contain trypsin inhibitors and chymotrypsin inhibitors.

DESCRIPTION: Straggling, climbing or trailing annual or short-lived perennial herb; stems 20–80 cm long, densely pubescent or pilose when young, often becoming glabrescent; leaves compound with 10–17 leaflets; leaflets linear, oblong or elliptic, 1–2.5 cm long, 1.5–6 mm broad, pilose; rachis terminated by a tendril; stipules semisaggitate, entire or denticulate; inflorescence 2–20-flowered; calyx-tube gibbous at base, the teeth unequal, the lower ones exceeding the tube; corolla 1–1.8 cm long, reddish-purple, with tip very dark; legumes straw-colored to brown, pubescent; outer surface depressed between the seeds, oblong to oblong-rhomboid, 2.5–4 cm long, 8–12 mm broad, flattened, usually densely pubescent or villous, but sometimes only on the suture, twisting slightly during tardy dehiscence; seeds 3–5 per pod, subspherical to sublenticular, smooth, brownish-gray with black mottles, ca. 5 mm long, 5 mm broad, 3.5 mm thick. Seeds 22,050/kg; wt. kg/hl = 77.

GERMPLASM: Assigned to the Mediterranean Center of Diversity, purple vetch or cvs thereof is reported to exhibit tolerance to high pH. ($2n = 12, 14$.)

DISTRIBUTION: Native to Mediterranean region of southern Europe from Azores and Portugal eastward. Now widely cultivated and often appearing as an escape or weed in other crops. Introduced in United States from southern Europe, and cultivated in Pacific Coast States. Found escaped in Kenya.

ECOLOGY: Usually found in fields of other crops growing as a weed; also in disturbed ground. In tropical East Africa (Kenya) grows at 2,400–2,800-m elevations. Vetches require a cool temperate climate for best development, even in the tropics. In the Pacific States, vetches make their growth during the fall, winter, and early spring and mature in late spring or early summer. Tolerates nearly any type of soil, but grows best in rich loams. Moderate moisture supply is necessary. Not drought resistant. Tolerates acid soil better than most legume crops; can be grown in most soils without addition of lime. Ranging from Warm Temperate Dry to Moist through Subtropical Moist Forest Life Zone, purple vetch is reported to tolerate annual precipitation of 3.1–16.6 dm (mean of 8 cases = 8.5), annual mean temperature of 10.8°–21.8°C (mean of 8 cases = 17.7) and pH of 4.8–8.2 (mean of 7 cases = 6.7).

CULTIVATION: All vetches are fall-sown in the southern United States and on the Pacific Coast. In northern states, all vetches except hairy should be sown early in spring; hairy vetch is sown in August or early September. Vetches may be broadcast or drilled. Less seed is necessary with the drill method. Broadcast seeding is usually disked in. On level, poorly drained land, vetch is generally broadcast in beds or ridges. Common practice is to broadcast seed in cotton middles, either by hand or by airplane. Then seed is covered with a 2-row or 3-row tractor-mounted middle-buster (listers). Operated at relatively shallow depths, these throw soil and seed into beds and at same time provide water furrows for drainage. On well-drained seedbed, seeding with grain drill is most satisfactory method. Depth of planting varies with soil and amount of surface moisture. In loam soils, good stands are obtained from plantings at depth of 10 cm; deeper plantings usually result in poorer stands. Shallower plantings give good results if sufficient moisture is present. Seed rate 56–67 kg/ha broadcast. Vetch is sown alone or with a supporting crop of small grain. Sowing with grain is common practice where crop is grown for forage. Oats, where successful, is the favorite grain to combine with vetch, though wheat, rye, and barley may be used. Oats is especially useful when vetch is grown for seed as the oat and vetch seeds are readily separable; separation is more difficult from rye, wheat, or barley. Wherever vetch is grown as a green-manure crop, it is nearly always sown alone. On the sandy lands of Michigan, where winters are severe, vetch is seeded in combination with rye. In many areas fertilizers usually are not necessary for a successful growth of vetch. In western Oregon, gypsum is commonly applied at rate of 85–167 kg/ha. In southern United States P and K are used in varying proportions, depending on soil needs. Fertilizers should be applied before or at time of seeding. Nitrogen fertilizers are seldom needed. Vetch contains 2.5–3.5% N, much of which is from the air. One ton of dry vetch contains about 27 kg of N. On very infertile, badly eroded lands of the south, it is desirable to include N along with P and K; about 17.5 kg N/ha should be sufficient.

HARVESTING: Vetches (common, hairy, winter, and other shattering types) are cut for seed as soon as lower pods are fully ripe. At this stage the upper pods are fully formed and yield of seed is maximum. Cutting later causes more seed shattering. Nonshattering species (purple and Hungarian) usually are allowed to ripen 75–90% pods before cutting. Vetches for seed are cut with grain binder and put in shocks, or mowed with a windrower attachment. Ordinary grain thresher used for vetch grown alone or intercropped with small grain. Combine harvesters are used satisfactorily in the northwest, especially for nonshattering and semishattering vetches. Fanning mills are used to clean and separate vetch seed from foreign matter and the seeds of grains. Vetches make good hay, alone or mixed with small grain, and are relished by livestock. Common and Hungarian vetches are not generally used, but hay from other vetches should be equally good, especially in the northeast. Vetch planted with small grain is often cut green and fed to cattle or other livestock. Succulent late-winter and early-spring feed can be supplied in mild climates in this way with little expense. Vetch is ordinarily cut for hay when the first pods are well developed. Crop cut with ordinary mower with windrower attachment when thin, or if heavier, cut and windrowed with a side-delivery rake. Hay may be cured in windrow or bunched with dump rake and allowed to cure in shocks. For pasture, vetches, alone or in mixture, extend the grazing season by supplying late fall and early spring feed. They stand trampling and are well suited for pasture. Common, hairy, and Hungarian vetches have been used for pasture. Usually vetch is pastured only when the ground is dry to avoid packing the soil and to reduce possible bloat in cattle and sheep. Hairy vetch is pastured in eastern United States, where it is more commonly grown. Vetches used as green manure in the south consistently increase the yields of cotton and corn crops that follow the vetches, mainly due to the nitrogen turned under in the green manure. The time for turning under vetch for soil improvement depends on the amount of vetch growth and the expected planting date for the succeeding crop. Vetch can be turned under with plows, preferably with rolling colters attached. In areas where the succeeding crop is planted on beds, vetch may be disked in the land rebedded with middle-busters. About 2–3 weeks should elapse between turning under and planting to permit decomposition of vetch. For rotation, in southeast United States corn, cotton, soybeans, and other crops follow vetch planted as a winter cover crop. Vetch can be utilized as a soil-improving crop with or without grazing and still fit well into southern rotations. In northern states, it is sometimes possible to cut fall-sown hairy vetch for hay early enough to plant late crop of millet. Other vetches, in the north, must be spring-sown and require the entire season for development. Under irrigation in the southwest, where winters are mild, vetches are grown alone or with a small grain crop for hay and pasture during the winter and followed by another crop.

YIELDS AND ECONOMICS: Vetches vary considerably in seed yield depending on cv or species, climatic conditions, method of growth of the crop, fall or spring-sown, and use for which a crop is intended. Common, Hungarian, monantha, and purple vetches average 25–30 bu/ha with 50–67 bu/ha as a maximum. Hairy and winter vetches yield 15–18 bu/ha with 30–38 bu/ha as maximum. Purple vetch seed used in the United States is produced in western Oregon, western Washington, and northwestern California. In other parts of California it is used as green manure crop and for hay.

BIOTIC FACTORS: Fungi reported on this vetch include the following species: *Ascochyta pisi, Erysiphe communis, Oidium erysiphoides,* and *Uro-*

Vicia ervilia (L.) Willd.

myces viciae-fabae. Nematodes isolated from this plant include: *Heterodera glycines*, *H. goettingiana*, *H. schachtii*, *H. trifolii*, and *Pratylenchus vulnus*. Since vetches are insect-pollinated, bee hives are recommended to insure fertilization and increase seed yield. Seven introductions of this vetch suggested that it is segregating for resistance to the vetch bruchid (*Bruchus brachialis* Fahraeus). *Vicia sativa*, *V. ervilla*, and *V. monantha* also were classified as resistant, *V. villosa* (incl. *V. dasycarpa*) as susceptible (Pesho et al., 1973).

REFERENCES:

Blum, A., and Lehrer, W., 1973, Genetic and environmental variability in some agronomical and botanical characters of common vetch (*Vicia sativa* L.), *Euphytica* **22**:89–97.

Gunn, C. R., 1971, Seeds of native and naturalized vetches of North America, *U.S. Dep. Agric., Agric. Handb.* **392**:1–42, illus.

Henson, P. R., and Schoth, H. A., 1968, Vetch culture and uses, *U.S. Dep. Agric., Farmers Bull.* **1740**:1–20, illus.

Hermann, F. J., 1960, Vetches in the US—Native, naturalized and cultivated, *U.S. Dep. Agric., Agric. Handb.* **168**:1–84, illus.

CONTRIBUTORS: J. A. Duke, C. F. Reed, J. K. P. Weder.

Figure 130. *Vicia ervilia* (L.) Willd.

Vicia ervilia (L.) Willd.

FAMILY: Fabaceae
COMMON NAME: Bitter vetch
SYNONYM: *Ervum ervilia* L.

USES: Plants of this species are not readily eaten by livestock, but it is an excellent winter green manure crop in California. Seed used for stock feed, especially for sheep.

FOLK MEDICINE: No data available.

CHEMISTRY: No data available.

DESCRIPTION: Annual herb; stems glabrate, erect, angular, 2–7 dm tall; leaves compound with 8–12 pairs of leaflets without tendrils; leaflets oblong-linear to linear, 4–15 mm long, 1–4 mm broad, apex rounded or shallowly emarginate; rachis terminating in a point; stipules small, acuminate, semihastate, more or less dentate; racemes 1–4-flowered, 1.5–4 cm long; flowers 7–9 mm long; corolla rose-colored to whitish with violet veins, nodding on short pedicels; calyx campanulate, the teeth acuminate, somewhat longer than the tube; legume straw to light brown, glabrous, sometimes with nectariferous glands, reticulate, with few to numerous knoblike processes, usually nearly moniliform, occasionally with pools of nectar in depressions, 1.2–2.5 cm long, 2–5 mm broad, terete to slightly flattened, oblong, abruptly acuminate, base long-acuminate, twisting loosely or tight during dehiscence; seeds pyramidal, smooth, pale straw to reddish-straw to lightly mottled with darker or blackish brown to densely mottled or streaked with blackish brown, 3.2–5.6 mm long, 2.8–5 mm broad, 2.7–4.8 mm thick; hilum small, about one-twelfth circumference of seed.

GERMPLASM: Assigned to the Near Eastern and Mediterranean Center of Diversity, bitter vetch or cvs thereof is reported to exhibit tolerance to high pH. ($2n = 14$.)

DISTRIBUTION: Native to southern Europe. Introduced and cultivated in the Pacific Northwest of the United States.

ECOLOGY: Crops produce good growth and good seed crops in western United States. In Cotton Belt it has made comparatively little growth and has often winterkilled, hence its use is limited. Ranging from Cool Temperature Steppe to Wet through Warm Temperate Thorn to Dry Forest Life Zones, bitter vetch is reported to tolerate annual precipitation of 3.6–11.6 dm (mean of 10 cases = 6.3), annual mean temperature of 8.3°–20.0°C (mean of 10 cases = 14.2), and pH of 5.6–8.2 (mean of 9 cases = 7.3).

YIELDS AND ECONOMICS: Extensively grown in Asiatic Turkey, and seed shipped in large quantities to England and other countries for stock feed, especially sheep. Bitter vetch seed produced in Mediterranean region. This crop not grown commercially in United States. However, in western United States it has made good growth and produced good seed crops, but not considered superior to other vetches in common use.

BIOTIC FACTORS: The following fungi have been reported on this vetch: *Aphanomyces eutriches, Erysiphe communis, Fusarium leguminosum, Mycosphaerella pinodes, Oidium* sp., *Pleospora herbarum, Uromyces fabae,* and *U. heimlerianus. Heterodera goettingiana* and *H. trifolii* have been isolated from this vetch.

CONTRIBUTORS: J. A. Duke, C. F. Reed.

Figure 131. *Vicia faba* L.

Vicia faba L.

FAMILY: Fabaceae

COMMON NAMES: Broadbean, Fava Bean, Horsebean, Windsorbean, Tickbeans (small types)

USES: Cultivated as a vegetable and used green or dried, fresh or canned, and for stock feed. Feeding value of broadbeans is high, considered in some areas superior to field peas or other legumes. Broadbean has been considered as a meat extender or substitute and as a skim-milk substitute. Sometimes grown for green manure, but more generally for stock feed. Large-seeded cvs used as a vegetable, and frequently grown as a home-garden crop, and for canning. One of the most important winter crops for human consumption in the middle East. Roast seed are eaten like peanuts in India.

FOLK MEDICINE: Said to be used for diuretic, expectorant, and tonic.

CHEMISTRY: Inhalation of the pollen or ingestion of the seeds may cause favism, a severe hemolytic anemia, perhaps causing collapse. It is an inherited enzymatic deficiency occasional among Mediterranean people (Greek, Italian, Semitics). The genetic disorder occurs in about 1% of whites, 15% of blacks. The favism-inducing toxins are believed to be divicine and isouramil, the aglycone moieties of vicine and convicine. Flesh of broadbeans contains ca. 0.61–2.38% vicine, common vetch contains 0.04%, peas contain traces, and soy flour is devoid of vicine. Injected intravenously in rabbits, broadbean extracts have produced haemoglobinuria and death. An ethanol–ether extract of broadbeans has estrogenic activity, 50 mg stimulates the nonpregnant uterus at dioestrus. The LD_{50} of the bean extract in mice was 19,000 mg/kg body weight. L-DOPA and epinene have been reported from the seeds. Among phytoalexins reported in broadbean are medicarpin, epoxide, and wyerone. Seeds are reported to contain trypsin inhibitors and chymo-

Vicia faba L.

trypsin inhibitors. The whole dried seeds contain (per 100 g): 344 calories, 10.1% moisture, 26.2 g protein, 1.3 g fat, 59.4 g total carbohydrate, 6.8 g fiber, 3.0 g ash, 104 mg Ca, 301 mg P, 6.7 mg Fe, 8 mg Na, 1123 mg K, 130 μg β-carotene equivalent, 0.38 mg thiamine, 0.24 mg riboflavin, 2.1 mg niacin, 162 mg tryptophane, and 16? mg ascorbic acid. Flour contains: 340 calories, 12.4% moisture, 25.5 g protein, 1.5 g fat, 58.8 g total carbohydrate, 1.5 g fiber, 1.8 g ash, 66 mg Ca, 354 mg P, 6.3 mg Fe, 10 μg β-carotene equivalent, 0.42 mg thiamine, 0.28 mg riboflavin, and 2.7 mg niacin. Taamiah, made from broadbean, contains 408 calories, 27.0% moisture, 10.0 g protein, 31.8 g fat, 26.3 g total carbohydrate, 1.2 g fiber, 4.9 g ash, 72 mg Ca, 153 mg P, and 6.1 mg Fe. Immature seeds contain: 75 calories, 76.3% moisture, 7.1 g protein, 0.4 g fat, 15.3 g total carbohydrate, 3.2 g fiber, 0.9 ash, 38 mg Ca, 127 mg P, 0.10 mg thiamine, 0.22 mg riboflavin, and 140 mg ascorbic acid. Germinated seeds contain: 111 calories, 64.5% moisture, 10.9 g protein, 0.3 g fat, 22.9 g total carbohydrate, 5.6 g fiber, and 1.4 g ash. Roasted seeds contain: 366 calories, 5.3% moisture, 26.4 g protein, 2.0 g fat, 63.3 g total carbohydrate, 1.7 g fiber, 3.0 g ash, 60 mg Ca, 479 mg P, 6.8 mg Fe, 0.21 mg thiamine, 0.35 mg riboflavin, 2.4 mg niacin, and 2 mg ascorbic acid. Green seeds contain 118 calories, 69.0% moisture, 9.3 g protein, 0.4 g fat, 20.3 g total carbohydrate, 3.8 g fiber, 1.0 g ash, 31 mg Ca, 140 mg Fe, 2.3 mg Fe, 60 μg vitamin A, 0.28 mg thiamine, 0.17 mg riboflavin, 1.7 mg niacin, and 28 mg ascorbic acid. Fried and salted seeds contain: 402 calories, 7.6% moisture, 26.4 g protein, 14.8 g fat, 47.4 g total carbohydrate, 3.8 g fiber, 3.8 g ash, 73 mg Ca, 331 mg P, 7.1 mg Fe, 994 mg K, 5 μg β-carotene equivalent, 0.10 mg thiamine, 0.05 mg riboflavin, and 1.0 mg niacin. The haulms contain: 85.0% moisture, 2.5% crude protein, 0.4% fat, and 4.9% crude fiber, 5.4% N-free extract, 1.8% ash, 0.22% Ca, 0.04% P, and 2.0% digestible crude protein. The amino acid content averages (mg/g N): isoleucine 250, leucine 443, lysine 404, phenylalanine 270, tyrosine 200, methionine 46, cystine 50, threonine 210, valine 275, arginine 556, histidine 148, alanine 259, aspartic acid 702, glutamic acid 942, glycine 258, proline 249, serine 280. The fatty acid composition of broadbean oil has been reported as 88.6% unsaturated (oleic 45.8, linoleic 30.0%, and linolenic 12.8%) and 11.4% saturated (8.2% stearic). Cholesterol (0.04%) and lipoxygenase are also reported.

DESCRIPTION: Coarse, upright, annual herb; stems large, unbranched 0.3–2 m tall, with 1 or more stems from the base; leaves compound, leaflets usually 6 large, broad, oval; flowers large, white with dark purple markings, horne on short pedicels in clusters of 1–5 in axils of leaves; 1–4 pods developing from each flower cluster; legumes greenish black, brown to black, glabrous, reticulate, 8–20 cm long, 10–30 mm broad, inflated, terete, flattened, oblong, obliquely acuminate at both ends, style usually permanent, 3–4-seeded, twisting loosely or tightly during dehiscence; seeds oblong or oval, flattened or rounded, smooth, bright reddish brown, light to dark greenish brown or light to dark purple, all obscurely mottled or dotted with colors similar to base colors, 6.517 mm long, 730.5 mm broad, 4.5–9 mm thick. *V. f.* var *major,* seeds 1,103 kg; *V. f.* var *equina,* seeds 6,615/kg. Wt. kg/hl = 77. Germination cryptocotylar.

GERMPLASM: Some of the many horsebean cvs developed and grown for their seeds, are 'Windsor,' 'Longpod,' 'Dwarf Fan,' 'Julienne,' 'Lorraine,' 'Black Spanish,' 'Mazagan,' 'Picardy,' and 'Winter' are a few of the U.S. cvs. 'Lindsay-Johnson Winter' bean is a large flat green-seeded cv. Small-seeded cvs are more often grown for green manure and forage. Other important cvs are: 'Albyn Tick,' 'Herz Freya,' 'Blue Rock,' and 'Maris Bead.' Smaller less productive tick beans are grown in Europe to feed pigeons and other livestock. 'Petite' tickbean yields ca. 2,500 kg/ha. Botanical vars *equina, faba, minor,* and *paucijuga* have been recognized in recent revisions, and subvarieties have been named. In this partially allogamous species, with considerable intrapopulational variation and no sterility barriers between subspecies, such fine-honed nomenclature may seem superfluous. Broadbean has been assigned to the Central Asian, Mediterranean, and South American Centers of Diversity. Cubrero (1974) postulated a Near Eastern center of origin, with four radii (1) to Europe (2) along the north African coast to Spain, (3) along the Nile to Ethiopia, and (4) from Mesopotamia to India. Secondary centers of diversity are postulated in Afghanistan and Ethiopia. The wild progenitor has not been discovered yet. Several wild species (*V. narbonensis* L. and *V. galilaea* Plitmann and Zohary) are taxonomically closely related to the cultivated crop, but they contain $2n = 14$ chromosomes. Numerous attempts to cross them with *Vicia faba* have failed. Broadbean or cvs thereof is reported to exhibit tolerance to high pH, insects, low pH, slope, and virus. ($2n = 12, 14$.)

DISTRIBUTION: Probably native to the Near East, broadbean is now widely introduced and cultivated in temperate North America, including Manitoba and Saskatchewan, South America, especially the Andes, and elsewhere (e.g., Burma, China, Sudan, and Uganda).

ECOLOGY: Broadbean requires a cool season for best development. Grown as a winter annual in warm temperate and subtropical areas. It can be grown anywhere it does not winterkill but where temperatures fluctuate rapidly. Well-adapted to wetter portions of cereal-growing areas of western Canada. Tolerates nearly any soil type; grows best on rich loams. Moderate moisture supply is necessary. Not drought resistant. Moisture requirement is highest ca. 9–12 weeks after establishment. More tolerant to acid soil conditions than most legumes. Can be grown in nearly all parts of the United States without liming. Hardier cvs tolerate winter temperatures of $-10°C$ without serious injury. Winter types fare well with average temperatures of 2°C, without severe frost. Growing season should have little or no excessive heat. Ranging from Boreal Moist to Wet through Tropical Desert to Dry Forest Life Zones, broadbean is reported to tolerate annual precipitation of 2.3–20.9 dm (mean of 95 cases = 8.0), annual mean temperature of 5.6°–27.5°C (mean of 95 cases = 12.1), and pH of 4.5–8.3 (mean of 87 cases = 6.6).

CULTIVATION: In localities having no hard frosts, most cvs can be sown in fall and survive the winter. In northern localities, or at high elevations farther south, fava bean should be planted in early spring, when ground can be worked, at the same time as the earliest ordinary spring crops. Large-seeded cvs are sown with planters used for lima beans; small-seeded cvs with a common corn planter. In some areas broadbeans are planted by hand. At seeding time the field is plowed shallow and seed dropped in every second or third furrow. Seed are usually sown 5–10 cm deep in rows ca. 75 cm apart, with seeds 15 cm apart in the rows. Rows 60 cm apart, or even closer, give good results under favorable conditions, but wider planting usually is preferable. Small-seeded cvs are planted at 90–122 kg/ha; large-seeded cvs, 78–90 kg/ha. In United Kingdom, 450 kg seed/ha produces maximum yields. Yields are economically optimal at 225–340 k/ha for large seeded cvs, and satisfactory at 190 kg/ha for small-seeded cvs. For green manure or forage small-seeded cvs are usually broadcast. Russian field trials showed that pretreatment of the seeds with 0.01% of vitamins of the B group increased seed yields by as much as 36%. Fertilizers and seed inoculation with proper legume bacteria are usually recommended. Inoculation is not always practiced (e.g., Britain, Europe in general, where nodulation with indigenous rhizobia is excellent). In southern United States, P and K are used before or at seeding. Early deep sowing into a well-drained firm seedbed gives best results. Broadbeans should be thoroughly cultivated throughout their growing period. When planted in 60 cm rows or closer, special machinery is necessary for cultivation. When planted in 90 cm rows, ordinary cultivators can be used. Zero tillage has depressed yields by 22%. In United Kingdom, thiram and captan are recommended as fungicides, chlorpropham plus diuron or fenuron, or simazine, as preemergence herbicides, dinoseb-acetate as a postemergence herbicide.

HARVESTING: Time of harvest depends on method—hand or mechanical. Beans mature 90–220 days after planting. Harvest can be delayed a little longer for hand than for mechanical harvest. In either case, crop should not be cut until the lower pods are matured and the upper ones fully developed. If harvest is delayed until the upper pods are ripe, loss from shattering is great. An ordinary mowing machine can be used, but the drop-rake reaper is more satisfactory and reduces shattering. Crop should be cut on cloudy day and maybe cut at night and shocked early the next day. Large-seeded cvs are threshed with a common bean thresher with special adjustments to the cylinder. Small-seeded types can be thrashed without difficulty. After threshing, seed are cleaned with ordinary fanning mills. For canning, beans are allowed to swell and then are picked by hand before they become hard. As a dried vegetable, they are prepared in same way as other common beans.

YIELDS AND ECONOMICS: Dry bean yields in the US average ca. 6,600 kg/ha; in Great Britain, ca. 3,000 kg/ha. Yields are closely correlated with number of pods per plants. In British field trials, maximum potential seed yield in 'Herz Freya' was 4,940 kg/ha, in 'Maris Bead' 6,710 kg/ha. Water

Vicia faba L.

could be more important in yield than solar radiation or plant competition. (Sprent *et al.*, 1977). In 1975, Asia produced 4,750,000 MT (1192 kg/ha), Europe 722,000 MT (1,433 kg/ha), Africa 688,000 (1,098 kg/ha), South America 135,000 MT (557 kg/ha), North America 43,000 MT (537 kg/ha). China was the largest producer with an estimated 4,660,000 MT (1189 kg/ha); Italy, 252,000 MT (1,222 kg/ha); Egypt, 234,000 MT (2125 kg/ha); Morocco, 213,000 MT (966 kg/ha); Ethiopia, 124,000 MT (886 kg/ha) (FAO, 1975). Switzerland reported highest yields (4,000 kg/ha), followed by Denmark (3,500 kg/ha) and West Germany (3,176 kg/ha). The United Kingdom yield was 2,395 kg/ha (FAO, 1976). Dry-matter yields reported in British field trials were 4,870 kg/ha for 'Triple White,' 6,320 for 'Gilletts Green Longpod,' 6,680 for 'Throws MS,' 7,280 for 'Minor' tick, 7,740 for 'E. W. King's Longpod,' and 8,060 for 'Strubes' broadbean. In Canadian studies of 16 cvs at 6 localities, mean seed yields ranged from 2,533 to 3,488 kg/ha, with the highest yield at over 7 MT/ha ('Klein Kronige'). In British studies, 'Albyn Tick' gave 6,765 kg seed/ ha, 'Blue Rock' 6,025, 'Herz Freya' 7,067, and 'Maris Bead' 6,602. The plants have been calculated to fix nearly 200 kg N/ha. Broadbeans for human consumption and horsebeans for green manure and stock feed are becoming important crops in southern United States and along the Pacific Coast. Broadbeans are grown in home gardens. Large-seeded green types are canned. Yields of fresh green broadbean for home consumption as a vegetable average 11–12.5 MT/ha in United Kingdom and 25 MT/ha was reported.

BIOTIC FACTORS: One study concluded that bees increase seed production by 15–20%. Honeybees were estimated to account for 80% of cross-pollination, bumblebees less than 20%, wild bees less than 1%. A closed-flower phenotype (recessive to normal) exists which lacks the typical scent and is avoided by bees (Poulsen, 1977). Many fungi attack broadbeans, depending on the area where they are grown. The following have been reported on broadbeans: *Alternaria brassicae* var *phaseoli*, *A. tenuis*, *A. tenuissima*, *Ascochyta boltshauseri*, *A. fabae*, *A. pinodella*, *A. pinodes*, *A. pisi* (*A. viciae*), *Aspergillus niger*, *Botrytis cinerea*, *B. fabae*, *Cercospora fabae*, *C. viciae*, *C. zonata*, *Cladosporium cladosporioides*, *C. herbarum*, *C. pisi*, *Clonostachys araucariae*, *Colletotrichum lindemuthianum*, *Corticium rolfsii*, *C. solani*, *Cunninghamella echinulata*, *Deplosporium album*, *Dothiorella fabae*, *Erysiphe pisi*, *E. polygoni*, many species of *Fusarium*, *Gibberella fujikuroi*, *G. saubinettii*, *Gloeosporium viciae*, *Helicobasidium purpureum*, *Leveillula taurica*, *Macrophomina phaseoli*, *Melanospora papillata*, *Mycosphaerella pinodes*, *Nectria anisophylla*, *Olpidium viciae*, *Peronospora fabae*, *P. lagerheimii*, *P. pisi*, *P. viciae*, *Phoma malaena*, *Phyllosticta fabae*, *Phymatotrichum omnivorum*, *Physoderma fabae*, *Phytophthora cactorum*, *Ph. cinnamoni*, *Pleospora herbarum*, *P. vulgaris*, *Pythium spp.*, *Rhizoctonia solani*, *Rhizopus nigricans*, *Sclerotinia fuckeliana*, *S. minor*, *S. sclerotiorum*, *Sclerotium rolfsii*, *Stagonospora carpathica*, *Stemphylium botryosum*, *S. consortiale*, *Trichothecium roseum*, *Uromyces appendiculatus*, *U. fabae*, *U. orobi*, and *U. viciae-fabae*. Broadbeans also attacked by the sweet pea streak, tooth-tumor swelling vein virus and broadbean wilt; red-clover vein mosaic (*Marmor trifolii*), virus 1-celery mosaic (a strain of cucumber mosaic virus: *Marmor cucumeris*), spotted wilt (*Lethum australiensis*). Bacteria causing diseases in broadbean include: *Bacterium phaseoli*, *B. viciae*, *Erwinia phytophthora*, and *Psuedomonas viciae*. Nematodes isolated from broadbean include: *Ditylenchus dipsaci*, *Heterodera glycines*, *H. goettingiana*, *H. rostochiensis*, *Longidorus maximus*, *Meloidogyne arenaria*, *M. artiellia*, *M. hapla*, *M. incognita*, *M. i. acrita*, *M. javanica*, *Pratylenchus brachyurus*, *P. coffeae*, *P. goodeyi*, *P. penetrans*, *P. pratensis*, *P. vulnus*, *P. zeae*, *Rotylenchulus reniformis*, *Tylenchorhynchus dubius*, *T. parvus*. The most serious insect pests are the broadbean weevil, *Bruchus rufimanus*, and aphids, especially the bean aphid, *Aphis fabae*. Broomrape (*Orobanche crenata*) may be a serious problem in the Middle East. Eptam, applied as a postemergence spray, was fairly effective, as was soil fumigation with dibromochloropropane, and oxak (terbutol), if deeply incorporated into the soil before sowing.

REFERENCES:

Bond, D. A., 1977, Field bean, in: *Evolution of Crop Plants* (N. W. Simmonds, ed.), pp. 179–182, Longman's, London, 339 pp.

Cubrero, J. I., 1973, Evolutionary trends in Vicia faba, *Theor. Appl. Genet.* **43**:59–65.

Evans, L. E., *et al.*, 1972, Horsebeans—A protein crop for Western Canada? *Can. J. Plant Sci.* **52**:657–659.

Ishag, H. M., 1973, Physiology of seed yield in field beans

(Vicia faba). II. Yield and yield components, *J. Agric. Sci. Camb.* **80:**181–189.

Lewis, W. H., and Elvin-Lewis, M. P. F., 1977, Medical Botany, Wiley, N.Y., 515 pp.

Poulsen, M. H., 1977, Obligate autonomy in Vicia faba L., *J. Agric. Sci.* **88**(1):253–256.

Sharaf, A., Kamel, S. H., and El-Shabrawy, O. A., 1972, Some pharmacological studies on Vicia faba cotyledons grown in Egypt, *Qual. Plant. Mater. Veg.* **22**(1):99–105.

Sprent, J. I., Bradford, A. M., and Norton, C., 1977, Seasonal growth patterns in field beans (Vicia faba) as affected by population density, shading and its relationship with soil moisture, *J. Agric. Sci.* **88**(2):293–301.

Yassin, T. E., 1973, Analysis of yield stability in field beans *(Vicia faba)* in the northern province of Sudan, *J. Agric. Sci. Camb.* **80**(1):119–124.

CONTRIBUTORS: J. A. Duke, A. Fyson, C. F. Reed, J. I. Sprent, J. K. P. Weder, D. Zohary.

Vicia monantha Retz.

FAMILY: Fabaceae
COMMON NAMES: Bard vetch
SYNONYM: *Vicia calcarata* Desf.

USES: In general growth habit and use bard vetch is similar to winter and common vetch. It is adapted only to southwest United States, where it succeeds well in irrigated Yuma and Imperial Valleys. Farther north in the West, it cannot compete with the other vetches, and in the Cotton Belt east of the Mississippi River it has never succeeded.

FOLK MEDICINE: No data available.

CHEMISTRY: No data available.

DESCRIPTION: Slender, glabrous annual herb; stems weak and viny, angular, prostrate or climbing, 2–4 dm long; leaves compound with 8–16 leaflets, these linear-lanceolate to oblong-elliptic; stipules entire, semisagittate, sometimes the lobes bifid; peduncles filiform, shorter than the leaves, aristate, generally 1–3-flowered; calyx sparingly appressed-villous, one-fourth length of corolla, teeth small, subequal, triangular, the upper pair somewhat connivent; corolla blue, 12–20 mm long; legume oblong, straw-colored, glabrous, reticulated, slightly depressed between seeds, 3.5–4.5 cm long, 12–20 mm broad, flattened, apex acuminate, base rounded, with 5–7 seeds, twisting loosely or tightly during dehiscence;

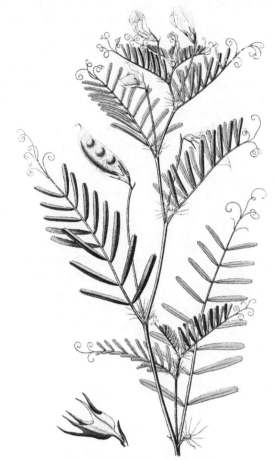

Figure 132. *Vicia monantha* Retz.

seeds subspherical to oval, smooth, light greenish, ochre, chestnut, brown, or black, mottled with brown, 4–6 mm long, ca. 4.3 mm broad, and 3.4 mm thick; hilum slightly wedge-shaped, occupying 9% of seed circumference.

GERMPLASM: Assigned to the Mediterranean, Near Eastern, and Eurosiberian Centers of Diversity, bard vetch or cvs thereof is reported to exhibit tolerance to high pH. ($2n = 14$.)

DISTRIBUTION: Native to southern Europe and eastern Asia. Cultivated only in southwest United States.

ECOLOGY: Adapted only to the Southwest where it succeeds well in irrigated areas. Ranging from Cool Temperate Steppe to Wet through Warm Temperate Moist to Thorn Forest Life Zones, bard vetch is reported to tolerate annual precipitation of 3.5–12.3

Vicia monantha Retz.

dm (mean of 6 cases = 6.3), annual mean temperature of 7.0°–16.0°C (mean of 6 cases = 11.2), and pH of 6.0–7.5 (mean of 5 cases = 6.9).

YIELDS AND ECONOMICS: Grown mainly in the irrigated areas of the Yuma and Imperial Valleys of the Southwest. Also grown to some extent in southern Europe and eastern Asia.

BIOTIC FACTORS: Fungi reported as attacking this vetch include the following species: *Aphanomyces eutreiches, Erysiphe polygoni, Pseudoplea trifolii, Sclerotinia* sp., *Pleosphaerulina briosiana,* and *Uromyces viciae-fabae.* Nematodes isolated from this vetch include: *Heterodera goettingiana, H. schachtii, Meloidogyne arenaria, M. incognita, M. i. acrita,* and *M. javanica.* Since vetches are insect-pollinated, nearby bee hives ensure fertilization and increase seed yield.

CONTRIBUTORS: J. A. Duke, C. F. Reed.

Vicia pannonica Crantz

FAMILY: Fabaceae
COMMON NAME: Hungarian vetch

USES: Grown in western Oregon, where it has recently become important commercially. Rather winter hardy, but its exact limitations have not been established. It is known to survive as far north as Washington D.C. Seed used to limited extent in mixed ground feeds. In Europe this species is grown for hay, pasture, and green manure, and to a lesser degree for cover crop.

FOLK MEDICINE: No data available.

CHEMISTRY: No data available.

DESCRIPTION: Annual herb; stems shortly soft- to shaggy-pubescent, 2 to several per plant, prostrate to ascending or climbing, rarely unbranched, 10–50 cm tall; leaves short-petioled to nearly sessile, the lower with 4, the upper with often 8 pairs of leaflets and mostly short tendrils; leaflets very short-petioled, linear to oblong; inflorescence very short-stalked, 2–4-flowered, rarely 1-flowered; flowers 1.3–2.2 cm long, the pedicel much shorter than the

Figure 133. *Vicia pannonica* Crantz

calyx; corolla about 3 times longer than the calyx, yellowish to purple; legume straw-colored, sericeous with numerous nectary glands and weakly reticulate, 2–3 cm long, 7–10 mm broad, flattened, oblong, obliquely acuminate at both ends, twisting loosely during dehiscence; seeds subpyramidal to compressed-spherical, smooth, rosy-brown, mottled and dotted with brown or black, 4.1 mm long, ca. 4.2 mm broad, 3.3 mm thick. Seeds 22,050/kg; wt. kg/hl = 77.

GERMPLASM: Hungarian vetch is assigned to the Near Eastern and Eurosiberian Centers of Diversity. ($2n = 12$.)

DISTRIBUTION: Native to Central Europe, especially to Hungary and adjacent territories. Introduced and cultivated, sometimes naturalized, in Pacific northwest United States. Also introduced into cultivation elsewhere in Europe but less extensively grown than hairy or common vetches.

ECOLOGY: Cultivated in regions where temperatures do not fluctuate sharply or where there is protection by snow. This species tolerates temperatures of −18°C or lower. Especially well adapted to heavy clay soils and does better in wet places than other vetches. In southern United States it has done poorly on sandy land. Ranging from Cool Temperate Steppe to Cool Temperate Wet Forest Life Zone, Hungarian vetch is reported to tolerate annual precipitation of 3.5–10.3 dm (mean of 8 cases = 6.1), annual mean temperature of 7.0°–13.1°C (mean of 8 cases = 9.4), and pH of 4.9–8.2 (mean of 7 cases = 6.9).

YIELDS AND ECONOMICS: Most Hungarian vetch seed used in the United States is produced in western Oregon and western Washington. It is produced to a limited extent in southern Europe also. Commercial use is confined almost wholly to the Pacific Northwest, where it is grown for hay, silage, green manure, pasture, and seed.

BIOTIC FACTORS: Since vetches are insect-pollinated, bee hives near fields insure fertilization of flowers and increase seed yield. Following fungi have been reported on Hungarian vetch: *Aphanomyces eutreiches, Ascochyta pinodella, Erysiphe pisi, Fusarium culmorum, Peronospora mayorii, Septoria viciae, Uromyces fischeri-eduardi. U. heinerlianus, U. pisi;* plants are resistant to *Kabatiella nigricans* and *Mycosphaerella pinodes. Meloidogyne* sp., a nematode, has been isolated from this vetch.

CONTRIBUTORS: J. A. Duke, C. F. Reed.

Vicia sativa L.

FAMILY: Fabaceae
COMMON NAME: Common vetch

USES: Common vetch can be grown as winter crop only in areas with mild winters. In western Oregon and western Washington, it is hardy in most winters; often winter-kills in the northern part of the Cotton Belt. Grown in Pacific Coast States, for hay, seed, green manure, silage, and pasture; in southeast United States, largely for green manure.

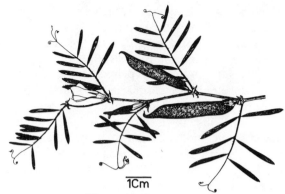

Figure 134. *Vicia sativa* L.

FOLK MEDICINE: Seeds are said to be astringent, detergent, and a remedy for diarrhea.

CHEMISTRY: Seeds of *Vicia sativa* are reported to contain (per 100 g): 343 calories, 9.5% moisture, 29.5 g protein, 0.8 g fat, 5.7 g fiber, 137 mg Ca, 240 mg P, 9.3 mg Fe, 0.15 mg thiamine, 0.10 mg riboflavin, and 2.7 mg niacin. Green forage of common vetch is reported to contain 79.5% moisture, 3.8 g protein, 0.5 g fat, 5.5 g fiber, 2.1 g ash. Hay of common vetch contains: 11.0% moisture, 13.3 g protein, 1.1 g fat, 25.2 g fiber, and 6.2 g ash. Essential amino acid contents of seeds from different countries ranged as follows (g/16 g N): 7.8–12.3 arginine, 1.3–5.3 histidine, 3.2–6.7 lysine, 0.8–1.6 tryptophan, 3.9–7.7 phenylanine, 0.9–2.6 methionine, 0.7–1.1 cystine, 3.2–5.2 threonine, 7.1–7.2 leucine, 3.8–5.7 isoleucine, and 4.2–7.9 valine. According to the Wealth of India, the seeds contain vicine and convicine, the toxic principles of *Vicia faba,* and betaine, cephalin, choline, cholesterol, citronic acid, guanine, lecithin, and two myoinositol α-D-galactosides. In addition, they are said to produce hydrocyanic acid and neurolathyrogens (β-cyano-α-alanine and γ-glutamyl-β-cyano-α-alanine).

DESCRIPTION: Straggling or ascending, robust, annual herb; stems 10–70 cm long, glabrescent to sparsely pubescent with short spreading hairs; leaves compound with 4–12 leaflets; leaflets variable in shape, linear, narrowly oblong to broadly obovate or obcordate, 0.3–4 cm long, 0.2–1.5 cm broad, acute, rounded or emarginate at apex, usually apiculate or cuspidate, mostly cuneate at base, thinly pubescent on both surfaces, or glabrescent; petiole short, up to 1 cm long; rachis of upper

Vicia sativa L.

leaves with a well-developed branched tendril; petiolules only 0.5 mm long, but distinct stipules semisaggitate, 3–8 mm long, 2–6 mm broad, with a distinct dark glandular median blotch, margins dentate; flowers solitary or paired, subsessile or short-pedicellate; calyx glabrescent or sparsely pubescent, tube 3.5–7 mm long, lobes linear-lanceolate, 3–8 mm long, often with a glandular blotch; corolla crimson, violet, bluish or white, the standard often palest; standard obovate, ovate or oblong, 0.7–2.5 cm long, 0.8–1.7 cm broad, emarginate; legumes light brown to straw, pubescent or puberulent, reticulate and depressed between the seeds, 4–8 cm long, 7–8 mm broad, terete to somewhat compressed, apex oblique, long-acuminate, almost falcate, base rounded, twisted loosely to tightly during dehiscence; seeds smooth, usually dark brown, sometimes marbled, rounded, reticulate, subglobose or slightly compressed, 4.5–6 mm long, 4.5–6 mm broad, 2.7–5 mm thick. Seeds 15,435/kg. Fl. July–Sept.; Fr. Aug.–Oct.

GERMPLASM: Many cvs of common vetch have been developed, the most commonly grown for seed in western Oregon is one of the most winter hardy, 'Willamette,' which is also well adapted to the fertile soils of southeastern United States. 'Pearl' vetch has light pink seeds and is grown occasionally in western Oregon as a spring-sown crop. 'Doark,' developed in Arkansas is slightly hardier than 'Willamette' and does well in Arkansas. 'Warrior,' developed in Alabama, is suitable grazing and green manure crop. Many botanical cvs have been described based on variations of leaves, seeds, and pubescence. Anatolian elements of ssp. *amphicarpa* are tolerant to cold, drought, and grazing. Some authors have treated *Vicia angustifolia* as a separate species, and it may be the weedy ancestral species to polymorphic *V. sativa,* with its primary center of diversity in the Near East. Within *angustifolia* there is wild ssp. *angustifolia* and the segetal subspecies, a weed in cereals, which readily crosses with *V. sativa,* and perhaps with its segetal convar *consentini.* With $2n = 14$, some ancestral type is suspected have given rise to the *angustifolia*–*sativa* complex with $2n = 12$, with a great deal of introgression, and further derivation of *V. cordata* Wulf. Assigned to the Near Eastern Centers of Diversity, common vetch or cvs thereof is reported to exhibit tolerance to frost, fungus, grazing, high pH, insects, low pH, nematodes, virus, and weeds. ($2n = 10$.)

DISTRIBUTION: Of ancient origin and now widespread as an escape from cultivation throughout the temperate regions of the World. Introduced in North America from Europe, and now cultivated and escaped along roadsides throughout the United States.

ECOLOGY: Common vetch fares well in loams and sandy loams and gravelly soils, especially with adequate lime, although excess lime is injurious. Ranging from Boreal Moist to Wet through Subtropical Thorn to Moist Forest Life Zone, common vetch is reported to tolerate annual precipitation of 3.1–16.3 dm (mean of 74 cases = 8.0), annual mean temperature of 5.6°–22.5°C (mean of 74 cases = 12.2), and pH of 4.5–8.2 (mean of 71 cases = 6.5).

YIELDS AND ECONOMICS: The most commonly grown vetch for the many uses mentioned above. Most common vetch seed used in the US is produced here in western Oregon and western Washington. Much is also produced in the more southern European countries and in the British Isles. World vetch production (excluding *Vicia faba*) was 1,-556,000 MT in 1975, averaging 997 kg/ha. The USSR was estimated to have produced 1,100,000 MT, averaging 1,048 kg/ha. Asia, excluding Russia, produced 227,000 MT (838 kg/ha), Turkey produced 150,000 MT (882 kg/ha). Europe produced 191,000 MT (1,007 kg/ha). Africa produced 38,000 MT (766 kg/ha). Spain followed Turkey, producing 60,000 MT, Syria 52,000, Yugoslovia 45,000, Morocco 36,000 MT. Yields were highest in Yugoslavia at 3,020 kg/ha, Malta at 2,558, Egypt at 2,381, Austria at 1947 Sweden at 1,635 kg/ha (FAO, 1975). Examining 24 lines, Blum and Lehrer (1973) found that DM yields ranged from 5,629 kg/ha ('26866') to 7,943 ('16642'); 'Blanchegraine' gave 6,759, 'Vedoc' 6,406. In Bulgarian trials of 7 cvs, 'Obrazet' gave the highest seed yield, and over 3 years averaged more than 2 MT/ha, outyielding the standard cv by 62.7%. Wet-matter yields were closer to 20 MT/ha.

BIOTIC FACTORS: Vetches are insect-pollinated, and bee hives near each field ensure pollination of flowers and increase seed yields. Many fungi attack this vetch: *Ascochyta fabae, A. pisi, A. viciae-pisiformis, A. vicina, Aspergillus glaucus, Botrytis cinerea, B. fabae, Cercospora viciae, Colletotrichum viciae-sativae, Erysiphe communis* f. *viciae-sativae, E. pisi, E. polygoni, Fusarium acuminatum, F. avenaceum, F. oxysporum, Kabatiella nigricans, Mycosphaerella pinodes, Ovularia sphaeroidea, Peronospora viciae, P. viciae-sativae, Phomopsis viciae, Phyllosticta fabae, Ph. viciae, Pleosphaerulina hyalospora, Pleospora americana, Rhizoctonia solani, Sclerotinia trifoliorum, Septoria viciae, Stagonospora viciae-pisiformis, Thielaviopsis basicola, Uromyces briardi, U. ervi, U. fabae, U. orobi, U. viciae-crassae,* and *U. viciae-fabae*. Plants are also attacked by the bacteria *Pseudomonas cannabina* and *Ps. viciae*, the virus curly top, and the parasitic plants *Cuscuta epithymum, C. pentagona, Orobanche barbata,* and *O. indica*.

CONTRIBUTORS: J. A. Duke, C. F. Reed, J. K. P. Weder.

Vicia sativa L. ssp. *nigra* (L.) Ehrh.

FAMILY: Fabaceae

COMMON NAMES: Blackpod vetch, Narrowleaf vetch

SYNONYMS: *Vicia abyssinica* Alef.
Vicia sativa var *abyssinica* (Alef.) Bak.
Vicia sativa var *angustifolia* L.

USES: Cultivated by few orchardists of Southern United States as a good crop which volunteers as a winter-cover and green manure crop. It volunteers in pasturelands and makes excellent pasture. Under cultivation, blackpod has seldom been successful.

FOLK MEDICINE: Seeds are said to be astringent and detergent. A seed decoction has been used for measles and smallpox.

CHEMISTRY: According to the Wealth of India, the seeds appear to play a role in human lathyrism. Ducks and monkeys, fed on the seeds, developed a syndrome causing deleterious effects on the nervous system. The seeds also contain the neurolathyrogens, β-cyano-l-alanine, and its γ-glutamyl derivative, also a cyanogenetic glucoside, vicianin, and vicine. Vicianin, known only from this genus, but not from *V. sativa*, is produced on hydrolysis. With an enzyme vicianase, yields HCN, benzaldehyde, and vicianose. Other items are arginine, pipecolic acid, and 2 unknown ninhydrin positive compounds.

Figure 135. *Vicia sativa* L. ssp. *nigra* (L.) Ehrh.

DESCRIPTION: Slender, annual herb; stems 10–50 cm tall; leaves compound, with 4–12 leaflets; leaflets 0.4–2 cm long, 1.5–6.5 mm broad, narrow, those of the lower leaves oblong and truncate, of the upper ones linear- to lance-attenuate, mucronate; flowers small, 1–1.8 cm long; standard mostly as brightly colored as the other petals, calyx 7–11 mm long; legumes slender oblong, 2.3–5 cm long,

Vicia sative L. ssp. *nigra* (L.) Ehrh.

4.5–6 mm broad, terete to plane, black to dark brown, short-pubescent to glabrescent, punctate or reticulate, twisting slightly during dehiscence; seeds 2.3–4 mm in diameter, rounded-squarish to subspherical, smooth, velvety purplish-black to straw or light reddish-brown or greenish-straw or reddish-straw, sometimes mottled, streaked or dotted with dark brown. Fl. Mar.–Oct.; Fr. May–Nov.

GERMPLASM: Several botanical varieties have been described; var *segetalis* (Thuill.) W. D. J. Koch, has leaflets of upper leaves oblong to oblong-obovate, truncate or emarginate mucronate at apex, found along roadsides, waste places and as a weed in wheat fields; var *uncinata* (Desv.) Rouy, has leaflets of upper leaves narrowly linear, truncate or abruptly narrowed, 1–2 mm broad, found locally from Newfoundland to Virginia. Blackpod vetch is assigned to the African Center of Diversity. ($2n = 12$.)

DISTRIBUTION: Native to tropical Africa (Sudan and Ethiopia to Congo, north to Mediterranean region, Sinai, Arabia), Europe and east to northwest India. Introduced to the United States from Africa and Europe, now occurring in the United States mostly as a weed. Abundant in grainfields of Spring Wheat Belt, and along roadsides and in waste places in Cotton Belt.

ECOLOGY: Grasslands, scrub, in dense herbage, bamboo, and other upland forests; ascending to 1,600–3,360 m in east tropical Africa. Ranging from Cool Temperate Steppe to Wet through Warm Temperate Dry to Moist Forest Life Zone, blackpod vetch is reported to tolerate annual precipitation of 3.5–11.6 dm (mean of 23 cases = 7.2), annual mean temperature of 6.5°–22.5°C (mean of 23 cases = 10.5), and pH of 4.9–8.2 (mean of 22 cases = 6.6).

YIELDS AND ECONOMICS: Sometimes in orchards in the South as a winter cover crop and a green manure crop. Also common in river-bottomlands of the South where Johnsongrass is established. There this vetch often occurs in abundance and makes up a good part of the hay from such areas. Seed sometimes saved from screenings of spring wheat crop in the Northwest.

BIOTIC FACTORS: Following fungi have been reported on this vetch: *Ascochyta fabae, A. viciae, A. pisi, Botrytis cinerea, Colletotrichum viciae, C. villosum, Erysiphe communis, Kabatiella nigricans, Leveillula taurica, Mycosphaerella pinodes, Peronospora mayorii, P. viciae, P. viciae-sativae, Phymatotrichum omnivorum, Trichocladia baumleri, Uromyces briardi, U. fabae,* and *U. viciae-fabae*. Nematodes isolated from this plant include the following species: *Heterodera goettingiana, Meloidogyne hapla, M. arenaria, M. incognita acrita,* and *M. javanica*.

CONTRIBUTORS: J. A. Duke, C. F. Reed.

Vicia villosa Roth

FAMILY: Fabaceae
COMMON NAMES: Winter vetch, Hairy vetch, Sand vetch, Woolypod vetch
SYNONYM: *Vicia dasycarpa* Ten.

USES: One of the oldest and most common vetches. Extensively cultivated for forage. Also makes good hay, silage, pasture, green manure, and cover crop. Because of its well known tolerance to adverse soil and climate, this species is used to replace fallow with forage in Spain. Winter vetch is the most hardy of the commercial vetches and is the only one recommended for fall planting in the North. Many times, ruderal races become serious weeds in cereals.

FOLK MEDICINE: No data available.

CHEMISTRY: Immature winter vetch hay (86.0–91.6% DM) contained (on a moisture-free basis): 20.9–25.0% crude protein (avg. of 24 cases = 22.6), 2.5–4.2% fat (avg. 3.2), 26.8–31.6% crude fiber (avg. 28.7), 8.7–11.2% ash (avg. 9.9), and 31.5–37.6% N-free extract (avg. 35.6). Based on 17 cases, the green forage (16.7–24.8% DM) contained: 15.6–27.3% crude protein (avg. 23.1), 2.1–3.8% fat (avg. 2.7), 22.1–30.6% crude fiber (avg. 27.5), 8.2–13.7% ash (avg. 12.1), and 32.7–41.4% N-free extract (avg. 34.6). Ca content averaged 1.1%, P 0.33% and K 2.25% on a moisture-free basis. The seed protein consists of 2 globulins as major fractions, glutelin and albumin as minor ones, and 0.3% lecithin and 0.08% cephalin, also γ-glutamyl-β-cyanoalanine, canavan-

Vicia villosa Roth

Figure 136. *Vicia villosa* Roth

ine, asparagine and a trace of arginine. The seeds are considered less toxic than those of common vetch, but are said to have caused poisoning in cattle, with extensive pain, bellowing, sexual excitement, and convulsions. Xanthophyll ranged from 1,813 to 2,477 mg/kg DM, carotene from 628 to 1,066 mg/kg DM.

DESCRIPTION: Straggling, climbing, prostrate or trailing annual, biennial or perennial herb; stems spreading-villous to glabrous, 0.3–2 m tall; leaves compound, with 8–24 leaflets; leaflets linear to elliptic or lanceolate, 0.3–3.5 cm long, 1–8 mm broad, pilose to glabrous, rachis terminated by a tendril; stipules semi-sagittate or falcate, entire; inflorescence 2–20-flowered; calyx-tube gibbous at base, the teeth unequal, glabrescent, the lower ones shorter than or exceeding the tube; corolla 1–2 cm long, violet, purple or blue, sometimes with white or yellow wings; legumes light to dark straw colored, narrowly oblong, obliquely short acuminate at both ends, almost rounded, 2–4 cm long, 4–10 mm broad, flattened, glabrous or appressed pubescent, twisting tightly to loosely during dehiscence; seed 2–8 per pod, spherical to nearly sublenticular, smooth, dull dark reddish brown to dull greenish brown, densely mottled or dotted with blackish brown, 3.5–4.9 mm long, 3.4–4.9 mm broad, 3.2–3.8 mm thick; hilum extending about one-seventh of the circumference of the seed. Seeds 22,000–44,000/kg; wt. kg/hl = 77. May–Oct.

GERMPLASM: Many cvs have been developed for particular areas. 'Madison Vetch' is a strain developed in Nebraska and well suited to conditions there. Although the seeds of winter vetch and those of *V. dasycarpa* are distinct, taxonomists have combined these 2 species on the basis of plant characteristics. "Dasycarpa" cvs include 'Auburn' (Ala.), 'Oregon' (Cal.), and 'Lana' (Ore.), which make good growth in late winter and early spring. Hence they are recommended as winter forage in southeastern United States. Much of what was called *V. dasycarpa* in the United States has been reidentified as *V. villosa*. Seed were originally produced in Michigan where cold-tolerant hairy forms prevailed. Then seed production was shifted to Oregon where a heat-tolerant glabrous ecotype soon prevailed and became the 'Winter' vetch for the southeastern United States. Ironically, the original 'Hairy' vetch was hairy but not tolerant of the northern winter (except as seed), while the derived 'Hairy' vetch, renamed 'Winter' vetch, was glabrous and tolerant of the southern winter. Assigned to the Near Eastern Center of Diversity, winter vetch is said to tolerate cold, drought, frost, high pH, poor soil, and virus. ($2n = 14$.)

DISTRIBUTION: Native to Europe and Asia, with the primary center of diversity probably in Western and Ante-Asia. Now widely cultivated in all countries in the Temperate Zone, mainly as a forage crop. Often escaped along roadsides and in waste places throughout the United States and Canada. In tropical eastern Africa it is found in grasslands and as a weed in maize, growing at 1,800–2,040 m altitude.

ECOLOGY: Cultivated and naturalized and escaped in fields, wastelands, and grasslands throughout temperate regions of the world, but faring best on sandy or sandy loam soils. Ranging from Boreal Moist to Wet through Subtropical Moist Forest Life

Vicia villosa Roth

Zone, winter vetch is reported to tolerate annual precipitation of 3.1–16.6 dm (mean of 42 cases = 8.1), annual mean temperature of 4.3°–21.1°C (mean of 42 cases = 12.0), and pH of 4.9–8.2 (mean of 40 cases = 6.6).

YIELDS AND ECONOMICS: Hairy and winter vetches yield 15–18 bu/ha, with 30–38 bu/ha as maximum. In irrigated trials in Caspian Russia a winter rye/winter vetch catchcrop yielded 24,640 kg WM/ha, compared with 6,290 for pure stands of vetch. The catchcrops improved the yields of subsequently planted rice by 1,380 kg mixed and 580 kg pure vetch, respectively. Mixed with wheat, it yielded 19,470–21,860 kg fresh forage/ha, while wheat alone gave only 15,450. The mixture yielded about 750 kg crude protein, pure wheat forage only 340 kg crude protein/ha. In Voronezh province, cvs 'Kurskaya' and 'Voronezhskaya,' the most cold-resistant cvs, yielded 25–26 MT fresh forage/ha, 6,000–7,600 kg hay. In Oklahoma, vetch/bermuda grass mixtures yielded 7 MT DM/ha. Most seed of winter and hairy vetch used in the United States are produced in Oregon, Oklahoma, Texas, Arkansas, and western Washington. It is also produced in quantity in European countries bordering the Baltic Sea and Hungary, and is imported into the United States from that region.

BIOTIC FACTORS: Since vetches are insect-pollinated, bee hives nearby ensure pollination of flowers and increase seed yields. Following fungi reported on this vetch include: *Ascochyta pisi, A. viciae, A. viciaevillosae, Aphanomyces eutreiches, Chaetomium globosum, Ch. funicola, Colletotrichum viciae, C. villosum, Corticium vagum, Erysiphe pisi, Fusarium oxysporum* f. *medicaginis, Gloeosporium americanum, Kabatiella nigricans, Mycosphaerella pinodes, Ovularia fallax, O. schwarziana, O. sphaeroidea, Phyllosticta fabae, Protocoronospora nigricans, Pseudopeziza medicaginis,* and *Septoria pisi*. Plants are also attacked by the virus, curly top. The pea seedborne mosaic virus can overwinter in seedlings of winter vetch. Nematodes isolated include: *Belonolaimus gracilis, B. longicaudatus, Criconemoides curvatus, C. lobatus, C. morgense, C. mutabile, C. xenoplax, Ditylenchus dipsaci, Heterodera glycines, H. goettingiana, H. schachtii, H. trifolii, Hoplolaimus galeatus, Meloidogyne arenaria, M. hapla, M. incognita acrita, M. javanica, Paratylenchus curvitatus, P. projectus, Pratylenchus crenatus, P. penetrans, P. pratensis, P. projectus, P. vulnus, Rotylenchulus reniformis, Rotylenchus robustus, Scutellonema brachyurum, Trichodorus christiei, Tylenchorhynchus claytoni,* and *Xiphinema americanum*.

CONTRIBUTORS: J. A. Duke, C. F. Reed.

Vigna aconitifolia (Jacq.) Marechal

FAMILY: Fabaceae
COMMON NAMES: Moth bean, Mat bean
SYNONYM: *Phaseolus aconitifolius* Jacq.

USES: In India, other parts of Asia, and Africa, moth bean is grown for human food. Green pods and ripe seeds, whole or split are cooked as vegetables. In India and southwestern United States plants are used for erosion control, green manure, pasture, or forage crops. Seeds may be processed for starch.

Figure 137. *Vigna aconitifolia* (Jacq.) Marechal

FOLK MEDICINE: Seed used medicinally in diets for fevers; roots are said to be narcotic.

CHEMISTRY: Beans contain 23% protein, 0.7% fat, and 43.5% carbohydrate. Whole dried seeds contain: 10.8% water, 23.6% protein, 1.1% fat, 56.5% carbohydrate, 4.5% crude fiber, and 3.5% ash, 0.22–0.32% Ca, 0.1–0.7% P, 1.0% K, 0.84% Mg, and 0.000% Fe. The main amino acids are lysine, leucine, isoleucine, phenylalanine, and tyrosine. Green forage is reported to contain 75.0% moisture, 3.0% crude protein, 0.4% fat, 7.7% crude fiber, 3.9% ash, 10.0% N-free extract, 0.9% Ca, 0.16% P, and 0.95% K. The approximate composition of the hay is: 10.0% moisture, 16.2% crude protein, 2.4% fat, 16.0% crude fiber, 14.0% ash, and 41.4% N-free extract. Miller (1958) reports 13.8% moisture, 17.2% crude protein, 1.7% fat, 29.4% crude fiber, 12.0% ash, and 39.7% N-free extract in hay. Seeds are reported to contain trypsin inhibitors and chymotrypsin inhibitors. The genus *Vigna* is reported to contain oxalic acid.

DESCRIPTION: Annual hairy herb, forming a short erect main stem 10–40 cm tall, producing up to 12 primary trailing branches 60–130 cm long, about 25 secondary branches 60 cm long, and finally 20 tertiary branches 30 cm long; leaves alternate, trifoliolate, with grooved petioles 5–10 cm long; stipules peltate, 1–2 cm long, with lanceolate lobes; leaflets 5–12 cm long with terminal leaflet slightly larger, with 5 or fewer acuminate narrow lobes, and the lateral leaflets with 5 or less lobes; inflorescence of axillary capitate racemes, on peduncles 5–10 cm long; flowers several, yellow, about 9 mm long; pods small, 2.5–5 cm long, 0.5 cm in diameter, nearly cylindrical, with brown short stiff hairs, short curved beak, 4–9-seeded; seeds light brown, whitish-green, or yellow-brown, rarely mottled, cylindrical to rectangular, distinctly elongate, subtruncate at both ends, 3–5 mm long, 2–2.5 mm wide; hilum 1.2–2.5 mm long, 0.4 mm wide, white, laterally compressed, narrowly obovate with base at micropyle, margin slightly concave, excentrically located; micropyle at base of hilum and elevated by the bulky radicle, pore not evident; weight 1 g/100 seed, germination phanerocotylar.

GERMPLASM: According to the currently prevailing concepts the genus (1) *Phaseolus sensu stricto* is represented by *Phaseolus vulgaris* and its allied species; (2) *Vigna* subgenus *Ceratotropis* (Piper) Verdc. includes all the yellow-flowered Asiatic species formerly under the genus *Phaseolus*, i.e., *P. aureus, P. mungo, P. angularis, P. aconitifolius, P. trilobus, P. calcaratus*, etc.; and (3) the genus *Macroptilium* (Benth.) Urban includes the deep purple flowered species formerly belonging to *Phaseolus*, e.g., *P. lathyroides, P. atropurpureus* and other allied species. S. Dana (personal communication, 1977) presents evidence from species crosses which do not support the above-mentioned revision. He reasons as follows: (1) He finds not a single report of a successful cross of any species included in presently reorganized *Vigna* subgenus *Ceratotropis* (Piper) Verdc. with any species under any other subgenus of *Vigna*. Within the subgenus crosses between *V. aconitifolia* and *V. radiata* have been obtained. Seedlings of the offspring plants were infertile; (2) Partially fertile hybrids have been reported from crosses of *Phaseolus vulgaris* with *P. mungo* and *P. angularis;* (3) A partially fertile hybrid has been reported from *Phaseolus aureus* × *P. lathyroides* cross. This hybrid had 23.8% pollen fertility. Assigned to the Hindustani and Indochina–Indonesian Centers of Diversity, moth bean or cvs thereof is reported to exhibit tolerance to drought, high pH, heat, and slope. ($2n = 22$.)

DISTRIBUTION: Moth bean is wild and cultivated from Ceylon to the Himalayas and Burma, in tropical regions up to 1,300 m in northwestern India. In cultivated forms it has spread to China, Africa, and southern United States.

ECOLOGY: Grown from sea level to 1,300 m, moth bean is a hot-weather crop, requiring uniformly high temperatures. Frost-sensitive, moth bean does best with soil temperatures ca. 27°C. It is very drought-resistant and does well with a well-distributed annual rainfall of 5–7.5 dm; heavy rainfall is harmful. Grows on many soil types, particularly suitable on dry light sandy soils, it is considered extremely hardy. Although reported as short-day, 19 lines proved to be day-neutral in Hawaii (Hartman, 1969). Ranging from Warm Temperate Moist to Thorn through Tropical Very Dry Forest Life Zone, moth bean is reported to tolerate annual precipitation of 3.8–17.3 dm (mean of 5 cases = 8.7), annual mean temperature of 21.1°–27.5°C (mean of 5 cases = 24.5), and pH of 5.0–8.1 (mean of 4 cases = 7.1).

Vigna aconitifolia (Jacq.) Marechal

CULTIVATION: Moth bean cultivated from seed, sown on a well prepared seedbed, at a seed rate of 3.4–4.5 kg/ha, in rows 75–90 cm apart for pure stands, at depth of 2.5–4.0 cm. When grown for forage in the United States, seed rate may be 7–34 kg/ha. In India seed is sown in October, 15 kg/ha, in rows 25 cm apart, or in southern India, in June or July with the onset of the monsoon. Germination is rapid at 27°C. In many areas moth bean is interplanted with cereals, or rotated as a green manure crop with cotton. Cultivation is the same as for other row crops for the first few weeks; after stems begin to lengthen, cultivation is stopped.

HARVESTING: Seeds mature ca. 90 days after planting. Decumbent habit makes crop difficult to harvest with a mower. Plants are usually cut with a sickle, dried for ca. 1 week, then threshed and winnowed.

YIELDS AND ECONOMICS: When planted in 90 cm rows for hay production, average hay yields are ca. 3,000 kg/ha. Grown for forage, yields 37–50 MT WM/ha, 7.5–10 MT DM. In India seed yields are 200–1,500 kg/ha, depending on climate and culture. In the United States yields are 34–45 MT WM/ha; 7–9 MT hay/ha; and 1,240–1,800 kg seed/ha. It is considered a first-class plant for food, forage, and green manure. It is grown to a limited extent as a pasture, forage, and green manure in southwestern United States, mainly in Arizona, Texas, and California. Although India is the leading producer of mothbeans (nearly 500,000 MT in 1970–1971), Thailand is the leading exporter (more than 40,000 MT/yr).

BIOTIC FACTORS: Flowers are self-fertile normally. Fungi causing diseases in moth bean include: *Ascochyta phaseolorum, Cercospora cruenta* (leaf spot), *Colletotrichum lindemuthianum* (anthracnose), *Gloeosporium phaseoli, Phyllosticta phaseolina,* and *Sphaerotheca humuli* (mildew). The yellow mosaic of mung virus attacks the plant. Also moth bean is parasitized by *Striga euphrasiodes* and *S. lutea* in India and Africa. Nematodes infesting moth bean: *Heterodera glycines* and *Meloidogyne* spp.

REFERENCES:

Hartman, R. W., 1969, Photoperiod responses of Phaseolus plant introductions in Hawaii, *J. Am. Soc. Hortic. Sci.* **94**:437–440.

Kennedy, P. B., and Madson, B. A., 1925, The mat bean, *Phaseolus aconitifolius, Calif. Agr. Exp. Stn., Bull.* **396**: 33 pp.

Piper, C. V., and Morse, W. J., 1914, Five oriental species of beans, *U.S. Dep. Agric., Bull.* **119**:28–30, illus.

CONTRIBUTORS: S. Dana, J. A. Duke, C. F. Reed, J. K. P. Weder.

Vigna angularis (Willd.) Ohwi & Ohashi

FAMILY: Fabaceae

COMMON NAMES: Adzuki bean, Adsuki bean

SYNONYMS: *Azukia angularis* (Willd.) Ohwi
Dolichos angularis Willd.
Phaseolus angularis (Willd.) Wight

USES: Adzuki beans are grown mainly in the Orient, dried and used for human food, either cooked whole or made into a meal used in soups, cakes, or confections. Beans may be popped like corn, used

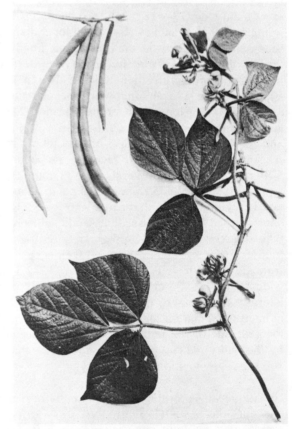

Figure 138. *Vigna angularis* (Willd.) Ohwi & Ohashi

as a coffee substitute, or candied by boiling in sugar (amanatto). Flour is also used for shampoos and to make facial creams. Adzuki bean meal is prepared by grinding dry beans and then removing seed coats with sieves. A more common wet process consists of 4 stages: (1) boiling beans until soft, usually after a presoaking; (2) crushing the cooked beans; (3) removing seed coats by forcing mass through sieves, or by putting bean paste in cold water, where seed coats are easily separated; and (4) drying bean paste. In Japan, fresh, undried bean paste is called *an,* and dried product *sarashian.* Sometimes seed coats are removed from the soaked and parboiled beans before they are crushed. On boiling, the red color of seed coats dissolves, necessitating a couple of changes of water, but the final bean meal is reddish. Crop grown for forage or green manure. Sprouted beans used as a vegetable.

FOLK MEDICINE: In traditional Chinese view, everything is composed of 2 complementary and antagonistic qualities, yin and yang. Food is considered to have a bipolar nature, and any dietary item may be placed on the yin–yang continuum. Imbalance of the primeval forces yin and yang in the body is considered the cause of disease, which may be treated by administration of food or medicine containing a yin–yang balance appropriate to the ailment. Thus if one is suffering from a yin sickness (excessive yin), a yang remedy is indicated to restore harmony. Foods that are strongly yin and yang are used medicinally. The adsuki is considered the most yang of beans. Thus it is used to treat yin sickness such as "kidney trouble, constipation, boils, abscesses, certain tumors, threatened miscarriage, difficult labor, retained placenta, and nonsecretion of milk." The leaves (yin) are said to lower fever and the sprouts treat threatened abortion caused by injury (Sacks, 1977).

CHEMISTRY: Dried seeds contain (per 100 g): 336 calories, 10.8% moisture, 19.9 g protein, 0.6 g fat, 64.4 g total carbohydrate, 7.8 g fiber, 4.3 g ash, 136 mg Ca, 260 mg P, 9.8 mg Fe, 0.06 mg thiamine, 0.09 mg riboflavin, 2.0 mg niacin, and 2 mg ascorbic acid. Another analysis (Kay, 1978) gives, per 100 g edible portion: 13.0% moisture, 25.3 g protein, 0.6 g fat, 57.1 total carbohydrate, 5.7 g fiber, 3.9 g ash, 253 mg Ca, 7.6 mg Fe, 15 IU vitamin A, 0.57 mg thiamine, 0.18 mg riboflavin, and 3.2 mg niacin.

The amino acid composition (mg/g N) is: 280 isoleucine, 490 leucine, 440 lysine, 340 phenylalanine, 210 tyrosine, 110 methionine, 70 cystine, 240 threonine, and 340 valine. Seeds boiled and sweetened contain: 218 calories, 45.3% moisture, 3.0 g protein, 0.1 g fat, 50.7 g total carbohydrate, 0.6 g fiber, 0.9 g ash, 26 mg Ca, 66 mg P, 3.9 mg Fe, 0.05 mg thiamine, 0.01 mg riboflavin, and 0.1 mg niacin/100 g. Immature raw seeds contain (per 100 g): 135 calories, 65.0% moisture, 9.6 g protein, 0.1 g fat, 24.9 g total carbohydrate, 0.6 g fiber, 0.4 g ash, 24 mg Ca, 80 mg P, 2.1 mg Fe, 0.01 mg thiamine, 0.01 mg riboflavin, and 0.2 mg niacin. Beans contain pipecolic acid. Seeds are reported to contain trypsin inhibitors and chymotrypsin inhibitors.

DESCRIPTION: Summer annual herb, usually bushy and erect, 27–90 cm tall, some later-maturing cvs slightly vining and some prostrate; leaves trifoliolate, on long petioles; leaflets ovate, usually entire, 5–9 cm long; inflorescence axillary, short, of 6–12 clustered flowers; flowers bright yellow, on short pedicels; pods cylindrical, 6–12.5 cm long, 0.5 cm in diameter, straw-colored, with blackish or brown forms, 4–12-seeded, somewhat constricted between the seeds; seeds cylindrical to cordate, often rounded-quadrilateral to subtrapeziform or rectangular, rounded at both ends; 5–7.5 mm long, 4–5.5 mm wide; hilum 2.4–3.3 mm long, 0.6–0.8 mm wide, slightly elongate to narrowly obovate at base of micropyle, excentrically located; seed coat smooth, wine red, occasionally buff, black or mottled (black on red, or black on beige); micropyle evident, pore countersunk; average weight 0.07 g/seed, 100 seeds weigh 10–20 g, depending on cv. Germination phanerocotylar. Roots possess root tubercles in abundance and fix nitrogen very efficiently.

GERMPLASM: More than 60 varieties and cvs have been recorded, differing in time of maturity, color of seed (straw, yellow, brown, green, blue-black, red, black, gray, maroon, mottled or speckled), and plant habit. Botanical variety: *Vigna angularis* var *nipponensis* (Ohwi) Ohwi & Ohashi (synonym: *Phaseolus nipponensis* Ohwi; *Ph. angularis* var *nipponensis* (Ohwi) Ohwi; *Azukia angularis* var *nipponensis* (Ohwi) Ohwi). Introduced cvs in the United States have come from north China, Manchuria, Siberia, Japan, Korea, and Hawaii. Early cvs were strictly bushy and mostly erect, while

Vigna angularis (Willd.) Ohwi & Ohashi

later cvs are slightly viny at the tips of stems and branches; some are decumbent. Later cvs have larger seeds than earlier ones. Unknown in the wild, adzuki bean is assigned to the China–Japanese and Indochina–Indonesian Centers of Diversity. China, Japan, and India have all been cited as centers of origin. Adzuki bean or cvs thereof is reported to exhibit tolerance to drought, frost, heat, slope, and virus. ($2n = 22$.)

DISTRIBUTION: Probably native to India or Japan; long established in China and Sarawak; cultivated for centuries in Japan, Korea, China, and Manchuria for human food. Has been introduced to Hawaii, southern United States, South America, Angola, India, Kenya, New Zealand, and Zaire.

ECOLOGY: Adzuki beans need about the same climatic conditions as soybeans. Although first classified as short-day, day-neutral responses are reported in Korean and Japanese lines grown in Hawaii. As in soybeans, there seems to be longitudinal adaptation in day-length sensitivity. In Japan, early maturing southern cvs are more sensitive to short-days than northern cvs. In Zaire, crop tolerates high temperatures, is fairly drought resistant, and does not tolerate waterlogging, frost, or cool conditions. Soil temperatures of above 16°C are required for germination, 15°–30°C for good growth. On volcanic soils, the bean is sensitive to B and Ni toxicities. Ranging from Warm Temperate Dry to Moist through Subtropical Dry to Moist Forest Life Zones, adzuki bean is reported to tolerate annual precipitation of 5.3–17.3 dm (mean of 8 cases = 12.5), annual mean temperature of 7.8°–27.8°C (mean of 8 cases = 17.1), and pH of 5.0–7.5 (mean of 8 cases = 6.1).

CULTIVATION: Seeds retain viability for over 2 yr. Seed should be planted 2.5 cm deep in rows 60–90 cm apart, when danger of frost is past and soil has become thoroughly warm. Plants are spaced 30 cm apart in the rows, but a closer spacing of 30 cm × 30 cm is used in Zaire. Seeding rates vary from 8 to 30 kg/ha. Cultivation is the same as for any other row crop. In tropics seeds may be planted wherever soil conditions are proper for germination. In Asia, seed may be sown directly in rice stubble at a high rate, reducing the weed problem. In Japan, adzuki beans are rotated with rice grown in rows. In India, applications of 300 kg/ha superphosphate, 100 kg/ha potassium sulfate, and 100 kg/ha ammonium sulfate are recommended on moderately fertile soils, followed by an additional application of ammonium sulfate. In New Zealand ca. 25 kg/ha N is added just before flowering.

HARVESTING: Adzuki beans are prolific seed producers; seed ripen evenly in 60–190 days. Pods do not shatter readily; crop is harvested with mower or bean harvester. Some pods are very thin, and some seeds germinate in the pods, especially during long-continued wet weather. Leaves persist until pods are fully mature. For hay, adzuki beans should be cut when pods are about half mature. When grown for seeds alone, cutting may be delayed until all pods are mature. In threshing, the ordinary grain separator is satisfactory when run about the same as for cowpeas, with blank concaves.

YIELDS AND ECONOMICS: Seed yields are 90–2,500 kg/ha; in Kenya, 500–600 kg/ha; in India, ca. 2,000–2,500 kg/ha; in the United States, ca. 100–700 kg/ha; in New Zealand, 1,350–2,250; in Taiwan, 1,450; in Japan, 1,900 kg/ha. In the Orient, adzuki bean is an important pulse food, grown mainly in China and Japan. In Japan, it is second only to soybeans, and commands higher prices than any other bean in Japan. Maroon-colored seed cvs are most used in Japan. Adzuki beans and adzuki bean flour are both important items of trade on Oriental markets. The Chinas, the Koreas, Colombia, and Thailand are the chief exporters, Japan being the only major importer.

BIOTIC FACTORS: Flowers are self-fertile when bagged, but cross-pollination between cvs is frequent, and form natural hybrids readily. Following fungi cause diseases in adzuki beans: *Aristastoma oeconomicum*, *Ascochyta phaseolorum* (leaf spot), *Cercospora canescens*, *C. cruenta*, *Clathrococcum nipponicum* (leaf blotch), *Colletotrichum phaseolorum*, *Corticium rolfsii*, *Fusarium* spp. *Macrophomina phaseoli*, *Macrosporium adzukiae*, *Phyllosticta phaseolina* (leaf spot), *Sclerotium bataticola*, *S. rolfsii*, *Sclerotinia sclerotiorum*, *Sphaerotheca fuliginea*, *Uromyces adzukicola* (rust), and *U. phaseoli*. The bacterium, *Xanthomonas phaseoli*, also infects adzuki bean. The viruses known to attack this bean include: adzuki mosaic virus, asparagus bean mosaic, and cucumber mosaic. Nematodes reported attacking adzuki beans include:

Heterodera glycines, H. schachtii, Meloidogyne sp., and *Pratylenchus penetrans*. In Japan, the butterbean borer *Ostrinia varialis* causes damage in Hokkaido. The adzuki podworm, *Matsumuraeses phaseoli;* the cutworm, *Spodoptera litura;* the corn seed maggot, *Hylema platura;* the chafer, *Anomala rufocuprea;* and the aphid, *Aphis craccivora,* are other insect pests in Japan; the caterpillar of the small looper moth, *Plusia,* may be troublesome in New Zealand. In New Zealand, preemergence application of dinoseb-amine and monolinuron has controlled weeds.

REFERENCE:
Sacks, F. M., 1977, A literature review of Phaseolus angularis—The adzuki bean, *Econ. Bot.* 31(Jan.–Mar.):9–15.

CONTRIBUTORS: J. A. Duke, C. F. Reed, J. K. P. Weder.

Vigna mungo (L.) Hepper

FAMILY: Fabaceae

COMMON NAMES: Urd, Black gram

SYNONYMS: *Azukia mungo* (L.) Masamune
Phaseolus mungo L.

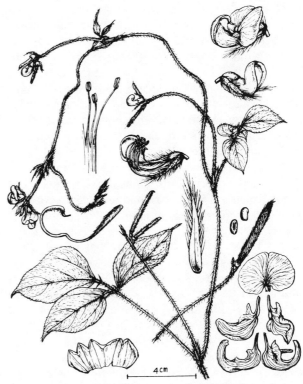

Figure 139. *Vigna mungo* (L.) Hepper

USES: Black gram is perhaps the most important pulse in India, where beans are eaten whole or split, boiled, or roasted, ground into flour and used to make cakes, breads, and porridge. Elsewhere green pods are eaten as vegetables. Plants are grown as green manure and cover crops, and for short-lived forage. Pulse and foliage are used as supplements in cattle feeds. The vines or haulms left after harvesting are used for animal feed. Seeds considered fattening; flour from seeds said to be excellent substitute for soap, leaving the skin soft and smooth.

FOLK MEDICINE: Medicinally, seeds considered cooling and astringent, and used as diet in fevers and for strengthening the eyes. Seeds are poulticed onto abscesses as a suppurative. In India said to serve as a snakebite remedy, and used for tumors.

CHEMISTRY: Dried seeds contain (per 100 g): ca. 344 calories, 10% water, 22–24% protein, 1–2% fat, 56 to 60% carbohydrate, and 0.9% crude fiber, 3.2% ash, 110 mg Ca, 382 mg P, 3.9 mg Fe, 45 g β-carotene equivalent, 0.48 mg thiamine, 0.21 mg riboflavin, and 2.3 mg niacin. Aspartic acid represents 12.7% of the total protein content, threonine 3.5, serine 5.1, glutamic acid 18.6, proline 4.3, glycine 4.4, alanine 4.5, valine 5.8, methionine 1.6, isoleucine 4.8, leucine 8.7, tyrosine 2.9, phenylalanine 6.0, lysine 7.2, histidine 3.0, and arginine 6.9. Oil composition of a Pakistan seed with 2.1% oil has: 14.1% palmitic acid, 4.3 stearic, 9.3 behenic, 3.8 linoceric, 20.8 oleic, 16.3 linoleic, and 35.7% linolenic. Trypsin inhibitors and hemagglutinins have been reported from the seed. Starch made from black gram contains (per 100 g): 370 calories, 12.2% moisture, 2.4 g protein, 3.8 g fat, 81.1 g total carbohydrate, 0.5 g ash, 50 mg Ca, 100 mg P, and 1.0 mg Fe. The haulms contains on a dry weight basis: 8.9% crude protein, 28.6% crude fiber, 12.6% ash, 2.8% fat, 47.1% N-free extract, 1.7% Ca, 0.2% P.

DESCRIPTION: Erect or decumbent, diffusely branched, herbaceous annual, 20–90 cm tall, up to 75 cm broad, sometimes trailing, with reddish-brown hairs; leaves trifoliolate, ovate, acuminate, 5–10 cm long, on long petioles; flowers yellow, 5–6

Vigna mungo (L.) Hepper

in elongated axillary racemes; pods erect, 4–7 cm long, 0.6 cm wide, very hairy, 4–10-seeded, brown when mature, with short hooked beak; seeds oblong, cylindrical to ovoid-oblong, slightly truncate at ends, 4.0–5.2 mm long, 3.5–4.1 mm wide; hilum 1.2–2.3 mm long, 0.7–1 mm wide, raised, compressed, narrowly obovate with base at micropyle, excentrically located, margin concave, elevated at micropylar end; micropyle obscured by caruncle and slanted toward hilum; seed coat usually black and mottled, but occasionally gray, dull brown, or dull green, covered with a mealy substance; 100 seeds weigh 1.5–4 g. Germination phanerocotylar. Strong taproot with many laterals. Fl. summer.

GERMPLASM: Many cvs are recognized in India; early- or late-maturing, small or large-seeded, with black or olive-green seed coat. 'T-9' (75–80 days) and 'Pusa 1' (85 days) are early-maturing cvs. 'Pusa-1,' 'H-10,' and 'G-1' are early-maturing cvs with considerable virus tolerance. Closely related *Vigna radiata* and *V. mungo* were both probably derived from ancestral *Vigna radiata* var *sublobata* (Roxb.) Verdc., (*Phaseolus sublobatus* Roxb., *P. trinervis* Wight & Arn.), which grows wild in India and Indonesia. Some taxonomists suggest that *Vigna radiata* and *Vigna mungo* are scarcely more than variants of 1 species. *Vigna mungo* is rich in γ-glutamylmethionine, white *V. radiata* is rich in γ-glutamylmethyl cysteine and its sulphoxide. Hybrids between the two contain both dipeptides. *Vigna mungo* has a yellow keel and narrow stipules, *V. radiata* a purple keel with largely oval, obtuse stipules. *V. mungo* has the pods nearly erect, with long hairs, and larger oblong seeds with a concave hilum; *V. radiata* has spreading or reflexed pods with short hairs, and globose seed with flat hilum. But unidirectional crossability, well manifested hybrid inviability and weakness and segregational hybrid sterility in crosses between these two species have been noted (Dana, 1966). The isolating barriers between these two species are well developed. Assigned to the Hindustani Center of Diversity, black gram or cvs thereof is reported to exhibit tolerance to drought, heat, high pH, heavy soil, low pH, slope, and virus. ($2n = 22, 24$.)

DISTRIBUTION: Urd is of ancient cultivation in India; not known in wild state. Although introduced in southern United States, the West Indies and other tropical areas, the crop is extensively cultivated only in India. In some countries it is grown as green manure.

ECOLOGY: Grows in India from sea level to 1,800 m in altitude. A crop of dry areas where rainfall is less than 9 dm annually. In areas of heavy rainfall it is planted after the rains cease. Quite drought resistant but intolerant of frost and prolonged cloudiness. Rain at flowering time is harmful. Crop is suited to clay soils, black cotton soils, or red light loams. Crops have been successfully grown on paddy soils just before or after the paddy crop. Ranging from Warm Temperate Thorn to Moist through Tropical Very Dry to Moist Forest Life Zone, black gram is reported to tolerate annual precipitation of 5.3–24.3 dm (mean of 13 cases = 12.4), annual mean temperature of 7.8°–27.8°C (mean of 13 cases = 22.0), and pH of 4.5–7.5 (mean of 12 cases = 6.4).

CULTIVATION: In India and Ceylon urd is grown both as a "summer" and "winter" crop, planted in mid-November and in mid-April; in Kenya and Nyasaland, planted after the rains start; in United States, planted in May–June. Land is prepared with one plowing then worked with harrow or cultivator. Seed may be broadcast and plowed in, or drilled 10–20 cm apart in rows 25–30 cm apart. Seed rates vary: in India, 13–22 kg/ha, in 25 cm rows; in Africa, 10–38 kg/ha, in rows 45 cm apart, and 15 cm apart in the rows; in eastern Africa, 11–17 kg/ha in United States, 58 kg/ha in rows 90 cm apart. Germinates within a week. Two cultivations may be necessary, first 20 days after planting, second 14 days later. Too fine a tilth may result in vegetative growth at the expense of seed. Little fertilizer required; usually a basal application of 125 kg/ha superphosphate and 30 kg/ha muriate of potash, or a mixed fertilizer of 16–8–8 at rate of 125 kg/ha initially at planting and later at flowering gives good results. Wood ash and dung are often applied in India. Gypsum applications may solve chlorosis problems. Crop may be grown in rotation with other crops, or intercropped with corn, cotton or sorghum. It may be grown as a midseason crop or a catch crop. Defective pod-setting has been corrected by floral applications of gibberellic acid, 1-naphthaleneacetic acid, 2-naphthooxyacetic acid, and *p*-chlorophenoxyacetic acid.

HARVESTING: Plants flower about 7 weeks after planting, and pods are ready for harvesting in about 75–140 days. Pods are harvested when mature but not open. Pods shatter easily, but can be combined. Pods are borne at top of foliage. Plants should be pulled early in the morning when wet to reduce shattering. In many areas plants are pulled by hand, with roots attached, stacked on threshing floors to dry about 1 week, and then threshed by flailing or trampled by oxen. Seeds are dried and cleaned; blemished, shrivelled, and wrinkled seed are removed by hand. In temperate regions, seed stores well for up to 3 yr; in tropics, best stored in airtight containers.

YIELDS AND ECONOMICS: Yields of dried seed average 340–560 kg/ha but can reach 1,500 kg/ha ('Pusa-1'); yields of dry forage usually weigh about 3 times as much. Greatest production is in India, where it is grown for local consumption.

BIOTIC FACTORS: Urd is self-pollinating, with up to 42% cleistogamy recorded. Following fungi are recorded as attacking urd: *Alternaria tenuis, Ascochyta phaseolorum* (blight), *A. pisi, Aspergillus niger, Cercospora canescens, C. cruenta* (leaf spot), *Colletotrichum lindemuthianum* (anthracnose), *Corticium solani, Elsinoe phaseoli, Erysiphe polygoni* (mildew, in warm moist weather), *Fusarium oxysporum,* f. *vasinfectum, Gloeosporium phaseoli, Macrophomina phaseoli* (dry root rot), *Myrothecium roridum, Nematospora coryli, Phomopsis* sp. (seedling blight), *Phyllosticta phaseolina, Protomycopsis phaseoli* (angular leaf spot), *Pythium aphanidermatum* (damping off), *Rhizoctonia bataticola, Rh. solani, Sclerotium bataticola, Sphaerotheca fuliginea, S. humuli, Synchytrium ajrekari, S. phaseoli-radiati, Uromyces appendiculatus* (rust), and *U. phaseoli*. Following viruses also attack urd: alfalfa mosaic (3 strains), alsike clover mosaic, bean chlorotic ringspot (4 strains), bean local chlorosis viruses (6 strains), bean mosaic, bean necrosis, leaf crinkle, leaf curl, leaf-spot, mosaic of Crotalaria mucronata, mosaic mottles, white clover mosaic, yellow mosaic of mung, and zonate leaf-spot. Trials of 118 cvs showed that 25 were resistant to yellow mosaic virus. Isolate SBMV 3 did not infect cv 'L41-13' or 'Krishna' or mungbean 'Krishna' while SBMV 1 and SBMV 2 infected 11 *V. mungo* and 7 *V. radiata* tested. Also the bacteria *Xanthomonas phaseoli* and *X. vignicola* have been isolated from black gram. Following nematodes attack plant: *Heterodera glycines, Meloidogyne arenaria, M. hapla, M. incognita, M. i. acrita, M. javanica, Pratylenchus brachyurus,* and *P. coffeae*. In Asia, the pod borer, *Apion ampulum;* the aphid, *Aphis craccivora;* and the hairy caterpillar, *Diacrisia obliqua,* are troublesome. In India the bean fly, *Ophiomyia phaseoli;* the common pulse beetle fly, *Madurasia obscurella,* and the white fly, *Bemisia tabaci,* are controlled during early urd growth, through soil applications of the granular insecticides, aldicarb, phorate, and monocrotophos. In eastern Africa, the bean aphid, *Aphis fabae* and the American bollworm, *Heliothis armigera,* are problems. Systemic insecticides like parathion have been used to control field infestations with the cowpea bruchid *Callosobruchus chinensis*. Dried neem leaves (*Azadiracta indica*) are occasionally burned to ward off storage insects. Preemergence applications of alachlor, chlorthal, diphenamid, nitrofen, and trifluralin have been recommended elsewhere for weed control.

REFERENCES:

Bose, R. D., 1932, Studies in Indian pulses. 5. Urd or black gram (*Phaseolus mungo* Linn. var. *roxburghii* Prain), *Indian J. Agric. Sci.* **2**:625–637.

Singh, R., and Singh, R., 1975, Studies on a mosaic disease of urd bean (*Phaseolus mungo* L.), *Phytopathol. Mediterr.* **14**(2/3):55–59.

CONTRIBUTORS: J. A. Duke, B. P. Pandya, C. F. Reed, J. K. P. Weder.

Vigna radiata (L.) Wilczek

FAMILY: Fabaceae

COMMON NAMES: Green gram, Golden gram, Mungbean

SYNONYMS: *Azukia radiata* (L.) Ohwi
Phaseolus aureus Roxb.
Phaseolus radiatus L.
Rudua aurea (Roxb.) Maekawa
Vigna aureus (Roxb.) Hepper

USES: Green gram is one of the very ancient crops of India, and is still a very important source of human food there. Highly nutritious seeds are eaten

Vigna radiata (L.) Wilczek

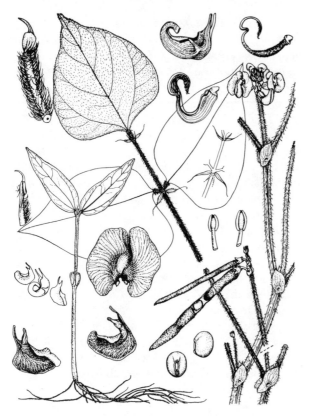

Figure 140. *Vigna radiata* (L.) Wilczek

split or whole, boiled or roasted. Green pods are eaten as a vegetable. In China and the United States sprouted mungbeans are a common vegetable in many dishes. For beansprouts, dry beans are soaked in water overnight, drained, and held in containers in a fairly warm, dark place. They are rinsed and drained every few hours and are ready for use in about 1 week. Various types of bean-sprouting chambers are available for home use. (A kilogram of dry mungbeans yields 6–8 kg sprouted beans.) Ripe roasted seeds are made into a flour (dahl) in many tropical areas. Green gram is grown as manure, hay, cover crop, and for forage. Husks are soaked and used for cattle feed.

FOLK MEDICINE: Medicinally, seeds are used both internally and externally, for paralysis, rheumatism, and affections of the liver, and for coughs and fevers. Roots are considered narcotic.

CHEMISTRY: Tyramine has been reported in mungbeans. Seeds are reported to contain trypsin inhibitors and chymotrypsin inhibitors. Whole seeds dried contain (per 100 g): 341 calories, 10.6% moisture, 22.9 g protein, 1.2 g fat, 61.8 g total carbohydrate, 4.4 g fiber, 3.5 g ash, 105 mg Ca, 330 mg P, 7.1 mg Fe, 6 mg Na, 1132 mg K, 55 g β-carotene equivalent, 0.53 mg thiamine, 0.26 mg riboflavin, 2.5 mg niacin, and 4 mg ascorbic acid. Mungbean amino acids average (mg/g N): aspartic acid 716, cystine 44, threonine 209, serine 296, glutamic acid 865, proline 229, glycine 210, alanine 242, valine 259, methionine 33, isoleucine 233, leucine 441, tyrosine 156, phenylalanine 306, lysine 504, histidine 182, and arginine 345. High protein content seems to be correlated with small seed size. The oil has been reported to contain: 28.1% palmitic acid, 7.8% stearic, 0.9% arachidic, 2.4% behenic, 6.3% cerotic, 6.4% oleic, 32.6% linoleic, and 14.4% linolenic. The unsaponifiable fraction may contain 0.023% stigmasterol. Mungbean starch is 28.8% amylose and 71.2% amylopectin. Total sugar content is reported as (per 100 g): 2.69–5.88% with monosaccharides 0.38–1.00, sucrose 1.06–2.19, raffinose 0.38–0.69, and stachyose 0.50–1.50. Raw sprouts contain (per 100 g): 30 calories, 90.1% moisture, 4.2 g protein, 0.2 g fat, 5.0 g total carbohydrate, 0.9 g fiber, 0.5 g ash, 15 mg calcium, 71 mg P, 1.2 mg Fe, 7 mg Na, 222 mg K, 20 g β-carotene equivalent, 0.11 thiamine, 0.10 mg riboflavin, 0.8 mg niacin, and 18 mg ascorbic acid.

DESCRIPTION: Erect or semierect, rather hairy, annual shrubby herb, 30–120 cm tall, diffusely branched, with tendency to twine at tips, deep-rooted; leaves alternate, trifoliolate, dark green, with scattered hairs on both sides, on long petioles, leaflets large, 5–10 cm long, ovate, entire or rarely trilobed; stipules prominent, peltate; stipels ca. 5–10 mm long, flowers pale yellow, crowded in axillary or terminal racemes in clusters of 10–25; pods black, gray or brownish, 2.5–10 cm long, 4–6 mm wide, spreading, reflexed, with short hairs; seeds 10–20, globose or oblong, 3.2–5 mm long, 3–3.8 mm wide, slightly flattened, usually green, but may be olive, yellow, brown, or purplish-brown, or marbled or mottled, etc.; hilum flat, 1.3–1.7 mm long, 0.5–0.6 mm wide, narrowly obovate, margin slightly concave; seed coat surface smooth or light-colored, farinose (mealy) and marked with fine crenulate lines; micropyle elevated, pore facing hilum; 100 seeds weigh 1.5–4 g; germination phanerocotylar. Fl. summer.

GERMPLASM: Over 2,000 strains and several botanical varieties of mungbean have been recognized. (1) *Vigna radiata* var *sublobata* (Roxb.) Verdcourt (Syn.: *Phaseolus sublobatum* Roxb.): *Ph. trinervis* Wight & Arn.; *Vigna opisotricha* A. Rich.; *V. brachycarpa* Kurz); (2) *Vigna radiata* var *glabra* (Roxb.) Verdcourt (Syn.: *Phaseolus glaber* Roxb.; *Ph. mungo* var. *glaber* (Roxb.) Bak. in Hook f.; *Ph. calcaratus* var *glaber* (Roxb.) Prain. Over 100 cvs of mungbean are grown in China and other Asiatic countries; they differ mainly in habit (erect or suberect, with some twining at tips), size of plant (usually 60 cm, but some 30 cm tall, others up to 150 cm), period of maturity (80–140 days), color of pods, size and color of seeds, cvs are being developed which can be gathered in one harvest. Main cvs in the United States are 'Golden' (sometimes called *aureus*), and 'Green' (sometimes called *typica* or *radiata*), the color refers to dry seed; both are grown for hay and green manure crops, but only green gram is grown for dry-bean production. In India, the black-seeded type may be called *grandis,* the brown-seeded type *bruneus*. 'Green' cvs with hard shining dark green seed coats are preferred for bean sprouts. 'Kopagaun,' 'Krishna-11,' 'Mung H70-16,' and 'Pusa Baisakhi' are short-duration high-yielding Indian cvs suitable for multiple cropping; 'Jawahar-45' is suitable for rainy season planting and for double cropping in rainfed and limited water areas. Assigned to the Indochina–Indonesian, China–Japanese, and Hindustani Centers of Diversity, mungbean or cvs thereof is reported to exhibit tolerance to drought, heat, high pH, laterite, low pH, salt, slope, and virus. ($2n = 22$.)

DISTRIBUTION: In India green gram is of ancient cultivation but does not grow wild. It was introduced early into southern China, Indochina, and Java. In recent times, it has been introduced into East and Central Africa, the West Indies, and the United States.

ECOLOGY: In India green gram is grown from sea level to 2,000 m, usually as a dryland crop following rice. It is drought-resistant but susceptible to frost, waterlogging, and salinity. Some cvs tolerate alkalinity and salinity. A well-distributed rainfall of 7–9 dm is recommended. Grown after rainy season where rains are heavy. Rain is detrimental at flowering. Usually grown on good loamy soil, or black cotton soils or those that tend to be slightly clayey, but can be grown on red light loams or alluvial soils, if they are sufficiently deep. Both short-day and long-day cvs are found in India. Photoinsensitive cvs are available, which flower only 30 days after planting. Ranging from Cool Temperate Moist through Tropical Thorn to Wet Forest Life Zones, mungbean is reported to tolerate annual precipitation of 2.8–41.0 dm (mean of 29 cases = 12.5), annual mean temperature of 7.8°–27.8°C (mean of 29 cases = 20.9), and pH of 4.3–8.1 (mean of 27 cases = 6.6).

CULTIVATION: Soil prepared with a shallow plowing and 2 diskings. Basal dressings of 400 kg/ha superphosphate and 100 kg/ha ammonium sulfate give good results on some tropical laterites. Studies suggest that 15–40 kg/ha phosphorus is the most economical level. Seed broadcast or drilled in 25–50 cm rows, in June–July or October. In India, seed rate is 13–17 kg/ha; in the United States, 6–9 kg/ha; but in some areas reported as high as 22 kg/ha. Germinates in 3–5 days. Usually 2 cultivations are necessary, the first ca. 20 days after germination, the second about 2 weeks later. In dry areas a couple of irrigations may be necessary. Stands may be pure or intercropped with taller, longer-aged crops like sorghum. Green gram is often grown as a green-manure crop with rice, or used as a catch midseason crop, or rotated before or after rice.

HARVESTING: For green pods 50–70 days are required; for ripe pods, 60–150 days, depending on cv. In some, fruiting extends over a long period, but in 'Pusa Baisakhi,' 75% of the crop can be harvested at the first picking, the remainder in about 10 days. Plants are pulled with their roots, stacked for a week in the sun or on a threshing floor, and then threshed by beating with sticks or trampled by oxen. Beans are then cleaned and bagged for market. In other areas these procedures are mechanized.

YIELDS AND ECONOMICS: Yields of 450–560 kg/ha of dried beans are common; Sri Lanka has reported yields up to 1,000 kg/ha; United States, 1,125 kg/ha. Yields of air-dried hay in United States are 2–7 MT/ha for golden gram and less for green gram. In India yields of beans average 100–200 kg/ha; good

Vigna radiata (L.) Wilczek

yields are 300–500 kg/ha. AVRDC trials with 20 elite cvs varied with season, but seed yields were 800–2,100 kg/ha. The Philippine line 'PHLV 18' averaged 1,800 kg/ha and was a relatively stable yielder. In summer and fall experiments, 'PHLV 18' produced 75% of its yield in 66 and 75 days, respectively. As a field crop, mungbean is comparable to cowpea or soybean. In 1974, India produced 494,000 MT (400 kg/ha); Thailand, 190,000 (800 kg/ha); Indonesia 17,000 MT (600 kg/ha); Philippines, 16,000 MT (400 kg/ha); Bangladesh, 14,000 (800 kg/ha); Sri Lanka, 6,000 (600 kg/ha); and Taiwan, 3,000, (700 kg/ha). Large acreages of green gram are grown for food in China, India, and Sri Lanka. About 11,250,000 kg of mungbeans are produced in the United States annually, mostly for canned bean-sprouts.

BIOTIC FACTORS: Many fungi cause diseases of mungbeans: *Alternaria tenuis, Ascochyta phaseolorum* (bean blight), *A. pisi, Botrytis cinerea, Cercospora canescens* (leaf spot), *C. caracallae, C. cruenta, Colletotrichum lindemuthianum* (anthracnose), *Corticium solani, Corynespora cassicola, Elsinoe phaseoli, Erysiphe polygoni* (downy mildew), *Fusarium bulbigenum, Helminthosporium anthyllids, Leveillula leguminosarum, Macrophomina phaseoli* (blight), *Macrosporium fasciculatum, Mycosphaerella phaseoli, Nematospora coryli, N. phaseoli, Phoma subcircinata, Phymatotrichum omnivorum, Pythium aphanidermatum, P. arrhenomanes, P. irregulare, P. rostratum, P. splendens, P. ultimum, Sclerotinia sclerotiorum, Sclerotium rolfsii* (root rot), *Sphaerotheca fuliginea, Sphaerotheca humuli,* and *Uromyces appendiculatus.* Plants are also attacked by the bacteria: *Corynebacterium glaucum, C. faciens,* and *Xanthomonas phaseoli.* Viruses isolated include: alfalfa mosaic, Brassica virus 2, Brazilian tobacco streak, clover yellow mosaic, cucumber mosaic, leaf crinkle, flower abortion, mosaic of Crotalaria mucronata, stunt, white clover mosaic, yellow mosaic, and zonate leaf-spot. Screening 157 cvs, Ahmad (1975) found none that were resistant to yellow mosaic, but 6601, 6602, MN1, MG1, MG588, and MG589 were tolerant. In the US, 'M-238' and 'M-330' are rather resistant to downy mildew and virus diseases. Plants are parasitized by *Striga lutea.* In Taiwan even a 70-day growing season does not allow mungbean to escape severe disease attack.

Leaf-spot *(Cercospora canescens)* can reduce yields by 47% during the warm wet season. Sixty-eight mungbean cvs proved resistant. Powdery mildew *(Erysiphe polygoni)* can be severe during the cool dry season. Ten cvs proved resistant. Seedlings often succumb to the combined attack of root rot fungi *(Fusarium solani* f. sp. *phaseoli, Pythium* sp., and *Rhizoctonia solani)* and root rot nematodes. Mungbean mottle virus has reduced yields as much as 79%. Nematodes isolated from green gram include: *Heterodera glycines, Hoplolaimus indicus, Meloidogyne arenaria, M. hapla, M. incognita, M. i. acrita, M. javanica, Paratylenchus minutus, Pratylenchus brachyurus,* and *P. coffeae.* Beanfly *(Melanagromyza phaseoli)* attacks leaves soon after germination, and larvae bore into tissues and traverse down stems, causing rot at the collar, this pest being especially serious in Sri Lanka. It is the most serious pest in southeastern Asia, sometimes causing complete losses. Granular carbofuran at planting is said to help. Cyolane, Cytrolane, and Solvirex are said to give satisfactory control of cutworms *(Chrysodeixis, Mocis,* and *Spodoptera),* the podborer, *Etiella zinckenella,* red spider mite, *Tetranychus cinnabarinus,* and aphids. Endrin is said to control the larvae of *Amsacta moorei* and *A. albistriga* as well as the hairy caterpillar, *Diacrisia obliqua.* Seeds are subject to serious attack by bruchids. The bruchid *(Callosobruchus chinensis)* sometimes attacks beans in the field but is principally a storage pest. Applying peanut or soybean oil helps prevent severe bruchid infestations without affecting seed viability. Flowers are fully self-fertile. For weed control, chlorthaldimethyl, diphenamid, and trifluralin have been recommended preplanting; alachlor, chloramben, chlorthal, dichlormate, linuron, nitrofen, and terbutryn, postplanting.

REFERENCES:

Ahmad, M., 1975, Screening of mungbean *(Vigna radiata)* and urdbean *(V. mungo)* germplasms for resistance to yellow mosaic virus, *J. Agric. Res. (Pakistan)* **13**(1):349–354.

AVRDC, 1976, *Mungbean Report for 1975,* Asian Vegetable Research and Development Center. Shanhua, Taiwan, Republic of China.

Biswas, M. R., and Dana, S., 1975, *Phaseolus aureus* X *P. lathyroides* Cross, *Nucleus* **18**:81–85.

Ligon, L. L., 1945, Mung Beans. A legume for seed and forage production, *Okla. State Univ. Bull.* **B-284:**1–12.

Matlick, R. S., and Oswalt, R. M., 1963, Mungbean varieties for Oklahoma, *Okla. State Univ. Bull.* **B612:** 1–15.

Wester, R. E., 1964, Growing mungbean sprouts, *U.S. Dep. Agric., Agric. Res. Serv., Rep.* **CA-4-59**(Jan.):1–2.

CONTRIBUTORS: J. A. Duke, C. F. Reed, J. K. P. Weder.

Vigna umbellata (Thunb.) Ohwi & Ohashi

FAMILY: Fabaceae

COMMON NAMES: Rice bean, Climbing mountain bean, Mambi bean, Oriental bean

SYNONYMS: *Azukia umbellata* (Thunb.) Ohwi
Phaseolus calcaratus Roxb.
Phaseolus pubescens Bl.

USES: Rice beans, widely grown and used as human food in Asia and the Pacific Islands, are eaten boiled, in soups, or with or instead of rice. Young pods and leaves are used as vegetables; sprouts are also consumed. Whole plants are used as forage for livestock, especially pigs. Grown as a cover crop and a green manure.

FOLK MEDICINE: No data avilable.

CHEMISTRY: The dried seeds contain per 100 g 13.3% moisture, 327 calories, 20.9 g protein, 0.9 g fat, 60.7 g total carbohydrate, 4.8 g crude fiber, 4.2 g ash, 200 mg Ca, 390 mg P, 10.9 mg Fe, 0.49 mg thiamine, 0.21 mg riboflavin, and 2.4 mg niacin. The amino acid composition is (mg/g N): arginine 462, histidine 380, leucine 606, isoleucine 387, lysine 769, methionine 769, cystine 44, phenylalanine 325, tyrosine 262, threonine 294, and valine 394. The fatty acid composition of the oil is (%): myristic 6.3–7.3, palmitic 5.6–6.0, stearic 2.1–4.4, behenic 4.6–5.8, arachidic 3.0–3.9, lignoceric 2.9–3.2, linoleic 7.5–9.7, and oleic 61.3–68.0. Seeds contain β-sitosterol. Philippine analyses (R. M. Lantican, personal communication, 1978) show much more fat than other analyses. The Philippine analyses showed 3.9% fat, 18.8% crude protein, and 50.8% starch. At the vegetative stage, the forage contains 16.0% dry matter, 18.0% crude protein, 1.1% fat, 31.5% crude fiber, 39.9% N-free extract, 9.5% ash, 1.4% Ca, and 0.35% P. At the flowering stage, it contains 24.0% dry matter, 14.5% crude protein, 1.0% fat, 32.1% crude fiber, 41.6 N-free extract, 10.8% ash, 1.2% Ca, and 0.4% P. Said to be free of cyanogenic glycosides, rice bean does contain trypsin inhibitors or chymotrypsin inhibitors.

DESCRIPTION: Annual or short-lived perennial; stems suberect, 30–75 cm tall; producing vining branches 1–2 m long, grooved, with short white hairs; leaves trifoliolate, petioles 5–10 cm long, stipules conspicuous, ovate-lanceolate, attached below the middle; leaflets ovate, 5–10 cm long, 2.5–6 cm broad, usually entire, sometimes 3-lobed; inflorescence of erect axillary racemes with 5–20 flowers, on peduncles 7.5–20 cm long; flowers 2–3 together at nodes, bright yellow, with bracteoles linear-lanceolate, with standard 1.5–2 cm in diameter; pods long and slender, 6–12.5 cm long, 0.5 cm broad, 6–12-seeded, glabrous, falcate, shattering; seeds oblong to strongly elongate, subtrapezoidal, 5–10 mm long, 2–5 mm wide; hilum 2.5–4.5 mm long, 0.6–1.5 mm wide, strap-shaped; micropyle usually hidden by caruncle; seed coat smooth, dark wine, green, yellow, brown, black, speckled, or mottled; 100 seeds weigh 8–12 g. Germination cryptocotylar. Fl. summer.

GERMPLASM: Cultivars differ in period of maturity and in color of seed coat. Late cvs are more vigorous in habit than earlier. Four botanical varieties are recognized: var *major*, hills of northern India and Burma, with larger flowers; var *rumbaiya*,

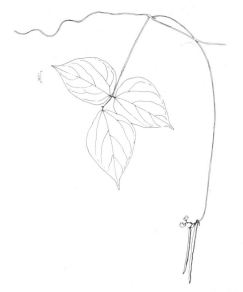

Figure 141. *Vigna umbellata* (Thunb.) Ohwi & Ohashi

Vigna umbellata (Thunb.) Ohwi & Ohashi

cultivated in Khasia Hills of Burma, as 'Rumbaiya,' with short erect or spreading stems; var *gracilis*, a wild form with slender smooth stems and rather narrow leaflets; var *glaber* has smooth stems and leaves. *Phaseolus torosus* Roxb., cultivated in Nepal, has torose pods, short and subcylindrical, pale-cream colored seeds. Cultivars have been introduced in the United States from China, Burma, India, Japan, and Java. Assigned to the Hindustani, Indochina–Indonesian, and China–Japanese Centers of Diversity, rice bean or cvs thereof is reported to exhibit tolerance to drought and slope. ($2n = 22$.)

DISTRIBUTION: Wild from the Himalayas and central China to Malaysia, spreading by cultivation to other parts of tropical Asia. Most widely cultivated in China, Korea, Japan, India, Java, Mauritius, and the Philippines, and to a limited extent in Southeast Asia, Africa, the West Indies, Australia, and the United States.

ECOLOGY: Wild forms grow on open sites and along roadsides; cvs fare best on fertile loams. Cultivars are well-adapted to practically the same areas as cowpeas, and can tolerate high temperatures and drought, but not frost. Best grown where temperatures average 18°–30°C, and rainfall 10–15 dm. Rice beans are short-day plants. In western Bengal, maximum temperatures for flower initiation were 25°–26°C, minimum, 10°–12°C (Chaudhuri and Prasad, 1972). Ranging from Subtropical Dry to Moist through Tropical Dry Forest Life Zone, rice bean is reported to tolerate annual precipitation of 7.0–17.3 dm (mean of 3 cases = 11.1), annual mean temperature of 23.5°–27.5°C (mean of 3 cases = 25.3), and pH of 6.8–7.5 (mean of 2 cases = 7.2).

CULTIVATION: Rice beans are usually broadcast at 68–90 kg/ha. Sown for row crops at 20–70 kg/ha. When planted in rows 90 cm apart, crop makes a dense growth to 60 cm deep. In Bengal, August planting is recommended. In Burma, crop is usually rotated with rice. In India, preplanting application of 50–60 kg/ha superphosphate is recommended. However, due to adverse effects of low temperatures (10°–12°C) prevailing during the short-day conditions of subtropical winter in West Bengal, August planting is favored. Flowering is delayed to 140–145 days under such conditions, presumably due to longer photoperiods.

HARVESTING: Rice beans mature in 60–140 days, depending on the cv. In the Philippines, days to flowering averages 64, to maturity, 92. Though very productive of seed, the vining habit of the plant and the shattering of seed make the crop difficult to harvest by machine. Harvesting in the morning, when pods are moist, reduces the loss. Forage rice bean may be harvested 70–80 days after sowing, but yields are higher at 120–130 days.

YIELDS AND ECONOMICS: Yields generally are ca. 225 kg/ha. However in Papua, New Guinea yields of 500–800 kg/ha were common. Yields exceeding 1 MT/ha are reported from western Bengal. Forage yields average 2,200–3,500 kg/ha. Rice beans are extensively grown for human food in Asia and the Pacific Islands but rarely enter international trade. In 1975, Japan imported ca. 7,000 MT from Thailand, 3,000 from China, 2,000 from Burma. Crop has been introduced in southern United States, but has not developed any economical importance.

BIOTIC FACTORS: Flowers are self-fertile. Following fungi cause diseases in rice bean: *Corticium solani*, *Myrothecium roridum*, and *Woroninella umbilicata*. In the Philippines, powdery mildew and rust occur sparingly. Cucumber mosaic virus also attacks the plant. Nematodes attacking rice bean include: *Heterodera glycines* and *Meloidogyne javanica*. Rice bean is recommended to follow rice, since the flooding reduces the nematode problem. Relatively speaking rice bean is a pest-free crop, even comparatively immune to most storage insects.

REFERENCE:
Chaudhuri, A. P., and Prasad, B., 1972, Flowering behaviour and yield of rice bean (*Phaseolus calcaratus* Roxb.) in relation to date of sowing, *Indian J. Agric. Sci.* **42**:627–630.

CONTRIBUTORS: S. Dana, J. A. Duke, T. N. Khan, R. M. Lantican, C. F. Reed, J. K. P. Weder.

Vigna unguiculata ssp. *cylindrica* (L.) Verdc.

FAMILY: Fabaceae
COMMON NAMES: Catjang, Hindu, Sowpea
SYNONYM: *V. cylindrica* (L.) Skeels

USES: In India and Sri Lanka, catjang is grown for seed crop; cultivated as a vegetable; tender green

Vigna unguiculata ssp. sesquipedalis (L.) Verdc.

Figure 142. *Vigna unguiculata* ssp. *cylindrica* (L.) Verdc.

pods consumed as green vegetable; dried seeds used whole or split; makes excellent forage; for making hay, often mixed with corn or sorghum for silage; for green manure (especially cvs with spreading habit and luxuriant growth).

FOLK MEDICINE: No data available.

CHEMISTRY: Seeds are reported to contain trypsin inhibitors and chymotrypsin inhibitors. Whole dried seeds contain (per 100 g) 342 calories, 23.1 g protein, 1.4 g fat, 61.4 g total carbohydrate, 4.8 g fiber, 3.3 g ash, 101 mg Ca, 383 mg P, 7.6 mg Fe, 70 μg β-carotene equivalent, 0.75 mg thiamine, 0.18 mg riboflavin, 2.5 mg niacin, and 1 mg ascorbic acid. Soaked and skinned seeds contain (per 100 g) 264 calories, 30.4% moisture, 16.9 g protein, 50.4 g total carbohydrate, 2.3 g ash, 0.57 mg thiamine, 0.10 mg riboflavin, and 1.8 mg niacin. Raw immature seeds contain (per 100 g) 49 calories, 83.6% moisture, 4.8 g protein, 0.4 g fat, 9.5 g total carbohydrate, 2.1 g fiber, 1.7 g ash, 151 mg Ca, 68 mg P, and 150 μg β-carotene equivalent. Raw leaves contain (per 100 g) 44 calories, 85.0% moisture, 4.7 g protein, 0.3 g fat, 8.3 g total carbohydrate, 2.0 g fiber, 1.7 g ash, 256 mg Ca, 63 mg P, 5.7 mg Fe, 7,870 μg β-carotene equivalent, 0.20 mg thiamine, 0.37 mg riboflavin, 2.1 mg niacin, and 56 mg ascorbic acid. Dried leaves contain (per 100 g) 277 calories, 10.6% moisture, 22.6 g protein, 3.2 g fat, 54.6 g total carbohydrate, 9.0 g ash, 1556 mg Ca, 348 mg P, 12.0 mg Fe, and 86 mg ascorbic acid.

DESCRIPTION: Herbaceous annual; stems erect, suberect or spreading, glabrous except at nodes, thin, rounded; leaves trifoliolate, on long petioles 5–15 cm long; leaflets large, dark green or purple-tinged, ovate, usually entire, apex acute, 6.5–16 cm long, 4–10 cm wide, lateral leaflets oblique; inflorescence axillary, a raceme on long peduncle, the flowering part nodose; corolla white, light pink or light blue, standard 2–3 cm in diameter, keel truncate; pods rounded, thin, 10–15 cm long; seeds vary in color and size, depending on cv. Germination phanerocotylar.

GERMPLASM: Numerous cvs: 'Cream Lady' does well in Puerto Rico; 'Bombay cowpea' is a grain variety, with erect vigorous growth habit, short growing period, nearly uniform maturity; pods do not shatter at maturity, relatively resistant to attack by aphids and bean fly, and fairly drought resistant. Assigned to the Indochina-Indonesian and Hindustani Centers of Diversity, catjang or cvs thereof is reported to exhibit tolerance to insect, poor soil, slope, virus, and weeds. ($2n = 22$.)

DISTRIBUTION: Native to India and Sri Lanka and widely cultivated in tropical areas, as East Africa. More common and with more cvs in Asia.

ECOLOGY: Does well on well-drained moderately light or medium loam soils; lime is essential. On rich soil, vegetative growth is excessive and seed yield low. Soil of medium fertility best suited for seed crop. Requires a reasonably warm growing season, fairly good distribution of rain, followed by bright weather during and after flowering. Wet areas reduce seed yield and cause excessive growth. Ranging from Warm Temperate Dry to Moist through Tropical Dry Forest Life Zone, catjang is reported to tolerate annual precipitation of 7.0–17.3 dm (mean of 8 cases = 12.2), annual mean temperature of 12.2°–27.5°C (mean of 8 cases = 21.3), and pH of 5.0–7.5 (mean of 7 cases = 6.1).

Vigna unguiculata ssp. *cylindrica* (L.) Verdc.

CULTIVATION: For catjang land should be worked to a fine tilth, 15–20 cm deep. Seeds maintain viability for 6–7 yr. When sown as pure crop, seed rate is 15–20 kg/ha; as mixed crop with sorghum or sesame, seed rate is lower. Germinates in 4–5 days. Plants should be spaced at 30-cm intervals in rows 45 cm apart. Heavy manuring is not recommended, as it causes excessive growth. Basal dressing of 1 cwt superphosphate and 0.25 cwt muriate of potash is applied during preparation. One intercultivation, when crop is 3 weeks old, usually adequate. Crop has heavy foliage that controls weeds.

HARVESTING: A limiting factor of cultivation for green pods is the cost of harvesting. Selections are made for erect stalks above foliage, and high yield. Pods mature in ca. 80 days. Crop is harvested in one operation by cutting plants close to ground. Harvested pods are dried, threshed, and cleaned by winnowing.

YIELDS AND ECONOMICS: Yield of dry grain is 1,390–1,440 kg/ha. In India some cvs yield only 900–1,000 kg dry seed/ha; others yield 2,400–2,500 kg/ha. Catjang is a very important crop in India and Sri Lanka, and in some tropical African areas. Although grown extensively in certain areas, it does not enter international trade.

BIOTIC FACTORS: Following fungi are recorded as attacking catjang: *Aristastoma oenconomicum*, *Cercospora cruenta*, *Rhizobium japonicum*, *Erysiphe polygoni* (powdery mildew), *Phytophthora* sp. (stem rot), *Uromyces phaseoli* (rust), *Rhizoctonia* sp. (root rot). Mosaic viruses (*Marmor* spp.) also attack this crop.

CONTRIBUTORS: J. A. Duke, C. F. Reed.

Vigna unguiculata ssp. *sesquipedalis* (L.) Verdc.

FAMILY: Fabaceae

COMMON NAMES: Yard-long bean, Asparagus bean, Pea bean

SYNONYMS: *Dolichos sesquipedalis* L.
Vigna sesquipedalis (L.) Fruwirth
Vigna sinensis ssp. *sesquipedalis* (L.) van Eseltine
Vigna sinensis var *sesquipedalis* (L.) Aschers. & Schweinf.

USES: Yard-long bean is grown for its immature pods, which are used as a vegetable, and for ripe dry beans. Young leaves and sprouting seed also serve as vegetables. Yard-long beans are good forage plants because of their ability to smother out weeds.

FOLK MEDICINE: Leaves, boiled with rice, have been used for earache. Pulped leaves with alum are applied to stop the flow of mothers milk. Juice from the leaves is used in traditional Malaysian and Indonesian medicine.

CHEMISTRY: Raw young green pods contain (per 100 g): 88.3% moisture, 37 calories, 3.0 g protein, 0.2 g fat, 7.9 g carbohydrate, 1.6 g fiber, 0.6 g ash, 44 mg Ca, 45 mg P, 0.7 mg Fe, 6 mg Na, 233 mg K,

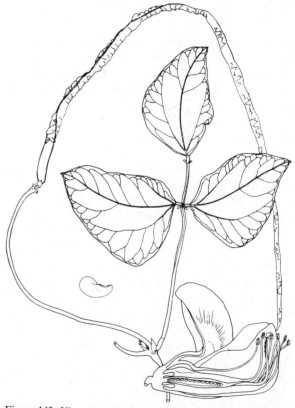

Figure 143. *Vigna unguiculata* ssp. *sesquipedalis* (L.) Verdc.

Vigna unguiculata ssp. *sesquipedalis* (L.) Verdc.

225 μg β-carotene equivalent, 0.12 mg thiamine, 0.11 mg riboflavin, 1.0 mg niacin, and 22 mg ascorbic acid. The protein content of pods can attain 9%, of mature seeds 24%. Raw leaves contain (per 100 g): 34 calories, 88.4% moisture, 4.2 g protein, 0.4 g fat, 5.8 g total carbohydrate, 1.7 g fiber, 1.2 g ash, 108 mg Ca, 106 mg P, 4.7 mg Fe, 25 mg Na, 366 mg K, 2400 μg β-carotene equivalent, 0.28 mg thiamine, 0.24 mg riboflavin, 1.2 mg niacin, and 35 mg ascorbic acid. Immature seeds contain (per 100 g): 34 calories, 89.0% moisture, 3.3 g protein, 0.19 fat, 6.9 g total carbohydrate, 1.3 g fiber, 0.7 g ash, 54 mg Ca, 68 mg P, 1.4 mg Fe, 235 μg vitamin A equivalent, 0.09 mg thiamine, 0.12 riboflavin, 1.1 mg niacin, and 26 mg ascorbic acid. Seeds are reported to contain trypsin inhibitors and chymotrypsin inhibitors.

DESCRIPTION: A strong climbing annual vine, twining counterclockwise to heights of 2–4 m; leaves trifoliolate; leaflets ovate, 7.5–12.5 cm long; flowers dirty yellow or violet, ca. 2.5 cm long, on long stalks; pods limp, 30–100 cm long, straight or irregularly twisted, pendent, more or less inflated when young, cylindrical; seeds 15–20, large, oblong to reniform, 8–12 mm long, usually black (African forms) or brown (in forms sold in US), wrinkled; hilum white. Weight 100 seeds, 10–25 g depending on variety. Germination phanerocotylar.

GERMPLASM: In Hong Kong, cvs are divided into two groups: Green podded forms 45–90 cm long; and 'White'-podded forms with very pale green pods, up to 45 cm long. 'Yard-Long' is a good cv; in Trinidad, the cv 'Long White' imported from Hong Kong has given good results. Bush cvs are being developed in the Philippines. Assigned to the African and Hindustani Centers of Diversity, yard-long bean or cvs thereof is reported to exhibit tolerance to laterite, photoperiod, poor soil, slope, and virus. ($2n = 22, 24$.)

DISTRIBUTION: Native to India or Africa, yard-long beans are mostly cultivated in the Far East (Bangladesh, India, Indonesia, Pakistan, Philippines). Also in the Caribbean, and to a lesser extent in Africa.

ECOLOGY: Produces vigorous growth in warm climates with mean monthly temperatures of 20°–30°C. Plants are sensitive to waterlogging and cold and killed by frost (cannot tolerate 4°C for long). Yard-long bean is grown in many soil types, from sandy loams to clay. Some cvs are sensitive to day length; other are day-neutral. Tolerates and requires a higher rainfall than cowpeas. Ranging from Warm Temperate Dry to Moist through Tropical Dry to Moist Forest Life Zones, yard-long beam or cvs thereof is reported to tolerate annual precipitation of 6.2–41.0 dm (mean of 16 cases = 16.7), annual mean temperature of 12.5°–27.4°C (mean of 16 cases = 21.7), and pH of 4.3–7.2 (mean of 15 cases = 5.7), (pH 5.5–6.0 recommended).

CULTIVATION: Seeds remain viable for several years. Seed sown when weather is warm, covered to twice the diameter of seed (or to 5 cm deep) and soil firmed over seed. Seed rates are 25–50 kg/ha. Seedlings are thinned to stand 30–50 cm apart in rows 75–100 cm apart, and must be kept moist. Seed may be planted, 4–6, in hills. Some type of support is necessary. Fertilizer in the United States consists of 600–1,000 kg/ha of 4–8–9 NPK mixture, applied in bands 5 cm below the seeds. In Chinese gardens, they are often interplanted with such vegetables as sweet potato, taro, and yam.

HARVESTING: Plants produce green pods in 50–90 days, ca. 1 kg/sq. m. Seed can be harvested in 60–90 days. Green and rip pods are usually harvested and processed manually.

YIELDS AND ECONOMICS: Yields vary with size of beans and frequency of picking. In the Philippines, yields of green pods are 4–5.5 MT/ha, but plants may produce 10 MT/ha. Yard-long beans form an important vegetable crop in southeastern Asia, particularly with Chinese market gardeners; grown also in the West Indies.

BIOTIC FACTORS: A heavy pollinating insect is required to depress the wings of the flower and expose the stamens and stigma. Flowers open in early morning, close before noon and fall by the end of the same day. Yard-long bean is bothered by many cowpea pests. In the Philippines, cutworms (*Agrotis* spp.) and bean flies (*Ophiomyia phaseoli*) are curbed by folidol, imidan, meptox, and sevin. Plants are damaged by nematodes *Meloidogyne incognita* and var *acrita,* and *M. javanica.* The yellow mosaic virus is particularly bad where aphids are prevalent. Plants are also attacked by

Vigna unguiculata ssp. *sesquipedalis* (L.) Verdc.

the fungi *Cercospora cruenta* and *Erysiphe polygoni,* and the bacterium, *Pseudomonas syringae.* Preemergence applications of the herbicides ametryne, diphenamid, nitrofen, and prometryne have increased yields in Trinidad.

CONTRIBUTORS: J. A. Duke, C. F. Reed, J. K. P. Weder.

Vigna unguiculata (L.) Walp. ssp. *unguiculata*

FAMILY: Fabaceae

COMMON NAMES: Cowpea, Crowder pea, Black-eyed pea, Southern pea

SYNONYM: *Vigna sinensis* (L.) Savi ex Hassk.

USES: Cultivated for the seeds (shelled green or dried), the pods and/or leaves that are consumed as green vegetables or for pasturage, hay, ensilage, and green manure. The tendency of indeterminate cvs to ripen fruits over a long time makes them more amenable to subsistence than to commercial farming. However, erect and determinate cvs, more suited to monocultural production systems, are now available. If cut back, many cvs continue to produce new leaves, that are eaten as a potherb. Leaves may be boiled, drained, sun-dried and then stored for later use. In the United States, green seeds are sometimes roasted like peanuts. The roots are eaten in Sudan and Ethiopia. Scorched seeds are occasionally used as a coffee substitute. Peduncles are retted for fiber in northern Nigeria. Crop used to some extent as pasturage, especially for hogs, and may be used for silage, for which it is usually mixed with corn or sorghum. Crop is very useful as a green manure, and leafy prostrate cvs reduce soil erosion.

FOLK MEDICINE: Cowpeas are sacred to Hausa and Yoruba tribes, and are prescribed for sacrifices to abate evil and to pacify the spirits of sickly children. Hausa and Edo tribes use cowpeas medicinally, 1 or 2 seeds ground and mixed with soil or oil to treat stubborn boils.

CHEMISTRY: Raw mature seeds typically contain (per 100 g): ca. 11.4% moisture, 338 calories, 22.5 g protein, 1.4 g fat, 61.0 g total carbohydrate, 5.4 g fiber, 3.7 g ash, 104 mg Ca, 416 g P, 0.08 mg thiamine, 0.09 mg riboflavin, 4.0 mg niacin, and 2 mg ascorbic acid. In results at IITA, based on several thousand distinct cvs, protein averaged 23–25% protein, ranged from 18 to 29%, with potential for perhaps 35%. The proteins consist of 90% water-insoluble globulins and 10% water-soluble albumins. The reported amino acid content is (mg/g N): isoleucine, 239, leucine 440, lysine 427, methionine 73, cystine 68, phenylalanine 323, tyrosine 163, threonine 225, tryptophan 68, valine 283, arginine 400, histidine 204, alanine 257, aspartic acid 689, glutamic acid 1027, glycine 234, proline 244, and serine 268. Although much variation occurs, cowpeas are deficient in cystine, methionine, and tryptophan. Total sugars range from 13.7 to 19.7% and include: 1.5% sucrose, 0.4% raffinose, 2.0% stachyose, 3.1% verbascose. Starch may vary from 50.6 to 67.0% with 20.9–48.7% amylose, 11.4–36.6% amylopectin. The fatty acid composition of Pakistani seed oil is reported as: linolenic 12.3, linoleic 27.4, oleic 12.2, 1.1 lignoceric, 4.0 behanic, 0.9 arachidic, 7.1 stearic, 33.4 palmitic.

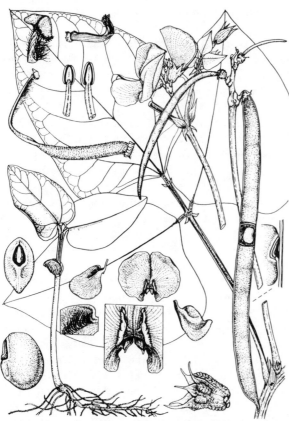

Figure 144. *Vigna unguiculata* (L.) Walp. ssp. *unguiculata*

Seeds also contain 0.025% stigmasterol. Immature pods contain (per 100 g): 85.3% moisture, 47 calories, 3.6 g protein, 0.3 g fat, 10.0 g total carbohydrate, 1.8 g fiber, 0.8 ash, 45 mg Ca, 52 mg P, 1.2 mg Fe, 170 µg vitamin A, 0.13 mg thiamine, 0.10 mg riboflavin, 1.0 mg niacin, and 22 mg ascorbic acid. Raw immature seeds contain (per 100 g): 66.8% moisture, 127 calories, 9.0 g protein, 0.8 g fat, 21.8 g total carbohydrate, 1.8 g fiber, 1.6 mg ash, 27 mg Ca, 175 mg P, 2.3 mg Fe, 2 mg Na, 541 mg K, 370 IU vitamin A value, 0.43 mg thiamine, 0.13 mg riboflavin, 1.6 mg niacin, and 29 mg ascorbic acid. Tender shoot apices, raw, contain (per 100 g): 89% moisture, 30 calories, 4.8 g protein, 0.3 g fat, 4.4 g total carbohydrate, 1.8 g ash, 73 mg Ca, 106 mg P, 2.2 mg Fe, 0.35 mg thiamine, 0.18 mg riboflavin, 1.1 mg niacin, and 36 mg ascorbic acid. The hay contains per 100 g: 9.6% moisture, 18.6 g crude protein, 23.3 g crude fiber, 2.6 g fat, 34.6 g N-free extract, and 11.3 g ash. Digestibility is improved by grinding the seeds into a fine powder. Seeds contain a trypsin inhibitor, a chymotrypsin inhibitor and a cyanogen in concentrations of ca. 2 mg/100 ml extract. Cooking improves the nutritive value, perhaps because the activity of trypsin inhibitors and/or the amount of other toxins are decreased by heat.

DESCRIPTION: Annual herb, erect or suberect, spreading, to 80 cm or more tall, glabrous; taproot stout with laterals near soil surface, roots with large nodules; stems usually procumbent, often tinged with purple; first leaves above cotyledons are simple and opposite; subsequent trifoliolate leaves are alternate, the terminal leaflet often bigger and longer than the two asymmetrical laterals; petiole, stout, grooved, 5–15 cm long; leaflets ovoid-rhombic, entire or slightly lobed, apex acute, 6.5–16 cm long, 4–11 cm wide, lateral leaflets oblique; inflorescence axillary, 2–4-flowered, crowded, near tips on short curved peduncles 2.5–15 cm long; calyx campanulate with triangular teeth, the upper 2 teeth connate and longer than rest; corolla dull white, yellow, or violet, with standard 2–3 cm in diameter, keel truncate; stamens diadelphous, the anthers uniform; pods curved, straight or coiled; seeds 2–12 mm long, globular to reniform, smooth or wrinkled, red, black, brown, green buff or white, as dominant color; full colored, spotted, marbled, speckled, eyed, or blotched; (5–30 g/100 seeds, depending on the cv). Germination phanerocotylar. Fl. early summer; Fr. mid and late summer, depending on cv sensitivity to local photoperiod and temperature conditions.

GERMPLASM: The most extensive collection of germplasm (and literature) is at IITA (over 7,000 cv accessions in 1975). Cowpea cvs may be grouped in the following manner for the United States: 'Crowder peas'; seeds black, speckled, brown, or brown-eyed, crowded in pods, seed usually globose, 'Brown Crowder' a good cv in Puerto Rico; 'Black-eyed': seeds white with black-eye around hilum, not crowded in pods. Extensively grown in California and southeastern United States and Puerto Rico; 'Cream' cvs: seed cream-colored, not crowded in pods; intermediates between 'Crowder' and 'Black-eyed' types, as 'Purple Hill' with deep purple mature pods and buff or maroon eyed seed; forage cvs: as 'New Era,' useful also for dry seeds in other geographical locations, e.g., western Africa. Other standard cvs are: 'Block,' 'Brabham,' 'Early Bluff,' 'Iron,' 'Taylor,' and 'Victor.' 'Gubgub' is an excellent table cv. Assigned to the African and Hindustani Centers of Diversity, cowpea or cvs thereof is reported to exhibit tolerance to aluminum, drought, high pH, heat, laterite, low pH, nematodes, poor soil, shade, slope, virus, weeds, and wilt. ($2n = 22$.)

DISTRIBUTION: Of ancient culture in Africa and Asia, and widespread in Africa, spreading by way of Egypt or Arabia to Asia and the Mediterranean. Now widely cultivated throughout the tropics and subtropics. Wild and cultivated forms readily cross. Steele (1972) and Sauer (1952) agree on a solely Ethiopian center of origin, followed by subsequent evolution predominantly in the ancient farming systems of the African Savanna.

ECOLOGY: Thrives on many kinds of soil, from highly acid to neutral; less well adapted to alkaline. Crop grows and yields at relatively low fertility levels, but often responds to P fertilization; N applications rarely effective on well-nodulated plants. Can withstand considerable drought and a moderate amount of shade, but is less tolerant of waterlogging than soybean. Some plants are indeterminant in growth, and continue to grow until killed by frost. In the tropics, such indeterminate plants may be weak perennials and continue growing as long as conditions are favorable. In some

Vigna unguiculata (L.) Walp. ssp. *unguiculata*

determinate types, the later flower initiation, the higher up the stem it is, the more flowers, and the greater the ultimate seed yield providing the growing season is sufficiently long. Dry matter production, seed yield, and root nodulation are reduced in photoperiods less than 12 hr, 13 min. Differences of as short as 12 min can affect flowering and seed yield. Cowpeas are short-day, warm-weather plants, sensitive to cold and killed by frost. They tolerate heat and relatively dry conditions and can be grown with less rainfall and under more adverse conditions than *Phaseolus vulgaris* or *P. lunatus*. Over a range of 21°C day/16°C night to 36°C day/31°C night dry matter production was greater at 27°C day/22°C night. Night temperatures strongly affect many phases of the life cycle and differences in day temperature during reproductive growth markedly affect crop duration and yield. Marked cv differences in environmental responses have been identified. Ranging from Warm Temperate Thorn to Moist through Tropical Thorn to Wet Forest Life Zones, cowpea is reported to tolerate annual precipitation of 2.8–41.0 dm (mean of 54 cases = 14.2), annual mean temperature of 12.5°–27.8°C (mean of 54 cases = 22.1), and pH of 4.3–7.9 (mean of 46 cases = 6.2).

CULTIVATION: Seeds remain viable for several years. Germination is epigeal. Should be planted after danger from frost is past. If seeded for hay or seed, crop should be sown early, but for green manure and pasture purposes, may be seeded late with good results. Rate of seeding varies with method: when planted in rows 10–40 kg/ha; for broadcasting, 90 kg/ha. Cowpeas may be planted in rows, broadcast or mixed with such other plants as cassava, corn, sorghum, sudangrass, johnsongrass, millets, peanuts, or soybeans. When grown for seed, it is planted in rows; for forage or green manure, broadcast. For hog feed or silage, cowpeas are planted with corn, either at the same time as or at the last cultivation of corn. In rows, cowpeas are spaced of 5–7.5 cm apart, in rows 75–90 cm apart. Two or more cultivations are necessary to control weeds. Ordinary corn cultivator equipment is satisfactory, and cultivation should stop when flowering begins. In United States, 600–1,000 kg/ha of a 4-8-8 NPK fertilizer may be applied in bands 5 cm below seeds when planting. Cowpeas are usually grown rainfed, rarely irrigated. For weed control, amines of 2,4-D and MCPA are said to be effective as preemergence sprays. Trifluralin at 0.56–1.12 kg/ha just before sowing is said to give good control. Cowpeas respond slightly to K application up to 45 kg/ha. Calcium ions in the soil aid inoculation. In the United States, application of ca. 1 MT of lime is recommended and favors seed increase more than hay increase. Superphosphate recommendations are 112–224 kg/ha in the United States. Sulfur can limit seed production and/or protein synthesis. Molybdenum recommendations are 20–50 g/ha, and Mn, Cu, Zn, and B are essential, in very small quantities, for effective nodulation and seed yield increases. The cowpea symbiosis has genetic potential for large seed yields; cowpea *Rhizobium* associations should require only nominal amounts of fertilizer N, if any.

HARVESTING: Early maturing cvs produce pods in 50 days, seed in 90 days; late cvs mature seed in 240 days. Crop ripens unevenly and proper stage for harvesting is difficult to determine. Usually flowers and green and ripe pods occur on vines at same time. Crop is cut for seed when one-half to two-thirds of pods are ripe. May be harvested by hand, with a special harvester or by self-rake reapers. For hay, crop cut when most pods are fully developed, and first ones have ripened. If cut too early, hay is difficult to cure; if cut too late, stems are long and woody and seed and leaves shatter badly. Ordinary mowing machine is used for harvesting cowpeas.

YIELDS AND ECONOMICS: Yields vary with cv, climate, soil, and culture. Yields are: in the tropics, 400–600 kg dried beans/ha; in the United States, 1,000–1,500 kg/ha, but up to 3,000 kg/ha in California. Hay yields are ca. 5 MT/ha. World production in 1975 from 5,170,000 ha was 1,097,000 MT seed, (212 kg/ha). Africa produced 1,003,000 MT (199 kg/ha); North America, 57,000 MT (715 kg/ha); Asia, 27,000 MT (629 kg/ha); Europe, 6,000 (803 kg/ha). Oceania produced 3,000 MT, averaging 633 kg/ha. Nigeria had the highest national production with 850,000 MT; Upper Volta, 75,000, Haiti, 37,000, Senegal, 23,000, and United States, 20,000. Highest reported national yields were Trinidad with 3,457 kg/ha; Egypt, 3,103, Sri Lanka, 1,107; Japan, 1,000; and Greece, 803 (FAO, 1976). With good cvs and integrated insect control, yields of 2,000 kg/ha would be feasible. The estimated annual N fixation is 73–354 kg/ha with a global average of 198 kg/ha (double that of soybeans). Cowpea is the second most important pulse in tropical Africa and in Tropical America, especially Venezuela (ca. 35,000 ha) and Brazil (ca. 400,000 MT). In the United States cowpeas ranked 36th in acreage in 1965 and 30th in value ($3.5 million). World production was estimated then at 272,000 MT annually (probably conservative). In southeastern United States cowpeas are grown mainly for green pods and green seeds in southeastern United States and for dry beans in California.

BIOTIC FACTORS: In either dry or humid areas of United States and Africa, cowpeas are almost entirely self-pollinated. Some outcrossing (10–15%) probably always occurs with ants, flies, and bees as the main vectors. Flowers open early in morning, close by noon, and fall off the same day. Pollen is sticky and heavy. Grasshoppers may be the worst pests of cowpeas but are not serious in western Africa; where *Maruca testulalis, Enarmonia pseudonectis* and some *Heliothis* spp. damage foliage, flowers, buds, and pods; thrips and leafhoppers damage foliage, and bruchids damage dry seeds. In southeastern United States, the cowpea curculio is the most serious pest. Other insects recorded as attacking cowpeas are: southern cowpea weevil *(Callosobruchus maculatus)*; fall armyworm *(Spodoptera frugiperda)*; green cloverworm *(Plathypena scabra)*; bean leaf beetle *(Cerotoma trifurcata)*; Mexican bean beetle *(Epilachna varivestis)*; spotted cucumber beetle *(Diabrotica undecimpunctata)*; white fringed beetle *(Graphognathus leucoloma)*; garden fleahopper *(Halticus apterus)*; and caterpillars of *Heliothis zea*. In West Africa, *Ootheca mutabilis, Maruca testulalis,* and *Enarmonia pseudonectis* have been controlled with 8 applications of a BHC/DDT mixture at 5-day intervals. Dieldrin has proved effective against *Maruca testulalis* and diazinon against *Ophiomyia phaseoli*. With effective integrated insect control, seed yield has increased 30-fold. Rust *(Uromyces phaseoli)*, canker *(Xanthomonas vignicola)*, wilt *(Fusarium oxysporum)*, mildew *(Erysiphe polygoni)*, and charcoal rot *(Sclerotium baticola)* are the principal diseases. Anthracnose *(Colletotrichum lindemuthianum)* is damaging in humid situations, and is very important in western Africa, where rust and leaf-spot prevail. Resistance to stemrot *(Phytophthora vignae)* is reported. Following fungi have been recorded as attacking cowpeas: *Aristastoma oenconomicum, Ascochyta pisi, Botrytis cinerea, Botryodiplodia theobromae, Cercospora canescens, C. cruenta, C. dolichi, C. vignicaulis, Chaetoseptoria vignae, C. wellmanii, Choanephora cucurbitarum, Cladosporium cladosporioides, C. vignae, Colletotrichum capsici, C. lindemuthianum, Corticium solani, Corynespora casiicola, Diaporthe phaseolorum, Diplodia natalensis, Erysiphe polygoni, Fusarium oxysporum* forma *tracheiphilum* (cowpea wilt, most destructive disease in California), *Glomerella cingulata, G. lindemuthianum, G. vignicaulis, Helminthosporium vignae, Leptosphaerulina vignae, Macrophomina phaseoli, Microsphaera alni, Mycosphaerella pinodes, Myrothecium roridum, Nematospora phaseoli, Pellicularia filamentosa, P. koleroga, Periconia byssoides, Peronospora trifoliorum, Phoma bakeriana, P. vignae, Phyllosticta phaseolina, Phymatotrichum omnivorum, Phytophthora cactorum, P. vignae* (stem rot), *Pythium artotrogus, P. debaryanum, P. splendens, P. ultimum, Rhizobium japonicum, Rhizoctonia solani, Rhizopus stolonifer, Sclerotinia sclerotiorum, Sclerotium bataticola,* (charcoal rot), *S. rolfsii, Septoria melanophthalmi, S. vignae, S. vignicola, Sphaerotheca fuliginea, Synchytrium dolichi, Stagonospora phaseoli, Thielaviopsis basicola, Thyrospora solani, Uromyces appendiculatus, U. phaseoli, U. vignae, Verticillium alboatrum,* and *Woroninella dolichi.* Cowpeas are infected by the following bacterial diseases: *Pseudomonas phaseolicola, Ps. solanacearum, Ps. tabaci,* and *Xanthomonas vignicola* (serious in West Africa). Virus disease of cowpeas are: alfalfa mosaic, Argentina sunflower, Brazilian tobacco streak, bushy stunt of tomato, carnation ringspot, clover mosaic, coconut wilt, cucumber mosaic, cucumber necrosis, eggplant mosaic, green mosaic, lucerne mosaic, of *Crotalaria hirsuta,* mosaic of *Crotalaria mucronata,* peach yellow bud mosaic, *Pelargonium* leaf curl, ringspot disease of red raspberry, subterranean clover stunt, tobacco etch, tobacco mosaic, tobacco necrosis, tomato aspermy, tomato black ring, tomato spotted wilt, cowpea mosaic (seed-borne, carried by bean-leaf beetle, *Cerotoma ruficornis*), white clover mosaic. Cowpeas are parasitized in Africa by *Striga gesnerioides, S. lutea,* and *S. senegalensis.* Despite nearly 75 years of efforts to find cvs resistant to root-knot nematodes (*Meloidogyne* spp.) and sting nematodes (*Belonolaimus gracilis*) the problems are still acute. Nematodes which infest cowpeas are: *Belonolai-*

mus gracilis, Criconemoides spp., Ditylenchus dipsaci, Helicotylenchus cavenessi, H. pseudorobustus, Hemicycliophora arenaria, Heterodera cajani, H. glycines, H. schachtii, H. vigna, Hoplolaimus seinhorsti, Meloidogyne arenaria, M. hapla, Meloidogyne incognita, M. i. acrita, M. javanica, M. thamesi, Paratylenchus minutus, Peltamigratus nigeriensis, Pratylenchus brachyurus, P. coffeae, P. goodeyi, P. minyus, P. penetraus, P. pratensis, P. scribaeri, Pratylenchus thornei, P. vulnus, P. zeae, Radopholus similis, Rotylenchulus renibormis, Scutellonema bradys, S. clathricaudatum, Trichodorus sp., Tylenchorhynchus brevidens, Xiphinema americanuum, X. basiri, X. ifacolum, and Zygotylenchus guevarai.

REFERENCES:
El Baradi, T. A., 1975, Pulses 1. Cowpeas, *Abstr. Trop. Agric.* **1**(12):9–19.
Lorz, A. P., *et al.*, 1955, Production of Southern Peas (Cowpeas) in Florida, *Bull. Fla. Agric. Exp. Stn.* **557**.
Lorz, A. P., 1961, Breeding southern peas for processing, *Proc. Fla. State Hortic. Soc.* **74**:282–284.
Philpotts, H., 1965, Effect of soil temperature on nodulation of cowpeas *(Vigna sinensis)*, *Aust. J. Exp. Agr. Anim. Husb.*, **7**:372–376.
Steele, W. M., 1972, Doctoral thesis, University of Reading, United Kingdom, 242 pp.
Summerfield, R. J., Huxley, P. A., and Steele, W. M., 1974, Cowpea *(Vigna unguiculata* (L.) Walp.), *Field Crop Abstr.* **27**(7):301–312.

CONTRIBUTORS: J. A. Duke, C. F. Reed, K. O. Rachie, R. J. Summerfield.

Vigna vexillata (L.) A. Rich.

FAMILY: Fabaceae
COMMON NAMES: Zombi pea, Aka sasage, Pois zombi, Pois poison
SYNONYMS: *Phaseolus pulniensis* Wight.
Phaseolus quadriflorus A. Rich.
Phaseolus sepiarius Dalz.
Phaseolus vexillatus L.
Plectrotropis hirsuta Schum. & Thonn.
Strophostyles capensis (Thunb.) E. Mey.
Vigna carinalis Benth.
Vigna crinita A. Rich.
Vigna golungensis Bak.
Vigna hirta Hook.
Vigna scabra Sond.
Vigna senegalensis A. Chev.
Vigna thonningii Hook. f.
Vigna tuberosa A. Rich.

USES: Tuberous roots eaten like sweet potatoes in Sudan and Ethiopia. Widely cultivated as green pasture, cover, and soil-improvement crop, as well as for erosion control.

FOLK MEDICINE: No data available.

CHEMISTRY: No data available.

DESCRIPTION: Perennial climbing or trailing herb, 0.6–3 m long, from a narrow woody rootstock; stems sparsely to very densely covered with brown hairs; leaves compound, petioles and peduncles rough and hispid; leaflets variable, ovate to lanceolate, varying from 3.7–6.5 cm long and 2.3 cm broad to 10–12.5 cm long and 1.3 cm broad, tapering to acute apex, green on both surfaces, sparsely hairy above, setose along veins beneath; peduncles 2–4-flowered at summit; flowers greenish-yellow tinged with purple; calyx tubular-campanulate, with

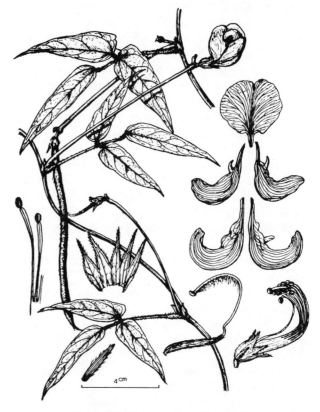

Figure 145. *Vigna vexillata* (L.) A. Rich.

5 acuminate segments somewhat longer than tube, vexillum broad, reflexed, ca. 2 cm long, with a folded claw, limb with inflexed ears at base; carina obliquely circularly incurved, spurred on one side; legume sessile, linear, straight, 7.4 cm long, 2.5–4 mm broad, 10–11-seeded, glabrescent or pubescent and covered with brown bristly hairs; seed oblong or oblong-reniform, buff to black, unspotted, or brown or dark red with black speckles, 2.5–4.5 mm long, 2–2.5 mm in diameter, sometimes densely brown-hispidulous. Root spindle-shaped, as a taproot or tuberous rootstock, sometimes divided, containing starch and edible. Fl. and Fr. July; Sept.–Oct. in Japan.

GERMPLASM: Many varieties have been recognized for this widely distributed species: var *linearis* Craib, var *pluriflora* Franch., var *stocksii* Bak., var *tsusimensis* Matsum., var *youngiana* F. M. Baill. and var *yunnanensis* Franch. Variations are based on pubescence, leaflet shape and size, and flower variations. In some cvs flowers are pink or purple and fade to yellow. Typical plants (var *vexillata*) are very densely ferruginous-pubescent, with ovate to lanceolate leaflets up to 16 cm long and 8 cm broad, and calyx-lobes up to 2 cm long; var *angustifolia* (Schumach. & Thonn.) Bak. is glabrescent or less hairy, with very narrow leaflets 4–8 cm long and 0.4–1.5 cm broad, and calyx-lobes short, 2–8 mm long; var *dolichonema* (Harms) Verdc. has calyx-lobes up to 2.2 cm long and leaflets up to 9.5 cm long and 5.5 cm broad, and is known only from Tanzania; var *hirta* (Hook.) Bak. f. is intermediate with narrow leaves and dense bristly indumentum, and grows in Uganda and Tanzania. All cvs freely interbreed. Assigned to the African and Hindustani Centers of Diversity, zombi pea or cvs thereof is reported to exhibit tolerance to low pH, poor soil, and slope. ($2n = 22$.)

DISTRIBUTION: Pantropical; throughout tropical Africa, extending to India, Malaysia, and Australia; spread to tropical South America and West Indies; often forming roadside thickets.

ECOLOGY: Grows native in grassy places, thickets, grasslands, bushland, forest, and abandoned cultivated areas from sea level to 2,250 m. Widely distributed from Tropics to warm temperate regions, often in humid and eroded areas. Ranging from Subtropical Dry to Moist through Tropical Moist to Wet Forest Life Zones, zombi pea is reported to tolerate annual precipitation of 6.4–42.9 dm (mean of 11 cases = 18.3), annual mean temperature of 18.3°–27.4°C (mean of 11 cases = 22.9°C), and pH of 5.0–7.8 (mean of 7 cases = 6.1).

CULTIVATION: Propagated by seed. Roots develop abundant nodules and plants serve as excellent pioneers for sterile land. Plants develop rapidly and quickly provide erosion control. Seed often planted in rows, but grown and treated much like potatoes.

HARVESTING: Tubers harvested at end of season by digging up entire plants and removing the tubers.

YIELDS AND ECONOMICS: An important tropical legume for food, soil improvement, and erosion control.

BIOTIC FACTORS: The nematodes, *Meloidogyne incognita* and *M. javanica,* have been isolated from this plant. Plants are very sensitive to herbicides.

CONTRIBUTORS: J. A. Duke, C. F. Reed.

Voandzeia subterranea (L). Thouars

FAMILY: Fabaceae
COMMON NAMES: Bambarra groundnut, Bambarra erdnuss, d'Angole, Voandzu
SYNONYMS: *Glycine subterra* L.
Voandzeia subterranea forma *sativa* Jacques-Felix

USES: Beans or seeds consumed as food, much like peanuts. Oil content is low, but the Azande of the Congo roast then pound seeds to extract the oil. Bambarra is not particularly rich in protein but is high in calories. Beans may be consumed fresh, grilled, boiled, or made into a flour to form cakes. Said to be intermediate in flavor between Brosimum and Pisum. Immature shelled or unshelled nuts are pounded and boiled to a stiff porridge, which keeps for a long time and can be used on journeys. Beans are used as coffee substitute. Seeds variously prepared may be served as relishes, appetizers, or served with cassava or plantain. Ghana produces more than 40,000 cans of seeds in gravy annually.

Voandzeia subterranea (L.) Thouars

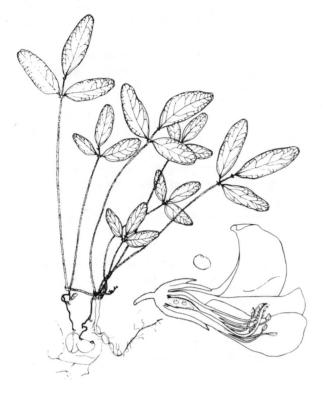

Figure 146. *Voandzeia subterranea* (L.) Thouars

Leaves, rich in nitrogen and phosphorus, are used for grazing and forage. Seeds may be pounded with capsicum and other ingredients, wrapped in musaceous or marantaceous leaves.

FOLK MEDICINE: No data available.

CHEMISTRY: The raw seeds contain per 100 g as purchased: 152 calories, 57.3% moisture, 7.8 g protein, 3.1 g fat, 30.0 g total carbohydrate, 3.0 g fiber, 1.8 g ash, 14 mg Ca, 258 mg P, and 1.2 mg Fe. The dried seeds contain (per 100 g) 367 calories, 10.3% moisture, 18.8 g protein, 6.2 g fat, 61.3 g total carbohydrate, 4.8 g fiber, 3.4 ash, 62 mg Ca, 276 mg P, 12.2 mg Fe, 10 μg β-carotene equivalent, 0.47 mg thiamine, 0.14 mg riboflavin, 1.8 mg niacin, and 0–8 mg ascorbic acid. The amino acid content is (mg/g N): leucine 494–510, lysine 400–430, valine 331–340, phenylalanine 219–360, isoleucine 275–280, threonine 219–240, methionine 113–120, and cystine 70–180. The fatty acid composition of the oil is palmitic, 19.4%; stearic, 11.8%, oleic, 24.4%; linoleic, 34.2%, arachidic, 5.3%; and behenic 4.9%. An early analysis compared the green seed (45.4% moisture, 1.28 ash, 2.35 cellulose, 2.41 fat, 20.49 sugar and starch, and 5.67 albuminoids) with the ripe seed (10.27% water, 3.23 ash, 4.63 cellulose, 5.03 fat, 46.8 sugar and starch, and 12.04 albuminoids) with fresh leaves (66.00 water, 3.32 ash, 10.35 cellulose, 1.45 fat, 14.63 sugar and starch, and 4.25 albuminoids) and dry leaves (10.76 water, 7.30 ash, 42.20 cellulose, 2.42 fats, 31.41 sugar and starch, and 5.91 albuminoids). The green shells are: 17.96 water, 0.18 ash, 1.23 cellulose, 0.02 fats, 2.66 sugar and starch, 0.35 albuminoids, and ripe shells 2.00 water, 0.71 ash, 5.23 cellulose, 0.14 fat, 8.60 sugar and starch, and 1.32 albuminoids. The seeds are reported to contain trypsin inhibitors and chymotrypsin inhibitors.

DESCRIPTION: Annual herb; short-creeping, much-branched stems, rooting at nodes, the internodes very short thus giving a bunchy appearance to the plant; leaves pinnately trifoliolate, petioles long, erect, grooved, thickened at the base, hairy, with stipules; petioles and stems pink, purple or bluish-green, depending on the cv; terminal leaflet subtended by two stipules, while the lateral ones have one each; margins entire, the lateral leaflets less regular in shape and shorter than the terminal one; flowers spreading out at ground level on hairy peduncles (which arise from the nodes of the stem) to the bulbous ends of which 1–3, usually 2, flowers may be attached by pedicels; petals in most cvs yellowish-white, in some cvs deeper yellow with reddish-brown striations, and in a few light pink; fruit a bivalved subterranean pod 1.25–2.5 cm long, indehiscent, white and smooth in the fresh mature state, becoming brownish and shriveled when dry, attached to the stem by long, flexible peduncle from which it detaches itself at maturity; seeds 1–2 per pod, rounded or flattened on one side, smooth, hard. Seed white or cream, brown, purple, black, and speckled in various shades.

GERMPLASM: There are many cvs of Bambarra groundnut (Johnson, 1968), differing in shape of leaves and size, hardiness, maturity period and color of seeds. Plants differ in growth habit, being bunched, semibunched or open; open plants are more like wild types. The species is both self-pollinating and cross-pollinating, ants being the main pollinators. Ants are thought to soften the soil, making penetration of the fertilized flowers easier. Center of origin: along the Nile, Northern Nigeria, northern Cameroons. *V. s.* var *spontanea*

(Harms) Hepper (Syn.: *V. s.* forma *spontanea* Harms) refers to the wild form found in n Nigeria and Cameroons. Assigned to the African Center of Diversity, Bambarra groundnut or cvs thereof is reported to exhibit tolerance to disease, drought, heat, insects, low pH, poor soil, sand, and savanna. ($2n = 22$.)

DISTRIBUTION: Widely cultivated throughout tropical Africa, as well as in Madagascar, Mauritius, India, Sri Lanka, Indonesia, Philippine Islands, Malaysia, Iowa, New Caledonia, northern Australia, tropical Central America, Surinam, and Brazil.

ECOLOGY: Best adapted to the savanna areas and rainforest/savanna transition (or derived savanna) of Africa. Thrives on high temperatures and in bright sunshine, faring best with average day temperatures of 20°–28°C. Gives a good crop under conditions of high temperature and low rainfall, but does better with frequent rains from sowing to flowering. Withstands heavy rainfall except during fruiting and harvesting, but can develop with as little as 5 dm/yr. Does well on sandy soils, pH 5.0–6.5, but grows on any well-drained soil, even on poor, worn-out soils (one of the best-suited crops for savanna ochrosols). Intolerant of calcareous soils. A short-day plant, requiring at least 100–120 days of frost-free weather. Thrives up to altitudes of 1,520 m. Ranging from Subtropical Dry to Moist through Tropical Very Dry to Dry Forest Life Zones, Bambarra groundnut is reported to tolerate annual precipitation of 5.2–41.0 dm (mean of 12 cases = 13.8), annual mean temperatures of 18.7°–27.4°C (mean of 12 cases = 22.7) and pH of 4.3–7.1 (mean of 10 cases = 5.5).

CULTIVATION: In the tropics two crops are often possible, most cvs maturing seeds in 100–180 days. Seeds are planted for the first crop in May–June, for the second, July–August, usually during the rainy season. However, planting dates vary with distances from the equator, being as early as February in Tanganyika and as late as mid-November in South Africa. Seed planted at rates varying from 25 to 75 (–190) kg/ha, spaces 10–45 cm apart in rows, which are 15–55 cm apart to permit weed cultivation or in double rows 20 cm apart on flat-topped ridges 60 cm apart. Whole pods may be planted, but usually the seeds alone are planted up to 7.5 cm deep. Scraped seeds germinate in about 6 days. Seedbed or field should be deeply plowed and worked until it is friable to enable plant to bury its seed pods after fertilization of flowers. In southern Africa plants are hand-weeded when 10 cm tall, and mounded up at flowering time to encourage development of the pods underground. Crop is best grown alone, but is often grown with grains, (millet, maize, or sugarcane) or root crops of other legumes. Addition of superphosphate at planting time and later at the rate of 60 hg/ha with 40 kg of sulfate (3 weeks after sowing) greatly increases the yield. On soils of low fertility, application of 225–450 kg/ha general fertilizer followed by 40–60 kg/ha ammonium sulfate, 3 weeks after planting, is recommended. Nitrogen-rich soils tend to encourage vegetative growth at expense of fruit production.

HARVESTING: Bunch types take ca. 90–120 days to mature, spreading types ca. 120–150 days. For eating, the seed should not be fully mature. The crop may be harvested when the lower half of the leaves are wilted or when foliage turns yellow and withers. It is uprooted by hand or with hoes, and the soil near the root zone is stirred in search of detached pods. For seed production, plants· are uprooted with a hoe, plowed out or pulled by hand, and allowed to dry in the field in windrows for 1 or 2 days. Dry weather is best for harvesting. Seeds will germinate in the pods if allowed to get wet. Pods may be shelled as soon as harvested, or they may be allowed to dry in the sun for 1–4 weeks. Threshing is usually manual, cracking the pods by various means. Seed is stored in gunny sacks, straw baskets, earthenware jars or in granaries. Since stored seed are subject to bruchid attack, ashes are sometimes added.

YIELDS AND ECONOMICS: Intercropped with sugar cane, Bambarra groundnut may yield nearly 2,400 kg/ha green nuts (50% moisture) and nearly 3 MT leaves. Seed yields average 300–800 kg/ha, depending on cv, weather conditions, spacing in the fields and size of seed. Yields average 650–850 kg/ha in Africa but can be as low as 56–112 kg/ha in Zambia and as high as 3,580 kg/ha in Rhodesia. Studying 14 cvs in Ghana, Karikari and Lavoe (1977) report that yields in bunch cvs are 808–1,100 kg/ha, while yields in semibunch cvs are 714–767 kg/ha. Highest reported yield was 4,400 kg/ha. The plant produces less biomass than Arachis. Bambarra groundnut does not enter world trade but is a very important local legume crop, especially for poor lands of tropical Africa, where it is second only to cowpeas and peanuts. Estimated production in Africa is 330,000 MT; Nigeria, 100,000; Upper Volta, 65,000; Niger, 30,000; Ghana, 20,000; Togo, 8,000; and the Ivory Coast, 7,000. In Ghana it ranks second only to cowpea in production and consumption. In addition to the seed for human consumption, the tops are also palatable to stock. Efforts are being made to select cultivars for local areas.

BIOTIC FACTORS: Although Bambarra groundnut is relatively free from fungal and insect pests, fungal diseases include: *Ascochyta phaseolorum* (attacks leaves), *Cercospora canescens, C. voandzeia, Colletotrichum capsici, Corticium solani, Elsinoe* sp., *Erysiphe polygoni, Fusarium oxysporum, Leptosphaerulina trifolii, Meliola vignaegracilis, Phaeolus manihotis* (causes root rot), *Phyllosticta voandzeiae, Sclerotium rolfsii, Rhizoctonia bataticola,* and *Sphaerotheca voandzeiae* (attack leaves), and *Synchytrium dolichi.* Viruses include: alfalfa mosaic (3 strains), alsike clover mosaic, bean chlorotic ringspot (4 strains), bean local chlorosis (5 strains), bean mosaic, bean necrosis, and white clover mosaic. *Meloidogyne javanica* (root-knot) and *Meloidogyne* sp. also attack this crop. Leafhoppers *(Empoasca facialis, Hilda patruelis)* and the larvae of *Diacrisia maculosa* and *Lamprosema indicata* may be somewhat troublesome. Bruchids *(Callosobruchus maculatus, C. subinnotatus)* may infest stored seeds, as does *Ctenocampa hilda* in Sierra Leone and *Piezotrachelus ugandanus* in western Africa.

REFERENCES:

Anonymous, 1910, The Bambarra groundnut, *Agric. News* 9(222):340–341.

Doku, E. V., and Karikari, S. K., 1971, Bambarra groundnut, *Econ. Bot.* **25**(3):255–262.

Doku, E. V., and Karikari, S. K., 1971, The role of ants in pollination and pod production of Bambarra groundnut, *Econ. Bot.* **25**(4):357–362, illus.

Doku, E. V., 1968, Flowering, pollination, and pod formation in Bambarra groundnut *(Voandzeia subterranea)* in Ghana, *Exp. Agric.* **4**:41–48.

Doku, E. V., 1969, Growth habit and pod production in Bambarra groundnut *(Voandzeia subterranea), Ghana J. Agric. Sci.* **2**.

Hepper, F. N., 1963, The Bambarra groundnut *(Voandzeia subterranea)* in West Africa, *Kew Bull.* **16**:398–407.

Holm, J. M., and Marloth, B. W., 1940, Bambarra groundnut or Njugobean, *Farming S. Afr. Bull.* **215**.

Johnson, D. F., 1968, The Bambarra groundnut *(Voandzeia subterranea), Rhod. Agric. J.* **65**(1):1–4.

Karikari, S. K., and Lavoe, S. K., 1977, Preliminary evaluation and utilization of fourteen cultivars of Bambarra groundnut *(Voandzeia subterranea* Thouars), *Acta Hortic.* **53**:195–199.

CONTRIBUTORS: J. A. Duke, B. N. Okigbo, C. F. Reed, J. K. P. Weder.

Appendix

TABLE 1.
Legume "Toxins": Their Toxicity and Generic Distribution[a]

Chemical	Oral LD_{50}[b]	Legume genera in which reported
Acetone	5,300	*Phaseolus*
Adenine	755	*Pachyrrhizus*
Albitocin	19	*Albizzia*
Anhaline		*Acacia*
p-Anisaldehyde		*Acacia*
Anthraquinone		*Acacia*
L-Arabinose (pectinose)		*Prosopis*
Benzaldehyde	1,000	*Acacia*
Benzyl alcohol	1,230	*Acacia*
Bufotenine		*Desmodium, Lespedeza, Piptadenia*
Butyraldehyde	2,490	*Acacia*
Caprylic acid (octanoic acid)		*Spartium*
Choline		*Canavalia, Cicer, Genista*
Chrysarobin		*Andira, Cassia*
Cinnamaldehyde	1,160	*Cassia*
Citric acid		*Tamarindus*
Coumarin	196	*Dipteryx, Melilotus*
p-Cresol	861	*Acacia*
Cuminic aldehyde (cuminaldehyde)	1,390	*Acacia*
Cyanocobalamin (vitamin B_{12})		*Medicago*
Cytisine	101	*Anagyris, Argyrolobium, Cladrastis, Coronilla, Cytisus, Euchresta, Genista, Gleditsia, Laburnum, Lotus, Piptanthus, Sophora, Spartium, Templetonia, Thermopsis, Ulex*
Decanal	3,730	*Acacia, Cassia*
Derris		*Derris*
Dimethyl tryptamine (indole)		*Acacia, Desmodium, Lexpedeza, Mimosa, Petalostylis, Piptadenia*
Donaxine (gramine)		*Desmodium, Lupinus*
DOPA	582	*Astragalus, Baptisia, Cytisus, Lupinus, Mucuna, Vicia*
Dopamine	165	*Mucuna*
Erysodine (HCl)	155	*Erythrina*
Erysopine (HCl)	18	*Erythrina*

(Continued)

Appendix

TABLE 1. (*Continued*)

Chemical	Oral LD$_{50}$[b]	Legume genera in which reported
Erysothiopine (Na)		*Erythrina*
Erythraline (HBr)	80	*Erythrina*
Erythramine (HBr)		*Erythrina*
β-Erythroidine		*Erythrina*
Folic acid (vitamin M)		*Lathyrus, Lens*
Fumaric acid		*Phaseolus*
Gallic acid	5,000	*Acacia*
Gramine (donaxine)		*Desmodium, Lupinus*
Guanidine		*Galega, Glycine, Vicia*
Heptanoic acid	160	*Acacia*
Histamine		*Trifolium*
Hydrocyanic acid	3.7	*Albizia, Bauhinia, Caesalpinia, Canavalia, Cassia, Castanospermum, Entada, Erythrophloeum, Gliricidia, Glycine, Indigofera, Lathyrus, Leucaena, Lotononis, Lotus, Medicago, Melilotus, Melolobium, Mundulea, Phaseolus, Tamarindus, Tephrosia, Trifolium, Vicia*
Hydroxysenkirkine		*Crotalaria*
Hypoxanthine		*Lupinus*
Indican		*Crotalaria, Indigofera, Robinia*
Indole	1,000	*Robinia*
Inositol-(meso) (myoinositol)		*Phaseolus*
Isobutyraldehyde	2,810	*Acacia*
Isobutyric acid	280	*Ceratonia*
Isochaksine		*Cassia*
Jacobine		*Crotalaria*
Lactose		*Ceratonia*
Lapachol	487	*Adenanthera?, Andira?, Intsia?*
Leucaenine		*Leucaena, Mimosa*
Longilobine (retrorsine)		*Crotalaria*
Lysine		*Lupinus, Pisum, Robinia, Vicia*
Malic acid		*Medicago, Tamarindus*
Malonic acid	1,310	*Anthyllis, Astragalus, Colutea, Lotus, Lupinus, Medicago, Melilotus, Ononis, Phaseolus, Sophora, Thermopsis, Trifolium, Trigonella*
Maltose		*Ceratonia, Glycine, Lathyrus, Phaseolus, Trifolium*
Melilotin (hydrocoumarin)	1,460	*Melilotus*
S-methyl-1-cysteine		*Astragalus, Phaseolus*
Methyl salicylate	887	*Acacia*
Monocrotalic acid		*Crotalaria*
Monocrotaline	66	*Crotalaria*
Nicotine	3	*Acacia, Mucuna*
Oxalic acid		*Cicer, Glycine, Lens, Medicago, Phaseolus, Pisum, Vigna*
Pachycarpine (D-sparteine)		*Ammodendron, Ammothamnus, Anagyris, Baptisia, Cytisus, Genista, Lupinus, Thermopsis*
Palmitic acid		*Acacia*
Pantothenic acid		*Lathyrus, Lens, Medicago*
Pectin		*Lupinus, Medicago, Phaseolus, Tamarindus*

TABLE 1. (*Continued*)

Chemical	Oral LD$_{50}$[b]	Legume genera in which reported
Phellandrene		*Caesalpinia*
Phenethylamine		*Acacia*
Physostigmine	4.5	*Dioclea, Mucuna, Physostigma, Vicia*
Piperidine	400	*Lupinus*
Piperonal (heliotropin)	2,700	*Robinia*
Quercitin	161	*Acacia, Bauhinia, Trifolium*
Quercitrin (flavone)		*Bauhinia, Lathyrus, Leucaena, Vicia*
Quinic acid		*Medicago*
Raton		*Gliricidia*
Retrorsine		*Crotalaria*
Ricinoleic acid		*Cassia*
Rotenone	132	*Cracca, Derris, Lonchocarpus, Milletia, Mundulea, Pachyrhizus, Piscidia, Spatholobus, Tephrosia*
Rutin		*Acacia, Baptisia, Bauhinia, Daviesia, Phaseolus, Sophora, Tephrosia*
Sanguinarine		*Ammodendron, Robinia*
Saponin		*Abrus, Acacia, Albizia, Astragalus, Cassia, Castanospermum, Cicer, Coronilla, Entada, Glycine, Glycyrrhiza, Lathyrus, Lens, Lotus, Lupinus, Medicago, Mora, Ononis, Pachyrhizus, Phaseolus, Pisum, Pithecellobium, Sesbania, Sophora, Spartium, Stryphnodendron, Swartzia, Trifolium*
Senecionine		*Crotalaria*
Seneciphylline		*Crotalaria*
Shikimic acid		*Caesalpinia, Lespedeza, Medicago, Vicia*
Sparteine (lupinidine)		*Ammodendron, Ammothamnus, Anagyris, Baptisia, Cytisus, Genista, Hovea, Lupinus, Piptanthus, Sarothamnus, Sophora, Spartium, Templetonia, Thermopsis*
Tannic acid		*Acacia, Glycyrrhiza, Lespedeza, Leucaena, Peltophorum*
Tartaric acid	5,000	*Tamarindus*
Terpineol	4,300	*Acacia*
Tonka absolute	1,380	*Dipteryx*
Trigonelline		*Acacia, Astragalus, Canavalia, Glycine, Lupinus, Medicago, Pisum, Trigonella*
Tryptophan		*Glycine, Lupinus, Medicago, Peltophorum*
Tyramine		*Acacia, Cytisus*
Vanillin	1,400	*Myroxylon*
Xanthine		*Lupinus, Vicia*

[a] Source: Duke (1977).
[b] All LD$_{50}$ entries in milligrams per kilogram body weight; all toxicity data derived from National Institute of Occupational Safety and Health (1975).

Appendix

TABLE 2.
Legume Genera and Their "Toxins"

Abrus[b] (Fabaceae): saponin
Acacia (Mimosaceae): anhaline, anisaldehyde, anthraquinone, benzaldehyde, benzyl alcohol, butyraldehyde, cresol, cuminic aldehyde, decanal, dimethyl tryptamine, eugenol methyl ether, gallic acid, heptanoic acid, hydrocyanic acid, indole, isobutyraldehyde, linalool, methyl salicylate, nicotine, palmitic acid, phenethylamine, quercitin, rutin, saponin, tannic acid, terpineol, trigonelline, tyramine
Adenanthera (Mimosaceae): lapachol
Adenocarpus (Fabaceae): sparteine
Albizia (Mimosaceae): albitocin, hydrocyanic acid, saponin
Ammodendron (Fabaceae): pachycarpine, sanguinarine, sparteine
Ammothamnus (Fabaceae): pachycarpine, sparteine
Anagyris (Fabaceae): cytisine, pachycarpine, sparteine
Andira (Fabaceae): chrysarobin, lapachol
Anthyllis (Fabaceae): malonic acid
Arachis (Fabaceae):
Argyrolobium (Fabaceae): cytisine
Astragalus (Fabaceae): dopa, malonic acid, methyl cysteine, saponin, selenium, trigonelline
Baptisia (Fabaceae): cytisine, dopa, pachycarpine, rutin, sparteine
Bauhinia (Caesalpiniaceae): hydrocyanic acid, quercitin, quercitrin, rutin
Caesalpinia (Caesalpiniaceae): hydrocyanic acid, phellandrene, shikimic acid
Canavalia (Fabaceae): choline, hydrocyanic acid, trigonelline
Castanospermum (Fabaceae): hydrocyanic acid, saponin
Ceratonia (Caesalpiniaceae): isobutyric acid, lactose, maltose
Cicer (Fabaceae): choline, oxalic acid, saponin
Cladrastis (Fabaceae): cytisine
Colutea (Fabaceae): malonic acid
Coronilla (Fabaceae): cytisine, saponin
Cracca (Fabaceae): rotenone
Crotalaria (Fabaceae): hydrocyanic acid, hydroxysenkirkine, indican, jacobine, longilobine, monocrotaline, retronecine, retrorsine, riddelline, senecionine, seneciphylline
Cytisus (Fabaceae): cytisine, dopa, pachycarpine, sparteine, tyramine
Daubentonia (Fabaceae):
Daviesia (Fabaceae): rutin
Delonix (Caesalpiniaceae): hydrocyanic acid
Derris (Fabaceae): derris, rotenone
Desmodium (Fabaceae): bufotenine, dimethyl tryptamine, donaxine
Dichrostachys (Mimosaceae):
Dioclea (Fabaceae): physostigmine
Dipteryx (Fabaceae): coumarin, tonka absolute
Dolichos (Fabaceae): hydrocyanic acid
Entada (Mimosaceae): hydrocyanic acid, saponin
Erythrina (Fabaceae): erysodine, erysopine, erysothiopine, erythraline, erythramine, erythroidine
Erythrophloeum (Mimosaceae): hydrocyanic acid
Euchresta (Fabaceae): cytisine
Galega (Fabaceae): guanidine
Genista (Fabaceae): choline, cytisine, pachycarpine, sparteine

TABLE 2. (*Continued*)

Gleditsia (Caesalpiniaceae): cytisine
Gliricidia (Fabaceae): hydrocyanic acid, raton
Glottidium (Fabaceae):
Glycine (Fabaceae): betaine, choline, guanidine, hydrocyanic acid, isovaleraldehyde, maltose, oxalic acid, saponin, trigonelline, tryptophane
Glycyrrhiza (Fabaceae): saponin, tannic acid
Gymnocladus (Caesalpiniaceae):
Hovea (Fabaceae): sparteine
Indigofera (Fabaceae): hydrocyanic acid, indican
Intsia (Mimosaceae): lapachol
Laburnum (Fabaceae): cytisine
Lathyrus (Fabaceae): folic acid, hydrocyanic acid, maltose, pantothenic acid, quercitrin, saponin
Lens (Fabaceae): choline, folic acid, oxalic acid, pantothenic acid, saponin
Lespedeza (Fabaceae): bufotenine, dimethyl tryptamine, shikimic acid, tannic acid
Leucaena (Mimosaceae): hydrocyanic acid, leucaenine, mimosine, quercitrin, tannic acid
Lotus (Fabaceae): cytisine, hydrocyanic acid, malonic acid, saponin
Lupinus (Fabaceae): dopa, hydrocyanic acid, hypoxanthine, lysine, malonic acid, pachycarpine, pectin, piperidine, saponin, sparteine, trigonelline, tryptophane, xanthine
Medicago (Fabaceae): choline, citric acid, hydrocyanic acid, limonene, malic acid, malonic acid, oxalic acid, pantothenic acid, pectin, quinic acid, saponin, shikimic acid, trigonelline, tryptophane
Melilotus (Fabaceae): coumarin, hydrocyanic acid, malonic acid, melilotin
Melolobium (Fabaceae): hydrocyanic acid
Milletia (Fabaceae): rotenone
Mimosa (Mimosaceae): anisaldehyde, dimethyl tryptamine, histamine, mimosine
Mora (Caesalpiniaceae): saponin
Mucuna (Fabaceae): dopa, dopamine, nicotine, physostigmine, serotonine
Mundulea (Fabaceae): hydrocyanic acid, rotenone
Myroxylon (Fabaceae): benzoic acid, peru balsam oil, phellandrene, vanillin
Ononis (Fabaceae): malonic acid, saponin
Oxytropis (Fabaceae): selenium
Pachyrhizus (Fabaceae): adenine, choline, rotenone, saponin
Parkinsonia (Caesalpiniaceae):
Peltophorum (Caesalpiniaceae): tannic acid, tryptophane
Petalostylis (Caesalpiniaceae): dimethyl tryptamine
Phaseolus (Fabaceae): acetone, choline, fumaric acid, hydrocyanic acid, inositol, malonic acid, maltose, methylcysteine, oxalic acid, pectin, rutin, saponin, succinic acid
Physostigma (Fabaceae): physostigmine
Piptadenia (Mimosaceae): bufotenine, dimethyl tryptamine
Piptanthus (Fabaceae): cytisine, sparteine
Piscidia (Fabaceae): rotenone
Pisum (Fabaceae): oxalic acid, saponin
Pithecellobium (Mimosaceae): saponin
Prosopis (Mimosaceae): arabinose

(*Continued*)

TABLE 2. (*Continued*)

Psoralea (Fabaceae):
Robinia (Fabaceae): chrysarobin, indican, indole, lysine, piperonal, sanguinarine
Sarothamnus (Fabaceae): sparteine
Sesbania (Fabaceae): saponin
Sophora (Fabaceae): anabasine, cytisine, malonic acid, pachycarpine, rutin, saponin, sparteine
Spartium (Fabaceae): caprylic acid, cytisine, saponin, sparteine
Spatholobus (Fabaceae): rotenone
Swartzia (Fabaceae): saponin
Tamarindus (Caesalpiniaceae): citric acid, hydrocyanic acid, malic acid, pectin, tartaric acid
Templetonia (Fabaceae): cytisine
Tephrosia (Fabaceae): hydrocyanic acid, rotenone, rutin
Thermopsis (Fabaceae): cytisine, malonic acid, pachycarpine
Trifolium (Fabaceae): histamine, hydrocyanic acid, malonic acid, maltose, quercitin, saponin
Trigonella (Fabaceae): choline, malonic acid, trigonelline
Ulex (Fabaceae):
Vicia (Fabaceae): choline, guanidine, hydrocyanic acid, lysine, physostigmine, quercitrin, shikimic acid, xanthine
Vigna (Fabaceae): oxalic acid
Wisteria (Fabaceae):

[a] Source: Duke (1977).
[b] Boldfaced genera were indexed by Kingsbury (1964). A generic entry not followed by the name of a toxin indicates that the genus contains toxic species, but that the toxin is not yet identified.

TABLE 3.
Ecosystematic Attributes of Legumes[a]

Scientific name	Precipitation (dm)				Temperature (°C)				pH			
	Min.	Mean	Max.	(No.)	Min.	Mean	Max.	(No.)	Min.	Mean	Max.	(No.)
Acacia farnesiana (L.) Willd.	0.9	11.6	42.9	(4069)	4.3	18.1	29.9	(4063)	4.2	6.5	8.7	(3278)
Acacia mearnsii de Wild.	6.4	14.0	40.3	(20)	14.7	24.1	27.8	(20)	5.0	6.8	8.0	(15)
Acacia nilotica (L.) Del.	6.6	12.6	22.8	(6)	14.7	22.6	27.8	(6)	5.0	6.5	7.2	(5)
Acacia pycnantha Benth.	3.8	12.0	22.8	(12)	18.7	24.1	27.8	(12)	5.0	6.9	8.0	(10)
Acacia senegal (L.) Willd.	6.6	13.0	22.8	(6)	12.8	19.9	27.8	(6)	5.0	6.4	7.2	(4)
Acacia seyal Del.	3.8	12.4	22.8	(9)	16.2	23.8	27.8	(9)	5.0	6.4	7.7	(7)
Aeschynomene indica L.	8.7	15.0	22.8	(7)	18.7	24.0	27.8	(7)	5.0	6.9	8.0	(5)
Albizia julibrissin Durazz.	9.7	12.0	14.3	(3)	14.9	22.0	27.3	(3)	6.5	6.8	7.0	(3)
Alysicarpus vaginalis (L.) DC.	9.7	14.4	20.9	(3)	14.4	20.4	26.5	(3)	5.6	6.2	7.0	(3)
Anthyllis vulneraria L.	9.1	15.7	42.9	(11)	18.3	24.6	27.4	(11)	5.5	7.0	8.7	(6)
Arachis hypogaea L.	4.4	7.0	13.6	(18)	6.6	9.1	18.6	(18)	4.8	6.8	8.0	(17)
Astragalus sinicus	3.1	13.8	41.0	(162)	10.5	23.5	28.5	(161)	4.3	6.5	8.7	(90)
Baphia nitida Lodd.	16.3	18.5	20.9	(4)	11.7	17.2	22.8	(4)	5.0	5.8	6.6	(4)
Bauhinia esculenta Burchell	13.6	23.8	40.3	(4)	23.5	25.4	26.6	(4)	5.0	5.1	5.3	(3)
Caesalpinia coriaria (Jacq.) Willd.	8.4	11.0	13.2	(3)	21.0	24.4	27.8	(3)	5.0	6.5	7.4	(3)
Caesalpinia echinata Lam.	5.9	15.1	42.9	(14)	14.7	25.0	27.5	(14)	4.5	6.7	8.7	(7)
Caesalpinia sappan L.	11.2	20.6	27.8	(3)	24.4	25.7	26.6	(3)	4.5	5.5	7.1	(3)
Caesalpinia spinosa (Mol.) Ktze.	7.0	23.3	42.9	(5)	24.2	26.1	27.5	(5)	5.0	6.5	7.5	(3)
Cajanus cajan (L.) Millsp.	6.6	10.5	17.3	(4)	14.7	23.0	27.5	(4)	6.8	7.1	7.5	(3)
Calopogonium mucunoides Desv.	5.3	14.5	40.3	(60)	15.8	24.4	27.8	(60)	4.5	6.4	8.4	(44)
Canavalia ensiformis (L.) DC.	8.7	22.4	42.9	(21)	18.7	25.0	27.4	(21)	4.3	5.7	8.0	(16)
Canavalia gladiata (Jacq.) DC.	6.4	17.1	42.9	(20)	14.4	23.4	27.8	(20)	4.5	6.1	8.0	(16)
Cassia alata L.	6.4	14.2	26.2	(7)	14.8	22.4	27.4	(7)	5.0	6.0	7.1	(6)
Cassia auriculata L.	6.4	19.5	42.9	(22)	14.7	23.3	29.9	(22)	4.3	6.1	8.0	(15)
Cassia fistula L.	3.8	15.6	42.9	(11)	16.2	23.1	27.5	(11)	5.0	6.5	7.7	(7)
Cassia occidentalis L.	4.8	14.1	27.2	(83)	18.0	25.5	28.5	(82)	5.5	7.0	8.7	(22)
Cassia senna L.	6.4	15.7	42.9	(32)	12.5	22.8	7.8	(32)	4.5	6.3	8.4	(26)
Cassia tora L.	6.4	13.3	19.4	(5)	15.8	22.6	27.1	(7)	5.0	5.8	7.1	(5)
Centrosema pubescens Benth.	8.4	14.7	23.3	(4)	24.3	25.3	26.2	(4)	4.5	6.2	7.4	(4)
Ceratonia siliqua L.	5.3	17.9	41.0	(7)	19.4	24.7	27.4	(7)	4.3	5.4	7.0	(7)
Cicer arietinum L.	3.1	10.5	40.3	(13)	12.7	20.3	26.5	(13)	5.0	7.3	8.3	(10)
Clitoria laurifolia Poir.	2.8	6.7	15.0	(41)	6.3	17.0	27.5	(41)	5.0	7.2	8.3	(35)
Clitoria ternatea L.	5.3	18.9	27.8	(5)	21.0	24.6	27.3	(5)	4.5	5.0	5.5	(5)
Copaifera officinalis (Jacq.) L.	3.8	14.3	42.9	(24)	19.4	24.4	27.9	(24)	4.5	6.7	8.7	(20)
Coronilla varia L.	6.4	20.9	40.3	(4)	2.3	24.6	26.6	(4)	4.5	5.4	7.0	(3)
Crotalaria anagyroides H. B. K.	5.2	8.6	13.6	(14)	7.0	11.5	22.5	(14)	4.8	6.5	8.3	(13)
Crotalaria brevidens Benth.	17.8	27.0	41.0	(3)	22.8	25.8	27.4	(3)	4.3	5.6	6.6	(3)
Crotalaria juncea L.	11.2	16.9	26.7	(4)	16.0	22.2	26.2	(4)	6.0	6.6	7.1	(2)
Crotalaria lanceolata E. Mey.	4.9	14.9	42.9	(29)	8.4	22.5	27.5	(29)	5.0	6.2	8.4	(24)
Crotalaria pallida Ait.	8.7	14.9	28.2	(10)	11.5	20.3	26.2	(10)	5.5	6.1	7.1	(8)
	8.7	18.3	36.1	(9)	16.0	22.2	26.2	(9)	4.5	6.0	7.1	(7)

(Continued)

TABLE 3. (Continued)

Scientific name	Precipitation (dm)				Temperature (°C)				pH			
	Min.	Mean	Max.	(No.)	Min.	Mean	Max.	(No.)	Min.	Mean	Max.	(No.)
Crotalaria retusa L.	9.7	14.1	21.4	(3)	20.3	22.9	26.2	(3)	6.2	6.8	7.1	(3)
Crotalaria spectabilis Roth	9.0	14.3	28.2	(17)	11.5	21.6	27.8	(17)	4.9	6.3	8.0	(14)
Cyamopsis tetragonoloba (L.) Taubert	3.8	7.5	24.1	(16)	7.8	22.0	27.5	(16)	5.3	7.2	8.3	(14)
Cytisus scoparius (L.) Link	5.7	6.7	7.6	(5)	7.4	8.5	9.7	(5)	6.0	6.8	7.0	(5)
Delonix regia (Boj. ex Hook.) Raf.	4.8	14.1	27.2	(82)	18.0	25.5	28.5	(81)	5.5	7.0	8.7	(21)
Derris elliptica (Roxb.) Benth.	13.5	24.7	41.0	(6)	23.3	25.7	27.4	(6)	4.3	6.2	8.0	(5)
Desmodium canum (J. F. Gmel.) Schinz & Thell.	10.3	12.5	16.0	(3)	21.8	23.8	26.2	(3)	6.8	7.2	7.7	(3)
Desmodium intortum (Mill.) Fawc. & Rendle	5.3	14.4	40.3	(20)	7.3	21.3	27.1	(20)	4.5	5.9	7.1	(17)
Desmodium tortuosum (Sw.) DC.	11.2	15.4	21.4	(3)	19.7	22.7	26.2	(3)	5.5	6.3	7.1	(3)
Desmodium uncinatum (Jacq.) DC.	5.3	12.0	24.3	(5)	18.3	21.4	27.3	(5)	4.5	5.3	5.6	(5)
Dipteryx odorata (Aubl.) Willd.	13.5	25.9	40.3	(4)	21.3	25.0	26.6	(4)	5.0	6.2	8.0	(4)
Genista tinctoria L.	5.2	8.0	11.6	(12)	7.0	9.3	14.7	(12)	4.5	6.2	7.4	(11)
Glycine max (L.) Merr.	3.1	12.8	41.0	(108)	5.9	18.2	27.8	(108)	4.3	6.2	8.4	(98)
Glycine wightii (R. Grah. ex Wight & Arn.) Verdc.	5.3	13.5	42.9	(15)	18.0	21.5	27.4	(15)	5.0	6.0	7.1	(13)
Glycyrrhiza glabra L.	4.0	6.5	11.6	(13)	5.7	13.9	25.0	(13)	5.5	7.1	8.2	(10)
Glycyrrhiza lepidota Pursh	3.5	4.4	5.3	(4)	6.9	7.5	8.4	(4)	7.3	7.4	7.5	(3)
Hedysarum coronarium L.	4.6	10.0	23.6	(12)	5.7	16.5	29.9	(12)	5.5	6.6	8.0	(7)
Indigofera arrecta Hochst. ex A. Rich.	8.7	20.8	42.9	(5)	18.7	23.6	27.4	(5)	5.0	5.8	6.5	(3)
Indigofera hirsuta L.	8.7	14.5	26.7	(13)	15.8	23.7	27.8	(13)	5.0	6.5	8.0	(10)
Indigofera spicata Forsk.	8.7	20.0	42.9	(7)	16.0	22.0	27.4	(7)	5.0	6.2	7.7	(4)
Indigofera tinctoria L.	6.4	15.2	41.0	(13)	10.5	23.1	27.4	(13)	4.3	6.5	8.7	(9)
Inga edulis Mart.	6.4	16.9	40.0	(9)	21.3	25.1	27.3	(9)	5.0	6.6	8.0	(7)
Lablab purpureus (L.) Sweet	7.0	14.5	40.3	(16)	14.5	21.9	27.5	(16)	5.0	6.2	7.5	(15)
Lathyrus hirsutus L.	3.2	8.2	12.9	(24)	4.3	13.6	23.8	(24)	5.5	6.6	8.2	(12)
Lathyrus odoratus L.	3.1	6.9	10.3	(5)	7.6	13.7	21.1	(5)	5.0	6.7	8.2	(5)
Lathyrus pratensis L.	5.2	6.6	8.3	(6)	7.4	8.4	9.6	(6)	6.0	6.7	7.0	(6)
Lathyrus sativus L.	3.2	7.4	13.6	(31)	4.3	13.0	27.5	(31)	4.5	6.8	8.3	(27)
Lens culinaris Medik.	2.8	7.9	24.3	(35)	6.3	14.5	27.3	(35)	4.5	6.8	8.2	(32)
Lespedeza bicolor Turcz.	12.1	13.8	16.3	(3)	12.2	13.7	14.9	(3)	5.0	6.0	6.8	(3)
Lespedeza cuneata (Dum.) G. Don	6.1	12.7	23.4	(14)	9.9	16.1	26.2	(14)	4.9	6.1	7.1	(12)
Lespedeza stipulacea Maxim.	4.9	11.7	16.0	(12)	8.4	16.1	26.2	(12)	5.5	6.3	7.3	(11)
Lespedeza striata (Thunb. ex Murr.) Hook. & Arn.	4.9	11.4	16.3	(14)	8.4	16.4	26.2	(14)	4.9	6.2	7.3	(13)
Leucaena leucocephala (Lam.) de Wit.	1.8	14.9	41.0	(30)	14.7	24.0	27.4	(30)	4.3	6.1	8.7	(21)
Lotus corniculatus L.	2.1	8.4	19.1	(59)	5.7	11.1	23.7	(59)	4.5	6.4	8.2	(56)
Lotus tenuis Waldst. & Kit. ex Willd.	4.4	7.0	11.6	(10)	7.0	10.2	16.9	(10)	5.6	6.7	8.0	(9)
Lotus uliginosus Schkuhr	3.5	8.5	13.6	(12)	5.9	12.0	21.3	(12)	5.5	6.6	8.2	(10)
Lupinus albus L.	3.6	8.4	17.8	(38)	5.7	12.7	26.2	(38)	4.8	6.4	8.2	(35)
Lupinus angustifolius L.	3.5	8.4	16.6	(39)	5.6	12.3	26.2	(39)	4.9	6.6	8.2	(38)
Lupinus luteus L.	3.5	8.5	20.9	(42)	6.6	13.0	26.2	(42)	4.5	6.5	8.2	(41)
Macroptilium atropurpureum (DC.) Urb.	3.8	11.8	28.2	(21)	12.5	22.0	26.8	(21)	5.0	6.2	8.0	(18)
Macroptilium lathyroides (L.) Urb.	9.7	15.4	23.3	(6)	21.8	24.7	27.3	(6)	4.5	6.5	7.7	(6)
Macrotyloma uniflorum (Lam.) Verdc.	7.0	19.8	42.9	(8)	23.4	25.3	27.5	(8)	5.0	6.4	7.5	(6)
Medicago arabica (L.) Huds.	3.1	7.7	12.9	(14)	7.0	14.5	26.2	(14)	5.6	6.9	8.2	(12)
Medicago falcata L.	3.1	6.9	13.6	(27)	5.7	10.4	19.9	(27)	4.9	6.8	8.2	(25)

Appendix

Species									
"Medicago hispida Gaertn."	2.0	8.9	16.0	(6)	9.5	16.6	21.8	(6)	5.6	6.8	8.2	(6)
Medicago lupulina L.	3.1	8.4	17.1	(52)	5.7	11.7	22.5	(52)	4.5	6.5	8.2	(48)
Medicago minima (L.) Bartal.	5.8	8.1	9.7	(3)	15.7	17.9	20.3	(3)	6.8	6.9	7.0	(3)
Medicago orbicularis (L.) Bartal.	4.4	8.0	12.8	(8)	14.2	18.7	21.8	(8)	6.5	7.3	8.0	(8)
Medicago polymorpha L.	3.1	8.6	19.1	(36)	10.5	16.3	27.5	(36)	5.3	6.8	8.2	(30)
Medicago sativa L.	0.9	10.3	27.2	(214)	4.3	17.7	28.5	(213)	4.3	6.5	8.7	(136)
Medicago scutellata (L.) Mill.	4.4	8.0	11.9	(7)	11.5	18.9	26.2	(7)	5.5	6.8	7.5	(7)
Melilotus alba Medik.	0.9	7.8	16.0	(54)	5.7	12.7	24.3	(54)	4.8	6.7	8.2	(50)
Melilotus indica (L.) All.	0.9	7.6	12.9	(18)	7.2	14.8	22.5	(18)	5.0	6.8	8.2	(17)
Melilotus officinalis Lam.	3.1	7.8	16.0	(47)	4.9	10.9	21.8	(47)	4.8	6.6	8.2	(43)
Melilotus suaveolens Ledeb.	4.9	8.3	11.6	(2)	8.4	10.5	12.5	(2)	5.6	6.5	7.3	(2)
Mimosa pudica L.	8.4	20.4	41.0	(10)	22.1	25.2	27.4	(10)	4.3	5.9	7.4	(10)
Mucuna deeringiana (Bort) Merr.	3.8	15.3	31.5	(12)	18.7	24.2	27.1	(12)	4.5	5.9	7.7	(10)
Mucuna pruriens (L.) DC.	8.7	13.7	22.1	(4)	18.7	23.5	27.3	(4)	5.0	6.0	6.8	(4)
Myroxylon balsamum (L.) Harms	13.5	26.4	40.3	(6)	23.1	25.4	27.3	(6)	5.0	6.1	8.0	(5)
Myroxylon balsamum var pereirae (Royle) Harms	13.5	24.1	40.3	(4)	23.3	24.6	26.5	(4)	5.0	6.9	8.0	(3)
Onobrychis viciifolia Scop.	3.5	6.6	10.7	(24)	5.9	10.2	18.6	(24)	4.9	7.1	8.2	(20)
Ononis arvensis L.	5.7	6.6	7.6	(3)	7.4	8.3	9.6	(3)	6.0	6.7	7.0	(3)
Ornithopus sativus Brot.	5.7	7.0	11.9	(11)	7.0	11.1	18.3	(11)	5.5	6.6	8.2	(11)
Pachyrhizus erosus (L.) Urban	6.4	21.0	40.3	(7)	21.3	24.2	27.4	(7)	5.0	6.0	6.8	(5)
Pachyrhizus tuberosus (Lam.) Spreng.	6.4	18.9	41.0	(5)	21.3	25.4	27.4	(5)	4.3	5.2	6.8	(4)
Peltophorum pterocarpum (DC.) Back. ex K. Heyne	6.4	17.2	40.3	(6)	18.7	23.1	26.6	(6)	5.0	6.0	7.4	(4)
Pentaclethra macrophylla Benth.	13.6	20.2	25.8	(4)	21.3	24.4	26.6	(4)	5.0	5.6	6.5	(3)
Phaseolus acutifolius A. Gray	6.4	10.6	17.3	(9)	17.0	22.2	26.2	(9)	5.0	6.3	7.1	(7)
Phaseolus coccineus L.	3.1	9.1	17.3	(31)	5.6	13.2	26.2	(31)	4.9	6.5	8.2	(27)
Phaseolus lunatus L.	3.1	15.5	42.9	(43)	8.7	20.4	27.5	(43)	4.5	6.2	8.4	(36)
Phaseolus vulgaris L.	0.9	12.8	42.9	(217)	5.7	19.3	28.5	(216)	4.2	6.4	8.7	(144)
Pisum sativum L.	0.9	9.2	27.8	(137)	5.0	12.9	27.5	(137)	4.2	6.3	8.3	(122)
Psophocarpus tetragonolobus (L.) DC.	7.0	19.3	41.0	(15)	15.4	25.2	27.5	(15)	4.3	5.8	7.5	(14)
Pterocarpus erinaceus Poir.	17.3	28.5	40.3	(3)	23.5	25.2	26.6	(3)	4.5	4.8	5.0	(2)
Pterocarpus santalinus L. f.	13.6	18.9	27.8	(3)	24.2	25.7	26.6	(3)	4.5	4.8	5.0	(2)
Pueraria javanica Benth.	9.6	14.7	24.3	(3)	19.4	22.8	27.3	(3)	4.5	5.6	6.8	(3)
Pueraria lobata (Willd.) Ohwi	9.7	13.5	21.4	(15)	12.2	18.1	26.7	(15)	5.0	6.1	7.1	(13)
Pueraria phaseoloides (Roxb.) Benth.	9.1	24.2	42.9	(18)	22.1	25.6	27.4	(18)	4.3	5.8	8.0	(15)
Rhynchosia minima (L.) DC.	3.8	10.3	16.0	(3)	23.3	25.4	26.7	(3)	7.1	7.5	7.7	(3)
Robinia pseudoacacia L.	6.1	11.2	19.1	(6)	7.6	13.0	20.3	(6)	6.0	6.6	7.0	(6)
Sesbania bispinosa (Jacq.) W. F. Wight	5.7	13.4	22.1	(4)	19.9	23.8	27.3	(4)	5.8	6.9	7.5	(3)
Sesbania exaltata (Raf.) Rydb.	6.2	13.1	23.2	(12)	12.5	21.6	27.8	(12)	4.5	6.2	7.2	(10)
Spartium junceum L.	2.9	7.5	13.7	(11)	8.1	13.3	18.7	(11)	4.5	6.6	8.2	(8)
Sphenostylis stenocarpa (Hochst. ex A. Rich.)	8.7	11.2	13.6	(2)	18.7	22.4	26.2	(2)	5.0	5.8	6.5	(2)
Stylosanthes erecta Beauv.	11.2	18.4	27.8	(4)	18.3	23.9	26.6	(4)	4.5	5.5	7.1	(4)
Stylosanthes guianensis (Aubl.) Sw.	5.3	15.6	41.0	(13)	18.3	23.5	27.4	(13)	4.3	5.7	7.1	(11)
Stylosanthes humilis H.B.K.	5.3	12.9	23.2	(11)	18.3	23.4	26.8	(11)	4.5	5.8	7.1	(10)
Tamarindus indica L.	0.9	14.5	42.9	(110)	17.3	25.3	28.5	(109)	4.5	6.8	8.7	(42)
Tephrosia candida (Roxb.) DC.	7.0	16.1	26.7	(9)	18.0	24.2	27.5	(9)	5.0	6.4	8.0	(7)
Tephrosia vogelii Hook. f.	8.7	14.5	26.7	(7)	12.5	19.8	26.2	(7)	5.0	5.6	6.5	(5)
Trifolium alexandrinum L.	3.8	8.7	16.6	(17)	7.0	16.0	26.7	(17)	4.9	6.8	7.8	(15)
Trifolium ambiguum Bieb.	4.9	9.3	11.6	(3)	8.4	10.2	12.5	(3)	4.5	5.8	7.3	(3)

(Continued)

Appendix

TABLE 3. (Continued)

Scientific name	Precipitation (dm)				Temperature (°C)				pH			
	Min.	Mean	Max.	(No.)	Min.	Mean	Max.	(No.)	Min.	Mean	Max.	(No.)
Trifolium arvense L.	5.2	7.3	11.8	(7)	7.4	9.0	12.8	(7)	6.0	6.6	7.0	(6)
Trifolium campestre Schreb.	5.7	8.3	11.8	(5)	7.4	11.6	20.3	(5)	6.0	6.8	7.0	(4)
Trifolium dubium Sibth.	5.7	8.2	11.8	(6)	7.4	11.3	20.3	(6)	6.0	6.8	7.0	(5)
Trifolium fragiferum L.	4.4	7.0	11.6	(20)	6.6	11.9	21.8	(20)	5.3	6.8	8.2	(19)
Trifolium glomeratum L.	4.8	9.8	12.8	(3)	12.8	14.9	17.6	(3)	6.1	6.3	6.5	(2)
Trifolium hirtum All.	4.4	7.5	13.6	(6)	14.1	17.4	19.7	(6)	5.5	6.8	8.2	(6)
Trifolium hybridum L.	3.5	8.2	17.6	(42)	4.9	8.4	19.7	(42)	4.8	6.3	7.5	(41)
Trifolium incarnatum L.	3.1	9.2	16.3	(30)	5.9	12.7	21.3	(30)	4.8	6.5	8.2	(26)
Trifolium medium L.	5.2	5.9	6.6	(2)	8.0	8.2	8.9	(2)	6.0	6.5	7.0	(4)
Trifolium nigrescens Viv.	6.2	11.0	13.3	(4)	12.5	16.4	21.3	(4)	5.6	5.9	6.5	(4)
Trifolium pratense L.	3.1	8.6	19.1	(91)	4.9	10.6	20.3	(91)	4.5	6.3	8.2	(84)
Trifolium repens L.	3.1	8.6	19.1	(101)	4.3	11.6	21.8	(101)	4.5	6.3	8.2	(89)
Trifolium resupinatum L.	2.8	7.8	13.6	(20)	6.3	12.7	20.3	(20)	4.8	6.7	8.2	(19)
Trifolium subterraneum L.	3.8	9.0	16.3	(34)	5.9	14.4	21.3	(34)	4.5	6.1	8.2	(31)
Trifolium variegatum Nutt.	5.6	9.2	12.8	(2)	8.9	13.3	17.6	(2)	6.3	6.4	6.5	(2)
Trifolium vesiculosum Savi	6.1	10.4	13.6	(3)	11.1	14.4	19.7	(3)	5.5	6.4	8.2	(3)
Trigonella foenum-graecum L.	3.8	7.0	15.3	(18)	7.8	15.9	27.5	(18)	5.5	7.3	8.2	(15)
Ulex europaeus L.	7.6	9.8	12.6	(6)	9.4	11.4	14.4	(6)	5.6	6.5	7.0	(3)
Vicia benghalensis L.	3.1	8.5	16.6	(8)	10.8	17.7	21.8	(8)	4.8	6.3	8.2	(7)
Vicia cracca L.	4.4	6.8	8.3	(7)	6.0	9.3	14.2	(7)	6.0	6.7	8.0	(7)
Vicia ervilia (L.) Willd.	3.6	6.3	11.6	(10)	8.3	14.2	20.0	(10)	5.6	7.3	8.2	(9)
Vicia faba L.	2.3	8.0	20.9	(95)	5.6	12.1	27.5	(95)	4.5	6.6	8.3	(87)
Vicia hirsuta (L.) S. F. Gray	5.7	7.9	11.8	(4)	7.4	9.4	12.8	(4)	6.0	6.7	7.0	(3)
Vicia monantha Retz.	3.5	6.3	12.3	(6)	7.0	11.2	16.0	(6)	6.0	6.9	7.5	(5)
Vicia pannonica Crantz	3.5	6.1	10.3	(8)	7.0	9.4	13.1	(8)	4.9	6.9	8.2	(7)
Vicia sativa L. ssp. *nigra* (L.) Ehrh.	3.5	7.2	11.6	(23)	6.5	10.5	22.5	(23)	4.9	6.6	8.2	(22)
Vicia sativa L. ssp. *sativa*	3.1	8.0	16.3	(74)	5.6	12.2	22.5	(74)	4.5	6.5	8.2	(71)
Vicia villosa Roth	3.1	8.1	16.6	(42)	4.3	12.0	21.1	(42)	4.9	6.6	8.2	(40)
Vigna aconitifolia (Jacq.) Marechal	3.8	8.7	17.3	(5)	21.1	24.5	27.5	(5)	5.0	7.1	8.1	(4)
Vigna angularis (Willd.) Ohwi & Ohashi	5.3	12.5	17.3	(8)	7.8	17.0	27.8	(8)	5.0	6.1	7.5	(7)
Vigna caracala (L.) Verdc.	6.9	6.9	6.9	(1)	17.4	17.4	17.4	(1)	7.2	7.2	7.2	(1)
Vigna lutea (Sw.) Gray	10.3	10.3	10.3	(2)	21.1	21.4	21.8	(2)	5.0	5.9	6.8	(2)
Vigna marina (Burm.) Merr.	23.6	23.6	23.6	(1)	29.9	29.9	29.9	(1)				(0)
Vigna mungo (L.) Hepper	5.3	12.4	24.3	(13)	7.8	22.0	27.8	(13)	4.5	6.4	7.5	(12)
Vigna radiata (L.) Wilczek	2.8	12.5	41.0	(29)	7.8	20.9	27.8	(29)	4.3	6.6	8.1	(27)
Vigna umbellata (Thunb.) Ohwi & Ohashi	7.0	11.1	17.3	(3)	23.5	25.3	27.5	(3)	6.8	7.2	7.5	(2)
Vigna unguiculata ssp. *cylindrica* (L.) Verdc.	7.0	12.2	17.3	(8)	12.2	21.3	27.5	(8)	5.0	6.1	7.5	(7)

[a] Source: Duke (1978).

TABLE 4.
Economic Legumes: Their Tolerances, Yields, Centers of Diversity, and Ecocenters

Common name	Scientific name	Tolerance[a]	Yield[b]	Center of diversity[c]	Ecocenter[d]
Acacia, sweet	*Acacia farnesiana* (L.) Willd.	Dr Hi Ht Lo Na Qz Sl Sv We		SA	1424
Ajipo	*Pachyrhizus tuberosus* (Lam.) Spreng.	In Lo	90,000 rt 600 sd	SA	1925
Alfalfa	*Medicago sativa* L.	Al *An Ba Di* Dr *Fu* Hf Hi Ht *In* Lo Mi My Na Ne Se Sl Sm So Vi We Wl *Wt*	74,000 hay	CE, NE, ME	1018
Alfalfa, yellow-flowered	*Medicago falcata* L.	Dr Fr Hi Na Sl		ES, NE	0710
Avaram	*Cassia auriculata* L.	Gr We		HI	1623
Babul	*Acacia nilotica* (L.) Del.	Hi Na Sv Wl		AF	1224
Balsam, Peru	*Myroxylon balsamum* var *pereirae* (Royle) Harms	Hi My Ps Sl	4 resin	MA	2425
Balsam, Tolu	*Myroxylon balsamum* (L.) Harms	My	4 resin/	SA	2625
Barbasco	*Lonchocarpus utilis* A.C. Sm.	In Sh	13,500 rt	SA	
Barwood	*Pterocarpus soyauxii* Taub.			AF	
Barwood, red	*Pterocarpus erinaceus* Poir.	Lo Sv		AF	2825
Bean, adzuki	*Vigna angularis* (Willd.) Ohwi & Ohashi	Dr Fr Ht Sl Vi	1,100 sd	CJ, II	
Bean, African locust	*Parkia filicoidea* Welw. ex Oliv.	Fi Sl Sv	5,000 sd		
Bean, broad	*Vicia faba* L.	Hi In Lo Sl Vi	6,600 sd	CE, ME, SA	0812
Bean, green	*Phaseolus vulgaris* L.	Al Ba *Di* *He* Hf Hi La Lo Pe Ph Sm So Vi	2,500 sd	MA, SA, CJ	1319
Bean, haricot	*Phaseolus vulgaris* L.	Al Ba *Di* Dr *He* Hf Hi La Lo Mn Pe Ph Sm So Vi Wl	5,000 fr	MA, SA, CJ	1319
Bean, ice cream	*Inga edulis* Mart.	Wl		MA	1725
Bean, lablab	*Lablab purpureus* (L.) Sweet	Al Di Dr Ht Ps Qz Sh Vi	1,400 sd	AF	1422
Bean, lima	*Phaseolus lunatus* L.	An Di Dr Hi Ht La Lo Ph Sm Vi	2,500 sd	MA, SA	1620
Bean, moth	*Vigna aconitifolia* (Jacq.) Marechal	Dr Hi Ht Sl	45,000 wm	HI, II	0924
Bean, mung	*Vigna radiata* (L.) Wilczek	Dr He Hi La Lo Sa Sl Vi	1,100 sd	II, CJ, HI	1221
Bean, rice	*Vigna umbellata* (Thunb.) Ohwi & Ohashi	Dr Sl	200 sd	HI, II, CJ	1125
Bean, scarlet runner	*Phaseolus coccineus* L.	An Qz Sl Vi	1,700 sd	MA	0913
Bean, tepary	*Phaseolus acutifolius* A. Gray	Di Dr *Ht* Qz Sl Vi	1,700 sd	MA	1122
Bean, tonka	*Dipteryx odorata* (Aubl.) Willd.	In Ps Sh Sl Wl	4 fr/	SA	2625
Bean, winged	*Psophocarpus tetragonolobus* (L.) DC.	Ht La Lo	4,000 rt	II	1925
Bean, yard-long	*Vigna unguiculata* ssp. *sesquipedalis* (L.) Verdc.	La Ph Ps Sl Vi	10,000 fr	AF, HI	1722

(Continued)

Appendix

321

Appendix

TABLE 4. *(Continued)*

Common name	Scientific name	Tolerance[a]	Yield[b]	Center of diversity[c]	Ecocenter[d]
Beggarlice	*Desmodium intortum* (Mill.) Fawc. & Rendle	Di Gr Lo Na Sh	8,300 dm	SA	1421
Brazilwood	*Caesalpinia echinata* Lam.	Lo Sl		SA	2126
Broom, Spanish	*Spartium junceum* L.	Dr Hi Lo Ps Qz Sh Sl We Wi	1,500 fib	ME	2413
Burclover, California	*Medicago polymorpha* L.	Ad Ak Hi Hs Sh Sl We	7,500 hay	ES, HI	0916
Burclover, spotted	*Medicago arabica* (L.) Huds.	Ad Ak Hi Hs Na Vi We	7,500 hay	ES	0814
Button clover	*Medicago orbicularis* (L.) Bartal.	Hi Sl Vi		ME, CE	0819
Camel's Foot	*Bauhinia esculenta* Burchell	Qz		AF	1124
Camelthorn	*Alhagi pseudalhagi* (Bieb.) Desv.	Dr Na Qz We		HI	
Camwood	*Baphia nitida* Lodd.	Sl Sv Vi		AF	2425
Canicha	*Sesbania bispinosa* (Jacq.) W. F. Wright	Hs Lo Na Qz We Wl	1,000 sd	HI	1324
Carob	*Ceratonia siliqua* L.	Dr Hi Ht In Li Ps Qz Sl Sm	2,500 fr	ME	1020
Catjang	*Vigna unguiculata* ssp. *cylindrica* (L.) van Eseltine	In Ps Sl Vi We	2,500 sd	II, HI	1221
Chickpea	*Cicer arietinum* L.	Di Dr Hi Na Ph Vi	2,000 sd	HI, CE, NE	0717
Clover, alsike	*Trifolium hybridum* L.	Ak Al An Di Fr Fu Gr Hs My Na Sl Vi We Wl	7,000 dm	ES	0808
Clover, alyce	*Alysicarpus vaginalis* (L.) DC.	Hs Lo Ps Sl We	45,000 wm	AF, HI, II	1625
Clover, arrowleaf	*Trifolium vesiculosum* Savi	Dr Fr Hs *Gr* Ne Sl		ME	1014
Clover, ball	*Trifolium nigrescens* Viv.	Gr Hi Sl		NE, ES	1116
Clover, berseem	*Trifolium alexandrinum* L.	Hi Vi	37,500 wm	ME	0916
Clover, crimson	*Trifolium incarnatum* L.	Gr Hi Hs Ne Qz Sl Vi We	12,000 dm	ME, ES	0913
Clover, hop	*Trifolium* sp.	Hi Lo		NE	0910
Clover, kura	*Trifolium ambiguum* Bieb.	Di Dr Fr In Lo Sl Uv		ME, NE, ES	0813
Clover, Persian	*Trifolium resupinatum* L.	Dr Fr Hi Hs Li Qz We		ES	0911
Clover, red	*Trifolium pratense* L.	Al *Di* Fr *Fu* Gr Hf Hi Lo Mi *Mw* My Ru Sl *Vi* We Wl	7,100 dm		
Clover, rose	*Trifolium hirtum* All.	Gr Hi *Lo* Ps Sl	3,000 dm	ME, NE	0817
Clover, seaside	*Trifolium willdenovii* Spring.	Ak Na Sl Uv Wl		ES, ME	0712
Clover, strawberry	*Trifolium fragiferum* L.	Ak Gr Hi *Na* Vi We *Wl*		ME, AU	0914
Clover, sub	*Trifolium subterraneum* L.	Al Di Gr Hi Lo Vi		ME, NE, ES	0912
Clover, white	*Trifolium repens* L.	Al *Di* Fr Gr He Hi Hs Ht Lo Mi My Na *Ne* Ps Sm Vi We Wt	7,000 dm		
Clover, whitetip	*Trifolium variegatum* Nutt.	Wl		NA	0913
Clover, zigzag	*Trifolium medium* L.	Fr Sh	4,000 hay		0608
Copaiba	*Copaifera officinalis* (Jacq.) L.	Qz Sl	17 resin/	SA	2125

Appendix

Common name	Scientific name	Characteristics	Yield	Region	Code
Crotalaria, lanceleaf	*Crotalaria lanceolata* E. Mey.	La Sl Wl		AF	1520
Crotalaria, showy	*Crotalaria spectabilis* Roth	Hi Hs Ne In Sl Vi We	50,000 hay	AF	1422
Crotalaria, slenderleaf	*Crotalaria brevidens* Benth.	Sl Wl		AF	1722
Crotalaria, smooth	*Crotalaria pallida* Ait.	Dr Sl		AF	1822
Crownvetch	*Coronilla varia* L.	Dr Fr Gr In Lo *Mi* My *Ps* Sl Sm We	12,500 hay 500 sd	ES	0912
Divi-divi	*Caesalpinia coriaria* (Jacq.) Willd.	In Ps Qz Sl Wl		MA	1525
Dryer's greenweed	*Genista tinctoria* L.	Li Lo My Ps Sh Sl Sm We		ES	0809
Fenugreek	*Trigonella foenum-graecum* L.	Di Dr Hi In Na Ps	3,000 sd	ME, NE	0716
Gram, black	*Vigna mungo* (L.) Hepper	Dr Hi Hs Lo Sl Vi	800 sd	HI	1222
Gram, horse	*Macrotyloma uniflorum* (Lam.) Verdc.	Dr Ps Qz	300 sd 12,000 wm	HI	2025
Groundnut, Bambarra	*Voandzeia subterranea* (L.) Thou.	Di Dr Ht In Lo Ps Qz Sv	4,200 sd	AF	1423
Groundnut, Kersting's	*Macrotyloma geocarpum* (Harms) Marechal & Baudet	Dr Ps Qz Sv		AF	1426
Guar	*Cyamopsis tetragonoloba* (L.) Taubert	Ba *Di* Dr *Fu* Hi Ht *Hs* Na Vi	2,000 sd	HI, AF	0822
Gum arabic	*Acacia senegal* (L.) Willd.	Ak Dr Fi Hi Ps Qz Sl		AF	1224
Hemp, sunn	*Crotalaria juncea* L.	Di Dr In La Ps Sl Vi We		HI	1522
Huisache	*Acacia farnesiana* (L.) Willd.	Dr Hi Ht Lo Na Qz Sl Sv We	3,800 dm 900 fib	SA	1424
Indigo	*Indigofera tinctoria* L.	La	13,000 wm	AF	1523
Indigo, hairy	*Indigofera hirsuta* L.	Di In Lo Ne Ps Sh Sl Vi We	17,000 wm	AF	1424
Indigo, Natal	*Indigofera arrecta* Hochst. ex A. Rich.	Sh We	92,000 wm	AF	2124
Indigo, spicate	*Indigofera spicata* Forsk.	Dr La Sl		AF	2022
Jackbean	*Canavalia ensiformis* (L.) DC.	Di Dr Fu In Lo Na Qz Sh Sl Vi Wl	50,000 wm	HI, SA, CJ	1723
Jackbean, oblique-seed	*Canavalia plagiosperma* Piper			SA	
Jicama	*Pachyrhizus erosus* (L.) Urban	Gr In La Vi	95,000 rt 600 sd	MA	2124
Kino, Bengal	*Butea monosperma* (Lam.) Taub.	Dr Fr Gr Hs Sl Vi We	40,000 wm	CJ	1418
Kino, Malabar	*Pterocarpus marsupium* Roxb.	Di Dr Fu Hs In La Lo Ps Sl We	50,000 wm	II	2426
Kudzu	*Pueraria lobata* (Willd.) Ohwi	Al Di Dr Fi Hi Hs In La Li Lo Na Sl We Wi Wl	80,000 wm	MA	1524
Kudzu, tropical	*Pueraria phaseoloides* (Roxb.) Benth.	Ak *Di* Dr Fr *Fu* Hi Hs Ht In Na Ps Sl Vi	1,700 sd	NE	0814
Lead tree	*Leucaena leucocephala* (Lam.) de Wit	Hi Li Ps Sl Vi We	7,500 hay	CJ	1116
Lentil	*Lens culinaris* Medik.	Ba *Di* Dr Fu Gr Li *Ne* Sl We	7,500 hay	CJ	1216
Lespedeza, common	*Lespedeza striata* (Thunb. ex Murr.) Hook. & Arn.	Gr Ps Sl We	9,000 wm	CJ	1316
Lespedeza, Korean	*Lespedeza stipulacea* Maxim.	Qz Sl We		NA	0408
Lespedeza, sericea	*Lespedeza cuneata* (Dum.) G. Don				
Licorice, American	*Glycyrrhiza lepidota* Pursh				

(Continued)

Appendix

TABLE 4. (Continued)

Common name	Scientific name	Tolerance[a]	Yield[b]	Center of diversity[c]	Ecocenter[d]
Licorice, common	*Glychrrhiza glabra* L.	Dr Gr Hi Hs Ht My Na Sl We Wl	22,000 rt	ME, ES	0614
Lucerne, Brazilian	*Stylosanthes guianensis* (Aubl.) Sw.	Al Dr Gr La Lo Ps Sl	6,000 dm	SA	1824
Lucerne, Nigerian	*Stylosanthes erecta* Beauv.	Lo Ps Qz Sl Sv		AF	1824
Lupine, European blue	*Lupinus angustifolius* L.	An Di Fr *Fu* Hi In Ph Qz Vi We	7,500 hay 1,000 sd	ME	0812
Lupine, European yellow	*Lupinus luteus* L.	Al Hi Ph Ps Vi	7,500 hay 1,000 sd	ME	0813
Lupine, white	*Lupinus albus* L.	Fr Fu Hi Lo Na Ph Vi *We* Wl	7,500 hay 1,000 sd	HI	0813
Medic, black	*Medicago lupulina* L.	Ba Hi Li Lo My Sl Vi We		ES	0812
Medic, snail	*Medicago scutellata* (L.) Mill.	Dr Fr Li		ME	0819
Milkvetch, cicer	*Astragalus cicer* L.	Sl	6,000 hay	ES	
Pea, asparagus	*Tetragonolobus purpureus* Moench	Dr Hi La Ph Sl Vi We		ME	
Pea, butterfly	*Clitoria ternatea* L.	Al Dr Hi Ht La Lo Ne Ps Sh Sl Vi We Wt	16,000 dm	II	1424
Pea, cow	*Vigna unguiculata* (L.) Walp. ssp. *unguiculata*	Al *Di* Fr *Fu* Hf Hi Ht La Lo *Mw* Sl Sm *Vi Wt*	2,500 sd	AF, HI	1422
Pea, garden	*Pisum sativum* L.	Dr Hi Hs Lo Ps Ru Vi Wl	1,800 sd	NE, ME, AF	0913
Pea, grass	*Lathyrus sativus* L.		1,000 sd 30,000 wm	CE, ME	0713
Pea, pigeon	*Cajanus cajan* (L.) Millsp.	Di Dr Fr Hi La Lo Na Ne Ph Qz Vi We Wi Wl Wt	12,800 fr	HI, AF	1423
Pea, rough	*Lathyrus hirsutus* L.	Hs Wl		CE, ME	0814
Pea, yam	*Sphenostylis stenocarpa* (Hochst. ex A. Rich.) Harms	Sv Wl		AF	1122
Pea, zombi	*Vigna vexillata* (L.) A. Rich.	Lo Ps Sl		Af, HI	1823
Peanut	*Arachis hypogaea* L.	Al Di Dr Fr *Fu* Hi Ht *In* La Li Lo Qz Sm Sv *Uv* Vi	5,000 sd	SA, AF	1424
Ringworm bush	*Cassia alata* L.	Hi In La Sm Wl		MA, HI	2023
Sainfoin	*Onobrychis viciifolia* Scop.	Ak Dr *Fr Gr* Hi Li Na Ps Sl	500 sd	NE, ES	0710
Sainfoin, Spanish	*Hedysarum coronarium* L.	Fr Li Sl Vi	112,000 wm	ME	1016
Sandalwood, red	*Pterocarpus santalinus* L. f.	Dr Fi Hi Ht La Lo Ps Sl Wi		HI	1926
Sappanwood	*Caesalpinia sappan* L.	Qz Sl		II	2326
Senna, Alexandrian	*Cassia senna* L.	Di Dr Hi In Ps Qz		AF	1323
Senna, coffee	*Cassia occidentalis* L.	Hi La Vi We		II	1316
Serradella	*Ornithopus sativus* Brot.		1,200 sd 8,000 dm	ME	0711
Sesbania, hemp	*Sesbania exaltata* (Raf.) Rydb.	Lo Vi We		NA	1322
Shittim wood	*Acacia seyal* Del.	Hi Hs My Na Ps Sl Sv Wl		AF	1524

Common name	Scientific name			
Siratro	*Macroptilium atropurpureum* (DC.) Urb.	Al Dr Fi Fr Gr Ht Lo Na Ne Qz Sl Wl	9,400 dm	MA, SA	1222
Soga	*Peltophorum pterocarpum* (DC.) Back. ex K. Heyne	Hi Lo		II	1723
Soybean	*Glycine max* (L.) Merr.	Al Ba Di Fr Fu Hf Hi Hs In La Li Lo My Ne *Ph* Pt Sm St Vi	22,000 hay 3,100 sd	CJ	1318
Soybean, perennial	*Glycine wightii* (Grah. ex Wight & Arn.) Verdc.	Gr Lo Na Sh Sl Sv We	7,000 hay	AF	1422
Stylo, Brazilian	*Stylosanthes guianensis* (Aubl.) Sw.	Al Dr Fi Gr La Fr Lo Ps Sl Wl		SA	1624
Stylo, Nigerian	*Stylosanthes erecta* Beauv.	Lo Ps Qz Sl Sv		AF	1824
Stylo, Townsville	*Stylosanthes humilis* H.B.K.	Al Dr Hi Lo Ps Sl	1,200 sd	MA, AF	1323
Sweetclover, Indian	*Melilotus indica* (L.) All.	Ak Dr Fr Hi In Li Na Sl Vi We		HI	0815
Sweetclover, sweet	*Melilotus suaveolens* Ledeb.	Hi, Qz	1,000 day	ES, CJ, II	0820
Sweetclover, white	*Melilotus alba* Medik.	Ak Ba Di *Dr Fr* Hi Ht In Li Mi My Ps Sl Sm So Vi We Wl	32,000 wm	ES, ME	0813
Sweetclover, yellow	*Melilotus officinalis* Lam.	Ak *Dr Fr* Hi Hs Ht In Li My Na Ps Sl Sm So Vi We	8,500 hay	ES, ME	0811
Swordbean	*Canavalia gladiata* (Jacq.) DC.	Dr Lo Na Qz Sh Sl Vi Wl	50,000 wm	II, CJ	1422
Tamarind	*Tamarindus indica* L.	Dr Gr Hi Lo Ps Sl Sv Wi Wl	20,000 pulp	AF	1425
Tarhui	*Lupinus mutabilis* Sweet.	Di Dr Fr Fu Lo Oz Sl	50,000 wm	SA	1023
Tara	*Caesalpinia spinosa* (Mol.) Ktze.	Dr Fi Gr In Lo Ps Wi	15,000 dm	SA	1420
Tephrosia, Vogel	*Tephrosia vogelii* Hook. f.	Dr Gr Lo Mi Ps Sh Sl Wl	15,000 dm	AF	1624
Tephrosia, white	*Tephrosia candida* (Roxb.) DC.	In Sh	13,500 rt	II	
Timbo, macaquinhou	*Lonchocarpus nicou* (Aubl.) DC.	In Sh	13,500 rt	SA	
Timbo, urucu	*Lonchocarpus urucu* Killip & A.C. Sm.			SA	
Tragacanth, gum	*Astragalus gummifer* Labill.	Dr Li Ps Sl		ME, NE	0812
Trefoil, big	*Lotus pedunculatus* Cav.	Di Fu Hi Wl		ES, CA	0811
Trefoil, birdsfoot	*Lotus corniculatus* L.	*Di Dr Fr Fu* Gr He Hi Hs *In* Lo Mi Na Ps Qz Sl Vi We Wl	600 sd	ES	
Trefoil, narrowleaf	*Lotus tenuis* Waldst. & Kit. ex Willd.	Ak Hi Hg Na We Wl	250 sd	ES	0710
Tuba, putch	*Derris elliptica* (Roxb.) Benth.	Hs In Ps Sh Sl	3,000 dry rt	II	2523
Tuba root	*Derris malaccensis* (Benth.) Prain	In Ps Sl	4,000 dry rt	II	
Urd	*Vigna mungo* (L.) Hepper	Dr He Hi Hs Lo Sl Vi	500 sd	II	1221
Velvetbean	*Mucuna deeringiana* (Bort) Merr.	Dr La Ps Qz Vi We	18,000 hay 2,000 sd	II	1524
Vetch, bard	*Vicia monantha* Retz.	Hi		ME, NE, ES	0611
Vetch, bitter	*Vicia ervilia* (L.) Willd.	Hi		NE, ME	0614

(*Continued*)

Appendix

TABLE 4. (*Continued*)

Common name	Scientific name	Tolerance[a]	Yield[b]	Center of diversity[c]	Ecocenter[d]
Vetch, common	*Vicia sativa* L. ssp. *sativa*	Fr Fu Gr Hi In Lo Ne Vi We	6,800 hay	NE, ME	0812
Vetch, kidney	*Anthyllis vulneraria* L.	Hi Ps Qz Sl		ES, ME	
Vetch, purple	*Vicia benghalensis* L.	Hi		ME	0818
Vetch, single-flowered	*Vicia articulata* Hornem.	Ps Qz	21,500 wm		0812
Vetch, winter	*Vicia villosa* Roth	Fr Hi Ps Vi	26,000 wm	NE	
Wattle, black	*Acacia mearnsii* de Wild.	Dr La Ps		AU	1323
Wattle, golden	*Acacia pycnantha* Benth.	Dr		AU	1313
Yambean, erosus	*Pachyrhizus erosus* (L.) Urban	Gr In La Vi	95,000 rt	MA	2124
Yambean, tuberous	*Pachyrhizus tuberosus* (Lam.) Spreng.	In Lo	95,000 rt	SA	1925

Source: Duke (1978).

[a] Tolerances: Ad, Adobe; Ak, Alkali; Al, Aluminum; An, Anthracnose; Ba, Bacteria; Bi, Bird; Bo, Boron; Bu, Bunt; Di, Disease; Dr, Drought; Fi, Fire; Fr, Frost; Fu, Fungus; Gr, Grazing; He, Herbicide; Hf, Hydrogen fluoride; Hg, Mercury; Hi, High pH; Hm, Heavy metal; Hs, Heavy soil; Ht, Heat; In, Insects; La, Laterite; Li, Limestone; Lo, low pH; Mi, Mine; Mu, Muck; Mw, Mildew; My, Mycobacteria; Ne, Nematode; Pe, Peat; Pg, Phage; Ph, Photoperiod; Ps, Poor soil; Pt, Pesticide; Qz, Sand; Ru, Rust; Se, Sewage sludge; Sh, Shade; Sl, Slope; Sm, Smog; So, SO_2; St, Smut; Sv, Savanna; Uv, Ultraviolet; Vi, Virus; We, Weed; Wi, Wind; Wl, Waterlog; and Wt, Wilt. Italicized tolerance abbreviations have been reported in a recent issue of Crop Science. (I have not thoroughly reviewed the journal.)

[b] Yields: a number indicates the yield in kg/ha. This will be followed by lower case code letters indicating what yield is recorded. Symbols: dm, dry matter; dr, drug; fib, fiber; fl, flower; fo, fodder; fr, fruit; hay, hay; inflo, inflorescence; lf, leaf; rt, root or rhizome; sd, seed; sh, shoot; st, stem; wm, wet matter or green biomass. Where hectare yields were not available, per-plant yield is indicated by the slash symbol (/) following the item for which the yield is recorded. For example, under aceituna, I only found per-plant yields (90 kg fr). This is recorded as follows: 90 fr/. It is not uncommon for yield data from different countries or different technologies to vary by a factor of 10. I have taken the higher figures, where credible, assuming that this represents a good target yield. Obviously this table is incomplete. One reason for issuing it now is to solicit additional yield and tolerance data on these and other economic plants.

[c] Centers of diversity: CJ, China–Japan; II, Indochina–Indonesia; AU, Australia; HI, Hindustani; CE, Central Asia; NE, Near East; ME, Mediterranean; AF, Africa; ES, Eurosiberian; SA, South America; MA, Middle America; and NA, North America.

[d] Ecocenter is a 4-digit number, the first 2 digits representing the mean annual precipitation (dm), the second 2 digits the mean annual temperature (°C). See Duke, 1978a.

TABLE 5.
Recommended Inoculants for Various Legumes[a]

Species	Inoculant	Species	Inoculant
Acacia decurrens	SPEC 1[b]	*Melilotus alba*	A
Alysicarpus vaginalis	EL	*Melilotus indica*	A
Anthyllis vulneraria	K	*Melilotus officinalis*	A
Arachis hypogaea	SPEC 1	*Melilotus suaveolens*	A
Astragalus gummifer	SPEC 1	*Mucuna deeringiana*	EL
Cajanus cajan	EL	*Onobrychis viciifolia*	F
Calopogonium mucunoides	SPEC 1	*Phaseolus acutifolius*	SPEC 1
Canavalia ensiformis	SPEC 2	*Phaseolus coccineus*	D
Canavalia gladiata	SPEC 1	*Phaseolus lunatus*	SPEC 2
Canavalia plagiosperma	SPEC 1	*Phaseolus vulgaris*	D
Cassia spp.	SPEC 1, EL	*Pisum sativum*	C
Centrosema sp.	EL	*Psophocarpus tetragonolobus*	SPEC 1
Cicer arietinum	SPEC 1	*Pueraria lobata*	EL
Clitoria ternatea	SPEC 1	*Pueraria phaseoloides*	SPEC 1
Coronilla varia	M	*Sesbania exaltata*	SPEC 1
Crotalaria juncea	EL	*Sphenostylis stenocarpa*	SPEC 1
Crotalaria lanceolata	EL	*Stylosanthes guianensis*	SPEC 1
Crotalaria spectabilis	EL	*Stylosanthes humilis*	SPEC 1
Cyamopsis tetragonoloba	EL	*Tephrosia vogelii*	SPEC 1
Desmodium intortum	EL	*Trifolium alexandrinum*	SPEC 1
Glycine max	S	*Trifolium ambiguum*	SPEC 3
Glycine wightii	EL	*Trifolium fragiferum*	SPEC 6
Hedysarum coronarium	SPEC 1	*Trifolium hirtum*	WR
Indigofera arrecta	EL	*Trifolium hybridum*	B
Indigofera hirsuta	EL	*Trifolium incarnatum*	R
Indigofera spicata	EL	*Trifolium nigrescens*	B
Indigofera tinctoria	EL	*Trifolium pratense*	B
Lablab purpureus	EL	*Trifolium repens*	B
Lathyrus hirsutus	C	*Trifolium resupinatum*	R
Lathyrus sativus	SPEC 3	*Trifolium subterraneum*	WR
Lens culinaris	C	*Trifolium variegatum*	
Lespedeza cuneata	EL	*Trifolium vesiculosum*	O
Lespedeza stipulacea	EL	*Trigonella foenum-graecum*	SPEC 1
Lespedeza striata	EL	*Vicia benghalensis*	C
Leucaena leucocephala	M	*Vicia ervilia*	SPEC 1
Lotus corniculatus	K	*Vicia faba*	Q
Lotus pedunculatus	SPEC 1	*Vicia monantha*	C
Lotus tenuis	K	*Vicia pannonica*	C
Lupinus albus	H	*Vicia sativa*	SPEC 2
Lupinus angustifolius	H	*Vicia sativa* ssp. *nigra*	C
Lupinus luteus	H	*Vicia villosa*	C
Macroptilium atropurpureum	EL	*Vigna aconitifolia*	EL
Medicago arabica	SPEC 1	*Vigna angularis*	EL
Medicago falcata	A	*Vigna mungo*	EL
Medicago lupulina	SPEC 1	*Vigna radiata*	EL
Medicago orbicularis	SPEC 1	*Vigna umbellata*	EL
Medicago polymorpha	SPEC 1	*Vigna unguiculata*	EL
Medicago sativa	A	*Voandzeia subterranea*	ARAC, SPEC 2
Medicago scutellata	SPEC 1		

[a] Source: J. C. Burton (personal communication, 1978).
[b] SPEC, Special Culture for the genus, e.g., SPEC 1 for Acacia is not the same as SPEC 1 for Arachis.

Appendix

TABLE 6. Zero-Moisture

Species	Part[a]	Calories	100/(100 − x)[b]	Protein (g)	Fat (g)	Total carb. (g)	Fiber (g)	Ash (g)
Abrus precatorius	L	295	4.76	26.7		65.2	17.6	8.1
Acacia sp.	L	307	5.38	43.0	3.2	48.4	30.1	5.4
Albizzia sp.	L	310	5.00	35.0	1.0	61.0	16.0	3.0
Arachis hypogaea	L	320	4.65	20.5	2.8	69.3	21.4	7.4
	S	587	1.07	24.8	47.9	24.6	3.1	2.7
Bauhinia esculenta	S	560	1.01	29.5	42.8	24.3	20.5	3.2
	L	307	3.75	16.1	0.4	74.6	25.5	9.0
Bauhinia malabirica	L	210	3.28	8.6	2.9	21.6	4.6	1.0
Bauhinia reticulata	L	295	4.61	22.1	0.5	66.4	31.3	11.1
Cajanus cajan	DS	383	1.11	21.6	1.4	72.7	8.1	4.2
	GS	383	3.22	24.1	1.9	69.6		4.5
	GF	320	2.81	24.4	1.7	68.8	10.1	5.1
Canavalia ensiformis	S	393	1.13	23.7	3.6	68.9	8.6	4.1
Canavalia ensiformis	S	389	1.12	27.4	2.9	66.1	8.3	3.6
	GF	324	8.77	21.0	2.6	71.0	15.8	5.2
	GF	377	8.77	23.7	1.8	69.3	15.8	5.2
Canavalia gladiata	S	375	1.18	32.0	0.7	63.5	13.7	4.2
	GF	315	9.26	25.9	7.4	67.6	13.9	4.6
	GS		8.77	23.7	1.8	56.1	13.2	5.2
Cassia obtusifolia	L	295	4.93	27.6	0.9	61.6	11.3	9.9
Cassia siamea	L	330	2.90	21.5	2.9	70.7	10.7	4.6
Cassia tora	L	320	4.33	19.1	4.3	66.2	11.3	10.3
Cassia (cf. *senna*)	S	384	1.11	20.0	1.4	74.3	7.2	3.8
Ceratonia siliqua	A	203	1.13	5.1	1.6	91.2	8.7	2.4
	S		1.13	19.9	2.8	66.8	8.6	3.4
	F		1.17	5.3	3.0	42.6	10.7	3.3
Cicer arietinum	S	396	1.11	21.7	4.1	70.5	7.4	3.4
	S	405	1.12	21.5	6.2	68.2	2.8	03.5
	SH		2.54	20.8				
Clitoria ternatea	F	330	5.00	19.0	2.0	75.0	24.0	4.0
Cordeauxia edulis	S	446	1.12	12.1	13.4	71.6	1.6	2.5
Crotalaria intermedia	L		3.92	34.5				6.3
Crotalaria longirostrata	L	304	5.43	38.0	4.3	49.4	10.9	8.1
Cyamopsis tetragonoloba	F		5.71	21.1	1.1		13.1	8.0
Dalbergia cultrata	L	309	3.91	21.5		71.9		6.6
Delonix regia	S	378	5.81	35.4	5.2	51.7		7.5
Desmodium cinereum		340	3.86	22.4	6.9	63.7	23.5	6.9
Dialium guineensis	P	356	1.21	4.1	0.1	94.9	2.7	1.2
Dolichos lignosus	S	374	1.10	27.1	0.9	67.0		4.8
Dolichos uniflorus	S	370	1.11	25.0	1.1	67.2	5.2	6.9
	S	389	1.10	31.8	4.5	59.2		4.6
	S		1.13	24.8	0.6	74.5	6.0	
Enterolobium cyclocarpa	S	389	1.01	31.6	3.9	60.8	2.0	3.2
Erythrina berteroana	L	304	6.33	27.8	1.3	63.3	15.2	7.6
Erythrina fusca	L	325	5.41	24.9	4.3	63.3	22.2	7.6
Gliricidia sepium	L	340	6.54	15.7	3.2	77.8	8.5	3.2
	I	328	7.14	17.1	1.4	77.1	11.4	4.3
Glycine max	GS	436	3.14	40.8	17.9	35.8	6.0	5.3
Yellow	S	444	1.11	39.0	19.6	35.5	4.7	5.5
Black	S	439	1.14	38.0	17.1	40.3	4.9	4.6
	S	445	1.10	37.1	19.7	37.3	5.2	5.5
	SP	335	5.41	41.7	9.7	43.3	3.8	5.4

Nutritional Analysis of Legumes

Ca (mg)	P (mg)	Fe (mg)	Na (mg)	K (mg)	β-carotene equiv. (μg)	Thiamine (mg)	Riboflavin (mg)	Niacin (mg)	Ascorbic acid (mg)
126.6	309								
500	452	19.9				1.08	0.91	45.73	264
2,100	310								
1,218	381	19.5			35,967	1.07	2.88	7.44	456
52	438	04.1			16	0.84	0.15	16.58	
196	479	6.6				0.08	0.98	1.92	
1,635	322								255
121	82	0.7			2,247	0.07	0.50	3.61	262
2,005	369								
179	316	16.6			61	0.80	0.16	3.22	
84	435	4.2	16	1,813	467	1.29	0.80	7.73	84
202	489	5.6	14	1,748	407	1.24	0.44	5.05	90
15.1	339	9.7	40	848		0.73	0.15	3.50	2
177	334	7.8				0.87	0.17	2.02	
526	350	17.5			219	0.88			
2,903	473	29.1			39,144	1.13	3.50	7.40	557
290	450	15.4			19,256	0.11	2.06	3.77	162
476	264	27.7			40,009	0.82	1.26	6.93	554
228	204					1.15			
397	91								
280	301	12.3			67	0.53	0.18	2.00	
128	433	02.5	011		16	0.51	0.22	1.30	
	120	2.0	020	1,545	3,350	0.20	0.90	7.00	1240
206	207	7.2							
35		3.1							
870									
1,558	391	25.5			9,986	1.79	2.66	10.86	543
	286	33.1			1,130				280
742	305								
109	383	6.3	006	2,295	1,743		1.16	4.65	
673	416	14.7			4,122	0.96	0.42	5.79	633
876	93	4.5							
237	427	4.3				0.19			
196	196								
		9.0							
323	441	8.6			80				
316									
	552	4.6			9	2.77	0.23	1.82	7
30	506	13.9			418	1.20	1.20	7.66	234
684	216	9.7			12,443	1.30	0.92	25.43	16
308	222	5.2				0.91	0.52	6.54	78
111	264	12.8			150	0.79	0.57	4.28	314
157	496	11.9	47	1,906	1,130	1.26	0.53	4.71	85
245	606	9.4		999	11	0.73	0.24	2.44	0
251	580	10.8		467	11	0.74	0.26	3.19	0
243	595	6.7			60	0.78	0.27	2.20	
201	313	6.0	162	1509	135	1.03	0.81	4.33	54
281									

(Continued)

Appendix

TABLE 6.

Species	Part[a]	Calories	100/(100 − x)[b]	Protein (g)	Fat (g)	Total carb. (g)	Fiber (g)	Ash (g)
Hedysarum coronarium	H		5.88	13.5	2.4		30.6	10.6
Hymenaea courbaril	F	361	1.17	6.9	2.5	88.1	15.7	2.3
Indigofera glandulosa	S		1.09	34.8	2.4		8.5	35.0
Indigofera spicata	H		5.13	21.0	3.1		24.1	11.3
Inga edulis	S	320	2.72	29.1	1.9	65.3	4.4	3.5
Inga sp.	S	386	1.14	21.5	2.4	71.7	3.9	4.0
Inga sp.	A	353	5.88	5.9	0.6	91.1	7.1	2.4
Kerstingiella (Macrotyloma) geocarpa	S	386	1.11	21.5	1.2	73.9	6.1	3.6
Lablab purpureus	GF	312	8.93	25.0	2.7	65.2	16.1	7.1
	S	382	1.13	25.1	1.7	68.9	7.8	4.0
	S	381	1.12	25.5	1.1	69.6	9.6	3.6
	S	380	1.14	24.5	1.4	69.9	7.8	4.3
	GF	312	8.00	24.8	2.4	65.6	15.2	7.2
	L	284	9.17	22.0	3.7	55.9	61.4	12.8
Lathyrus sativus	S	379	1.09	29.9	1.2	65.2	8.0	3.6
Lens culinaris	S	383	1.11	27.6	1.3	67.8	4.3	3.2
	S	387	1.14	23.0	0.7	74.1		2.4
Leucaena leucocephala	L	332	4.88	14.2	3.9	74.7	8.8	7.3
	GF	306	5.18	43.5	4.7	45.6	19.7	6.2
Lupinus albus	S		1.12	36.7	10.0			2.9
Lupinus angustifolius	S		1.16	36.0	7.4		13.3	2.8
Lupinus mutabilis	S	440	1.08	47.8	17.8	30.4	7.7	3.6
	RS	513	1.86	32.2	32.6	32.2	7.1	3.0
	GS	441	3.50	48.0	17.8	31.5	2.1	2.8
Medicago denticulata	L	284	7.46	35.8	2.2	50.0	6.7	11.9
Medicago sativa	SH	301	5.78	34.7	2.3	54.9	17.9	8.1
Mimosa pudica	L	314	14.29	20.0	0.0	72.9		7.1
Mucuna deeringiana	S	398	1.13	24.0	5.0	67.2	5.2	4.0
Mucuna pruriens	S	385	1.12	33.0	3.6	59.2	8.7	3.9
Mucuna sp.	S	403	1.12	26.9	6.2	62.3	4.9	3.7
Mucuna utilis	S	392	1.18	28.3	3.5	64.9		3.5
Neptunia oleracea	L	283	8.33	43.3	1.7	45.0	15.0	10.0
Pachyrhizus erosus	R	359	7.63	10.7	0.8	83.9	04.6	4.6
	R	357	6.49	11.0	0.6	84.4	9.1	3.9
	GF		7.35	19.1	2.2	73.5	21.3	5.1
Parkia sp.	S	456	1.08	34.6	21.1	40.1	4.4	4.4
	P	348	1.15	3.9	0.5	92.1	14.5	3.3
Parkia speciosa	F	443	3.41	27.3	27.6	40.5	1.7	4.4
Pentaclethra macroloba	S	597	1.07	24.2	49.5	24.2	2.7	2.5
Phaseolus acutifolius	S	378	1.07	20.7	1.3	72.5	5.1	3.3
Phaseolus coccineus	S	385	1.14	23.1	2.1	70.7	5.5	3.9
	GS		1.52		0.5		18.5	4.3
	F		2.40	17.8	2.4	71.5	4.6	3.8
Phaseolus lunatus	S	384	1.15	25.0	1.6	70.3	4.9	3.9
	S	388	1.12	22.2	1.5	73.2		3.4
	GS	377	3.17	26.6	1.6	66.6	3.2	5.1
	SP	306	2.78	36.1	2.2	55.6	1.7	6.1
	L	286	35.70	21.4	0.0	60.7		17.9
Phaseolus trilobus	S	365	1.13	27.7	0.7	64.6	8.7	7.3
Phaseolus vulgaris	G	383	1.14	24.7	1.7	69.4	5.0	4.1
	GS	417	8.33	14.2	26.7	45.8	7.5	13.3
	F	318	8.85	22.1	1.7	69.9	15.9	6.2
	L	273	7.58	27.3	3.0	50.0	21.2	19.7

Appendix

(*Continued*)

Ca (mg)	P (mg)	Fe (mg)	Na (mg)	K (mg)	β-carotene equiv. (μg)	Thiamine (mg)	Riboflavin (mg)	Niacin (mg)	Ascorbic acid (mg)
32	167	3.7			1	0.27	0.20	4.80	13
522	348	08.4			0	0.52	0.35	32.6	35
145	292	5.5			0	0.36	0.03	1.60	3
123	118	5.3			1	0.24	0.35	2.35	53
114	435	16.6				0.84	0.21	2.55	0
509	473	8.9	18	2,545	3,108	0.80	0.98	8.04	179
82	472	5.8				0.70	0.20	2.37	
101	367	10.1				0.60	0.16	2.58	1
112	393	4.4				0.46	0.14	2.05	0
600	400	9.6	16	2,232	1,280	0.64	1.04	4.80	128
1,100	523	155.9			28,839	2.57			146
138	447	10.9							
71	331	7.7			66	0.46	0.21	2.44	1
78	370	7.8	33	866		0.52	0.38	1.48	0
2,699	248								
710	57	47.7			24,501		0.47	27.97	41
336	258	10.1	10	761					
97	588	6.8			1	0.30	0.54	2.81	
100	487	4.3			0	0.30	0.54	2.05	9
98	588	6.7			31	0.31	0.53	2.80	
992	462	380.0		2,342	18,015	1.34	2.09	5.22	731
69	295	31.2			11,825	0.75	0.81	2.89	936
500	757	20.0			93,028	1.71	0.71		257
149	361					0.14	0.11	3.39	2
						0.57	0.27	1.90	
146	236	2.4			45	0.35	0.20	2.83	0
466	300	32.5	892	2,307	29,613	1.00	1.17	9.16	175
97	266	5.2			0	0.26	0.26	0.19	65
889	287	9.6			2,536	0.81	0.66	5.88	
314	415	35.9				0.32	0.22	3.24	6
143	184	4.1			2,794	1.21	0.82	1.15	278
259	283	2.4				0.38	0.04	3.41	20
203	184	17.1				0.07	0.34	0.96	
120	332					0.35	0.13	2.99	0
130	404	10.3				0.57	0.21	2.62	2
93	421	6.2				0.81	0.21	3.50	3
120	384	6.2			82	0.81	0.45		65
133	445	5.6				0.38	0.18	2.42	1
101	269	6.3	20	330	1	0.51	0.23	1.57	0
79	377	7.0	6	2,368	285	0.50	0.50	4.75	95
303	1,061	22.8			83	0.47	0.38	5.56	19
285	1,285	82.1							
137	368	9.3			11	0.42	0.18	2.74	1
350	300	6.7							
381	425	12.4			6,638	0.71	1.06	4.42	239
2,076	568	69.7			24,559	1.36	0.45	9.85	834

(*Continued*)

Appendix

TABLE 6.

Species	Part[a]	Calories	100/(100 − x)[b]	Protein (g)	Fat (g)	Total carb. (g)	Fiber (g)	Ash (g)
Pisum macrocarpum	F	317	8.13	24.4	1.6	69.1	13.8	4.9
Pisum sativum	GS	382	4.55	28.7	1.8	65.5	9.1	4.1
	GS	326	3.36	25.5	1.3	70.6	10.1	2.7
	RS	391	1.14	25.6	2.3	70.0	5.4	2.8
	RS	384	1.13	27.2	1.5	68.1	5.5	2.9
Pithecellobium dulce	F	351	4.55	13.5	1.8	81.9	5.4	2.7
	A		4.52	0.32	2.7	89.9	5.4	3.2
	S		1.16	20.4	19.8		9.0	3.0
	L		1.00	29.0	4.4		17.5	5.6
Pithecellobium jiringa	F	388	4.22	26.2	8.0	71.3	5.5	2.1
Prosopis africana	S	361	1.04	16.0	1.7	78.5	7.6	3.4
Prosopis chilensis	L		1.08	28.4	9.2	34.3	26.8	1.5
Psophocarpus tetragonolobus	GF	321	12.80	26.9	3.8	62.7	21.8	6.4
	GF	324	9.52	18.1	1.0	75.2	15.2	5.7
	S	450	1.11	36.4	18.8	40.5		4.4
	S	447	1.18	35.0	17.8	43.4	11.1	4.1
	R	370	4.07	11.4	2.4	82.6	6.1	3.7
	L	313	6.67	33.4	3.3	56.7		6.7
Pterocarpus sp.	L	340	2.60	36.1	8.8	50.2	12.2	4.9
Pueraria thunbergiana	LC	327	9.09	3.6	0.9	88.2	70.0	7.3
	R	359	3.18	6.7	0.3	88.4	2.2	4.4
Sesbania grandiflora	L	321	4.17	36.3	7.5	47.1	9.2	9.2
	I	345	9.09	14.5	3.6	77.3	10.9	4.5
Sesbania roxburgii	L	321	7.14	25.7	2.9	51.0	27.8	6.4
Sesbania sp.	S	393	1.10	35.2	5.0	56.3	12.5	3.4
Sophora japonica	S	435	1.05	17.2	12.2	67.0	10.0	3.9
Sphenostylis stenocarpa	S	391	1.10	21.1	1.2	74.1	5.7	3.2
	R	366	2.84	10.8	0.6	86.3	1.1	2.3
Strophostylis helveola	S	383	1.14	23.3	1.3	71.8	6.3	3.8
Tamarindus indica	I	375	5.00	12.5	9.0	75.0	6.0	3.5
	L	342	4.55	23.3	4.0	70.7	5.7	2.6
	L	332	4.55	14.1	3.6	75.1	18.7	7.3
	F	343	5.05	10.1	1.0	83.8	42.4	5.0
	P	349	1.63	0.37	0.3	92.4	3.1	3.4
Trigonella foenum-graecum	L	282	8.06	37.1	1.6	49.9	11.3	11.3
	S	400	1.08	30.4	6.4	58.9	8.6	3.9
Vicia faba	S	380	1.16	29.0	1.4	66.0	5.9	3.6
Vigna mungo	S	385	1.12	23.5	1.8	71.0	4.9	3.8
Vigna radiata	S	381	1.12	25.6	1.3	69.2	4.9	3.9
	SP	303	10.10	42.4	2.0	50.5	9.1	5.0
Vigna umbellata	S	390	1.16	21.5	1.2	71.6		2.3
Vigna unguiculata ssp. *sesquipedalis*	F	316	8.55	25.6	1.7	67.5	13.7	5.1
	L	293	8.62	36.2	3.4	50.0	14.7	10.3
Vigna unguiculata	GS	382	3.01	27.1	2.4	65.6	5.4	4.8
	RS	384	1.12	25.5	1.7	69.1	4.9	3.9
	S	384	1.13	25.7	1.8	68.9	4.7	3.6
	F	319	8.85	32.7	5.3	54.9	10.6	7.0
	SH	273	9.09	43.6	2.7	40.0		16.4
Voandzeia subterranea	S	412	1.11	17.8	6.7	72.2		3.3

[a] A, Aril; DS, dry seed; F, fruit; GF, green fruit; GS, green seed; H, hay; I, inflorescence or flower; L, leaves; LC, leaves cooked; P, pulp of fruit; R, root; RS, ripe seed; S, seed; SH, shoot; SP, sprout.
[b] x, Percentage water; $100/(100 - x)$, conversion factor for converting to zero-moisture basis.

(*Continued*)

Ca (mg)	P (mg)	Fe (mg)	Na (mg)	K (mg)	β-carotene equiv. (μg)	Thiamine (mg)	Riboflavin (mg)	Niacin (mg)	Ascorbic acid (mg)
398	439	8.9	73	1,098	1,991	1.38	0.81	7.32	293
118	528	8.6	09	1,437	1,747	1.59	0.64	13.20	123
81	417	6.7			252	1.28	0.47	7.39	87
91	331	6.6			17	0.65	0.19	3.42	1
72	384	5.8	40		81	0.84	0.32	3.39	
58	189	2.2	86	999	324	0.33	0.45	2.70	598
59	244	6.3				0.99	0.27	0.90	120
999	350								
97	160	3.0			1,667	0.59	0.04	1.69	34
438	329								
					4				
504	457	1.9	29	19,513	237	1.81	0.76	9.52	200
88	222	2.2							
894	540	41.4			20,977	1.87			
606									
318	181	44.5			0	2.27	8.27	7.27	0
47	57	1.9							
1,684	258		21	2,005	25,979	1.00	1.04	9.17	242
145	290	5.4	291	1,400	636	0.91	0.72	14.54	473
364	400	58.5			14,316	1.85		20.00	171
299	553								
266	273				1,069				
61	437					0.76			
28	227								
265	220	7.0	25	1,270	9,675	0.35	0.40	6.00	25
105	228	8.8	35	1,198	11,020	0.44	0.48	6.58	26
2,302	100								
303	490								
132	140	2.1	5	929	16	0.34	0.13	1.79	5
1,209	387								
238	387	26.1			59	0.35	0.32	1.62	
121	461	04.9	9	1,150	75	0.52	0.22	27.80	0
123	390	9.4		192	22	0.65	0.22	2.58	2
118	370	7.9	7		62	0.59	0.29	2.80	4
152	717	12.1	71	2,242	202	1.11	1.01	8.08	182
93	444	5.8			0	0.35	0.24	0.28	10
376	385	6.0	51	1,992	1,924	1.03	0.94	8.55	188
931	914	40.5	215	3,155	20,688	2.41	2.07	10.34	302
81	518	6.9	6	1,628	668	1.29	0.39	4.82	87
83	477	6.5	39		20	1.29	0.24	2.46	
124	432	7.3	7	777	11	0.67	0.25	2.60	1
478	522	12.4	18	1,947	4,027	1.24	0.89	8.85	212
664	964	20.0				3.18	1.64	10.00	327
94	293	4.7			0	0.20			0

Appendix

TABLE 7. Amino Acid Composition of

Species	% Protein	% Oil	Lys.	Met.	Cys.	Arg.	Gly.	His.	Ile.
Abrus precatorius	19.4	2.8	5.4	1.2	—	4.8	3.9	3.2	3.4
Acacia farnesiana	55.0	7.8	4.7	0.9	—	9.2	3.4	2.3	3.5
Acacia willardiana	35.0	21.0	5.3	0.9	—	5.4	3.1	3.1	2.9
Alysicarpus vaginalis	34.0	6.6	5.2	1.1	—	8.5	4.2	2.5	3.5
Amicia zygomeris	23.0	21.0	4.8	1.0	—	9.5	3.7	2.5	3.5
Arachis hypogaea	23.4	45.3	3.4	0.9	0.9	11.7	—	2.2	3.2
Astragalus crassicarpus	45.0	5.4	3.1	0.7	—	8.5	4.0	1.8	2.5
Astragalus mexicanus	39.0	4.6	3.8	0.8	—	10.2	4.6	2.3	3.0
Astragalus panduratus	52.5	8.5	2.8	0.6	—	8.1	4.0	1.9	2.3
Baptisia leucantha	37.5	10.2	4.7	0.8	—	10.9	4.7	2.0	3.3
Cajanus cajan	21.9	1.5	6.8	1.2	—	5.9	3.7	3.4	3.8
Calliandra eriophylla	39.0	16.0	6.0	0.9	—	7.1	3.9	2.4	3.6
Calycotome villosa	31.3	3.0	5.9	0.9	—	8.2	5.2	2.3	4.1
Canavalia ensiformis	30.0	2.4	5.1	1.0	—	4.5	3.3	2.4	3.5
Cassia emarginata	25.0	2.5	5.3	1.6	—	10.2	5.1	2.8	3.3
Cassia marilandica	17.0	3.6	5.6	1.7	—	8.6	4.1	2.3	3.6
Cassia occidentalis	20.6	2.7	6.2	1.9	—	7.8	4.2	2.3	3.9
Ceratonia siliqua	47.0	3.9	5.6	1.0	—	11.8	5.3	2.5	3.5
Cicer arietinum	19.4	5.5	7.2	1.4	—	8.8	4.0	2.3	4.4
Clitoria ternatea	47.0	12.0	6.1	1.0	—	7.4	4.1	2.4	4.2
Colutea arborescens	71.0	3.0	2.7	0.6	—	8.2	2.8	1.8	2.4
Coronilla varia	32.0	8.0	4.1	0.8	—	6.5	4.8	2.4	3.0
Coursetia glandulosa	56.0	17.0	4.3	0.8	—	6.6	3.1	2.4	2.6
Crotalaria intermedia	31.0	2.7	4.7	0.9	—	8.5	5.2	2.3	4.2
Crotalaria spectabilis	31.0	2.6	5.8	1.3	—	9.7	4.1	2.3	3.9
Cyamopsis tetragonoloba	33.0	4.0	4.0	1.4	—	12.5	5.1	2.5	3.2
Dalea nutans	26.9	10.9	4.9	0.8	—	10.6	4.4	2.1	3.1
Dalea oaxacana	37.0	6.8	5.0	1.1	—	11.0	4.9	2.1	3.5

Various Legumes (g/16 g N)

Leu.	Phe.	Tyr.	Thr.	Trp.	Val.	Ala.	Asp.	Glu.	Hyp.	Pro.	Ser.
5.9	3.6	3.0	3.5	—	3.9	3.5	8.2	13.7	0.4	8.6	4.4
7.5	3.5	2.8	2.5	—	3.9	4.3	8.8	12.6	0.0	5.1	4.1
5.8	3.0	2.6	2.5	—	3.3	3.1	7.3	14.3	0.3	3.5	3.6
6.5	3.5	3.0	3.3	—	3.8	3.4	10.8	15.0	0.1	4.1	4.5
6.2	4.0	2.9	3.1	—	3.9	3.5	10.3	16.9	0.5	4.9	4.6
5.6	4.2	2.5	3.0	1.1	5.0	2.5	11.0	19.0	—	14.6	5.8
4.2	2.4	2.1	2.6	—	3.0	2.6	6.4	13.2	0.1	2.9	3.4
5.0	3.0	2.5	2.9	—	3.5	3.3	7.7	16.2	0.6	3.5	4.2
3.9	2.3	2.2	2.2	—	2.6	2.7	6.2	12.8	0.0	2.9	3.2
5.7	3.3	3.2	2.6	—	3.4	2.8	8.4	20.0	0.7	3.6	4.0
7.2	10.0	3.1	3.6	—	4.5	4.3	9.8	20.1	0.0	4.4	4.7
7.1	3.8	3.6	3.2	—	4.3	3.9	10.0	17.6	0.5	4.4	4.2
7.3	4.1	3.5	3.5	—	4.6	3.6	9.5	19.2	0.3	3.8	4.5
6.4	4.0	3.1	3.9	—	4.0	3.7	9.0	9.1	0.3	3.6	4.3
4.7	3.8	2.8	3.1	—	4.2	3.6	8.2	20.0	1.0	4.2	4.2
6.4	5.0	2.9	3.7	—	4.7	3.9	9.3	17.0	0.7	4.1	5.0
6.9	5.1	3.1	3.9	—	5.1	4.6	10.1	18.0	0.0	3.7	5.3
6.5	3.2	3.5	3.6	—	4.4	4.1	9.0	28.0	0.1	4.0	5.0
7.6	6.6	3.3	3.5	—	4.6	4.1	11.7	16.0	0.0	4.3	5.2
7.4	3.6	—	2.2	1.2	4.4	3.5	9.3	15.6	—	3.3	5.0
4.0	2.6	2.1	2.3	—	3.8	2.4	5.9	13.0	0.1	2.7	3.0
5.6	3.1	—	2.9	1.3	3.9	3.8	8.6	13.7	—	3.2	4.4
4.8	3.4	2.7	2.2	—	3.0	2.6	6.7	12.7	0.0	3.5	3.3
5.9	3.0	2.8	3.4	—	3.9	3.6	9.3	18.8	0.0	3.9	4.3
6.1	2.8	2.5	2.9	—	4.5	3.4	9.2	19.9	—	3.2	5.2
5.9	3.7	—	2.8	1.9	4.2	4.2	10.2	20.1	—	3.1	4.9
5.3	3.3	3.0	2.7	—	3.7	3.3	10.5	15.3	0.0	3.6	3.9
5.7	3.9	2.9	2.8	—	4.1	3.6	8.4	19.4	0.1	4.1	4.2

(Continued)

Appendix

TABLE 7.

Species	% Protein	% Oil	Lys.	Met.	Cys.	Arg.	Gly.	His.	Ile.
Delonix regia	58.0	11.0	4.5	1.1	—	12.0	3.8	2.2	3.2
Desmodium cinerascens	38.0	12.0	6.2	1.2	—	6.7	3.2	2.4	3.4
Ebenus laguroides	51.3	12.7	5.2	1.5	—	13.0	4.1	—	3.7
Genista monosperma	53.0	11.0	4.5	0.7	—	8.5	3.4	2.4	3.5
Glycine max	35.1	17.0	6.3	1.3	1.3	6.8	4.1	2.7	4.6
Hedysarum fontanesii	55.6	10.0	3.9	1.0	—	10.3	3.6	2.5	3.2
Hedysarum varium	50.0	8.3	3.6	0.8	—	5.4	3.9	2.0	2.6
Lablab purpureus	23.4	1.1	6.8	0.9	—	6.6	4.6	3.2	4.4
Lathyrus ornatus	33.0	1.4	5.6	0.7	—	6.7	3.5	2.3	3.2
Lathyrus sylvestris	33.4	0.8	5.8	0.7	—	7.5	3.7	2.6	3.7
Lens culinaris	26.9	0.8	6.7	0.6	—	7.8	3.7	2.1	3.8
Lespedeza stipulacea	52.0	7.0	5.3	1.0	—	11.9	4.1	2.6	3.9
Lotus rigidus	38.0	6.8	3.6	1.0	—	11.3	5.1	2.2	2.9
Lotus scoparius	35.0	8.4	3.9	1.0	—	9.1	5.2	2.2	3.0
Lupinus luteus	43.0	5.0	4.8	0.5	—	9.9	4.0	2.5	3.9
Lysiloma desmostachya	41.0	9.1	4.4	0.6	—	3.8	2.6	—	2.4
Medicago lupulina	33.0	5.0	4.7	1.2	—	7.6	4.2	2.6	3.5
Melilotus albus	39.4	5.9	5.7	1.3	—	9.3	4.4	2.7	3.9
Melilotus indicus	33.0	5.4	5.5	1.2	—	7.9	4.6	2.8	3.9
Milletia ovalifolia	27.0	37.0	5.1	0.8	—	4.2	4.0	1.6	
Mucuna deeringiana	29.4	5.6	6.3	1.1	—	6.6	4.2	2.3	4.7
Onobrychis aurantiaca	57.5	13.4	4.9	1.7	—	10.6	5.0	4.3	3.3
Onobrychis viciifolia	41.0	7.6	5.6	1.3	—	10.4	4.1	3.7	3.2
Onobrychis vulgaris	41.0	6.8	5.5	1.3	—	9.9	4.0	3.5	3.1
Ormosia jamaicensis	12.5	9.3	4.1	0.5	—	4.0	2.6	1.4	2.3
Oxytropis uralensis	27.0	4.3	4.6	1.0	—	9.3	4.2	2.2	3.7
Pachyrhizus erosus[a]			5.5	1.1	0.7	6.2	3.4	2.9	3.5
Phaseolus vulgaris	20.3	1.2	8.0	1.0	0.7	6.1	3.4	4.8	6.2

(Continued)

Leu.	Phe.	Tyr.	Thr.	Trp.	Val.	Ala.	Asp.	Glu.	Hyp.	Pro.	Ser.
6.2	3.9	2.8	2.6	—	4.2	3.7	8.3	21.1	0.0	3.6	4.2
6.2	3.2	2.5	3.1	—	3.7	3.4	9.7	13.0	0.2	3.9	3.9
6.1	4.2	2.5	3.4	—	4.0	4.2	9.6	18.3	0.0	4.5	4.6
6.4	3.3	3.2	2.9	—	3.5	3.1	9.0	17.5	0.2	3.9	3.8
7.9	5.5	2.6	4.0	1.2	4.7	4.5	11.6	19.0	—	5.3	4.9
5.0	3.3	2.5	2.8	—	3.4	3.1	8.1	14.3	0.0	3.2	3.7
4.3	2.4	2.2	2.7	—	2.9	3.1	7.3	9.1	0.1	2.7	3.0
8.5	4.9	3.6	4.2	—	5.2	4.5	13.0	15.7	0.6	4.3	5.4
5.5	3.4	2.5	3.0	—	3.4	3.1	8.4	14.8	0.2	3.4	3.6
6.4	3.8	2.6	3.4	—	4.4	3.9	10.2	14.8	0.0	3.6	4.3
6.6	4.2	3.9	3.3	—	4.2	3.5	10.9	14.5	0.0	3.5	4.4
7.2	4.3	—	3.0	1.4	4.2	3.8	11.3	16.7	—	3.7	5.9
4.9	3.4	2.6	2.7	—	3.3	3.2	8.6	15.1	0.2	3.3	4.3
5.1	3.4	2.8	2.8	—	3.3	3.2	8.3	13.9	0.1	3.5	4.2
7.5	3.6	—	3.1	1.0	3.5	3.4	10.1	24.4	—	3.2	4.9
4.9	2.5	2.5	2.1	—	2.7	2.6	6.2	8.6	0.2	3.3	3.0
5.9	4.0	2.8	3.4	—	4.0	3.6	9.4	13.1	0.1	3.6	4.2
6.9	4.8	3.2	3.1	—	4.4	3.6	9.9	15.5	0.0	4.2	4.4
6.2	4.5	2.9	3.2	—	3.9	3.8	10.3	14.2	0.0	3.7	4.2
6.2	4.0	2.7	3.1	—	3.9	3.1	8.7	11.4	0.0	3.8	4.0
7.0	4.4	4.8	4.0	—	4.9	3.4	11.6	11.8	0.0	5.5	4.1
5.4	3.0	2.8	3.0	—	3.7	3.3	10.7	14.2	0.0	3.1	4.0
5.6	3.0	2.5	3.0	—	3.7	3.7	9.0	15.6	0.4	3.7	4.1
5.4	2.9	2.4	3.1	—	3.6	3.2	9.4	15.7	0.1	3.8	4.1
4.7	2.5	2.6	2.3	—	2.6	2.4	5.6	10.7	0.4	2.7	2.5
6.2	3.6	3.0	2.8	—	4.4	4.0	9.5	15.6	—	3.4	4.3
5.4	3.7	3.2	3.8	—	5.7	4.0	34.0	8.6	—	3.5	4.8
7.9	5.5	2.6	4.2	1.2	6.1	2.7	9.8	17.6	—	3.8	5.3

(Continued)

Appendix

TABLE 7.

Species	% Protein	% Oil	Lys.	Met.	Cys.	Arg.	Gly.	His.	Ile.
Pisum "arvense"	25.6	0.9	7.0	0.8	—	8.8	4.2	2.2	3.9
Pisum sativum	22.5	1.4	7.2	0.6	0.8	9.3	4.2	2.4	4.4
Psophocarpus tetragonolobus[a]			7.7	1.3	2.3	4.7	4.7	3.2	4.4
Retama raetum	25.6	2.8	4.9	0.8	—	8.0	4.0	2.3	3.5
Robinia neomexicana	41.0	13.0	3.6	0.7	—	8.0	3.6	2.0	2.4
Scorpiurus subvillosa	35.0	2.0	4.5	1.3	—	10.9	4.5	2.2	3.2
Sesbania exaltata	40.0	4.0	4.7	1.3	—	7.6	4.3	2.4	4.0
Sesbania macrocarpa	38.0	6.4	4.9	0.9	—	8.2	4.3	2.6	3.2
Spartium junceum	34.4	12.4	5.3	1.0	—	8.8	4.6	2.3	4.0
Sphenostylis stenocarpa			7.7	1.7	1.8	5.8	4.9	4.7	4.6
Tephrosia leiocarpa	44.0	11.0	5.0	0.9	—	7.2	3.2	2.2	3.0
Trifolium incarnatum	44.0	4.0	5.4	1.4	—	9.8	4.1	2.7	4.2
Trigonella arabica	28.1	3.3	5.4	1.4	—	11.4	5.6	2.6	3.9
Trigonella foenum-graecum	26.0	5.0	6.0	1.3	—	9.2	4.4	2.0	4.5
Vicia angustifolia	47.5	6.2	3.4	0.9	—	10.4	4.6	1.8	2.9
Vicia faba	25.0	1.2	6.1	0.6	—	7.9	3.8	2.4	3.8
Vicia gigantea	38.1	1.2	4.1	0.6	—	5.4	2.7	1.7	2.9
Vigna angularis	21.1	1.0	7.3	1.3	0.9	6.3	3.4	3.3	3.9
Vigna radiata	22.9	1.2	9.4	2.0	0.5	6.3	3.3	2.9	4.1
Vigna "sesquipedalis"	22.7	1.6	6.6	1.4	0.5	8.7	3.8	5.0	6.7
Vigna umbellata	18.5	1.0	12.3	2.5	0.7	7.4	—	6.1	6.2
Vigna unguiculata	26.0	1.4	6.4	1.2	—	7.3	4.1	3.2	4.0
Vigna vexillata (?unguilata?)[a]			6.1	1.6	1.6	14.2	4.7	3.6	3.3

[a] Meal analyses of roots, presented as grams amino acid per 100 g recovered amino acid by Evans *et al.*, 1977.

(Continued)

Leu.	Phe.	Tyr.	Thr.	Trp.	Val.	Ala.	Asp.	Glu.	Hyp.	Pro.	Ser.
6.8	4.3	3.3	3.5	—	4.3	3.8	11.2	15.7	0.0	4.2	4.5
6.7	4.5	2.2	3.7	0.9	5.2	4.5	9.8	16.0	—	5.6	5.1
7.0	5.0	3.5	4.8	—	6.1	4.8	17.7	10.0	—	7.0	5.9
6.4	3.6	3.2	2.8	—	3.7	3.1	8.2	19.8	0.6	4.2	3.5
4.8	2.8	2.3	2.4	—	3.2	2.8	6.5	12.4	0.1	3.0	3.5
5.0	4.9	3.3	3.0	—	3.4	3.5	9.2	14.2	0.0	3.0	4.1
6.2	3.9	—	2.8	1.7	4.2	3.2	8.2	14.6	—	3.0	4.0
5.7	3.7	2.7	2.9	—	3.5	3.2	7.9	14.3	0.1	3.8	4.0
7.2	4.1	3.8	3.3	—	4.3	3.6	9.6	18.9	0.4	4.2	4.1
7.7	4.6	3.2	4.3	—	5.6	5.5	12.5	12.0	—	7.7	5.8
6.2	3.7	2.6	2.6	—	3.3	3.0	8.9	13.2	0.1	3.8	3.8
7.3	3.9	—	3.1	1.8	4.4	4.1	11.0	16.5	—	3.4	4.7
6.3	4.2	2.9	3.4	—	4.6	3.8	10.4	16.3	0.0	4.4	4.5
6.8	3.8	—	3.0	1.6	3.4	4.0	10.9	15.8	—	4.6	5.2
5.0	3.6	2.4	2.6	—	3.4	3.0	8.4	14.4	0.0	3.5	4.0
6.7	3.9	3.2	3.5	—	4.3	3.7	10.1	14.9	0.1	3.9	4.3
4.9	—	—	2.6	—	3.3	2.9	7.3	11.1	0.2	2.7	3.1
7.2	5.4	3.4	3.4	1.7	4.4	4.0	9.8	17.2	—	4.7	4.3
7.0	5.5	2.4	3.2	1.9	4.3	3.5	10.7	13.6	—	3.5	4.0
8.5	5.2	2.6	4.0	1.4	6.8	4.0	12.8	15.8	—	6.6	5.3
9.7	5.2	4.2	4.7	1.0	6.3	—	—	—	—	—	—
7.2	6.0	3.4	3.6	—	4.7	4.4	10.6	16.9	0.5	3.4	4.5
6.2	3.6	3.2	4.9	—	4.3	4.4	14.1	11.9	—	6.8	5.6

General References

Abramova, E. P., and Chernikov, M. P., 1964, Amount of proteinase inhibitors in the seeds of certain legumes, *Prob. Nutr.* **23**:13.

Acosta Solis, M., 1947, Commercial possibilities of the forests of Ecuador, mainly Esmeraldas Province, *Trop. Woods* **89**:1–47.

Adams, C. D., 1972, *Flowering Plants of Jamaica*, University of the West Indies, Mona, Jamaica, 848 pp.

Ahn, P., 1961, Soil vegetation relationships in the western forest area of Ghana, *Tropic Soils and Vegetation Procedures of the Abidjan Symposium*, p. 78., Oct. 20–24, 1959.

Backer, C. A., and van den Brink, R. C. Bakhuizen, Jr., 1965, *Flora of Java*, Noordhoff, Groningen, The Netherlands.

Bailey, L. H., 1922, *Standard Cyclopedia of Horticulture*, Macmillan, New York, 6 vols.

Bailey, L. H., and Bailey, E. Z., 1941, *Hortus Second*, Macmillan, New York, 778 pp.

Bailey, L. H., and Bailey, E. Z., 1976, *Hortus Third*, Macmillan, New York, 1290 pp.

Baker, E. G., 1929, *The Leguminosae of Tropical Africa*, Erasmus Press, Ghent, 3 vols.

Berliz, H. D., Wassner, H. P., and Weder, J., 1968, Proteinasen-Inhibitoren in Lebensmitteln. I. Vorkommen und Thermostabilität von Trypsin- und Chymotrypsin-Inhibitoren in verschiedenen Erbsen und Bohnen, *Z. Lebensm. Unters. Forsch.* **137**:211.

Black, J. M., 1948, *Flora of South Australia*, 2nd ed., Part II, p. 414, illus., W. L. Hawes, Government Printer, Adelaide, Australia.

Blombery, A. M., 1955, *Native Australian Plants, Their Propagation and Cultivation*, Sydney, Australia.

Bogdan, A. V., 1977, *Tropical Pasture and Fodder Plants (Grasses and Legumes)*, Longmans, London, 475 pp.

Borchers, R., and Ackerson, C. W., 1947, Trypsin inhibitor. IV. Occurrence in seeds of the Leguminosae and other seeds, *Arch. Biochem.* **13**:291.

Brenan, J. P. M., 1959, *Flora of Tropical East Africa, Leguminosae Subfamily Mimosoideae*, Whitefriars Press, London, 173 pp.

Brenan, J. P. M., 1967, *Flora of Tropical East Africa, Leguminosae Subfamily Caesalpinioideae*, Whitefriars Press, London, 230 pp.

Britton, N. L., and Wilson, P., 1924, *Scientific Survey of Puerto Rico and the Virgin Islands*, 5(3), N.Y. Academy of Sciences, New York.

Bunting, B., and Milsum, J. N., 1928, Cover crops and green manures, *Malay. Agric. J.* **16**:256–280, illus.

Burkill, I. H., 1935 (repr. 1966), *A Dictionary of the Economic Products of the Malay Peninsula*, Art Printing Works, Kuala Lumpur, 2 vols.

Cerighelli, R., 1955, *Cultures Tropicales. Planta Vivières*, Vol. 1, Librairie J. B. Baillière & Fils, Paris.

Chandler, W. H., 1950, *Evergreen Orchards*, Lea & Febiger, Philadelphia, Pa.

Chattopadhyay, H., and Banerjee, S., 1953, Effect of germination on the biological value of proteins and the trypsin-inhibitor activity of some common Indian pulses, *Indian J. Med. Res.* **41**:185.

Chopra, R. N., Nayer, S. L., and Chopra, I. C., 1956, *Glossary of Indian Medicinal Plants*, Council of Scientific and Industrial Research, New Delhi.

Chuang, Ching-Chang, and Huang, Chia, 1965, *The Leguminosae of Taiwan for Pasture and Soil Improvement*, Joint Commission on Rural Reconstruction, Taipei, 286 pp.

Clapham, A. R., Tutin, T. G., and Warbur, E. F., 1962, *Flora of the British Isles*, Cambridge University Press, Cambridge.

Claus, E. P., Tyler, V. E., and Brady, L. R., 1970 (repr. 1971), *Pharmacognosy*, 6th ed., Lea & Febiger, Philadelphia, PA, 518 pp.

Corner, E. J. H., and Watanabe, K., 1969, *Illustrated Guide to Tropical Plants*, Tokyo, 1147 pp.

Correll, D. S., and Johnston, M. C., 1970, *Manual of Vascular Plants of Texas*, Renner, Texas, 1881 pp.

CSIR, 1948–1976, *The Wealth of India*, 11 vols., Publications and Information Directorate, Council of Scientific & Industrial Research, New Delhi, India.

Dalziel, J. M., 1937, *The Useful Plants of West Tropical Africa*, Secretary of State for the Colonies, London, 612 pp.

Dastur, J. F., 1952, *Medicinal Plants of India and Pakistan*, D. B. Taraporevala Sons & Co., Ltd., Bombay.

De Geus, J. G., 1973, *Fertilizer Guide for the Tropics and Subtropics*, 2nd ed., Centre d'Etude de l'Azote, Zurich, 774 pp.

General References

de Sornay, P., and Flattely, F. W., 1916, *Green Manures and Manuring in the Tropics,* J. Bali, Sons & Danielson, London, 466 pp.

Dickson, J. G., 1956, *Diseases of Field Crops,* McGraw-Hill, New York.

Duke, J. A., 1972, *Isthmian Ethnobotanical Dictionary,* Harrod Press, Baltimore, MD, 96 pp.

Duke, J. A., 1977a, Vegetarian vitachart, *Q. J. Crude Drug Res.* **15**:45–66.

Duke, J. A., 1977b, Phytotoxin tables, *Crit. Rev. Toxicol.* **5**(3):189–237.

Duke, J. A., 1978a, Ecosystematics, in: *Biosystematics in Agriculture* (J. A. Romberger, ed.), pp. 53–68, Beltsville Symposia in Agricultural Research, Allanheld, Osmun & Co., Montclair, NJ, 340 pp.

Duke, J. A., 1978b, The quest for tolerant germplasm, in: *Crop Tolerance to Suboptimal Land Conditions* (G. A. Jung, ed.), pp. 1–61, ASA Special Publ. No. 32, American Society for Agriculture, Madison, WI, 343 pp.

Duke, J. A., 1978c, Ecological amplitudes of legumes, *Trop. Grain Leg. Bull.,* **13/14**:3–8.

Duke, J. A., and Terrell, E. E., 1974, Crop diversification matrix: Introduction, *Taxon* **23**(5/6):759–799.

Duke, J. A., Hurst, S. J., and Terrell, E. E., 1975, Ecological distribution of 1000 economic plants. Informacion al Dia Alerta. IICA-Tropicos, *Agronomia* **1**:1–32.

Erdman, L. W., 1967, Legume inoculation: What it is; what it does. *U.S. Dep. Agric., Farmers Bull.* **2003**:1, 7.

Evans, I. M., Boulter, D., Eaglesham, A. R. J., and Dart, P. J., 1977, Protein content and protein quality of tuberous roots of some legumes determined by chemical methods, *Qual. Plant. Plant Foods Hum. Nutr.* **27**(3/4):275–285.

Everett, T. H., 1960, *New Illustrated Encyclopedia of Gardening,* Vol. 8, pp. 1381–1384, illus., Greystone Press, New York.

FAO, 1976, *Production Yearbook,* Vol. 29, 1975, Rome, 555 pp.

Firminger, T. A., 1947, *Manual of Gardening for India,* 8th ed., Thacker, Spink & Co., Calcutta.

Franke, W., 1976, *Nutzpflanzenkunde,* Georg Thieme Verlag, Stuttgart, 474 pp.

Geithens, T. S., 1948, *Drug Plants of Africa,* African Handbook No. 8, University of Pennsylvania Press, Philadelphia, PA.

Gibbs, R. D., 1974, *Chemotaxonomy of Flowering Plants,* McGill-Queen's University Press, Montreal and London.

Gillett, J. B., Polhill, R. M., and Verdcourt, B., 1971, *Flora of Tropical East Africa. Leguminosae* Part 4, *Subfamily Papilionoideae,* Whitefriars Press, London, 2 vols.

Gleason, H. A., 1958, *The New Britton and Brown's Illustrated Flora of the Northeastern United States and Adjacent Canada,* New York Botanical Gardens, New York.

Graham, E. H., 1941, Legumes for erosion control and wildlife, *U.S. Dep. Agric., Misc. Publ.* **412**:153 pp.

Grieve, M., 1931, Repr. 1974, *A Modern Herbal,* Hafner Press, New York, 916 pp.

Grisebach, A. H. R., 1860–1864, *Flora of the British West Indian Islands,* Lovell Reeve & Co., London, 789 pp.

Grist, D. H., 1936, *An Outline of Malayan Agriculture,* Malayan Planting Manual, No. 2, 377 pp.

Guenther, E., 1952, *The Essential Oils,* Vol. 5, pp. 220–225, Van Nostrand-Reinhold, New York, 6 vols.

Hafenrichter, A. L., Schwendiman, J. L., Harris, H. L., MacLauchlan, R. S., and Miller, H. W., 1968, Grasses and legumes for soil conservation in the Pacific Northwest and Great Basin States, *U.S. Dep. Agric., Handb.* **339**:69 pp.

Harborne, J. B., Boulter, D., and Turner, B. L. (eds.), 1971, *Chemotaxonomy of the Leguminosae,* Academic Press, London and New York, 612 pp.

Haring, E., 1967, *The Complete Book of Growing Plants from Seed,* Diversity Books, Grandview, MO, 240 pp.

Haselwood, E. L., and Motter, G. G., 1966, *Handbook of Hawaiian Weeds,* Hawaiian Sugar Planter's Association, Honolulu, 479 pp.

Heath, M. E., Metcalfe, D. S., and Barnes, R. F. (eds.), 1973, *Forages,* 3rd ed., Iowa State University Press, Ames, 755 pp.

Hedrick, U. P., 1972, *Sturtevant's Edible Plants of the World,* p. 152, Dover, New York, 686 pp.

Heiser, C. B., 1965, Cultivated plants and cultural diffusion in nuclear America, *Am. Anthropol.* **67**:930–949.

Herklots, G. A. C., 1972, *Vegetables in South-East Asia,* Hafner Press, New York, 525 pp.

Hilebrand, W., 1888, *Flora of the Hawaiian Islands,* Westermann, New York, 673 pp.

Hooker, J. D., 1876–1879, *The Flora of British India,* Vol. 2, 1st Indian Reprint, 1973, M/s Bishen Singh Mahendra Pal Singh, Dehra Dun, 792 pp.

Hopkinson, D., 1969, Leguminous cover crops for maintaining soil fertility in sisal in Tanzania: I. Effects for growth and yield, *Exp. Agric.* **5**(4):283–294, illus.

Horsfal, J. G., *et al.,* 1972, *Genetic Vulnerability of Major Crops,* National Academy of Science, Washington, D.C.

Howell, T., 1903, *A Flora of Northwestern America,* Vol. 1, Portland, Oregon, 792 pp.

Hutchinson, J., and Dalziel, J. M., 1958, *Flora of West Tropical Africa,* 2nd ed. (rev. by R. W. J. Keay), Crown Agents, Millbank, 3 vols.

ICAR, 1970, *Pulse Crops of India,* Indian Council of Agricultural Research, New Delhi, 334 pp.

INEAC, 1952–1954, *Flore du Congo Belge et du Ruandi Urundi,* Vols. 3–6, l'Institut National Pour l'Etude Agronomique du Congo Belge, Brussels.

Irvine, F. R., 1961, *Woody Plants of Ghana,* Oxford University Press, London, 868 pp.

Irvine, F. R., 1968, *West African Agriculture,* 3rd. ed., Vol. 2, *West African Crops,* Oxford University Press, London, 272 pp.

Jaffe, W. G., 1950, Protein digestibility and trypsin inhibitor activity of legume seeds, *Proc. Soc. Exp. Biol. Med.* **75**:219.

Kay, D. E., 1979, *TPI Crop and Product Digest.* Vol. 3: *Food Legumes,* Tropical Products Institute, Ministry of Overseas Development, London.

Kearney, T. H., and Peebles, R. H., *et al.,* 1969, *Arizona Flora,* 2nd ed., University of California Press, Berkeley, CA, 1085 pp.

Kennard, W. C., and Winters, H. F., 1960, Some fruits and nuts for the Tropics, *U.S. Dep. Agric., Agric. Res. Serv, Misc. Publ.* **801**:135 pp.

Kernick, M. D., 1961, *Agricultural and Horticultural seeds,* FAO Agricultural Studies, No. 55, Rome.

King, G., 1936, Materials for a flora of the Malaysian Peninsula, *J. Asiat. Soc. Bengal* **2**(1):544 pp.

Kingsbury, J. M., 1964, *Poisonous Plants of the United States and Canada*, Prentice-Hall, Englewood Cliffs, NJ, 626 pp.

Kipps, M. S., 1970, *Production of Field Crops*, 6th ed., McGraw-Hill, New York, 790 pp.

Khan, T. N., and Erskine, W., 1975, Production and improvement of food legumes in Papua New Guinea and its implication to Malaysian food self-sufficiency, Proceedings of the Conference on Malaysian Food Self-Sufficiency, University of Malaysia, Kuala, Lampur.

Komarov, V. L. (ed.), 1945, *Flora USSR* **11**:139–148 (English ed., pp. 106–114, 1971), Israel Program for Scientific Translations, Jerusalem.

Kretschmer, A. E., Jr., Bolman, J. B., Snyder, G. H., and Gascho, G. J., 1973, Production of six tropical legumes each in combination with three tropical grasses in Florida, *Agron. J.* **65**:890–892.

Kundu, B. C., 1967, Some edible rhizomatous and tuberous crops of India, *Proc. Int. Symp. Trop. Root Crops* **1**.

Leese, B. M., Jr., 1958, Identification of Asiatic species of *Phaseolus* by seed characters, *Am. Midl. Nat.* **60**:132–144.

Leffel, R. C., 1973, Other legumes, in: *Forages* (M. E. Heath, D. S. Metcalfe, and R. F. Barnes, eds.), Chapter 20, 3rd ed., Iowa State University Press, Ames, 755 pp.

Leiner, I. E. (ed.), 1969, *Toxic Constituents of Plant Foodstuffs*, Academic Press, New York.

Lewis, W. H., and Elvin-Lewis, M. P. F., 1977, *Medical Botany*, Wiley, New York, 515 pp.

Little, E. L., Jr., and Wadsworth, F. H., 1964, Common trees of Puerto Rico and the Virgin Islands, *U.S. Dep. Agric. For. Serv., Agric. Handb.* **249**:548 pp.

Litzenberger, S. C. (ed.), 1974, *Guide for Field Crops in the Tropics and Subtropics*, AID, Washington, D.C., 321 pp.

Macbride, J. F., 1943, Flora of Peru, *Field Mus. Nat. Hist., Bot. Ser.* **13**:[Part 3(1)]507 pp.

McDougall, W. B., 1973, *Seed Plants of Northern Arizona*, Museum of Northern Arizona, Flagstaff, 594 pp.

MacGillivray, J. H., 1953, *Vegetable Production with Special References to Western Crops*, McGraw-Hill, New York, 397 pp.

McGregor, S. E., 1976, *Insect Pollination of Cultivated Crop Plants*, ARS, USDA, Washington, D.C., 411 pp.

MacMillan, H. F., 1946, *Tropical Planting and Gardening*, 5th ed., Macmillan, London, 560 pp.

Magness, J. R., Markle, G. M., and Compton, C. C., 1971, Food and feed crops of the United States, *N.J. Agric. Exp. Stn., Bull.* **828**:255 pp.

Maheshwari, P., and Singh, U., 1965, *Dictionary of Economic Plants in India*, Indian Council of Agricultural Research, New Delhi.

Maiden, J. H., 1889, *The Useful Native Plants of Australia*, The Technological Museum of New South Wales, Sydney, Turner and Henderson, 696 pp.

Mansfeld, V., Ziegelhoffer, A., Horakova, Z., and Hladovec, J., 1959, Isolierung der trypsin-inhibitoren aux einigen hulsenfruchten, *Naturwissenschiften* **46**:172.

Mappledoram, B. D., and Theron, E. P., 1972, Notes on the adaptability of several tropical legumes to different environments in Natal, *Proc. Grassl. Soc. S. Afr.* **7**:84–86.

Martin, F. W., and Ruberte, R. M., 1975, *Edible Leaves of the Tropics*, Antillian College Press, Mayaguez, Puerto Rico, 235 pp.

Masefield, G. B., et al., 1969 (repr. 1971), *The Oxford Book of Food Plants*, Oxford University Press, London, 206 pp.

Menninger, E. A., 1970, *Flowering Vines of the World*, Hearthside Press, New York, 410 pp.

Merrill, E. D., 1923, An enumeration of Philippine flowering plants, *Bur. Sci. Manila* **2**(3):241–323.

Metcalf, C. L., and Flint, W. P., 1939, *Destructive and Useful Insects*, McGraw-Hill, New York, 981 pp.

Miller, D. F., 1958, Composition of cereal grains and forages, *NAS–NRC, Publ.* **585**:663 pp.

Millspaugh, C. F., 1887, *American Medicinal Plants*, repr., Dover, New York.

Mors, W. B., and Rizzini, C. T., 1966, *Useful Plants of Brazil*, Holden-Day, San Francisco, CA, 166 pp.

Mortensen, E., and Bullard, E. T., 1968, *Handbook of Tropical and Sub-Tropical Horticulture*, rev. ed., Department of State, AID, Washington, D.C., 186 pp.

Morton, J. F., 1962, *Wild Plants for Survival in South Florida*, Hurricane House, Miami, FL, 76 pp.

Muencher, W. C., 1939, *Poisonous Plants of the United States*, Macmillan, New York, 266 pp.

Munz, P. A., and Keck, D. D., 1968, *A California Flora*, University of Calif. Press, Berkeley, CA, 1681 pp.

NAS, 1975, *Tropical Legumes: Resources for the Future*, National Academy of Sciences, Washington, D.C., 331 pp.

Neal, M. C., 1948, In gardens of Hawaii, *Bernice P. Bishop Mus. Spec. Publ.* **40**:398–399.

NIOSH, 1975, *Registry of Toxic Effects of Chemical Substances*, United States Department of Health, Education and Welfare, Rockville, MD.

Ochse, J. J., 1931, *Vegetables of the Dutch East Indies*, Printed and edited by Archipel drukkerij, Buitenzorg, Java, 1005 pp.

Ohwi, J., 1965, *Flora of Japan*, Smithsonian Institution, Washington, D.C., 1067 pp.

Parodi, L., 1964, *Encyclopedia Argentina de agricultura y jardenaria*, Acme, Buenos Aires, 2 vols.

Pearson, R. S., and Brown, H. P., 1932, *Commercial Timbers of India*, Calcutta, 2 vols.

Perdue, R. E., and Hartwell, J. L. (eds.), 1975, Proceedings of the 16th Annual Meeting of the Society of Economic Botany: "Plants and cancer," *Cancer Treatment Rep.* **60**(8):975–1215.

Perez-Arbelaez, E., 1956, *Plantas utiles de Colombia*, 3rd ed., Libreria Colombiana, Bogota, 832 pp.

Piper, C. V., and Morse, W. J., 1914, Five oriental species of beans, *U.S. Dep. Agric., Bull.* **119**.

Polunin, O., 1969, *Flowers of Europe, A Field Guide*, Oxford University Press, London, 662 pp.

Polunin, O., and Huxley, A., 1966, *Flowers of the Mediterranean*, 1st American ed., Houghton-Mifflin, Boston, MA, 257 pp.

Pulle, A. (ed.), 1939, *Flora of Surinam (Papilionaceae*, by G. J. H. Amshoff), Vol. 2, Part 2, J. H. deBussy, Amsterdam, 257 pp.

Purseglove, J. W., 1968, *Tropical Crops, Dicotyledons*, Wiley, New York, 2 vols.

General References

Rachie, K. O., 1978, Productivity potentials of edible legumes in the lowland Tropics, in: *Crop Tolerance to Suboptimal Land Conditions* (G. A. Jung, ed.), pp. 71–96, American Society of Agronomy, Special Publication 32, ASA, Madison, WI, 343 pp.

Rachie, K. O., and Roberts, L. M., 1974, Grain legumes of the lowland tropics, *Adv. Agron.* **26**:1–132.

Radford, A. E., Ahles, H. E., and Bell, C. R., 1968, *Manual of the Vascular Flora of the Carolinas*, University of North Carolina Press, Chapel Hill, NC, 1183 p.

Record, S. J., and Mell, C. D., 1924, *Timbers of Tropical America*, Yale University Press, New Haven, CN, 610 pp.

Reed, C. F., and Hughes, R. O., 1970, Selected weeds of the United States, *U.S. Dep. Agric., Agric. Res. Serv., Agric. Handb.* **336**:463 pp.

Rydberg, P. A., 1918, *Flora of the Rocky Mountains and Adjacent Plains*, published by the author, New York Botanical Gardens, New York, 1110 pp.

Sampson, M., 1928, Cover crops in tropical plantations, *Kew Bull.* **1928**:163.

Sauer, C. O., 1952, *Agricultural Origins and Dispersals*, American Geographical Society, New York, 110 pp.

Schery, R. W., 1972, *Plants for Man*, 2nd ed., Prentice-Hall, Englewood Cliffs, NJ, 657 pp.

Senewiratne, S. T., and Appadurai, R. R., 1966, *Field Crops of Ceylon*, Colombo, Ceylon, 376 pp.

Shaw, E. J. F., and Khan, A. R., 1931, Studies in Indian pulses, *Dep. Agric. India, Bot. Ser., Mem.* **19**.

Sievers, A. F., 1948, Production of drug and condiment plants, *U.S. Dep. Agric., Farmers Bull.* **1999**.

Sievers, A. F., and Higbee, E. C., 1942, Medicinal plants of tropical and subtropical regions, *U.S. Dep. Agric., For. Agric. Rep.* **6**.

Simmonds, N. W., 1976, *Evolution of Crop Plants*, Longman Group Ltd., London, 339 pp.

Smartt, J., 1976, *Tropical Pulses*, Longman Group Ltd., London, 348 pp.

Smith, D., 1977, *Forage Management in the North*, Kendall-Hunt, Dubuque, Iowa.

Smith, J. S., 1896, Fodder and forage plants, *U.S. Dep. Agric., Bull.* **2**.

Sohonie, K., and Bhandarkar, A. P., 1955, Trypsin inhibition in Indian foodstuffs. II. Inhibitors in pulses, *J. Sci. Ind. Res. (New Delhi), Sect. C* **14**:100.

Souci, S. W., Fachmann, W., and Kraut, H., 1974, Die zusammensetzung der lebensmittel, *Wiss. Verlagsges mbH Stuttg.*

Standley, P. C., 1937, Flora of Costa Rica, Part II, *Field Mus. Nat. Hist. Publ., Bot. Ser. 392*, Chicago, **18**(2):487–559.

Stanton, W. R., 1966, *Grain Legumes in Africa*, pp. 113–114, FAO, Rome, 182 pp.

Stecher, P. G. (ed.), 1968, *The Merck Index*, 8th ed., Merck, Rahway, NJ, 1714 pp.

Stehle, H., and Stehle, M., 1960, *Flore agronomique des Antilles Françaises*, Vol. III, Guadelupe.

Steinmetz, E. F., 1947, *Codex Vegetabilis*. IV, published by the author, Amsterdam, pagination unusual.

Thompson, H. C., and Kelly, W. C., 1957, *Vegetable Crops*, 5th ed., McGraw-Hill, New York, 611 pp.

Tidestrom, I., and Kittell, T., 1941, *Flora of Arizona and New Mexico*, Catholic University of America Press, Washington, D.C., 897 pp.

Tindall, H. D., 1965, *Fruits and Vegetables in West Africa*, FAO, Rome, 259 pp.

Tutin, T. G., Heywood, V. H., Burges, N. A., Moore, D. M., Valentine, D. H., Walters, S. M., and Webb, D. A. (eds.), 1968, *Fora Europaea*, Vol. 2, University Press, Cambridge, 455 pp.

Unwin, A. H., 1920, *West African Forests and Forestry*, Unwin, London, 527 pp.

Uphof, J. C. Th., 1968, *Dictionary of Economic Plants*, Richard Mayr, Wurzburg, 591 pp.

U.S. Department of Agriculture, Legume inoculation: What it is, what it does, *U.S. Dep. Agric., Farmers Bull.* **2003**.

Van Etten, C. H., Kwolek, W. F., Peters, J. E., and Barclay, A. S., 1967, Plant seed as protein sources for food or feed evaluation based on amino acid composition of 379 species, *Agric. Food Chem.* **15**(6):1077–1089.

Van Etten, C. H., Miller, R. W., Wolff, I. A., and Jones, Q., 1961, Amino acid composition of twenty-seven selected seed meals, *Agric. Food Chem.* **9**(1):79–82.

Van Etten, C. H., Miller, R. W., Wolff, I. A., and Jones, Q., 1963, Amino acid composition of seeds from 200 angiospermous plant species, *Agric. Food Chem.* **11**(5):399–410.

Van Rensburg, H. S., 1967, *Pasture Legumes and Grasses in Zambia*, Zambia Min. Agricultural Center Research Station, Mount Makulu, Miscellaneous Bulletin, Lusanka, 40 pp.

Van Wijk, H. L. G., 1971, *A Dictionary of Plant Names*, Asher, Vaals-Amsterdam, 2 vols.

Verdcourt, B., 1970, Studies in the Leguminosae—Papilionoideae: IV, *Kew Bull.* **24**:507–569.

Watt, B. K., and Merrill, A. L., 1963, Composition of foods, *U.S. Dep. Agric., Agric. Res. Serv., Agric. Handb.* **8**:189 pp.

Watt, G., 1889–1893, *A Dictionary of the Economic Products of India*, Supt. of Government Printing, Calcutta, 6 vols.

Watt, J. M., and Breyer-Brandwijk, M. G., 1962, *The Medicinal and Poisonous Plants of Southern and Eastern Africa*, 2nd ed., E. & S. Livingstone, Edinburgh, 1457 pp.

Weast, R., 1973, The development of new honey sources, *Am. Bee J.* **113**(2):60–62.

Westcott, C., 1971, *Plant Disease Handbook*, 3rd ed. Van Nostrand-Reinhold, New York, 843 pp.

Westphal, E., 1974, *Pulses in Ethiopia. Their Taxonomy and Agricultural Significance*, Centre for Agricultural Publishing and Documentation Wageningen, 261 pp.

Wheeler, W. A., 1950, *Forages and Pasture Crops*, Van Nostrand-Reinhold, New York, 752 pp.

Whistler, R. L., and BeMiller, J. N., 1959, *Industrial Gums, Polysaccharides and Their Derivatives*, Academic Press, New York, 766 pp.

White, F., 1962, *Forest Flora of Northern Rhodesia*, Oxford University Press, London, 455 pp.

Whiteman, P. C., 1968, The effects of temperature on the vegetative growth of six tropical legume species, *Aust. J. Exp. Agric. Anim. Husb.* **8**:528–532.

Willaman, J. J., and Li, Hui Lin, 1970, Alkaloid-bearing plants and their contained alkaloids, *Lloydia* **33**(3A):1–286.

Willaman, J. J., and Schubert, B. G., 1961, Alkaloid-bearing plants and their contained alkaloids, *U.S. Dep. Agric., Agric. Res. Serv., Tech. Bull.* **1234,** USGPO, Washington, D.C., 287 pp.

Williams, L., 1936, Woods of Northeastern Peru, *Field Mus. Nat. Hist. Publ., Bot. Ser.* **377,** Chicago, 587 pp.

Williams, L. O., 1960, Drug and condiment plants, *U.S. Dep. Agric., Agric. Res. Serv., Agric. Handb.* **172.**

Williams. R. O., 1934, *Flora of Trinidad and Tobago,* Trinidad Government Printing Office, Port-of-Spain, 2 vols.

Williams, R. O., 1949, *The Useful and Ornamental Plants of Zanzibar and Pemba,* St. Ann's Press, Timperley, 497 pp.

Williams, R. O., and Williams, R. O., Jr., 1951, *The Useful and Ornamental Plants in Trinidad and Tobago,* Guardian Commercial Printery Port-of-Spain, 355 pp.

Wu Leung, Woot-Tsuen, Busson, F., and Jardin, C., 1968, *Food Composition Table for Use in Africa,* FAO and U.S. Department HEW, 306 pp.

Wu Leung, Woot-Tsuen, Butrum, R. R., and Chang, F. H., 1972, *Proximate Composition, Mineral, and Vitamin Contents of East Asian Foods in Food Composition Table for Use in East Asia,* Part I, FAO and U.S. Department HEW, 334 pp.

Wu Leung, Woot-Tsuen, and Flores, M., 1961, *Tabla de Composicion de Alimentos Para Uso en America Latina,* National Institutes of Health, Bethesda, MD, 132 pp.

Yegna Narayan Aiyer, A. K., 1954, *Field crops of India,* Bangalore Printing, Bangalore City, 618 pp.

Zafar, R., Deshmukh, V. K., and Saoji, A. N., 1975, Studies on some Papilionaceous seed oils, *Curr. Sci.* **44**(9):311–312.